PROGRESS IN BRAIN RESEARCH

VOLUME 145

ACETYLCHOLINE IN THE CEREBRAL CORTEX

Volume 117: Neuronal Degeneration and Regeneration: From Basic Mechanisms to Prospects for Therapy, by F.W. van Leeuwen, A. Salehi, R.J. Giger, A.J.G.D. Holtmaat and J. Verhaagen (Eds.) – 1998, ISBN 0-444-82817-6.

Volume 118: Nitric Oxide in Brain Development, Plasticity and Disease, by R.R. Mize, T.M. Dawson, V.L. Dawson and M.J. Friedlander (Eds.) – 1998, ISBN 0-444-82885-0.

Volume 119: Advances in Brain Vasopressin, by I.J.A. Urban, J.P.H. Burbach and D. De Wied (Eds.) – 1999, ISBN 0-444-50080-4.

Volume 120: Nucleotides and their Receptors in the Nervous System, by P. Illes and H. Zimmermann (Eds.) – 1999, ISBN 0-444-50082-0.

Volume 121: Disorders of Brain, Behavior and Cognition: The Neurocomputational Perspective, by J.A. Reggia, E. Ruppin and D. Glanzman (Eds.) – 1999, ISBN 0-444-50175-4.

Volume 122: The Biological Basis for Mind Body Interactions, by E.A. Mayer and C.B. Saper (Eds.) – 1999, ISBN 0-444-50049-9.

Volume 123: Peripheral and Spinal Mechanisms in the Neural Control of Movement, by M.D. Binder (Ed.) – 1999, ISBN 0-444-50288-2.

Volume 124: Cerebellar Modules: Molecules, Morphology and Function, by N.M. Gerrits, T.J.H. Ruigrok and C.E. De Zeeuw (Eds.) – 2000, ISBN 0-444-50108-8.

Volume 125: Volume Transmission Revisited, by L.F. Agnati, K. Fuxe, C. Nicholson and E. Syková (Eds.) – 2000. ISBN 0-444-50314-5.

Volume 126: Cognition, Emotion and Autonomic Responses: The Integrative Role of the Prefrontal Cortex and Limbic Structures, by H.B.M. Uylings, C.G. Van Eden, J.P.C. De Bruin, M.G.P. Feenstra and C.M.A. Pennartz (Eds.) – 2000, ISBN 0-444-50332-3

Volume 127: Neural Transplantation II. Novel Cell Therapies for CNS Disorders, by S.B. Dunnett and A. Björklund (Eds.) – 2000, ISBN 0-444-50109-6.

Volume 128: Neural Plasticity and Regeneration, by F.J. Seil (Ed.) – 2000, ISBN 0-444-50209-2.

Volume 129: Nervous System Plasticity and Chronic Pain, by J. Sandkühler, B. Bromm and G.F. Gebhart (Eds.) – 2000, ISBN 0-444-50509-1.

Volume 130: Advances in Neural Population Coding, by M.A.L. Nicolelis (Ed.) – 2001, ISBN 0-444-50110-X.

Volume 131: Concepts and Challenges in Retinal Biology, by H. Kolb, H. Ripps and S. Wu (Eds.) – 2001, ISBN 0-444-50677-2.

Volume 132: Glial Cell Function, by B. Castellano López and M. Nieto-Sampedro (Eds.) – 2001, ISBN 0-444-50508-3.

Volume 133: The Maternal Brain. Neurobiological and neuroendocrine adaptation and disorders in pregnancy and post partum, by J.A. Russell, A.J. Douglas, R.J. Windle and C.D. Ingram (Eds.) – 2001, ISBN 0-444-50548-2.

Volume 134: Vision: From Neurons to Cognition, by C. Casanova and M. Ptito (Eds.) – 2001, ISBN 0-444-50586-5.

Volume 135: Do Seizures Damage the Brain, by A. Pitkänen and T. Sutula (Eds.) – 2002, ISBN 0-444-50814-7.

Volume 136: Changing Views of Cajal's Neuron, by E.C. Azmitia, J. DeFelipe, E.G. Jones, P. Rakic and C.E. Ribak (Eds.) – 2002, ISBN 0-444-50815-5.

Volume 137: Spinal Cord Trauma: Regeneration, Neural Repair and Functional Recovery, by L. McKerracher, G. Doucet and S. Rossignol (Eds.) – 2002, ISBN 0-444-50817-1.

Volume 138: Plasticity in the Adult Brain: From Genes to Neurotherapy, by M.A. Hofman, G.J. Boer, A.J.G.D. Holtmaat, E.J.W. Van Someren, J. Verhaagen and D.F. Swaab (Eds.) – 2002, ISBN 0-444-50981-X.

Volume 139: Vasopressin and Oxytocin: From Genes to Clinical Applications, by D. Poulain, S. Oliet and D. Theodosis (Eds.) – 2002, ISBN 0-444-50982-8.

Volume 140: The Brain's Eye, by J. Hyönä, D.P. Munoz, W. Heide and R. Radach (Eds.) – 2002, ISBN 0-444-51097-4.

Volume 141: Gonadotropin-releasing Hormone: Molecules and Receptors, by I.S. Parhar (Ed.) – 2002, ISBN 0-444-50979-8.

Volume 142: Neural Control of Space Coding and Action Production, by C. Prablanc, D. Pélisson and Y. Rossetti (Eds.) – 2003, ISBN 0-444-50977-1.

Volume 143: Brain Mechanisms for the Integration of Posture and Movement, by S. Mori, D.G. Stuart and M. Wiesendanger (Eds.) – 2004, ISBN 0-444-51389-2.

Volume 144: The roots of Visual Awareness, by C.A. Heywood, A.D. Milner and C. Blakemore (Eds.) – 2004, ISBN 0-444-50978-X.

PROGRESS IN BRAIN RESEARCH

VOLUME 145

ACETYLCHOLINE IN THE CEREBRAL CORTEX

EDITED BY

LAURENT DESCARRIES

Departments of Pathology and Cellular Biology and of Physiology, Centre de Recherche en Sciences Neurologiques, Université de Montréal, C.P. 6128, Succ. Centre-ville, Montreal, QC H3C 3J7, Canada

KREŠIMIR KRNJEVIĆ

Anaesthesia Research Unit and Department of Physiology, McGill University, McIntyre Building, 3655 Promenade Sir William Osler, Montreal, QC H3G 1Y6, Canada

MIRCEA STERIADE

Laboratory of Neurophysiology, Department of Anatomy and Physiology, Faculty of Medicine, Université Laval, Pavillon Ferdinand-Vandry, Quebec, QC G1K 7P4, Canada

ELSEVIER
AMSTERDAM – BOSTON – HEILDELBERG – LONDON – NEW YORK – OXFORD
PARIS – SAN DIEGO – SAN FRANCISCO – SINGAPORE – SYDNEY – TOKYO
2004

ELSEVIER B.V.
Sara Burgerhartstraat 25
P.O. Box 211, 1000 AE Amsterdam, The Netherlands

First edition 2004

Library of Congress Cataloging in Publication Data
A catalog record from the Library of Congress has been applied for.

British Library Cataloguing in Publication Data
A catalogue record from the British Library has been applied for.

ISBN: 0-444-51125-3 (volume)
ISBN: 0-444-80104-9 (series)
ISSN: 0079-6123
⊗ The paper used in this publication meets the requirements of ANSI/NISO Z39.48-1992 (Permanence of Paper).
Printed in The Netherlands.

123103/14668 u8

List of Contributors

E.X. Albuquerque, Departamento de Farmacologie Básica e Clínica, Instituto de Ciências Biomédicas, Centro de Ciências da Saúde, Universidade Federal do Rio de Janeiro, Rio de Janeiro, RJ 21944, Brazil

M. Alkondon, Department of Pharmacology and Experimental Therapeutics, University of Maryland School of Medicine, Baltimore, MD 21201, USA

N. Aznavour, Departments of Pathology and Cell Biology and Physiology, Université de Montréal, C.P. 6128 Succ. Centre-ville, Montreal, QC H3C 3J7, Canada

S.B. Backman, Department of Anaesthesia, Royal Victoria Hospital, 687 Pine Avenue W., Montreal, QC H3A 1A1, Canada

D.P. Calderon, Department of Neuroscience, Albert Einstein College of Medicine, 1300 Morris Park Avenue, Bronx, NY 10461, USA

N. Champtiaux, Laboratoire de Neurobiologie Moléculaire, Centre de la Recherche Scientifique, Unité de Recherche Associée 2182 'Récepteurs et Cognition', Institut Pasteur 75724, Paris Cedex 15, France

J.-P. Changeux, Laboratoire de Neurobiologie Moléculaire, Centre de la Recherche Scientifique, Unité de Recherche Associée 2182 'Recepteurs et Cognition', Institut Pasteur 75724, Paris Cedex 15, France

P.B.S. Clarke, Department of Pharmacology and Therapeutics, McGill University, 3655 Promenade Sir Wiliam Osler, Montreal QC, H3G 1Y6 Canada

L. Descarries, Departments of Pathology and Cell Biology and Physiology, Université de Montréal, C.P. 6128 Succ. Centre-ville, Montreal, QC H3C 3J7, Canada

P. Fiset, Department of Anaesthesia, Royal Victoria Hospital, 687 Pine Avenue W., Montreal, QC H3A 1A1, Canada

D.D. Flynn, Department of Molecular and Cellular Pharmacology, University of Miami School of Medicine, P.O. Box 016189, Miami, FL 33101, USA

E. Hamel, Laboratory of Cerebrovascular Research, Department of Neurology and Neurosurgery, Montreal Neurological Institute, McGill University, 3801 University Street, Montreal, QC H3A 2B4, Canada

M.E. Hasselmo, Department of Psychology, Center for Memory and Brain and Program in Neuroscience, Boston University, 2 Cummington Street, Boston, MA 02215, USA

C. Hsieh, Department of Neurobiology and Behavior, University of California, Irvine, 2205 McGaugh Hall, Irvine, CA 92697-4550, USA

B.E. Jones, Department of Neurology and Neurosurgery, McGill University, Montreal Neurological Institute, Montreal, QC H3A 2B4, Canada

S. Kar, Douglas Hospital Research Center, Department of Psychiatry, McGill University, Montreal, QC H4H 1R3 Canada

K. Krnjević, Anaesthesia Research Unit and Physiology Department, McIntyre Building Room 1215, McGill University, 3655 Promenade Sir William Osler, Montreal, QC H3G 1Y6, Canada

A.I. Levey, Department of Neurology and Center for Neurodegenerative Disease, Emory University School of Medicine, Whitehead Biomedical Research Building, Suite 505, 615 Michael Street, Atlanta, GA 30322, USA

J.-S. Liang, Department of Molecular and Cellular Pharmacology, University of Miami School of Medicine, P.O. Box 016189, Miami, FL 33101, USA

M.H. McCollum, Department of Molecular and Cellular Pharmacology, University of Miami School of Medicine, P.O. Box 016189, Miami, FL 33101, USA

J. McGaughy, Department of Psychology, Center for Memory and Brain and Program in Neuroscience, Boston University, 2 Cummington Street, Boston, MA 02215, USA

N. Mechawar, Departments of Pathology and Cell Biology and Physiology, Université de Montréal, C.P. 6128 Succ. Centre-ville, Montreal, QC H3C 3J7, Canada

M.-M. Mesulam, Cognitive Neurology and Alzheimer's Disease Center, Departments of Neurology and Psychiatry, Feinberg Medical School, Northwestern University, Chicago, IL 60611, USA

R. Metherate, Department of Neurobiology and Behavior, University of California, Irvine, 2205 McGaugh Hall, Irvine, CA 92697-4550, USA

A. Nordberg, Karolinska Institute, Neurotec Department, Division of Molecular Neuropharmacology, Huddinge University B84, S-141 86 Stockholm, Sweden

A. Peinado, Department of Neuroscience, Albert Einstein College of Medicine, 1300 Morris Park Avenue, Bronx, NY 10461, USA

E.K. Perry, Development in Clinical Aging, MRC Building, Newcastle General Hospital, Westgate Road, Newcastle upon Tyne, NE4 6BE, UK

R.H. Perry, Development in Clinical Aging, MRC Building, Newcastle General Hospital, Westgate Road, Newcastle upon Tyne, NE4 6BE, UK

G. Plourde, Department of Anaesthesia, Royal Victoria Hospital, 687 Pine Avenue W., Montreal, QC H3A 1A1, Canada

L.T. Potter, Department of Molecular and Cellular Pharmacology, University of Miami School of Medicine, P.O. Box 016189, Miami, FL 33101, USA

R. Quirion, Douglas Hospital Research Center, Department of Psychiatry, McGill University, Montreal, QC H4H 1R3, Canada

K. Semba, Department of Anatomy and Neurobiology, Faculty of Medicine, Dalhousie University, Halifax, NS B3H 1X5, Canada

Z. Shao, Laboratory of Signal Transduction, National Institute of Environmental Health Sciences, N.I.H., P.O. Box 12233, 111 T.W. Alexander Drive, Research Triangle Park, NC 27709, USA

O.K. Steinlein, Institute of Human Genetics, University Hospital Bonn, Friedrich-Wilhelms-University, Wilhelmstr. 31, D-53111 Bonn, Germany

M. Steriade, Laboratoire de Neurophysiologie, Faculté de Medecine, Université Laval, Laval, QC G1K 7P4, Canada

L.A. Volpicelli, Department of Neurology and Center for Neurodegenerative Disease, Emory University School of Medicine, Whitehead Biomedical Research Building, Suite 505, 615 Michael Street, Atlanta, GA 30322, USA

K.C. Watkins, Departments of Pathology and Cell Biology and Physiology, Université de Montréal, C.P. 6128 Succ. Centre-ville, Montreal, QC H3C 3J7, Canada

P.J. Whitehouse, Case Western Reserve University, 12200 Fairhill Road, Suite C357, Cleveland, OH 44120-1013, USA

J.L. Yakel, Laboratory of Signal Transduction, National Institute of Environmental Health Sciences, N.I.H., P.O. Box 12233, 111 T.W. Alexander Drive, Research Triangle Park, NC 27709, USA

XXIVth International Symposium of the Centre de recherche en sciences neurologiques
Acetylcholine in the Cerebral Cortex, held at the Université de Montréal,
May 6 and 7, 2002

Speakers from left to right (see number key below). 1. Jean-Pierre Changeux; 2. Paul B.S. Clarke; 3. Kazue Semba; 4. Michael E. Hasselmo; 5. Laurent Descarries; 6. Raju Metherate; 7. Manickavasagom Alkondon; 8. Barbara E. Jones; 9. Agneta Nordberg; 10. Ortrud K. Steinlein; 11. M.-Marsel Mesulam; 12. Edith Hamel; 13. Elaine K. Perry; 14. Rémi Quirion; 15. Krešimir Krnjević; 16. Lincoln T. Potter; 17. Mircea Steriade; 18. Allan I. Levey; 19. Peter J. Whitehouse; 20. Robert H. Perry; 21. Steven B. Backman; 22. Jerrel L. Yakel; 23. Alejandro Peinado.

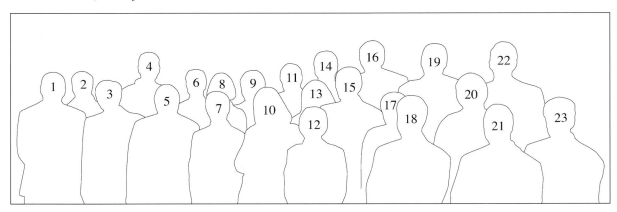

Preface

Each year, the *Centre de recherche en sciences neurologiques* (*CRSN*) of the *Université de Montréal* hosts an international symposium gathering leaders of the scientific community on a particular topic of the neurosciences. In 2002 (May 6 and 7), the XXIVth International Symposium of the *CRSN*, organized jointly by Laurent Descarries (Université de Montréal), Krešimir Krnjević (McGill University) and Mircea Steriade (Université Laval) was devoted to 'Acetylcholine in the Cerebral Cortex'. The two-day meeting consisted of 22 presentations on all aspects of this fascinating subject, ranging from its most elementary, at the molecular and cellular levels, to its systemic and holistic implications, including its role in cognition and involvement in human diseases and therapeutics. This effort to integrate current knowledge at all levels of organization of the nervous system, from the basic to the applied neurosciences, was greatly appreciated by an audience of students, scientists and clinicians from diverse disciplinary horizons.

It was most appropriate to hold this meeting in Montreal, where so much of the history of 'Acetylcholine in the Cerebral Cortex' has been written. For example, one can cite the pioneering studies demonstrating ACh release in the cortex. The 'cortical cup' technique, originally developed in 1949 by H.H. Jasper and K.A.C. Elliott at the Montreal Neurological Institute, allowed the first convincing demonstrations by Elliott and then F.C. MacIntosh that ACh is released in the brain *in situ*. Working in Sir Henry Dale's lab in London, MacIntosh had confirmed that ACh release in ganglia is a physiological, and not a pathological process. In separate experiments, both he and Jasper demonstrated a clear correlation between changes in cortical ACh release and the behavioral (waking–sleep) state of the animal. This opened up the possibility that ACh was a key element controlling thalamocortical interactions and their putative role in sleep–waking, a major focus of further studies by Jasper, J.P. Cordeau and their many collaborators in Montreal.

Acetylcholine has been a canonical neurotransmitter/modulator in both senses of the word. First to be discovered, hence the most ancient, it also served as a model, setting the rules for other chemical messengers. As illustrated by the present monograph, the constitutive elements of the cholinergic system are among the best known in the central nervous system. Following their immunocytochemical identification in brain, some 20 years ago, many of the molecular and cellular processes, properties and mechanisms governing their functioning have been elucidated. Their involvement in cortical functions has led to the beginning of an understanding of complex and refined behaviors and capacities previously unamenable to neurobiological exploration and characterization. Cholinergic neurons, receptor sites and effector mechanisms have been recognized as major players in pathological ageing, and as targets for new diagnostic or therapeutic procedures. Natural or synthetic compounds capable of mimicking or modifying their actions have been developed, raising the hope of being one day able to improve cognitive performance!

Like the presentations at the meeting, the chapters in this book have been grouped under four headings: I. The Acetylcholine Innervation of Cerebral Cortex; II. Modes of Action of Acetylcholine in the Cerebral Cortex; III. Cortical Properties and Functions Modulated by Acetylcholine, and IV. Clinical, Pathological and Therapeutic Implications. These somewhat arbitrary subdivisions should not detract from the effort made by all authors (and the editors) to integrate their topics in the broader framework of a global perspective on the subject. It is a pleasure to thank them all for participating in such a challenging endeavor.

Laurent Descarries, Krešimir Krnjević and Mircea Steriade

Acknowledgments of sponsorship and support

This monograph is the 24th in a series of reports based on the proceedings of an annual international symposium organized by the *Centre de recherche en sciences neurologiques (CRSN)* of the Université de Montréal (UdeM). The cost of this annual meeting is partly defrayed by an operating grant of the University to the Groupe de recherche sur le système nerveux central (GRSNC). We also gratefully acknowledge the complementary financial support received from the Fonds de la recherche en santé du Québec (FRSQ), the Ministère de la Recherche de la Science et de la Technologie du Québec, the Canadian Institutes of Health Research (CIHR), the Savoy Foundation/Epilepsy, Merck-Frosst Canada, Servier Canada, AstraZeneca Canada, and the Fonds Michel Bergeron. For their help in the organization of the symposium, we also wish to thank the secretarial and technical staffs of the CRSN and of the Department of Physiology at UdeM, and especially Chantal Nault and Daniel Cyr.

Contents

xiv

insectivores, or bats (this conclusion is yet to be confirmed using ChAT immunohistochemistry). A simple form of the basal nucleus is, however, identified in rodents, as a group of magnocellular neurons that lie ventral to the lentiform nucleus and are scattered in the anterior perforated substance without forming a cytoarchitectonic unit. The degree of complexity varies among different rodents. The nucleus basalis of lagomorphs such as rabbit is less developed than in some rodent species.

In ungulates including lamb, horse and donkey, the magnocellular basal nucleus is as well defined as in the rodents with the most-developed basal nucleus, but with a further increase in cell number. Many cells are located lateral to the globus pallidus, and sparse cells ventral to the globus pallidus. Among the carnivores, the nucleus basalis of raccoon and bear is distinctly larger than that of dog and cat. In sirens such as the manatee, the nucleus basalis is further developed in size and complexity. In primates, it shows the highest degree of development with the greatest number of cells and structural differentiation as well as compactness as a nucleus. Interestingly, an extensive development of this nucleus is also seen in cetaceans in which many magnocellular cells are located lateral to the globus pallidus. This phylogenetic trend toward increased size and complexity corresponds to the development of the neocortex, in particular the temporal lobe, which reaches its greatest development in the primates and cetaceans.

In the 1960s and 1970s, AChE histochemistry was used to identify cholinergic BF neurons (reviewed by Parent, 1986), confirming most of Gorry's earlier findings based on Nissl stain. However, applying AChE histochemistry to nonmammalian species is problematic because AChE is not a definitive marker for cholinergic neurons. This problem has been overcome by the development of antibodies that recognize ChAT across species including nonmammalian taxa. Thus, ChAT immunohistochemistry has made it possible for the first time to study the evolution of the cholinergic BF neurons across vertebrate classes (Fig. 1).

This section addresses the following questions: How common is the cholinergic BF system among living vertebrates, and what might have been the earliest ancestor that had a cholinergic BF system? The presence of intrinsic neurons in the cortex or pallium will also be reviewed. As mentioned above, the cholinergic BF system here is defined as a cluster of large cholinergic neurons in the BF or ventral telencephalon that project to the cerebral cortex or pallium.

Basal forebrain cholinergic neurons and their cortical projections

Mammals

As already mentioned, early studies using Nissl, myelin and Golgi stain indicated that the magnocellular basal nucleus or nucleus of Meynert shows phylogenetic progression in structural growth and complexity among mammalian species from rodents to human, in parallel with the evolution of the neocortex (Gorry, 1963). Obviously, the cholinergic nature of these magnocellular cells was not identified, and the observations in some species will have to be confirmed using ChAT immunohistochemistry and other methods. However, the temporal cortex was already recognized as an important area of innervation by these neurons, and this is significant because the piriform cortex is now known to receive among the densest cholinergic innervation in the cortex (Lysakowski et al., 1986).

The cholinergic basal nuclear complex has been identified in a number of placental mammalian species using ChAT immunohistochemistry. Of these, the cholinergic BF system shows the highest degree of development in the human brain (Hedreen et al., 1984; Saper and Chelimsky, 1984), in particular with the growth and differentiation of the caudal aspects of the cholinergic basal nuclear complex, i.e., the nucleus of Meynert (Mesulam and Geula, 1988; Butcher and Semba, 1989; Fig. 2A, B). The number of cholinergic neurons in the nucleus of Meynert (or Ch4) is estimated to be approximately 200,000 in each hemisphere (Arendt et al., 1985). Compared to the nucleus of Meynert, the medial septum is simple, and its cholinergic neurons may contain lesser amounts of ChAT mRNA (Kasashima et al., 1998). The cholinergic BF system has been identified in a number of nonhuman primates, including New World monkeys such as blackcapped capuchin (*Cebus apella*; Kordower et al., 1989) and common marmoset (*Callithrix jacchus*; Everitt et al., 1988), and Old World monkeys such as rhesus monkey

6

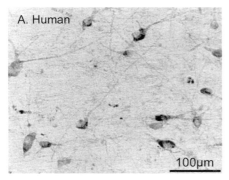

A. Human

100μm

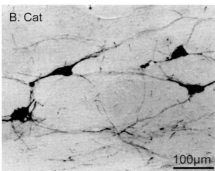

B. Cat

100μm

C. Turtle

100μm

Fig. 1. ChAT-ir neurons in the substantia innominata-magno-cellular basal nucleus region in human (A), cat (B) and turtle (C). Note the considerable morphological maturity of immunolabeled neurons in the turtle compared to the human and cat counterparts. A, B and C are adapted from Mesulam and Geula (1988), Vincent and Reiner (1987a) and Powers and Reiner (1993), respectively. Reproduced by courtesy of the original authors and permission of Wiley-Liss (A) and Karger (C).

(*Macaca mulatta*; Mesulam et al., 1983, 1984) and baboon (*Papio papio*; Satoh and Fibiger, 1985). The nucleus of Meynert in these nonhuman primates is highly differentiated and complex, but not to the same degree as in human.

Among carnivores, the cholinergic basal nuclear complex has been identified in the ferret (Henderson, 1987), cat (Kimura et al., 1981; Vincent and Reiner, 1987), raccoon (*Procyon lotor*; Brückner et al., 1992; Brauer et al., 1999) and mongrel dog (*Canis familiaris*; St.-Jacques et al., 1996). The magno-cellular basal nucleus in these animals is not as well developed as in primates, but is larger and more differentiated than in rodents.

The cholinergic BF system of the rat is by far the best studied (Schwaber et al., 1982; Houser et al., 1983; Semba and Fibiger, 1989; Semba, 2000); Fig. 2C, D). The cholinergic BF neurons have also been mapped in other rodents: the guinea pig (Maley et al., 1988) and mice of the CD-1 (Mufson and Cunningham, 1988; Kitt et al., 1994), BALB/c ByJ (Kitt et al., 1994), C57BL/6 and DBA/2 strains (Albanese et al., 1985). The cholinergic BF system is fairly similar among these rodents, and is generally less complex than in carnivores.

The European hedgehog (*Erinaceus europaeus*) is an insectivore with one of the structurally simplest and smallest mammalian brains. Although inform-ation on the entire cholinergic BF system of this species is lacking, its BF neurons project to the cortex and many of them are ChAT-immunoreactive (ir) (Dinopoulos et al., 1988). No information is available for the BF cholinergic system of nonplacental mammals, i.e., marsupials and monotremes.

Collectively, these studies show that the choliner-gic BF system varies among mammals, in terms of the number of neurons in each sector of the basal nuclear complex as well as differentiation within each sector. The projections of BF cholinergic neurons, and colo-calization of neuropeptides in these cholinergic neu-rons also vary among selected mammals that have been studied (reviewed in Semba, 2000). Furthermore, there is evidence for cholinergic neurons in the lateral septum of the rat (Kimura et al., 1990), cat (Kimura et al., 1981), raccoon (Brauer et al., 1999) and monkey (*Macaca Mulatta*; Kimura et al., 1990); these lateral septal neurons, however, tend to stain only lightly for ChAT.

Despite these variations, the basic scheme of the cholinergic basal nuclear complex, as defined by a continuous distribution of cholinergic neurons from the septum to the magnocellular basal nucleus, and their projections to the cerebral cortex, is similar

The acetylcholine innervation of cerebral cortex

Progress in Brain Research, Vol. 145
ISSN 0079-6123

CHAPTER 1

Phylogenetic and ontogenetic aspects of the basal forebrain cholinergic neurons and their innervation of the cerebral cortex

Kazue Semba*

Department of Anatomy and Neurobiology, Faculty of Medicine, Dalhousie University, Tupper Medical Building, 6850 College Street, Halifax, NS B3H 1X5, Canada

Introduction

Acetylcholine (ACh) is best known as a neurotransmitter. However, ACh probably evolved long before the appearance of a nervous system, since the machinery to synthesize and degrade ACh is present in bacteria, fungi, protozoa and plants. Even in so-called higher organisms, ACh is present in nonneural tissues such as the placenta, suggesting that ACh has cellular functions other than neurotransmission.

As a neurotransmitter, ACh was among the first to be identified. ACh was identified as a possible mediator of cellular function by Hunt in 1907. In the 1910s, it was shown by Sir Henry Dale to mimic the effect of parasympathetic nerve stimulation and to have different actions depending on the tissue. The latter observation suggested the presence of separate 'receptive substances'. ACh release upon nerve stimulation was demonstrated in the 1920s by Otto Loewi, in elegantly simple experiments using two frog hearts.

Despite this impressive beginning, a definitive anatomical mapping of central cholinergic pathways was not available until the 1960s, when Shute and Lewis published a series of studies using acetylcholinesterase (AChE), providing the first overall picture of cholinergic pathways (Lewis and Shute, 1967; Shute and Lewis, 1967). However, the results of these landmark studies remained somewhat tentative, because AChE is present not only in cholinergic neurons, but also in noncholinergic neurons such as certain noradrenergic and dopaminergic neurons. In the early 1980s, antibodies to choline acetyltransferase (ChAT), the biosynthetic enzyme for ACh, became available (Kimura et al., 1981; Crawford et al., 1982; Eckenstein and Thoenen, 1983; Wainer et al., 1984) and these antibodies made it possible, for the first time, to identify cholinergic pathways definitively. Thus, central cholinergic pathways were mapped in selected mammalian species (e.g., Wainer et al., 1984). Further studies combining ChAT immunostaining with retrograde tracing confirmed many of the pathways revealed by AChE histochemistry, with the exception of the ventral mesencephalic pathway, which was found to be noncholinergic. Recently, antibodies to the vesicular ACh transporter became available. In addition, in situ hybridization techniques have been developed to detect mRNA for ChAT. These markers are complementary to ChAT immunoreactivity and are useful as additional anatomical markers for cholinergic neurons.

These advances in understanding the anatomy of central cholinergic pathways paralleled those in ACh receptors. In the early 1980s, the nicotinic ACh

---

Tel.: (1)-902-494-2008; Fax: (1)-902-494-1212;
E-mail: semba@dal.ca

DOI: 10.1016/S0079-6123(03)450001-2

receptor became the first neurotransmitter receptor whose primary structure was determined (Changeux et al., 1984) and to be cloned (Sakmann et al., 1985). This was followed by reports of the primary structure of muscarinic receptors in the 1980s and 1990s. Thus, ACh is presently one of the best-studied neurotransmitters.

Of the cholinergic cell groups in the mammalian brain, the cholinergic neurons of the basal forebrain (BF) have received special attention probably for three reasons. First, cortical ACh plays an important role in cortical arousal, attention, plasticity, learning and memory and BF cholinergic neurons are the primary source of ACh to the cerebral cortex (revized by Rasmusson, 2000; Semba, 2000). Second, nerve growth factor was identified as a trophic factor for BF cholinergic neurons in the 1980s and the septohippocampal pathway, in particular, has been used as a model system to study the role of trophic factors in neuronal development and regeneration (e.g., Korsching, 1993). Finally, BF neurons degenerate massively in Alzheimer's disease and this fact has stimulated many studies with the goal to develop pharmacotherapies for cognitive dysfunction associated with this neurodegenerative disease (e.g., Bartus et al., 1982).

As mentioned above and discussed in more detail below, in all species examined to date, the primary source of cortical ACh is the intracortical axon terminals of cholinergic BF neurons. In the rat, an additional source of ACh is provided by the cholinergic neurons in the mesopontine tegmentum, in particular those in the laterodorsal tegmental nucleus. However, this projection is limited to the medial prefrontal cortex and the presence of a homologous projection in other species is unknown (reviewed by Semba and Fibiger, 1989). In the rat and a few other vertebrates, intrinsic cholinergic neurons also exist in the cortex, presumably supplying a minor source of ACh. In addition to the prominent cortical projections, cholinergic BF neurons send projections to a number of subcortical structures. However, axonal collateralization between targets does not appear to occur often, and for most cortically projecting cholinergic BF neurons, the cortex is likely the only target to innervate (Semba, 2000). The BF also contains noncholinergic (GABAergic and putatively glutamatergic) neurons, and some of them project to the cortex (reviewed in Semba, 2000).

The goal of this review is three-fold. The first is to provide an overview of the cholinergic BF system and its homologues across vertebrates in which this system has been identified. The second goal is to review the ontogeny of the cholinergic BF complex and its projection to the cerebral cortex in mammals. Finally, an attempt will be made to integrate the phylogenetic and ontogenetic data in order to understand the neurobiological significance of the cholinergic BF system.

The cholinergic BF system is defined in this chapter as a cluster of cholinergic neurons in the BF or ventral telencephalon that project to the cerebral cortex or pallium (e.g., Semba and Fibiger, 1989; Semba, 2000). Following the nomenclature by Butcher and Semba (1989), BF cholinergic neurons refer to those cholinergic neurons that are distributed, in a rostral to caudal direction, in the medial septum, vertical and horizontal limbs of the diagonal band of Broca, ventral pallidum, magnocellular preoptic area, substantial innominata and magnocellular basal nucleus, also known in primates as the nucleus of Meynert. These cholinergic neurons form a longitudinal column that is distributed in association with these nuclei, but largely ignore their classically defined boundaries (Schwaber et al., 1987; Butcher and Semba, 1989). Previous reviews are available on the phylogeny of magnocellular basal nucleus based on Nissl and AChE stain (Parent, 1986), and the ontogeny of central cholinergic neurons (Semba, 1992).

Phylogenetic aspects

Comparative anatomical studies in the 1960s and earlier used Nissl, myelin and Golgi stain to identify homologues of the magnocellular basal nucleus in various mammalian species (reviewed in Gorry, 1963). These observations provided useful insights into the evolution of this nucleus, although some conclusions later turned out to be incorrect in the light of ChAT immunohistochemical data. Thus, after examining representatives from 11 orders of mammals, Gorry (1963) concluded that the magnocellular basal nucleus exhibits a phylogenetic trend, with increasing numbers of cells and differentiation as a nucleus, in parallel with the general evolution of the brain. Specifically, the basal nucleus was not identifiable in marsupials,

Human

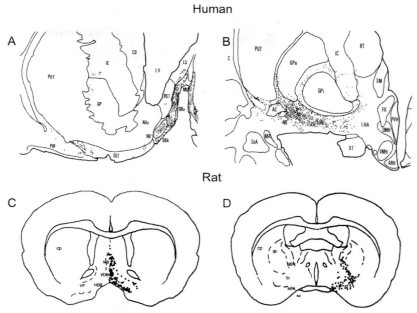

Fig. 2. Distribution of ChAT-ir neurons in the human (A, B) and rat (C, D) basal forebrain. In the human brain, ChAT-ir neurons are seen in the medial septum (MS) and the vertical and horizontal limbs of the diagonal band of Broca (DBv, DBh). The number increases caudally, in the substantia innominata (SI) and nucleus basal (NB). In the rat, ChAT-ir neurons are present in the medial septum (MS), vertical and horizontal limbs of the diagonal band (VDB, HDB), magnocellular preoptic area (MPA), substantia innominata (SI) and magnocellular basal nucleus (MBN). A and B, and C and D are adapted from Saper and Chelimsky (1984) and Semba (2000), respectively. A and B are reproduced by courtesy of the original authors.

among the mammalian species thus far studied. Within this basic scheme, the cholinergic system displays a phylogenetic trend of increased size and complexity, in particular with respect to the magnocellular basal nucleus, which reaches its highest degree of complexity and differentiation in human. The reason for prominent growth and complexity in cetaceans is unclear, but might be related to the fact that, like primates, they have highly differentiated cerebral cortex.

Birds

The avian telencephalic hemisphere consists of the pallium and the underlying basal ganglia, including the striatum, pallidum and olfactory bulb (Northcutt and Braford, 1980; Reiner et al., 1998). The only bird in which the distribution of ChAT-ir cell bodies and fibers has been studied is the pigeon (*Columba livia*; Medina and Reiner, 1994). In the pigeon

telencephalon, the largest group of ChAT-ir cell bodies is found in its ventral aspect, encompassing both the pallidal and striatal structures (Fig. 3A, B). Thus, many large ChAT-ir neurons are present in the nucleus of the fasciculus diagonalis Brocae, the ventral pallidum, the paleostriatum primitivum (dorsal pallidum in birds) and the intrapeduncular nucleus. The distribution of these neurons continues into the medial part of the paleostriatum primitivum (striatum). These ChAT-ir neurons in the ventral telencephalon appear to form a single field of cholinergic neurons. They are intensely immunoreactive, and display long dendrites. Their axons can be followed to the dorsal ('neocortex-like') telencephalon or pallium. The pallium contains numerous ChAT-ir fibers, but no ChAT-ir cell bodies. Some ChAT-ir neurons in the ventral telencephalon also appear to project to the striatum and the medial habenula. Many small ChAT-ir cells are present in the lateral septum, but unlike the other ventral

8

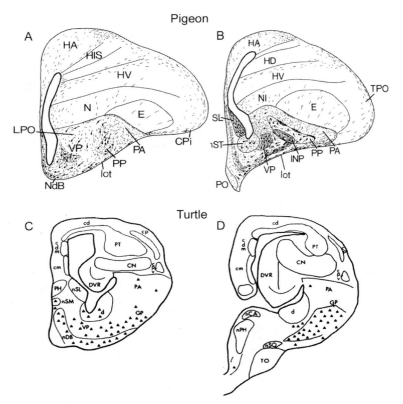

Fig. 3. Distribution of ChAT-ir neurons in the basal forebrain of pigeon (*Columba livia*; A, B) and turtle (*Pseudemys scripta* or *Chrysemys picta*; C, D). In the pigeon telencephalon, ChAT-ir neurons are seen in the nucleus of the diagonal band of Broca (NdB), ventral paleostriatum (VP, ventral pallidum) and nucleus intrapeduncularis (INP, avian basal ganglia), as well as in the paleostriatum primitivum (PP; avian striatum). In the turtle telencephalon, immunolabeled neurons are seen in the nucleus of the diagonal band of Broca (nDB), ventral paleostriatum or pallidum (VP) and globus pallidus (GP). The distributions of ChAT-ir neurons are remarkably similar to those in mammals. A and B are adapted from Medina and Reiner (1994), and C and D from Powers and Reiner (1993). Reproduced by courtesy of the original authors and permission of Wiley-Liss (A, B) and Karger (C, D).

telencephalic ChAT-ir neurons, they are lightly stained. The olfactory tubercle and the bed nucleus of the stria terminalis also contain ChAT-ir cells.

These observations suggest that the basal telencephalic cholinergic system in the pigeon closely resembles the cholinergic basal nuclear system in mammals in terms of location and projections to the cerebral cortex or pallium.

Reptiles

Species from three orders of reptiles have been studied: crocodiles, lizards and turtles. Crocodiles share a common ancestor with birds as the Archosauria, and therefore, the crocodilian brain generally bears closer resemblance to the avian brain than to the lizard brain (Brauth et al., 1985). In the crocodylian *Caiman crocodilus*, a large field of ChAT-ir neurons is present in the ventral paleostriatum, a region considered to be comparable to the substantia innominata and ventral pallidum of mammals (Brauth et al., 1983). A few ChAT-ir neurons are also present in the medial septal nucleus and the nucleus of the diagonal band. Although the projections of these cholinergic neurons to the cortex, or elsewhere, remain unknown, these observations suggest that the cholinergic BF complex is present in the caiman ventral telencephalon. In addition, ChAT-ir cell bodies are present in the region of the ventral telencephalon (the large-celled component of the rostral portion of the ventrolateral area of the telencephalon) that is comparable to the mammalian caudate-putamen, and these neurons appear to be interneurons providing a ChAT-ir plexus in the same region. These observations suggest that not only cholinergic BF neurons, but also striatal intrinsic cholinergic neurons similar to those in mammals exist in the caiman (Brauth et al., 1985).

In the gekkonid lizard *Gekko gecko*, one of the two species of lizard studied, most of the ChAT-ir neurons in the telencephalon are present in its ventral aspect, forming a continuum encompassing the septum, vertical and horizontal limbs of the diagonal band of Broca, bed nucleus of the medial forebrain bundle and lateral preoptic area (Hoogland and Vermeulen-VanderZee, 1990). The size of these ChAT-ir neurons varies, but they generally have a multipolar appearance. ChAT-ir fibers can be traced from these cell bodies through the septum and the alveus to cortical destinations, including parts of the medial and dorsal cortices, with regionally different laminar patterns. These observations suggest the presence of a cholinergic basal nuclear complex with cortical projection in the gecko brain. However, the organization of *Gekko gecko* striatum appears to be slightly different from the mammalian striatum, in that the region of cholinergic cell bodies is segregated from the region of cholinergic neuropil (Hoogland and Vermeulen-VanderZee, 1990).

Similar to gecko, the largest telencephalic group of ChAT-ir neurons in the lacertid lizard *Gallotia galloti* (Lacertidae) is in the basal ganglia (Medina et al., 1993). These neurons are seen in the olfactory tubercle, nucleus accumbens, striatum and ventral pallidum. More caudally, large ChAT-ir cells form a continuum encompassing the nucleus of the diagonal band of Broca, and regions around the lateral and medial forebrain bundles, including some within the globus pallidus. Dense ChAT-ir plexuses are present in the medial and dorsal cortices, and a few ChAT-ir cells in the septum appear to project to the medial cortex, the reptilian homologue of the hippocampus. Thus, on the basis of location, large cell size and cortical projection, the ventral telencephalic group of large cholinergic neurons in lizards appears to be homologous to the mammalian cholinergic BF system.

Turtles belong to the phylogenetically most conservative lineage of all living reptiles. They are also the phylogenetically oldest amniotes in which cholinergic neurons have been studied. Despite this archaic status, the distribution of cholinergic neurons in the two species of emydid turtles studied, red-eared (*Pseudemys scripta*) and painted turtles (*Chrysemys picta*), is remarkably similar to that of mammalian cholinergic BF neurons (Mufson et al., 1984; Powers and Reiner, 1993; Fig. 3C, D). Many ChAT-ir cell bodies are present in a continuous zone of the basal telencephalon encompassing the medial septum, the nucleus of the diagonal band, the ventral paleostriatum (which resembles the mammalian ventral paleostriatum or pallidum), the globus pallidus and the striato-amygdala area. There are regional differences in cell density; the medial septum and the vertical limb regions contain only a few ChAT-ir cells, whereas increasing numbers of ChAT-ir cells are seen more caudally, in particular in the ventral paleostriatum or pallidum (Powers and Reiner, 1993). In addition, these caudal cholinergic neurons are labeled intensely and are multipolar, whereas the rostral cholinergic neurons in the septum are stained more lightly and are oval or fusiform (Mufson et al., 1984). These observations suggest that, based on location and morphology, turtles have cholinergic neurons in the ventral telencephalon that are homologous to the mammalian cholinergic basal nuclear complex. However, the cholinergic neurons in the turtle medial septum appear to be relatively few compared to the mammalian counterpart (Powers and Reiner, 1993). It is possible that the septohippocampal system is not well developed in turtles.

ChAT-ir cells are present in the olfactory tubercle in both red-eared and painted turtles (Mufson et al., 1984; Powers and Reiner, 1993). However, there is a disagreement concerning the presence of cholinergic neurons in the turtle homologue of the mammalian striatum. Mufson et al. (1984) reported that ChAT-ir cell bodies were absent from the striatum of the red-eared turtle. Later, Powers and Reiner (1993) reported that ChAT-ir cell bodies are present in the paleostriatum augementation and area 'd' of the painted, as well as red-eared, turtle. These areas correspond to the striatum and nucleus accumbens, respectively, of mammals. The reason for this discrepancy might be technical, as the two studies used different antibodies.

Amphibians

The telencephalon of amphibians is distinctly less complex than that of reptiles, and is similar to that of some fishes, which might be related to the fact that amphibians are terrestrial anamniotes (Northcutt and Braford, 1980; Sarnat and Netsky, 1981). The distribution of ChAT-ir neurons has only been studied in anurans and urodeles (Marín et al., 1997). In the anurans, *Rana perezi* (Iberian green frog) and *Xenopus laevis* (South African clawed frog), numerous large and intensely ChAT-ir neurons, often with long processes, are present in the medial septum, the diagonal band of Broca, the region of the medial forebrain bundle, and the medial and lateral amygdala (Fig. 4A, B). Inclusion of the amygdala in this group is consistent with the view that the substantia innominata is part of the extended amygdala in mammals (Alheid and Heimer, 1988), even though the mammalian amygdala proper contains only a few small cholinergic neurons (e.g., Vincent and Reiner, 1987). In both frogs, some intensely labeled cells are also present in a ventrolateral telencephalic region lateral to the lateral forebrain bundle. The ChAT-ir neurons scattered throughout the ventral telencephalon form a continuous field. No ChAT-ir cells are present in the olfactory bulb or olfactory tubercle. The ventral telencephalic field with ChAT-ir cells also contains moderate to dense ChAT-ir fibers and varicosities. In addition, dense plexuses of ChAT-ir fibers and varicosities are seen in the lateral pallium, and less dense varicose fibers in the medial pallium; the dorsal pallium contains only a few ChAT-ir fibers. The striatum contains a few weakly stained ChAT-ir cells in *Rana*, but not in *Xenopus*.

The distribution of ChAT-ir cells in the urodele *Pleurodeles walti* (Iberian ribbed newt) is considerably less extensive than in anurans (Marín et al., 1997). ChAT-ir neurons are present mainly in the medial amygdala, with a few cells extending rostrally. ChAT-ir neurons appear to be absent from the medial septum. In the pallium, ChAT-ir fibers are seen mostly in the rostral part of the dorsal pallium,

and to a lesser extent, in the medial and lateral pallia. No ChAT-ir cells are present in the striatum or the olfactory tubercle.

These observations suggest that in both amphibian orders, there are cholinergic neurons in those regions of the basal telencephalon that correspond to the mammalian BF cholinergic cell regions. Unlike in

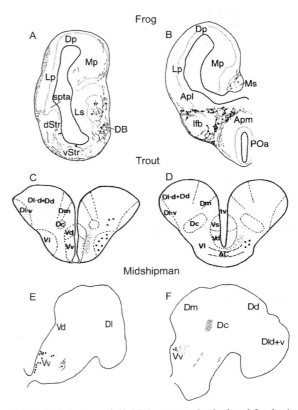

Fig. 4. Distribution of ChAT-ir neurons in the basal forebrain of an anuran amphibian (*Rana perezi*; A, B), and two teleosts, rainbow trout (*Oncorhynchus mykiss*; C, D) and midshipman (*Porichthys notatus*; E. F). In the frog, labeled cells are present in the diagonal band region (DB), and the medial amygdala (Apm). In the trout, labeled cells are seen in the Vl, which is considered homologous to the olfactory tubercle or the septum of mammals. In midshipman, labeled cells are in the Vv, which is considered homologous to the mammalian septum. Although homologies are less obvious at the first glance, the basic topographic relationship of ChAT-ir neurons in amphibians and bony fishes is similar to that seen in tetrapods. Adapted from Marín et al. (1997), Pérez et al. (2000) and Brantley and Bass (1988). Reproduced by courtesy of the original authors and permission of Wiley-Liss.

mammals, birds (pigeon) and reptiles, the distribution of these cholinergic neurons extends into the amphibian amygdala. Tract-tracing studies have shown that the amphibian pallium receives afferents from cells located in a region encompassing the diagonal band, the medial septum and the medial amygdala (Neary, 1990; Sassoè-Pognetto et al., 1991). Thus, it is highly likely that the cholinergic neurons in the basal telencephalon project to the pallium, and on this basis, the ChAT-ir in the basal telencephalon of amphibians might represent the amphibian homologue of the mammalian cholinergic BF system. These cholinergic neurons also appear to project to the striatum, which is largely devoid of intrinsic cholinergic neurons in both anuran and urodelean amphibians.

Fishes

As in amphibians, the telencephalon of the fish can be divided into the pallium, homologous to the cortex, and the subpallium (Nieuwenhuys, 1965; Northcutt, 1981; Sarnat and Netsky, 1981). The pallium in the teleosts is outwardly bowed, giving it a strikingly different appearance compared to the other fishes and terrestrial vertebrates.

The subpallium of the fish consists of several nuclei, and the interpretation of their homologies varies among different fishes, as well as among different authors. For example, according to Northcutt and his colleagues (Northcutt and Braford, 1980; Northcutt and Davis, 1983), the dorsal (Vd) and ventral (Vv) subpallial nuclei of teleosts are homologous to the lateral and medial septal nuclei of tetrapods, respectively, and the lateral subpallial nucleus (Vl) is homologous to the tetrapod olfactory tubercle. Using neurochemical and molecular markers, Wullimann and Rink (2002) recently proposed that the ventral tier of the subpallium including the Vl and Vv represents the septal formation, whereas the dorsal tier including the Vd and the Vc (central subpallial nucleus) corresponds to the striatal formation, of terrestrial vertebrates. In either scheme, therefore, the Vv is considered homologous to the septum, whereas the Vl is considered homologous to the olfactory tubercle by Northcutt and his colleagues, but to the septum by Wullimann and Rink (2002).

Of the numerous species of living fishes, cholinergic neurons of the ventral telencephalon have been studied in one to several species each of teleosts, chondrosteans, elasmobranches and cyclostomes.

Teleosts. In a cyprinid teleost, *Phoxinus phoxinus* (European minnow), there is a small population of small ChAT-ir neurons in the area ventralis telencephali pars lateralis (corresponding to the Vl above; Ekström, 1987). This is the only ChAT-ir cell group seen in the telencephalon. ChAT-ir neurons are also seen in adjacent diencephalic regions, including the periventricular hypothalamus and the suprachiasmatic nucleus. ChAT-ir fibers and terminals are present in a small medial zone of the telencephalic area dorsalis, i.e., the teleostean cortex. As mentioned above, the area ventralis telencephali pars lateralis (Vl) is considered by some to be homologous to the olfactory tubercle of tetrapods (Northcutt and Davis, 1983). The olfactory tubercle is not part of the BF cholinergic system in mammals, nor in nonmammalian species in which this system has been identified (see above). However, the Vl is also considered to be homologous to the septum of tetrapods (Wullimann and Rink, 2002). Furthermore, most cells in the lateral part of the ventral telencephalic area project to dorsal telencephalic areas in the sebastid teleost *Sebastiscus marmoratus* (coastal rockfish; Murakami et al., 1983). This suggests that the cholinergic neurons in the minnow area ventralis telencephali pars lateralis (Vl) might represent a teleostean homologue of the mammalian BF cholinergic system. Positive identification of their projection to the pallium would support this hypothesis.

Two species of salmonid teleosts, the rainbow trout (*Oncorhynchus mykiss*) and the brown trout (*Salmo trutta*), have been studied (Pérez et al., 2000; Fig. 4C–F). As in the cyprinid teleost *Phoxinus phoxinus*, there is a cluster of ChAT-ir cell bodies in the lateral part of the ventral telencephalic area or the Vl. Most of these cells are medium-sized with bipolar appearance, but some are large, in particular at the rostral level. The area Vl is suggested to be homologous to the olfactory tubercle (Northcutt and Davis, 1983) or the septum (Wullimann and Rink, 2002) of tetrapods. Unlike olfactory tubercle cells in tetrapods, most Vl cells project to the pallium in teleost (Murakami et al., 1983). Interestingly, scarce medium-sized ChAT-ir neurons are seen in the pallium, and they are continuous with the more ventrally located ChAT-ir cells in the Vl. Large neurosecretory neurons of the magnocellular preoptic area are also distinctly ChAT-ir. These observations suggest that the cholinergic population of the trout Vl might be homologous to the mammalian BF cholinergic system.

The midshipman (*Porichthys notatus*) belongs to the Batrachoididea family, the toadfishes, and is distantly related to the European minnow and the salmonid trouts discussed above. In the midshipman, ChAT-ir neurons are present in the ventral nucleus of the area ventralis (Vv) of the telencephalon (Fig. 4E, F). This region faces the ventricle medially, and a particularly dense population of ChAT-ir cells is found along its medial border (Brantley and Bass, 1988). The Vv is considered homologous to the tetrapod septum (Wullimann and Rink, 2002), particularly the medial septum (Northcutt and Braford, 1980; Northcutt and Davis, 1983), on the basis of topographical relationship and connectivity. The caudal part of the area dorsalis telencephali, comparable to the pallium of tetrapods, contains ChAT-ir fibers, but no cell bodies (Brantley and Bass, 1988). In view of these observations, the cholinergic neurons in the Vv of the ventral telencephalon of the midshipman might be homologous to the mammalian and nonmammalian cholinergic BF neurons. Examination of the projections of these cholinergic neurons could substantiate this possibility.

These studies in three species of teleosts suggest that a small number of cholinergic neurons are present in parts of the ventral telencephalon (i.e., the Vl and Vv) that are considered homologous to the septum, and the olfactory tubercle in the case of the Vl, of tetrapods. The inclusion of the teleostean olfactory tubercle might represent a rostral extension of the cholinergic BF complex. Cells in the Vl of teleosts are known to project to the pallium, and ChAT-ir fibers are seen in the pallium. Thus, the cholinergic neurons in the teleostean ventral telencephalon might be homologous to the BF cholinergic complex of tetrapods.

Chondrosteans, elasmobranchs and cyclostomes. Like teleosts, chondrostean fishes are ray-finned fishes; however, they are derived earlier than teleosts from a common ancestral lineage. In a chondrontean fish, the Siberian sturgeon (*Acipenser baeri*), there are no ChAT-ir cell bodies in the telencephalon, either in the pallium or subpallium, although the adjacent diencephalic regions including the magnocellular preoptic regions and the suprachiasmatic nucleus contain many ChAT-ir neurons (Adrio et al., 2000).

The elasmobranchs originate early from a common ancestral lineage with the ray-finned fishes, and retain more ancestral features of the fish lineage than the teleostean or chondrostean fishes. In an elasmobranch, the lesser spotted dogfish (*Scyliorhinus canicula*, shark), there are no ChAT-ir neurons or fibers in the basal telencephalon (Anadón et al., 2000). However, ChAT-ir cells are present in the olfactory bulb, and more caudally, in a region that might correspond to the olfactory tubercle or the Vl. These latter neurons appear to project to the olfactory bulb. They might correspond to the cholinergic cell group in the Vl of teleosts, but this is unclear. In addition, there are numerous small ChAT-ir cells in the superficial dorsal pallium of the telencephalon. These cells are bipolar, tripolar or multipolar with several dendrites. The presence of terminal-like labeling in the same region suggests that they are intrinsic neurons. The abundance of cholinergic neurons in the dogfish homologue of the cerebral cortex is striking, and contrasts with the absence of such cells in teleosts and most tetrapods thus far examined.

Lampreys are cyclostomes, which are the closest living species to the common ancestor of all vertebrates. In the *Petromyzon marinus*, a few ChAT-ir cells are present in the 'striatum' of the dorsal subpallium, but such cells are absent in the *Lampetra fluviatilis* (Pombal et al., 2001). No other ChAT-ir cells are present in the entire telencephalic region, including the olfactory bulb. As in anurans, neurons projecting to the neurohypophysis are ChAT-ir.

Summary. In teleosts, cholinergic neurons are present in ventral telencephalic regions that are considered homologous to the septum or the olfactory tubercle of tetrapods. Although the projections of these cholinergic neurons have not been studied, unidentified neurons in these regions have been

13

shown to project to the pallium. In sturgeons, dogfish and lampreys, which are derived earlier than teleosts from the common lineage of vertebrates, the basal telencephalon is devoid of cholinergic perikarya. Thus, the teleosts are the phylogenetically oldest species that are known to have BF cholinergic neurons.

Intrinsic cholinergic neurons in the cortex

Unlike cholinergic neurons in the ventral telencephalon, cholinergic neurons in the cerebral cortex or pallium have been confirmed only in selected species across various lineages of vertebrates. In the human, a small number of neurons in selected layers and regions of the cerebral cortex are weakly ChAT-ir or express ChAT mRNA (Kasashima et al., 1998, 1999). These are medium-sized or large pyramidal neurons. Cholinergic neurons are absent from the cortex of nonhuman primates including marmoset (Everitt et al., 1988) and *Cebus apella* (Kordower et al., 1989).

Reports on the existence of intrinsic cholinergic neurons in the feline cortex are mixed (Vincent and Reiner, 1987; Avendaño et al., 1996). Intrinsic cholinergic neurons are present in the rat cortex (reviewed by Semba and Fibiger, 1989; see also Kasashima et al., 1999). Unlike other cholinergic neurons, these cortical cholinergic neurons lack AChE (Levey et al., 1984). Cholinergic intrinsic neurons are absent from the cortex of the BALB/c ByJ mouse (Kitt et al., 1994); reports are mixed for CD-1 mouse (Mufson and Cunningham, 1988; Kitt et al., 1994). In another rodent, the guinea pig, cholinergic neurons are absent from the cortex (Maley et al., 1988). When present, cholinergic neurons in the cerebral cortex of nonhuman mammals are small, and fusiform or bipolar, in contrast to cholinergic neurons in the human cerebral cortex which are pyramidal neurons.

Cholinergic neurons are absent from the pallium of the pigeon (Medina and Reiner, 1994). Among reptiles, cholinergic neurons are absent from the pallium of the *Gekko gecko* (Hoogland and Vermeulen-VanderZee, 1990), red-eared (*Pseudemys scripta*) and painted turtles (*Chrysemys picta*) (Powers and Reiner, 1993). However, they are present in parts of the lateral, dorsal and medial cortices of the lizard *Gallotia gallotti* (Medina et al., 1993). Cholinergic neurons are absent from the pallium of amphibians (Marín et al., 1997).

Among fishes, cholinergic neurons are present in the pallium of the dogfish, an elasmobranch (Anadón et al., 2000), but not of the sturgeon, a more recently evolved, ray-finned fish (Adrio et al., 2000) or lampreys, a cyclostome, i.e., the phylogenetically oldest, living vertebrate (Pombal et al., 2001).

In summary, intrinsic cholinergic neurons are found in the cerebral cortex or its homologue in the human, rat, possibly cat and one strain of mouse, lizards and the dogfish, but not in many other species thus far studied. It is possible that such sporadic occurrence reflects technical problems with the sensitivity of antibodies. It is also possible, however, that it is the result of brain evolution, as discussed below.

Conclusions

Conservation of the cholinergic basal forebrain system

With the exception of mammals, the number of species examined so far for each class is small. However, evidence to date suggests that a BF cholinergic system, consisting of a cluster of cholinergic neurons in the ventral telencephalon and their projections to the cortex or pallium, is present in tetrapods (mammals, birds, reptiles and amphibians) and bony fishes (Fig. 5). In contrast, the cholinergic BF system appears to be absent in selected orders of fishes that were derived earlier from the common ancestral lineage of vertebrates. These include chondrosteans as represented by sturgeon, elasmobranchs as represented by dogfish, and cyclostomes as represented by lampreys. The conservation of the BF cholinergic system among tetrapods and bony fishes indicates that it probably appeared in a common ancestor of the osteichthyan (bony fish) lineage.

Despite this conservation, variation exists among terrestrial vertebrates. First, the size and complexity of the cholinergic BF system exhibit a phylogenetic graduation, in parallel with general evolution of the brain. Specifically, the number of cholinergic BF neurons appears to be relatively small in fishes, whereas it is the highest in humans, and possibly in

14

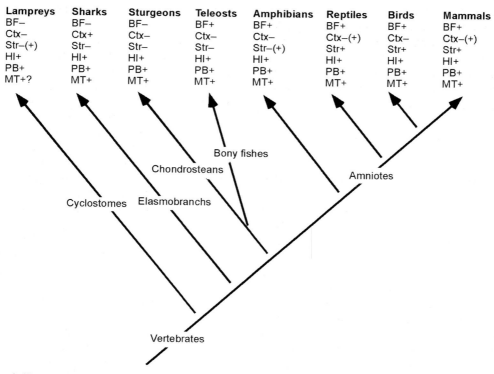

Fig. 5. Schematic illustration of the evolution of the cholinergic BF system, and intrinsic cholinergic neurons in the cortex (Ctx) and the striatum (Str), as well as three other main central cholinergic pathways: habenulopeduncular (HI) pathway, parabigeminal (PB) pathway and mesopontine tegmental (MT) pathway. The presence or absence of a given system is indicated by (+) or (−). Exceptions to the rule in each class are indicated in parenthesis: Lampreys, Str(+) in *Petromyzon marinus*; Amphibians, Str (+) in *Rana perezi*; Reptiles; Ctx(+) in *Gallotina gallotti*; Mammals, Cortex (+) in human, rat and possibly cat and mouse.

cetaceans. Similarly, the area of innervation, i.e., the cortex, increases in size, and also in complexity as reflected by the formation of multiple layers. It is possible that the phylogenetic trend toward a larger and more complex cholinergic BF system reflects an adaptation in support of the 'higher' functions that a more complex cortex can handle. However, this trend might also simply be a reflection of the fact that a larger brain can accommodate more neurons and more axons. It will be important, in future studies, to compare the relative size of the cholinergic BF system as a function of total brain size. Increase in the relative size of the cholinergic BF system would suggest a significant role of this system in the 'higher' functions of more evolved brains.

The relative size of different sectors of the cholinergic BF complex appears to change with the general evolution of the brain. For example, in

human brain, the magnocellular basal nucleus, located relatively caudally within the BF, is the largest sector. In contrast, in rodents and lower species, more rostral sectors, such as the medial septum and diagonal band regions, are as large as the caudal sector. Although a large proportion of the cholinergic neurons in urodelean amphibians is located caudally in the amygdala, this appears to be an exception. In some reptiles and in teleosts, the olfactory tubercle, which only contains cholinergic interneurons in mammals, appears to contain some cholinergic neurons projecting to the pallium; this projection might be related to the primary importance of olfaction in these species. The primary function of the fish telencephalon is olfaction. The telencephalon of 'higher' species assumes an increasing number of nonolfactory functions. It is possible that, as the brain evolved, the rostral cholinergic neurons, and particularly those in the olfactory structures, lost their original projection feature, in parallel with the growth of the more caudal cluster of BF cholinergic neurons. These newly emerging caudal cholinergic neurons may have thus prevailed as the main supplier of cholinergic input to the pallium, where ACh is used in functions uniquely associated with a more complex cerebral cortex, such as attention, learning and memory.

Comparison with other central cholinergic systems

BF cholinergic neurons are by no means the only cholinergic cell group in the brain of any species. Many of the studies quoted above have in fact examined ChAT immunoreactivity in the whole brain. Although the discussion of these other cholinergic pathways is beyond the scope of this chapter, a brief summary may help to set the BF findings in a broader perspective (see the original articles above, and reviews by Rodríguez-Moldes et al., 2002; Wullimann and Rink, 2002).

As already discussed, it is possible that cholinergic neurons intrinsic to the cerebral cortex exist only in certain species, perhaps as the result of recent independent evolution of these species. Alternatively, it is possible that this feature was once a common feature of all ancestors of living vertebrates, but was subsequently lost in most species. In any case, this lack of commonality of intrinsic cholinergic neurons in the cortex is in sharp contrast to the more highly conserved, cholinergic BF system.

The intrinsic cholinergic neurons in the striatum also appear to be less well conserved than the BF cholinergic neurons, although they are not as sporadic as those in the cortex (see also Reiner et al., 1998). They have been shown to be present in all mammals, birds and reptiles examined to date, but only occasionally in amphibians (i.e., present in *Rana perezi*, but absent in *Xenopus* and *Pleurodeles*), and fishes (i.e., absent in teleosts, chondroteans, elasmobranchs and the lamprey *Lampetra fluviatilis*; but very few of these cells are found in the lamprey *Petromyzon marinus*). The conservation of intrinsic cholinergic neurons in the striatum among amniotes, but apparently not among anamniotes, is consistent with the phylogenetic constancy of the striatal

complex in birds, reptiles and mammals (Reiner et al., 1984).

The habenulo-interpeduncular system is present and is also cholinergic in all the vertebrates thus far studied. Magnocellular preoptic neurosecretory cells are cholinergic in most fishes, but not in tetrapods. The parabigeminal nucleus contains a compact group of cholinergic neurons that project to the superior colliculus (i.e., the tectum). It is known as the nucleus isthmi in 'lower' species. This cholinergic projection system is highly conserved across all the species discussed in this chapter, i.e., both jawed vertebrates, and jawless vertebrates such as lampreys. Neurons of the cranial motor nuclei are also cholinergic in all vertebrates.

The mammalian brainstem contains an extensive cholinergic projection system originating in the mesopontine tegmentum, specifically, the pedunculopontine and laterodorsal tegmental nuclei. These neurons project widely to the forebrain regions, such as the thalamus and the BF, and to the brainstem reticular formation. This cholinergic projection system plays an important role in the control of sleep and wakefulness in mammals (Semba, 1999). This cholinergic pathway is well conserved across all the tetrapods and fishes studied to date. In the lampreys, these cholinergic neurons are clearly present, but their projections have not been identified.

To summarize, the cholinergic BF system is a feature of all amniotes and bony fishes, but not of 'lower' fishes. This conclusion is consistent with the observation that the contents of ACh, ChAT and AChE in the telencephalon relative to the whole brain increase rapidly from the anamniotes to reptile, birds and mammals (Hebb and Ratkovic, 1964; Wachtler, 1981). Better conserved, the mesopontine tegmental cholinergic system is common to all vertebrates, with the possible exception of cyclostomes. In contrast, three other main cholinergic pathways, i.e., the habenulo-interpeduncular system, the parabigeminal system, and the cranial motor nuclei and their peripheral projections, are well conserved across all vertebrates.

Unsettled issues

Three issues require consideration. One is the difference in the sensitivity of antibodies used in the previous comparative studies, which might have contributed to the variations seen among different species, and between different studies investigating the same species. This is particularly problematic when the content of the enzyme is low, as may be the case with cholinergic intrinsic neurons in the cerebral cortex. Furthermore, it is possible that ChAT occurs in slightly different molecular forms in different species (Wainer et al., 1984; Goldbach et al., 1998; Oda, 1999), perhaps reflecting the evolution of the enzyme itself. Antibodies might therefore not recognize these different forms with the same sensitivity. It is reassuring in this regard that Western blot analysis of protein extracts from brains indicated the presence of similar bands at ~ 68–72 kDa and a degradation form at 55 kDa in the rat, trout, sturgeon and dogfish (Anadón et al., 2000). However, not all antibodies have been tested for cross-species reactivity. Furthermore, although virtually all comparative studies have used ChAT as a marker for cholinergic neurons, differential distributions of ChAT, vesicular ACh transporter and their mRNAs have been reported in adult rat brain (Ichikawa et al., 1997). No information is currently available as to relative abundance of these molecules in nonmammalian species. It will, therefore, be important to use additional cholinergic markers, particularly when ChAT appears to be absent.

A second issue is the constancy of transmitter phenotype of the same neuronal system. The magnocellular neurosecretory neurons are cholinergic in some fishes and frogs, but they are noncholinergic in mammals. By the same token, it is possible that BF neurons changed their neurotransmitter phenotype during evolution. Thus, cortically projecting BF neurons in the 'lower' orders of fishes (chondrosteans, elasmobranches and lampreys) might not be cholinergic, even though the cholinergic phenotype appears to be well conserved from the 'higher' fishes (teleosts) upwards across amniotes. It might then be relevant to note that, in several mammals examined, cholinergic BF neurons are mixed with some noncholinergic neurons that also project to the cerebral cortex. These noncholinergic BF neurons projecting to the cortex might perhaps represent an older form of BF projection neuron, and this noncholinergic phenotype might dominate BF neurons projecting to the pallium in the 'lower' fishes.

If so, these neurons cannot be detected by ChAT immunohistochemistry. A combination of tract tracing techniques and immunohistochemistry will be required to address this issue.

Finally, it is not known whether ACh has the same action on cortical neurons, or whether ACh release in the cortex has the same behavioral consequence, across vertebrates. Functional studies of ACh action in different vertebrates will be useful for investigating these issues.

Ontogenetic aspects

How the cholinergic BF system develops is an important question because of the role played by ACh in the adult cortex and its potential role in cortical maturation. The cholinergic BF system is also a useful model system to study mechanisms of axonal guidance, the formation of specific innervation patterns, and the contribution of trophic factors in these events. The ontogenetic development of the BF cholinergic system has been studied, initially, using AChE histochemistry, and more recently, using ChAT immunohistochemistry and in situ hybridization for ChAT mRNA. Tract tracing techniques have also provided complementary information to the neurochemical results, by identifying emerging projections that do not express detectable amounts of cholinergic enzymes.

Another marker extensively used for studying developing cholinergic BF neurons is the p75 nerve growth factor receptor (p75-NGFR). In the 1970s, NGF was recognized as a trophic factor for BF cholinergic neurons, and its role in neuronal development and in regeneration following axotomy was investigated using the septohippocampal cholinergic pathway as a model system (Korsching, 1993). BF neurons express p75 NGFR abundantly. In the 1980s, an antibody to this receptor became available. This 192-IgG antibody has since been used frequently as an anatomical marker for studying the development of BF cholinergic neurons, due to its superior sensitivity compared to many ChAT antibodies.

This section will review the ontogenetic development of cholinergic projections from the BF to the cerebral cortex. The time of origin, migration and cellular differentiation of the cholinergic BF neurons will be discussed first, and then the axonal outgrowth and innervation of the cerebral cortex by these neurons. The development of noncholinergic BF neurons and their projections to the cortex will also be considered, as well as the development of intrinsic cholinergic neurons in the cortex. The gestation periods in the various species discussed in this section are: 21–22 days (rat); 20–21 days (mouse); 65–68 days (cat); 42 days (ferret); 165 days (rhesus monkey).

Origin and migration of basal forebrain cholinergic neurons

Germinal source

AChE is expressed in immature BF neurons, and therefore, using this enzyme as a marker, the origin and migration of cholinergic BF neurons have been studied in the *cat* (Krnjević and Silver, 1966). BF neurons appear to be derived from the ganglionic eminence (Krnjević and Silver, 1966), a bulge on the ventricular wall that is present during early stages of telencephalic development. The ganglionic eminence also gives rise to the striatum, the globus pallidus

and other subcortical structures (see discussion in Krnjević and Silver, 1966). These progenitor cells move back and forth between the ventricular zone and the subventricular zone to repeat mitosis, but after going through their final mitotic cycle, they leave the ventricular wall and migrate to 'the lenticular nucleus' and more rostrally, to the septum. The study appears to be inconclusive as to whether the septum arises from the lenticular nucleus or is derived independently from the latter. From each of these two structures, developing AChE-positive neurons migrate further to their final destinations (i.e., those regions that are now recognized as housing the cholinergic basal nuclear complex), where they differentiate into mature neurons with specific phenotypes (Krnjević and Silver, 1966).

Very similar observations have been reported in the *rat* using AChE histochemistry (Fine, 1987). In this study, AChE staining was first detected at E14 in the ventricular wall (Fig. 6). Over the next few days, AChE was detected in what appears to the precursor of the basal nucleus, and at E15, was present in the rostral cell group comprising the medial septum and the vertical limb of the diagonal band. At this stage, no continuity was detected between this rostral cell group and the more caudal,

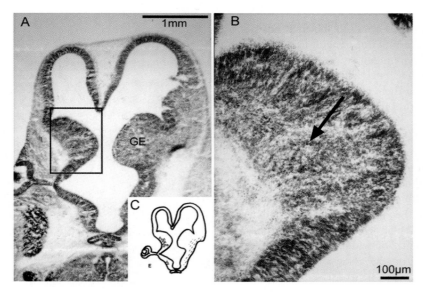

Fig. 6. The presence of AChE-positive cells (arrow in B) in the subventricular zone of the ganglionic eminence (GE) in an E14 rat. The boxed area in A is shown at a higher magnification in B. C is a diagram of the section shown in A; triangles represent AChE-positive cells. E: eye. Adapted from Fine (1987). Reproduced by courtesy and permission from Dr. Alan Fine who retains the copyright.

18

basal nucleus. Similarly, a study using p75-NGFR immunoreactivity (Koh and Loy, 1989) reported the earliest detection of immunoreactivity in the intermediate zone of the ventrolateral telencephalic wall at E13, followed by the detection of a second cell group more rostrally in the dorsomedial wall at E15. These two groups of neurons were thought to be precursors for the basal nucleus and medial septum-diagonal band, respectively, and by E15, the two groups merge, forming a continuum and displaying an adult-like distribution pattern by E17 (Koh and Loy, 1989).

These observations in the cat and rat indicate that all cholinergic BF neurons are derived from the embryonic striatal ventricular zone, and are of telencephalic origin. However, whether they all arise from a single precursor population is unclear. There are some indications, in fact, to suggest that the caudal BF (basal nucleus) and the rostral BF (medial septum and the vertical limb of the diagonal band) might arise separately, with the rostral group arising slightly later than the caudal group, although the two groups then quickly merge to form a continuum. This issue remains to be investigated.

Time of origin

The time at which cells undergo the final mitosis is referred to as the time of origin. Traditionally, this timing has been studied by using incorporation of tritiated thymidine into cells that are in mitotic cycles. If cells leave their mitotic cycle shortly after thymidine incorporation, the radiolabel remains in the cells, and these cells can be detected autoradiographically. It is also possible to examine cholinergic phenotype of thymidine-pulsed cells, for example, by combining with ChAT immunohistochemistry. Because the cholinergic neurons in the BF are distributed in a longitudinal continuum from the medial septum to the magnocellular basal nucleus, one obvious question is whether these neurons are generated concurrently or sequentially. This will also have implications for the issue of single or dual germinal source for these neurons.

In *rodents*, studies that combined tritiated thymidine autoradiography with ChAT immunohistochemistry or AChE histochemistry have revealed that cholinergic BF cells leave mitotic cycles in a caudal to rostral gradient. In the *rat*, which has a gestation period of 21 to 22 days, this wave of neurogenesis occurs from E12 to E16 (Fig. 7; Schwab et al., 1988; Semba and Fibiger, 1988; Brady et al., 1989). Similar observations have been made for AChE-positive neurons in the *mouse*, with neurogenesis occurring from E11 to E15 (Sweeney et al., 1989). These results are consistent with the time of neurogenesis of large, potentially cholinergic neurons of the BF in Nissl-stained material from the rat (Bayer, 1979, 1985), mouse (Creps, 1974) and hamster (ten Donkelaar and Dederen, 1979).

Not all forebrain neurons are generated in a caudal to rostral gradient. For example, in the *rat* striatum, cholinergic interneurons are generated synchronously with BF cholinergic neurons in the same gradient, whereas all somatostatin-containing interneurons are generated simultaneously, with a single peak at E15 (Semba et al., 1988). Comparison of the ChAT and AChE results with those of Nissl staining indicates that cholinergic neurons are among the first neurons to become postmitotic in the BF (Creps, 1974; ten Donkelaar and Dederen, 1979).

The pattern of neurogenesis in *primates* might be different from that in rodents, according to a study using Nissl stain (Kordower and Rakic, 1990).

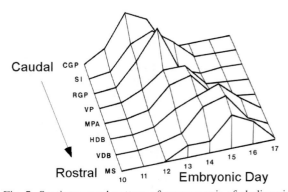

Fig. 7. Spatiotemporal pattern of neurogenesis of cholinergic neurons in the rat. Note the caudal to rostral progression of neurogenesis. Adapted from Semba and Fibiger (1988). CGP: caudal globus pallidus; HDB: horizontal limb of the diagonal band of Broca; MPA: magnocellular preoptic area; MS: medial septum; RGP: rostral globus pallidus; SI: substantia innominata; VDB: ventral limb of the diagonal band of Broca; VP: ventral pallidum.

Specifically, the neurogenesis of large hyperchromic neurons in the BF of rhesus monkey occurs during a 2-week period within the first trimester of the 165-day gestation period. This occurs in two stages. The initial brief burst occurs simultaneously at all levels of the BF around E30. Following a short quiescent period, a second phase of neurogenesis occurs in a caudal-to-rostral gradient, similar to that seen in rodents, with peaks at E33-43. The first burst appears to generate about 20% of total magnocellular neurons, and they are among the earliest to be generated in the entire telencephalon. A total of 10–30% of magnocellular neurons in the primate BF are thought to be noncholinergic, and it is possible that these initially generated neurons are noncholinergic (Kordower and Rakic, 1990).

Migration

Following final mitosis, newly generated BF neurons migrate to their final destinations in a caudal to rostral gradient in the *mouse* (Schambra et al., 1989). A similar caudal to rostral gradient in the migration of postmitotic cholinergic BF neurons is suggested by a study using p75-NGFR immunohistochemistry in the *rat*. Thus, p75-NGFR immunoreactivity is first detected in the intermediate zone of the ventrolateral telencephalic wall at E13, and in the dorsomedial telencephalic wall at E15 in the rat (Koh and Loy, 1989). The onset of NGF expression follows the time of last mitosis with a delay of approximately one day. How this neurotrophin expression is induced is not clear, but the induction does not appear to be dependent on target-derived factors, because these neurons have not yet innervated their target, the cerebral cortex.

Summary

The cholinergic BF neurons are derived from the ganglionic eminence of the embryonic telencephalic wall. These neurons are generated in a caudal to rostral gradient, and are among the first neurons to leave mitotic cycles in the telencephalon of both rodents and primates. There is a hint for separate precursors for the caudal and rostral BF cell groups, and this requires further investigation. Following

final mitosis, migration appears to occur in a caudal to rostral gradient. The early maturation of the cholinergic BF neurons places them in a strategic position to influence the subsequent development of noncholinergic neurons within the BF, and of the entire cortical mantle.

Expression of cholinergic phenotype in basal forebrain neurons

AChE histochemistry

AChE staining is first detected at or near the ventricular surface of the striatal eminence during the fourth gestation week in the *cat* (Krnjević and Silver, 1966). These cells are localized mostly around or between columns of migrating cells. The AChE activity at this stage appears to be nonselective. Once these neurons migrate laterally from the striatal ependyma and form the 'lenticular nucleus', they express a more selective and intense AChE activity. The intense AChE activity is also detectable in the septum, located more medially, at this stage. This intense AChE activity probably represents mature forms of AChE (Krnjević and Silver, 1966).

In the *rat*, AChE-positive neurons are detectable in the developing magnocellular basal nucleus and the horizontal limb of the diagonal band by E14-15, and in the medial septum and the vertical limb of the diagonal band, by E16 (Fig. 6; Fine, 1987; see also Ishii, 1957; Milner et al., 1983), suggesting a caudal to rostral gradient of AChE expression. In the *mouse*, AChE-positive neurons are detected in the magnocellular nucleus at E18 (Höhmann and Ebner, 1985). These findings are consistent with results based on enzymatic assays of the rat BF (Thal et al., 1991). Specifically, in the medial septum, diagonal band and magnocellular preoptic area, AChE activity is barely detectable at E14, but reliably detected at E18. AChE activity in these regions then increases rapidly between P5 and P30, followed by small decreases to adult levels (Thal et al., 1991). This overshoot appears to be unique to the cell body region of BF cholinergic neurons, and no overshoot is seen in their target areas, the neocortex and hippocampus, that contain their axon terminals (Thal et al., 1991). Postnatally, the intensity of AChE staining in the BF has been reported to increase earlier than that of ChAT immunoreactivity (Gould et al., 1991).

In the *ferret*, a species with protracted development compared to rodents (the gestation period of the ferret is 42 days), the earliest detectable AChE activity, at E28, is associated with the nigrostriatal pathway (Henderson, 1991). However, AChE-positive cell bodies in the BF are seen by E35 in the magnocellular basal nucleus, and slightly less intensely, in the septum and diagonal band nuclei. The staining intensity of these neurons generally increases by the time of birth (Henderson, 1991).

In the *human* fetus, AChE staining is seen in the basal nucleus anlage at 9 weeks of gestation, and this is reported to be the earliest AChE activity detectable in the telencephalon (Candy et al., 1985; Kostović, 1986). AChE staining is first seen in the undifferentiated basal nucleus, and then expands to occupy the sublenticular, diagonal band and septal areas (Kostović, 1986). At 15 weeks, distinctly AChE-stained perikarya are visible in clusters corresponding to the basal nucleus, diagonal band region and medial septum (Kostović, 1986). At 15 weeks, magnocellular basal nucleus neurons are the most differentiated cells in the entire telencephalon (Kracun and Rösner, 1986).

In summary, BF neurons express an immature, nonselective form of AChE activity shortly after final mitosis. However, this immature pattern is replaced by more intense and selective, mature form of AChE activity by the time they reach their final destinations, by E16 in the rat. A caudal to rostral gradient in the onset of AChE expression is observed in the human and rat BF. The intensity of AChE activity in the BF increases rapidly postnatally, and an overshoot is reported in rodents, occurring at the end of the third postnatal week.

ChAT immunohistochemistry

In general, the maturation of ChAT activity appears to be somewhat delayed compared to that of AChE. In addition to ChAT immunohistochemistry, a few studies used in situ hybridization or northern blot analysis to investigate the development of ChAT mRNA expression in rat septal complex. However,

no in situ hybridization results are available for other BF regions of the rat, or for the BF of other species.

In the *rat*, ChAT mRNA is first detected at E20 in the septum (Li et al., 1995; Bender et al., 1996). This is about the time septohippocampal fibers reach the hippocampus (Linke and Frotscher, 1993). Although cholinergic neurons are among the first BF cells to become postmitotic, ChAT mRNA is not detected until three days after the first detection of GAD67 mRNA at E17, the earliest time point studied (Bender et al., 1996). ChAT mRNA grain counts over individual neurons are relatively low at P4, but show a dramatic, 3.4-fold increase between P4 and P11, followed by a decrease by P30 (Li et al., 1995). This postnatal overshoot in ChAT mRNA appears to precede a similar overshoot in AChE activity (see above). The number of neurons expressing ChAT mRNA and the intensity of expression increase steadily, reaching adult levels at P16 (Li et al., 1995; Bender et al., 1996). Similar trends are seen with northern blot analysis of ChAT mRNA (Li et al., 1995). This ChAT mRNA expression parallels TrkA mRNA expressed by the same neurons, suggesting a role of trophic factors in the expression of ChAT mRNA (Li et al., 1995).

Specific ChAT immunoreactivity in the *rat* BF is first detected in the septum-diagonal band region at E17; at this time, only a few neurons are seen in the magnocellular basal nucleus (Armstrong et al., 1987; Dinopoulos et al., 1989; Fig. 8). This is in contrast to AChE activity, which is first detected in the basal nucleus (see above). ChAT immunoreactivity has been reported to be present in mitotic cells in the ventricular zones of developing forebrain as early as E13.5 in the *mouse* (Schambra et al., 1989), but this finding remains to be confirmed; there has been no other report of mitotic cells expressing neurotransmitters in the central nervous system (Rothman et al., 1980; Reiner et al., 1988). In the *rat*, ChAT-immunoreactive neurons are present in all BF regions by birth (Gould et al., 1991), although some studies did not find ChAT-ir neurons until later, at P7 (Sofroniew et al., 1987) or P8 (Bender et al., 1996). This discrepancy is probably due to differences in the antibodies used. Interestingly, similar to the first detection, the most intense ChAT immunoreactivity at P1 is still found in the medial septum, which also contains the highest percentage of AChE-positive

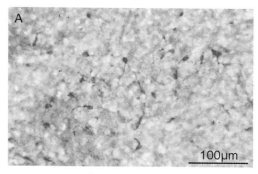

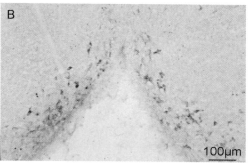

Fig. 8. ChAT-ir neurons in the developing basal forebrain of the rat. The morphology of these neurons is quite mature although they are small. A: magnocellular basal nucleus at E20. B: diagonal band region at P0.

neurons that are also ChAT-immunoreactive (Gould et al., 1991). The number of ChAT-ir neurons, and the staining intensity for ChAT in the septal complex increase until the end of the third postnatal week (Bender et al., 1996). These findings indicate that, in the rat, the maturation of intense ChAT immunor-eactivity in the BF occurs in a rostral to caudal direction, opposite to the gradient of the time of origin of the same neurons. This disparity suggests that the timing of phenotypic expression of the biosynthesizing enzyme does not correlate with the age of the neurons (Gould et al., 1991).

The late detection of ChAT compared to AChE might be, in part, due to the sensitivity of antibodies used. Thus, according to a study using the enzymatic assay of the rostral regions of the BF in the rat (Thal et al., 1991), ChAT activity in the medial septum, diagonal band and magnocellular preoptic area becomes detectable at E14, which is about the same time as the first detection of AChE activity in the same regions (no assay was conducted on the more

caudal, magnocellular basal nucleus). This is fol-lowed by a 10- to 15-fold increase by E18, and a rapid increase during the first four postnatal weeks, to reach peak values at P30. This increase is followed by a small decrease to adult levels at P60 (Thal et al., 1991). These patterns closely parallel the develop-ment of AChE activity (see above). There has been no study using enzymatic assay to investigate the development of ChAT activity in the basal nucleus.

In the *ferret*, ChAT staining is first detectable at E39, followed by a gradual increase during the first four postnatal weeks, to reach full intensity by P22–P29 (Henderson, 1991). Unlike in the rat, the inten-sity of ChAT immunoreactivity develops earlier in the magnocellular basal nucleus than in the septum and diagonal band region; no overshoot is reported in the ferret (Henderson, 1991).

An increase followed by a decrease in ChAT staining has been reported in the magnocellular basal nucleus of the *human* fetus, with ChAT activity increasing during gestation, peaking at about the 16th week, followed by a decrease (Perry et al., 1986).

In summary, previous studies are generally consistent regarding the prenatal onset of ChAT expression and its increase over the first three postnatal weeks in the rat. In many but not all of the species studied, this increase is followed by a subsequent gradual decrease to adult levels. However, information is incomplete with respect to possible regional differences in this overshoot. There are also a number of apparent inconsistencies regarding the spatial pattern of development of ChAT and AChE in the BF. Specifically, the spatial pattern of appearance of ChAT expression in the rat disagrees with that for AChE in the same species, and is also discrepant with the ChAT results of the ferret. The veracity of these differences remains to be clarified. One approach would be to use more sensitive ChAT antibodies.

Summary

The expression of AChE appears to begin shortly after future BF neurons undergo their last mitotic cycles, and to proceed as they migrate toward their final destinations. Intense AChE staining is seen as early as E14–15 in the rat, and there is a rapid increase postnatally with a peak at the end of the third week, followed by a small decrease to adult levels. In the rat the onset of AChE expression follows the neurogenetic gradient. The maturation of ChAT expression appears, by and large, to follow that of AChE, rather than preceding it, with a delay of approximately two days in the rat septal complex. However, evidence is inconclusive as to the pattern of the onset of ChAT expression across the BF regions. This requires further investigation.

Morphological maturation of basal forebrain cholinergic neurons

Golgi staining

The advantage of Golgi staining is that it offers detailed morphology of individual neurons, the disadvantage being that the neurotransmitter phenotype cannot be determined. Analyzing Golgi-stained neurons in the *rat* basal nucleus that display morphological features of cholinergic neurons, Gould et al. (1989) reported that basal nucleus neurons increase in cell body area, number of primary dendrites and length of dendrites, with a peak at P18. All of these values, however, decrease by P27, with the exception of dendritic length, which continues to increase to reach adult levels. For example, mean somatic areas are 218, 400, 328 and 260 μm^2 at P10, P18, P27 and in adult, respectively.

Also using Golgi staining in the rat but focusing on the diagonal band region, Dinopoulos et al. (1992) reported that the basic form (soma shape, number of primary dendrites, dendritic branching and dendritic varicosity) of large, potentially cholinergic neurons is already established by P4. However, the somatic size and total dendritic length increase further and stabilize to adult levels at P14, with no or little overshoot. The density of dendritic spines increases until P14, followed by a substantial decrease by P24 and a further gradual decrease to reach near-adult levels at P30.

These results of Golgi staining indicate rapid maturation of perikarya during the first two weeks after birth. In the basal nucleus, this is followed by rapid perikaryal shrinkage, readily noticeable by P24–30, and then a gradual tapering to reach

adult-like morphology. This overshoot in size, however, is not obvious in the diagonal band region. There is no clear explanation for this regional difference, but the lack of information on neurotransmitter phenotype of Golgi-stained neurons might be confounding the results.

p75-NGFR immunohistochemistry

Because most cholinergic neurons in the BF express p75-NGFR, immunoreactivity to this neurotrophin receptor has been used to study the development of BF neurons. In the *rat*, p75-NGFR-ir cells become discernible at E15, with their short neurites appearing over the next 4–5 days in a clear caudal to rostral gradient; at birth, these neurons display basic features of adult-like neuronal morphology with extensive dendrites (Koh and Loy, 1989). However, dendritic growth and branching, as well as increase in perikaryal size, continue during the first three postnatal weeks, which is followed by a decrease in dendritic arborization and sharp shrinkage in the size of perikarya to reach near-adult levels by P30. This period of hypertrophy is particularly noticeable in the basal nucleus (Koh and Loy, 1989). These observations roughly correspond to the Golgi observations of basal nucleus cells described above.

In the *ferret*, cell bodies immunoreactive for p75-NGFR are present at E28, two weeks before birth (Henderson, 1991). The intense labeling reveals both cell bodies and dendrites clearly. ChAT-immunoreactive neurons are not seen until approximately 10 days later. The precedence of p75-NGFR over ChAT is consistent with the hypothesis that NGF supports differentiation of cholinergic neurons.

ChAT immunohistochemistry

Consistent with the results with p75-NGFR immunoreactivity, ChAT-ir neurons in both the medial septum and basal nucleus in postnatal *rat* increase in size and dendritic complexity toward the end of the third postnatal week, followed by shrinkage and simplification to reach adult-like morphology by 45 days (Sofroniew et al., 1987; Gould et al., 1989, 1991). For example, somatic area grows rapidly over the first 2–3 postnatal weeks to 2–5 times the size at birth, and this is followed by a decrease by approximately 50% to reach adult size (mean somatic areas of 94, 380 and 270 μm² at P1, P18 and in adult, respectively, in the basal nucleus; Gould et al., 1991; Fig. 9). In addition, the intensity of ChAT immunoreactivity in individual cell bodies follows the same pattern with a transient increase followed by a decrease (Armstrong et al., 1987), although an increase without a subsequent decrease has also been reported (Gould et al., 1991). The peak of hypertrophy has been reported to progress in a caudal to rostral direction (Sofroniew et al., 1987), which is opposite to the rostral to caudal onset of ChAT expression (see above). In the diagonal band region, the marked growth of soma and dendrites precedes or coincides with the first detection of ChAT-ir fibers in the visual cortex, an increase in cortical and BF ChAT activity, and the period of maximal expression of NGF and NGF mRNA in the cortex (Dinopoulos et al., 1992).

In the *ferret* magnocellular basal nucleus, the size of ChAT-ir cell bodies peaks at P46, followed by a slight decrease to reach adult levels. In the diagonal band region, adult size is already attained by P35, with no further growth in size (Henderson, 1991).

Retrograde labeling

Using retrograde tracing with the lipophilic carbocyanine dye DiI on fixed postnatal (P1–14) rat brains, Calarco and Robertson (1995) reported that BF neurons retrogradely labeled from the visual cortex at P0 are located in the diagonal band region, and that this distribution is remarkably similar to the pattern seen in adults. The morphology of the labeled BF neurons is also remarkably mature already at P0, except that they are small. The size of cell bodies and the extent of dendritic fields increase steadily, as do dendritic branching and the number of dendritic spines (Calarco and Robertson, 1995). Similar results have been reported using wheatgerm agglutinin-conjugated horseradish peroxidase in vivo for the visual (Dinopoulos et al., 1989), medial prefrontal and somatosensory cortices (K. Semba and H. C. Fibiger, unpublished observations). As a caveat, it should be noted that the transmitter content of the retrogradely labeled neurons was not determined in

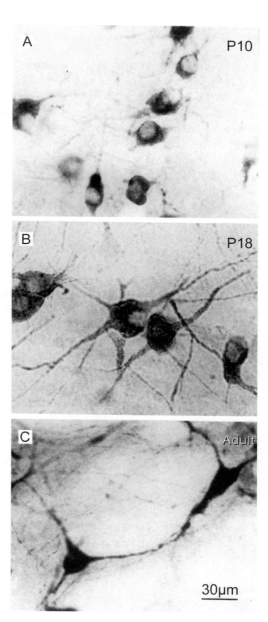

Fig. 9. Changes in the size and shape of ChAT-ir neurons in the rat basal forebrain during the first month after birth. ChAT-ir neurons at P18 are larger and display more dendrites than those at P10 or in adult. Adapted from Semba (1992). Reproduced by courtesy of Dr. Larry L. Butcher.

these studies. However, by inference from the available data with adult rats (Gritti et al., 1997; Manns et al., 2001), at least one-third are expected to be cholinergic.

Number of basal forebrain cholinergic neurons

In the *mouse*, the number of ChAT immunoreactive neurons in the medial septum and vertical limb of the diagonal band increases by 60% from P6 to P15 (Fagan et al., 1997; Ward and Hagg, 1999). In the magnocellular basal nucleus, this increase appears to be 2- to 3-fold from P0 to P60 (Villalobos et al., 2001). Because these neurons are known to be postmitotic by E18 (Semba and Fibiger, 1988), and because the intensity of immunolabeling of individual neurons increases during the same period (see above), it is likely that the increase in the number of ChAT-ir neurons is due to increased detectability.

Also in the mouse, the total number of neurons in the magnocellular basal nucleus that are retrogradely labeled from the medial prefrontal cortex increases rapidly from P4 to P8, followed by a decrease by P10 (Villalobos et al., 2000). At P13, the total cell counts drop to about half of the maximal values achieved at P8, and stabilize to adult levels at about P16. Regionally analyzed, this overshoot followed by a decline is apparent in the intermediate and posterior, but not the anterior, regions of the magnocellular basal nucleus. The increase in the number of retrogradely labeled neurons is likely due to increased ability of BF neurons to take up and transport tracers, and/or possible increases in the size and density of terminals in the cortex. The subsequent decrease in the number of retrogradely labeled neurons may reflect loss of noncholinergic neurons projecting to the medial prefrontal cortex, because there is no decrease in the number of ChAT-ir neurons in the same study.

Summary

The results with a variety of staining techniques suggest that cholinergic BF neurons begin to display a neuron-like form with short processes prenatally, at E15 in the rat. At birth, these neurons show considerable maturity in morphology, except that cell bodies are smaller and dendrites are shorter than in adult neurons. They grow rapidly during the first three weeks after birth.

The maturation of BF cholinergic neurons shows both regional and neurochemical differences. In terms of size, there is consistent evidence, based on Golgi, AChE, ChAT and p75 NGFR data, that the caudal, magnocellular basal nucleus neurons show hypertrophy followed by shrinkage by P30 in the rat. Evidence for hypertrophy for the more rostral, septal and diagonal band region neurons is less consistent. The transient increase in the activity of cholinergic enzymes, particularly AChE, in caudal BF neurons occurs slightly later than this morphological hypertrophy, with a peak between P30 and P60 in the rat.

The mechanisms of transient hypertrophy of many, if not all, cholinergic neurons are not clear. However, it is possible that trophic factors such as NGF and neurotrophin-3 have a role. In support of this possibility, NGF and NGF mRNA levels in the cerebral cortex reach a peak at P 21 in the rat (Large et al., 1986). In addition, maturation of morphology and cholinergic phenotype are sensitive to thyroid hormone (Gould and Butcher, 1989) and glucocorticoids (Hu et al., 1996). Loss or decease in these trophic factors and hormones might be responsible for a subsequent decrease in neuronal size and dendritic pruning.

Outgrowth of cholinergic axons of basal forebrain neurons and innervation of the cortex

As neurons begin to differentiate, their axons emerge and grow toward their target regions. The mechanisms underlying this directed outgrowth and formation of synaptic contact with target neurons are not well understood, although chemoaffinity (Sperry, 1963) and contact guidance (Horder and Martin, 1978; Székely, 1990) have been suggested as potential mechanisms. The growth of axons of cholinergic BF neurons into the cerebral cortex has been studied using immunohistochemistry for p75 NGFR and ChAT, AChE histochemistry, and retrograde tracing techniques. Not surprisingly, biochemical evidence indicates that AChE and ChAT activities develop first in the region of cell bodies, followed by their target areas (Thal et al., 1991).

The neocortex

AChE histochemistry. In the *rat*, early reports indicated that AChE-positive fibers are seen in the external capsule, internal capsule and in the deep part of layer 6 of the cortex at P4 (Gould et al., 1991; Kiss and Patel, 1992). Over the next several days, AChE-positive fibers innervate increasingly superficial layers of the cortex, with some regional differences, reaching all cortical layers by P7. The adult-like densities of AChE-positive fibers are seen by about 28 days after birth (Kiss and Patel, 1992). AChE levels in the neocortex rapidly increase between P5 and P30 (Thal et al., 1991).

Using a more sensitive AChE staining technique, De Carlos et al. (1995) compared the time course and spatial distribution of AChE-positive fibers arising from the thalamus and the BF. The great majority of AChE-positive fibers in the embryonic neocortex, in fact, appear to be thalamocortical axons from the sensory relay nuclei in the thalamus (De Carlos et al., 1995) expressing mature molecular forms of AChE (Gorenstein et al., 1991). However, these thalamo-cortical axons show laminar distributions, temporal characteristics and morphological features that are different from those of BF axons (Robertson et al., 1991). The BF afferents also follow a different trajectory than thalamocortical fibers (De Carlos et al., 1995). The first group of AChE-positive fibers derived from the BF arise in the magnocellular basal nucleus and substantia innominata, and reach the developing neocortex via the internal capsule at E18, which is approximately two days after thalamocortical fibers arrive in the subplate. AChE-positive fibers from more rostral BF regions do not reach the neocortex until several days later (De Carlos et al., 1995).

A similar pattern of outgrowth of AChE-positive fibers from the BF is seen in the *mouse*. AChE-stained fibers emerge from the magnocellular basal nucleus at E15, and enter the lateral cortex by P1 (Höhmann and Ebner, 1985).

In the *cat*, AChE-positive fibers reach the neocortex just before birth (Krnjević and Silver, 1966). In the visual cortex, the density of AChE-positive fibers rapidly increases, reaching a level higher than in adult by four weeks. This is followed by a decrease starting at 8 weeks to reach adult-like laminar patterns after three months of age (Bear et al., 1985). The period of high density of AChE-positive fibers corresponds to the critical period of visual plasticity in the cat (Hubel and Wiesel, 1970).

However, the possibility that some of this AChE expression belongs to thalamocortical afferents as in the rat has not been investigated.

In the *ferret*, AChE -positive fibers are seen to penetrate the cortical subplate at E35, and begin to invade the deep layers of the cerebral cortex a week before birth, initially forming a broad band in the middle of the cortical plate (Henderson, 1991). An adult-like distribution is assumed by P37 in the visual cortex, slightly later than in the medial, lateral and frontal cortices; transient increase in AChE staining is not mentioned in this study. Retrograde labeling also indicates that the BF projections to the cortex are adult-like at birth in terms of distribution of labeled cells (Henderson, 1991). AChE-positive fibers also reach the developing *human* neocortex just before birth (Kostović, 1986).

In summary, AChE-positive fibers reach the developing cortex prenatally in rat (E18) and ferret (E35), and just before birth in cat and human. The time course of maturation of laminar pattern shows regional differences, and a transient increase in density is reported in the feline visual cortex.

ChAT immunohistochemistry. In the *rat*, early studies detected ChAT-ir fibers in the cortex first during the second postnatal week; the intensity and density reached adult levels at about P35 (Dori and Parnavelas, 1989; Gould et al., 1991). The density of ChAT-ir puncta in the cerebral cortex first increased and then decreased slightly as a function of postnatal age (Oh et al., 1991).

A more recent postnatal study in the rat using a highly sensitive antibody reported significantly earlier arrival of ChAT-ir fibers, also revealing more morphological details (Mechawar and Descarries, 2001). Thus, a few ChAT-ir fibers bearing growth cones are present in the cortical subplate at birth, probably arriving via the anterior cingulate gyrus and, to a lesser extent, the external capsule. These fibers are mostly unbranched and have only few varicosities, presumably corresponding to sites of ACh storage and release. They then invade the cortex rapidly during the next two weeks, becoming more varicose and branching more often, and finally reaching all layers by P8 (Fig. 10). Maturation to adult-level densities of varicosities and fibers occurs first in the parietal cortex, then in the frontal cortex

26

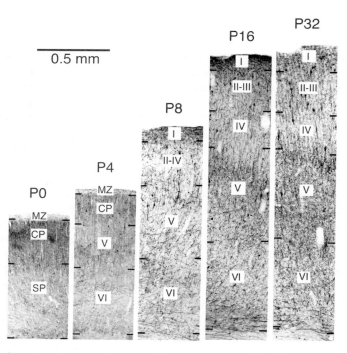

Fig. 10. Ingrowth of ChAT-ir fibers into the rat parietal cortex. Note the presence of ChAT-ir fibers in the subplate and deep aspect of the cortical plate at P0. This is followed by progressive ingrowth of ChAT-ir fibers and a rapid increase in density particularly during the first two weeks. Adapted from Mechawar and Descarries (2001). Reproduced by courtesy of the original authors.

and finally in the occipital cortex (Mechawar and Descarries, 2001; Fig. 11). For example, at P4, the density of varicosities is twice as high in the frontal and parietal cortices than in the occipital cortex. There is no overshoot in these values in any of the areas examined (Mechawar and Descarries, 2001). The early and rapid innervation of the cortex by ChAT-ir fibers, which were detected by a sensitive antibody, is in contrast to the earlier reports of much later development (see also Chapter 2).

The ChAT-ir terminals in the rat primary somatosensory area at P8, P16 and P32 have also been examined using electron microscopy (Mechawar et al., 2002). The pattern of connectivity is very similar to previous observations in adult rat, and classically defined synapses are observed only in a small fraction (17%, compared with 14% in adult) of varicosities. These results suggest a more diffuse mode of transmission, or volume transmission, in addition to classical synaptic transmission (Umbriaco et al., 1994), and this is obvious already at early stages of development. A similar observation has

been reported for noradrenergic terminals in the rat cortex, with a small trend for an increase in classical synapses (Latsari et al., 2002). These results suggest that both ACh and noradrenaline use a diffuse mode of transmission in the cortex both during development and in adulthood (see also Papadopoulos et al., 1989; Séguéla et al., 1990, for adult).

In the *cat*, the visual cortex has been studied much more extensively than other cortical regions, probably due to evidence for a role of ACh in experience-induced plasticity in ocular dominance in kittens (Bear and Singer, 1986). ChAT-ir fibers in the primary visual cortex are first detected at E54; although sparse, they are distributed throughout all cortical layers (Stichel and Singer, 1987). The fiber density then increases, beginning in layer 6 and continuing into more superficial layers, and the mature pattern of cholinergic innervation is established by 13 weeks postnatally (Stichel and Singer, 1987). This inside-out pattern parallels the pattern of neurogenesis in the primary visual cortex (Luskin and Shatz, 1985). In kittens, the critical period of visual

A.

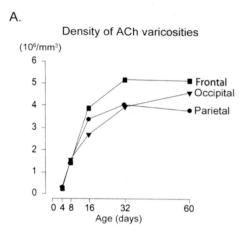

Density of ACh varicosities

(10⁶/mm³)

- Frontal
- Occipital
- Parietal

Age (days)

B.

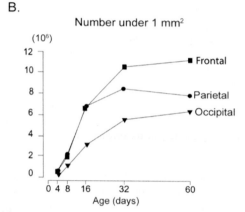

Number under 1 mm²

(10⁶)

- Frontal
- Parietal
- Occipital

Age (days)

C.

ChAT-ir axons in layer V
in occipital cortex at P8

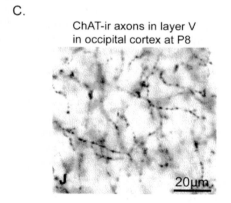

20μm

Fig. 11. Postnatal increase with age of the density of ChAT-ir varicosities and the number of varicosities under 1 mm² of cortical surface in three regions of the rat cerebral cortex. Note the rapid increase during the first two weeks, followed by a plateau. Adapted from Mechawar and Descarries (2001); their Figs. 6–8.

plasticity extends from 4 weeks to 3 months of age (Hubel and Wiesel, 1970), and there is interest in finding whether the development of cholinergic innervation has a role in this plasticity. Indeed, during this period, there is a transient increase in ChAT and AChE activities (Potempska et al., 1979) and muscarinic receptor binding (Shaw et al., 1984) in the cat striate cortex. However, these transient increases do not appear to be paralleled by the linear and gradual increase displayed by in-growing cholinergic fibers (Stichel and Singer, 1987).

In the *ferret*, ChAT-ir axons appear in subcortical fiber tracts at P22, and are present in the cerebral cortex by P46 (Henderson, 1991). Adult-level intensities of fiber staining in the cortex are not reached until P56. For comparison, tyrosine hydroxylase-ir fibers are already visible in the subcortical plate of the neocortex by E53, and these results were taken to conclude that cholinergic innervation of the cortex is late compared to catecholaminergic innervation (Henderson, 1991). However, the relatively late (P22) first detection of ChAT-ir fibers in this study might suggest a compromise due to the antibody used.

In summary, recent studies using sensitive ChAT antibodies have indicated that cholinergic fibers reach the cortex prenatally. This occurs at E54 for the kitten primary visual cortex. ChAT-ir fibers are also already present in the subplate at birth in the rat, but the prenatal outgrowth of these fibers has not been investigated using sensitive antibodies. The systematic information on the spatiotemporal patterns of arrival of cholinergic afferents in the cortex would have important implications for understanding the mechanisms of ingrowth of cholinergic fibers and the role of ACh in cortical maturation.

p75-NGFR immunohistochemistry. In the *rat*, p75-NGFR-ir fibers emerge from the magnocellular basal nucleus at E15, and penetrate the lateral aspect of the embryonic cortex at P1 (Koh and Loy, 1989). p75-NGFR-ir fibers are also found in the subplate as early as E15, but these fibers are likely to be derived from subplate neurons (Koh and Loy, 1989), which are known to express p75-NGFR (De Carlos et al., 1995). In the *ferret*, diffuse, noncellular p75-NGFR immunoreactivity is seen in the subplate of the emerging cortex at E28, and clearly delineated immunoreactive axons, presumably representing BF afferents, are seen in the cortical plate by P7 (Henderson, 1991).

Retrograde and anterograde tracing. Retrograde tracing results with DiI on fixed prenatal *rat* brains indicate that a very small number of neurons in the basal nucleus and substantia innominata are retrogradely labeled from the rostral cortex at E17, which is at least one day later than the first appearance of labeling in the thalamus; labeling in the diagonal band and ventral pallidum is not seen until near-birth (De Carlos and O'Leary, 1992; De Carlos et al., 1995). The labeling of BF neurons at E17 is also confirmed using in utero injections of Fast Blue or DiI into the rostral cortex at E17 (De Carlos et al., 1995). Anterograde labeling in fixed brains in the same study also indicates that the first BF neurons innervating the neocortex, which are caudal BF neurons, project through the internal capsule, as do thalamocortical axons, but more rostral BF neurons subsequently send their axons through the external capsule and cingulate bundle (De Carlos et al., 1995). In the adult rat, BF neurons project through these two latter pathways, and not through the internal capsule (Saper, 1984).

Postnatal studies confirm that BF projections to the rat visual (Dinopoulos et al., 1989; Calarco and Robertson, 1995), somatosensory and medial prefrontal cortices are present at P0 (K. Semba and H. C. Fibiger, unpublished observations; Fig. 12). Deeper placement of the tracer in the cortex results in larger numbers of retrogradely labeled neurons, suggesting that density of axons from the BF is greater in deep, compared to superficial, layers of the visual cortex (Calarco and Robertson, 1995). Anterograde tracing with DiI reveals that labeled fibers are seen in the subcortical white matter and subplate (layer 6) of the occipital cortex at P0, but not in the cortical plate nor in the most superficial, marginal zone (Calarco and Robertson, 1995; Fig. 13). Growth cones are typically seen at the tip of growing axons. Labeled fibers begin to reach the newly differentiated layer 6 of the cortical plate by P3. By P6, all cortical layers have been differentiated, and labeled fibers are seen in all these layers, although they are still denser in deep layers. As in the adult rat (Saper, 1984), most BF axons course dorsally, turning around the genu of the corpus callosum,

and then travel caudally to the visual cortex, whereas some fibers travel laterally through the striatum to reach the external capsule and travel caudally therein (Calarco and Robertson, 1995). It should be noted that the transmitter content of labeled fibers was not identified, and the possibility of labeling of passing fibers cannot be excluded (Calarco and Robertson, 1995). Nevertheless, the time course and pattern of distribution of anterogradely labeled axons closely resemble those of ChAT-immunoreactive neurons reported by Mechawar and Descarries (2001); see also Chapter 2).

Summary. The results based on AChE histochemistry, ChAT immunohistochemistry and tract tracing techniques indicate that the axons of developing BF cholinergic neurons begin to emerge from cell bodies prenatally, at E15–16 in the rat, starting with the most caudal neurons, and that these axons grow toward the developing cerebral cortex via three pathways. The first set of fibers reach the cortex prenatally, at E18 in the rat, and this is followed by the arrival of more fibers and rapid ingrowth into the deep to superficial layers of the cortex, with increasing densities, during the first two weeks after birth. These in-growing axons from the BF begin to express AChE prenatally, at E18 in the rat. ChAT is expressed in these fibers by birth at the latest. In parallel with increasing densities of cholinergic fibers in the neocortex, significant increases in the levels of ACh, choline and ChAT occur during the first month, in particular during the second and third postnatal weeks in the rat (Ladinsky et al., 1972).

The mechanisms for laminar differences in cholinergic innervation remain to be investigated. Initially, during the first two weeks after birth in the rat, both the density of varicosities and the number of varicosities per unit length of axon, as well as axonal branching, increase in parallel in all layers and regions in the cortex (Mechawar and Descarries, 2001; Fig. 11). It is possible, therefore, that the final laminar and regional differences in the density of ACh innervation might be due to regulatory factors that are distributed differentially across cortical layers and act subsequent to ingrowth of these axons (Conner and Varon, 1997; Sanes and Yamagata, 1999).

The machinery for cholinergic transmission is not complete without postsynaptic receptors and

PO   Adult

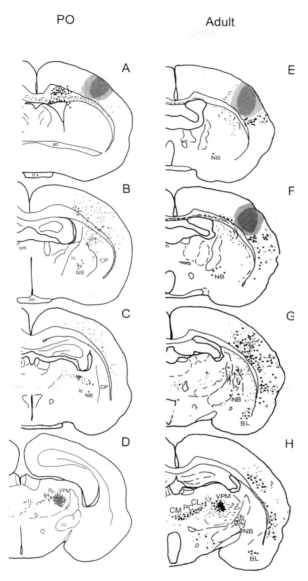

Fig. 12. Distributions of retrogradely labeled neurons (large dots) and labeled fibers and terminals (fine lines and dots) after wheatgerm agglutinin-conjugated horseradish peroxidase (1%, 0.05 μl; shaded areas for core and halo) injection into the barrel region of the somatosensory cortex in P0 (A–D) and adult rats (E–F). Both rats were perfused 24 h after tracer injections. Note the remarkable similarity in the distribution of labeled neurons in the nucleus basalis (NB), and the medial ventroposterior nucleus and intralaminar nuclei of the thalamus. Some cells in the corpus callosum are labeled at P0 (A) possibly due to phagocytosis of the tracer by glial cells. ac: anterior commissure; BL: basolateral amygdala; CL: centrolateral thalamic nucleus; CM: central medial thalamic nucleus; CP:

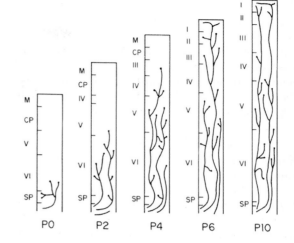

Fig. 13. A schematic representation to illustrate progressive ingrowth of axons anterogradely labeled with DiI from the diagonal band region of the fixed rat BF. The pattern is similar to that seen with ChAT immunoreactivity shown in Fig. 10. Adapted from Calarco and Robertson (1995). Reproduced by courtesy of the original authors and permission of Wiley-Liss.

recycling mechanisms for the transmitter. According to a study using pharmacological binding in vitro and autoradiography in the rat, the densities of vesicular ACh transporter and nicotinic binding sites in selected neocortical regions are already about 92% and 36% of adult levels, respectively, at E20 (Aubert et al., 1996). In contrast, the densities of M1 and M2 binding sites are only at 11 and 8%, respectively, at E20, and both reach adult levels at P14. The sites for high affinity choline uptake (plasma membrane transporter) develop more slowly, becoming detectable only at P7, and not reaching adult levels until at P60. These results are largely consistent with previous reports in the mouse based on biochemical and pharmacological assays for ACh and vesicular ACh transporter (Coyle and Yamamura, 1976; Brown and Brooksbank, 1979) and immunoreactivity for muscarinic receptor subunit (m1, m2, and m4; Höhmann et al., 1995). Together, these findings suggest that ACh released by cholinergic fibers entering the neocortex will have the machinery

caudate-putamen; ic: internal capsule; IL: intralaminar thalamic nucleus; NB, nucleus basalis; ox: optic chiasm; PC: paracentral thalamic nucleus; Po: posterior thalamic nucleus; sm: stria medullaris; VPM, medial posteroventral thalamic nucleus.

necessary to produce fast excitatory actions through nicotinic receptors by birth. Mechanisms for the slower muscarinic actions including autoinhibition will take more time to develop, with the uptake mechanisms for choline after hydrolysis of ACh maturing at about two months of age in the rat.

The hippocampus

AChE histochemistry and p75-NGFR immunohistochemistry. The cholinergic input to the hippocampus arises mainly from the medial septum and the nucleus of the vertical limb of the diagonal band. In the *rat*, p75-NGFR-ir fibers begin to emerge from the medial septum-diagonal band region at E15, and reach the most rostral pole of the hippocampus by E17, and other parts of the hippocampus by P1 (Koh and Loy, 1989). Postnatal studies using AChE histochemistry reported that AChE-positive fibers are seen consistently in the rostral parts of the dentate gyrus, hippocampus and subiculum by P2–P3, and by P5–P11 these fibers innervate the entire hippocampal formation, assuming adult-like patterns by P14–21 (Mellgren, 1973; Matthews et al., 1974; Milner et al., 1983; Henderson, 1991; Nyakas et al., 1994). Densities of AChE fibers in the hippocampus reach adult levels by the end of the fourth postnatal week (Gould et al., 1991). Similarly, biochemical assays indicate that the AChE levels in the hippocampus increase rapidly between P5 and P30 (Thal et al., 1991).

Makuch et al. (2001) investigated the timing of the arrival of cholinergic afferents from the septum in relation to the development of their target neurons, using the dentate gyrus as a model system. In adult rat, the highest density of cholinergic fibers in the hippocampal formation is seen in the external blade of the dentate gyrus, and AChE staining in this region is a selective marker for cholinergic afferents from the septum. These AChE-positive fibers innervate the external blade of the dentate gyrus two days after granule cells have differentiated to form a clearly defined layer in the external blade and begin to express mRNA for NGF and neurotrophin-3. These results are consistent with the notion that target cell development is required for ingrowth of septal afferents to the external blade of the dentate gyrus.

In the *cat* (Krnjević and Silver, 1966) and *human* (Kostović, 1986), AChE-positive fibers from the medial septum grow prenatally toward and along the medial wall of the cerebral hemisphere, and just before birth, these fibers penetrate the hippocampus. This event appears to occur at the same time as the penetration of AChE-positive fibers into the neocortex (see above).

ChAT immunohistochemistry. Biochemical studies have reported a lag between development of AChE and ChAT activities in the *rat* hippocampus (Nadler et al., 1974). At birth, ChAT activity in the hippocampal region of the rat is about 10% of adult levels, increasing three-fold by the end of the first postnatal week and reaching adult levels by the end of the third week. In contrast, AChE activity is already about 25% of adult levels in neonates, and matures more gradually than ChAT, reaching adult levels by P32. For comparison, high-affinity choline uptake, a presynaptic marker for cholinergic axons, is detectable at P6 (the earliest day examined) in the fascia dentata of the hippocampal formation, increasing three-fold by the third postnatal week, followed by an overshoot by 60% before decreasing to adult levels (Shelton et al., 1979).

The results with ChAT immunohistochemistry are discrepant from these biochemical results. In the *rat*, ChAT immunoreactivity is not detected in growing cholinergic axons until after they form synaptic contacts (Henderson, 1991; Thal et al., 1991). ChAT-positive fibers begin to emerge from the medial septum and ventral limb of the diagonal band toward the end of the first or the beginning of the second postnatal week. Stained fibers grow into the fornix and fimbria, and begin to appear in the hippocampus by P10, followed by a steady increase to reach adult-level densities in the entire hippocampal formation by P35 (Gould et al., 1991). These timings are considerably later than those indicated by biochemical techniques, or histological staining for p75 NGFR or AChE. They are also later than the appearance of the same markers in the neocortex (see above). It is likely that relatively late detection of ChAT-ir fibers in the hippocampus is due to lower sensitivity of the antibodies used.

Retrograde and anterograde tracing. Retrograde tracing studies in the *rat* indicate that the projection from the medial septum to the hippocampus is present at E20, and the distribution of retrogradely labeled neurons is similar to that seen in adult (Milner et al., 1983).

Results using anterograde tracing with DiI on fixed embryonic and postnatal rat brains are similar to those of p75-NGFR staining (see above); septohippocampal fibers reach the anterior pole of the hippocampus by E17, and the fimbrial pole of the hippocampus at E18 (Linke and Frotscher, 1993). The majority of the septohippocampal fibers arrive in the hippocampus shortly after birth, and an adult-like pattern of septohippocampal innervation is seen at P10 (Linke and Frotscher, 1993). In the dentate gyrus, results with anterograde labeling with DiI in the septal region indicate the presence of few labeled fibers in the external blade of the dentate gyrus, although intense fiber labeling is present in the fornix, at P2 (Makuch et al., 2001). At P6–8, modest labeling is seen in the external blade, with more prominent labeling in the internal blade. The densities increase in the external blade between P7 and P11. The time course and pattern of these anterogradely labeled fibers are remarkably similar to those of AChE-positive fibers (Makuch et al., 2001). However, the possibility of labeling fibers of passage in the septum cannot be ruled out in these studies, and the transmitter content of retrogradely labeled neurons and DiI-labeled fibers remains to be identified.

Summary. The results using p75NGFR immunohistochemistry and AChE histochemistry as well as tract tracing techniques indicate that septohippocampal fibers emerge from their cell bodies in the medial septum and diagonal band region at E15, and reach the hippocampal formation by E17 in the rat. By P1 all hippocampal regions are innervated by septohippocampal fibers, and an adult-like pattern of cholinergic innervation is seen by 2–3 weeks after birth. Thus, the development of the cholinergic septohippocampal cholinergic projections fairly closely parallels that of the cholinergic innervation of the neocortex, at least in the rat. There is some evidence that the penetration of a target region in the hippocampal formation follows the cellular maturation of the target region. The development of cholinergic receptors in the hippocampus also appears to take similar time courses as for the neocortex, with considerable maturation of nicotinic

receptors at birth, followed by development of muscarinic receptors over the first 2 weeks after birth (Höhmann et al., 1995; Aubert et al., 1996). Incoming cholinergic axons to the hippocampus are probably capable of releasing ACh exocytotically in an action potential-induced calcium-sensitive manner by P3, with inhibitory autoreceptors on these fibers expressed by P15–16 (Goldbach et al., 1998).

Other target areas

Although the neocortex and hippocampus are the main targets of cholinergic BF neurons, BF cholinergic neurons also project to other structures, including the thalamus, amygdala and olfactory bulb. These projections may have roles in different aspects of vigilance (Semba, 2000). In particular, the cholinergic BF projection to the thalamus is mainly directed to the reticular thalamic nucleus, which plays a critical role in regulating cortical activation (Steriade et al., 1997). The BF can, therefore, influence cortical activity via this indirect route as well as through direct projections. Currently, little information is available on the development of cholinergic BF projections to these additional targets. Nonetheless, a brief summary of what is known might be useful.

The ChAT activity determined by enzymatic assay in the midbrain-thalamus is higher at birth, and reaches adult levels earlier, than that in the cortex of the *rat* (Coyle and Yamamura, 1976). The ChAT activities specifically in the lateral geniculate nucleus and pulvinar also mature later than those in the cortex in the *rat* and *cat* (Kvale et al., 1983; Fosse et al., 1989; Carden et al., 2000). However, most of the ChAT activities measured in these studies are probably not of BF origin, because, of the nuclei examined or included in the analysis, only the reticular nucleus in both rat and cat, and the mediodorsal, anterior and ventromedial nuclei in cat receive cholinergic projections from the BF (reviewed in Semba, 2000).

There is some evidence that the cholinergic innervation of the amygdala by BF neurons precedes that of the cortex. The ChAT activity in the amygdala, which mostly represents cholinergic input from the BF, is already about 50% of adult level at P8, reaching near adult levels at P25 in the *rat* (Brown and Brooksbank, 1979). In the *ferret*, ChAT-ir axons appear in the basolateral nucleus of the amygdala at P22, before they appear in the cortex at P46 (Henderson, 1991).

The most of the AChE activity in the olfactory bulb is attributable to cholinergic fibers from the BF. The AChE and ChAT activities in the *rat* olfactory bulb are about 20% of adult values at birth, and remain similar through P10, followed by a rapid increase to near adult values at P25 (Brown and Brooksbank, 1979; Meisami and Firoozi, 1985; Rea and Nurnberger, 1986). Similarly, AChE-positive and ChAT-ir fibers are seen in the *rat* olfactory bulb at birth, and their densities increase gradually to reach adult levels at about a month postnatally (Le Jeune and Fourdan, 1991).

In summary, cholinergic BF projections to the amygdala mature earlier than those to the cerebral cortex, which includes the neocortex, hippocampus and olfactory bulb. The development of the cholinergic BF projections to the thalamus remains to be investigated.

Development of noncholinergic basal forebrain neurons

Cholinergic neurons are by no means the only neuronal type in the BF, but noncholinergic neurons are intermixed with the cholinergic neurons and many of them project to the cerebral cortex (Semba, 2000). Some of these neurons are GABAergic, whereas others are suggested to be glutamatergic (Gritti et al., 1997; Manns et al., 2001). Relatively little is known about the development of these noncholinergic BF neurons.

Comparing the time of neurogenesis of ChAT-ir neurons and Nissl-stained neurons, cholinergic neurons have been suggested to be among the first neurons to leave mitotic cycles in the rat BF (Bayer, 1985; Semba and Fibiger, 1988). It appears, therefore, that most noncholinergic BF neurons are generated later than cholinergic neurons. This implies that the more mature, cholinergic neurons might influence the development of noncholinergic neurons.

The neurogenesis data suggest that, as noncholinergic neurons, GABAergic BF neurons are not generated earlier than the cholinergic neurons.

However, GAD67 mRNA is detected in the rat septal complex already at E17, three days before ChAT mRNA becomes detectable at E20 (Bender et al., 1996). The reason for this accelerated differentiation of GABAergic, compared to cholinergic, neurons is unknown. By E20, many GAD67 mRNA-expressing neurons are seen in the medial septum and diagonal band region, and the staining intensity increases postnatally, reaching near adult levels by P22; the number of labeled neurons does not change (Bender et al., 1996). In adult rat, approximately 40% of all cells in the medial septum express GAD67 mRNA (Bender et al., 1996) and GAD67 is a prominent isoform of GAD in the medial septum (Esclapez et al., 1994).

GABAergic neurons in the septum that project to the hippocampus contain the neuropeptide parvalbumin in the adult rat (Freund, 1989), and this colocalization provides an opportunity to study the time course of development of the classical transmitter and a colocalized peptide in the same neurons. The first GAD-ir neurons in the BF are found as early as E16 (Lauder et al., 1986). In contrast, the immunoreactivity for parvalbumin is first detected in the septum at E21 (Solbach and Celio, 1991; but see Bender et al., 1996). Both events occur before the arrival of septohippocampal fibers in the hippocampus at E19 (Linke and Frotscher, 1993). Postnatally, parvalbumin-ir neurons increase in number first, and then in staining intensity, both reaching adult levels by P20 (Bender et al., 1996). These results suggest that septohippocampal neurons initially express the colocalized neuropeptide several days later than their primary transmitter GABA, but that both molecules reach maturation about the same time, i.e., three weeks postnatally.

The fact that septal neurons express GAD67 mRNA prior to the arrival of septohippocampal fibers in the hippocampus (Linke and Frotscher, 1993) suggests the possibility that, unlike the cholinergic phenotype, expression of the GABAergic phenotype might not be regulated by target-derived factors such as neurotrophins (Linke and Frotscher, 1993; Bender et al., 1996). However, GABAergic septal neurons include both septohippocampal neurons and local interneurons, and it is possible that the earliest GAD expression is attributable to interneurons, and that septohippocampal GABAergic neurons do not express GAD until later. In fact, the removal of hippocampus at P5 or P10 reduces the number of cholinergic septal neurons to 60% and that of parvalbumin-ir septal neurons (which represent GABAergic septohippocampal neurons) to 62% (Plaschke et al., 1997). These findings suggest that the development of both cholinergic and GABAergic septohippocampal neurons is dependent on factors derived from the hippocampus, at least between P5-10 in the rat.

In conclusion, GABAergic neurons in the septal complex appear to mature earlier than their cholinergic counterparts, despite the fact that their time of origin is not earlier than that of cholinergic neurons. The reason for the fast development of GABAergic neurons compared to cholinergic neurons is unclear.

Development of cholinergic interneurons in the cortex

In the *rat*, intrinsic cholinergic neurons are present in the adult neocortex and hippocampus (Semba and Fibiger, 1989). These neurons express ChAT protein but probably very small amounts of ChAT mRNA (Butcher et al., 1992; Lauterborn et al., 1993). Developmentally, faint ChAT-ir neurons are detected first at P4 in the frontal, parietal and occipital cortices, and these neurons become strongly immunoreactive at P8 (Mechawar and Descarries, 2001).

In fetal *monkey* (*Macaca mulatta*) cortex, ChAT-ir neurons are present in layers 5 and 6 as well as in the subjacent white matter of selected cortical regions examined, at 110–150 days of gestation (Hendry et al., 1987). However, ChAT-ir neurons appear to be absent from adult monkey cortex (see above).

In the *human* cerebral cortex, AChE-positive neurons are expressed in layers 3 and 5 after childhood and during adulthood (Mesulam and Geula, 1991). However, these neurons are not cholinergic. This AChE might have a role in neural differentiation, neurite outgrowth and morphogenesis (Mesulam and Geula, 1991).

Conclusions and unsettled issues

The following summarizes the main features of the development of cholinergic BF neurons and their innervation of the cerebral cortex. It is primarily based on the findings in the rat, but is consistent with the available information for the mouse, cat, ferret and primates including human.

(1) Cholinergic BF neurons emerge from the ganglionic eminence of the telencephalic wall. They become postmitotic in a caudal to rostral gradient, and are among the first cells in the telencephalon to differentiate. They migrate laterally and rostrally to reach their final destinations.

(2) AChE is detected first, followed by ChAT, in the migrating neurons. p75 NGFR is expressed concurrently with AChE, and the onset of AChE/p75 NGFR expression follows the neurogenetic gradient; evidence for a gradient of ChAT expression is incomplete. The expression of AChE and ChAT in the cell bodies increases rapidly during the first three weeks after birth. This is followed by a small decrease in the rat, and this overshoot is selective to the cell bodies and is not seen in the cortex, the terminal areas of these cholinergic neurons.

(3) Cholinergic BF neurons begin to grow dendrites prenatally, and show considerable maturity in their morphology at birth, although they are small and their dendrites are relatively short. Postnatally (during the first three weeks after birth in the rat), these neurons grow rapidly in size, and their dendrites grow longer. Transient hypertrophy of cholinergic BF neurons is evident particularly in caudal regions such as the magnocellular basal nucleus.

(4) The axons of cholinergic BF neurons emerge prenatally, and reach the neocortex and hippocampus shortly thereafter. Initially, these axons express minimal amounts of cholinergic enzymes. By the time of birth, these axons stain clearly for AChE and ChAT and reach the cerebral cortex, which is followed by rapid ingrowth and further increase in the enzymatic expression. The adult-like pattern of cortical innervation is seen by 2–3 weeks after birth in the rat, with regional variations.

(5) The incoming cholinergic axons in the developing cerebral cortex are capable of releasing ACh shortly after birth. Sufficient apparatus appears

to be present by the time of birth for excitatory actions through nicotinic receptors (Aubert et al., 1996). Muscarinic M1 receptors follow, and M2 receptors and high-affinity choline uptake develop more slowly to complete inhibitory and recycling mechanisms. Cholinergic transmission is in operation probably at a level close to adult levels by a month postnatally, and is totally mature by two months after birth (Aubert et al., 1996). ACh in the developing cortex has been thought to have roles in synaptic plasticity (Bear and Singer, 1986; Gu and Singer, 1993), morphological development such as cellular differentiation, and refinement of cortical connectivity (Höhmann et al., 1988; Aramakis et al., 2000; reviewed by Höhmann and Berger-Sweeney, 1998; Lauder and Schambra, 1999).

(6) Although the mechanisms for the maturation of cholinergic neurons and their axons are beyond the scope of this chapter, selected findings discussed in this chapter indicate that various neurotrophins such as BDNF, neurotrophin-3 and NGF present in the cerebral cortex have roles in cellular maturation, phenotypic expression and axonal growth and guidance of cholinergic, and possibly also GABAergic, BF neurons. In addition, it is also possible that afferents and target cortical neurons interact and influence each other during establishment of connections (Makuch et al., 2001).

While the above information represents significant advances made over the last decade in our knowledge of the ontogeny of cholinergic BF neurons, there are a number of unresolved questions that require further attention:

Are cholinergic BF neurons derived from a single or dual germinal source? It is not clear whether cholinergic BF is derived from a single or dual germinal source, despite an apparently continuous time of origin. Previous studies using AChE, p75 NGFR and ChAT did not report clear continuity between septal neurons and more caudal neurons in the substantial innominata and basal nucleus. The possibility that the septal cholinergic neurons are derived from a separate progenitor has been raised on the basis that, in trisomy 16 mice, rostral cholinergic neurons were generated two days earlier than in euploid littermates, while no difference was seen with caudal cholinergic neurons (Sweeney et al., 1989). It was proposed that the rostral neurons are derived from the embryonic lateral ventricle, whereas the caudal neurons originate in the embryonic third ventricle (Sweeney et al., 1989). Cholinergic neurons in the septum show a number of neurochemical features that are not fully shared by more caudal cholinergic neurons (Semba, 2000), and these differences might possibly be related to their different origins. Molecular genetics markers might be useful to address this lineage question.

What is the time course of expression of ChAT, AChE and vesicular ACh transporter in embryonic BF cell bodies? Information on the spatiotemporal pattern of ChAT expression in BF neurons at the cell body level is incomplete. Available data are also inconsistent with the spatial pattern of neurogenesis and AChE expression. While there may be real differences between these enzymes, it is also possible that this is due to technical problems, in particular the antibodies used. In fact, it has been suggested that ChAT protein undergoes posttranslational changes and that most available antibodies may not detect immature forms of ChAT (Bender et al., 1996). The results by the Descarries group for postnatal studies using a sensitive ChAT antibody (Mechawar and Descarries, 2001; Mechawar et al., 2002; see also Chapter 2) closely resemble previous results with retrograde labeling. It is, therefore, possible that the use of such an antibody would produce information on the prenatal development of ChAT expression. In situ hybridization for ChAT mRNA might also be useful. It will also be important to use vesicular ACh transporter antibodies to study the development of BF cholinergic neurons. These three proteins, ChAT, AChE and vesicular ACh transporter, have different functions in cholinergic transmission, and comparison of their development might provide useful insights as to how cholinergic transmission matures.

What is the time course of development of noncholinergic BF neurons innervating the cerebral cortex? There is a substantial population of noncholinergic BF neurons projecting to the cortex, and some of these are GABAergic. Although some information is available on the organization and function of these noncholinergic BF neurons in adulthood (reviewed by Semba, 2000), little information is available on their development. It will be important to investigate and compare the functions of these two components of the BF system both in adulthood and during development. One intriguing question might be whether the relative proportions change between the two phenotypic components, and if so, what mechanisms regulate such changes. Comparison of these results with those from comparative analysis across vertebrates will provide further insights into the function of the BF and its cortical projections.

How does the arrival and ingrowth of cholinergic axons into the cortex compare to those of other cortical afferents? The cholinergic fibers are not the only fibers arriving in the developing cerebral cortex during the perinatal period. To understand their function in cortical development, it is important to know how the pattern of arrival and ingrowth of cholinergic afferents compare to those of other cortical afferents. Noradrenergic fibers reach the anterior parietal cortex around E20, and occipital cortex at birth (Verney et al., 1984). Dopaminergic fibers enter the cortical plate just before birth (Kalsbeek et al., 1988), whereas serotonergic fibers and terminals are seen in the cortex by P1 (Seiger and Olson, 1973; Lidov and Molliver, 1982). Thalamocortical afferents arrive in layer 4 of the rat visual cortex by P2 (Calarco and Robertson, 1995). Although a delay in cholinergic innervation has been reported earlier (see above), a more recent study using sensitive antibodies to ChAT has indicated considerably earlier innervation, comparable to some of the other cortical afferents (Mechawar and Descarries, 2001). The information regarding the spatiotemporal pattern of arrival and ingrowth of cholinergic fibers in relation to that of other cortical afferents would provide useful information as to the role of cholinergic neurons in cortical maturation.

Conclusions

Both phylogenetic and ontogenetic approaches have been used in the past to understand the formation and organization of the brain. The phylogenetic approach aims to understand how the brains of living animals evolved through interaction of both genetic and environmental factors during phylogeny. The ontogenetic approach aims to uncover cellular and molecular processes that underlie the formation of the brain. The famous principle of Haeckel stated that 'ontogeny recapitulates phylogeny' (cited in Sarnat and Netsky, 1981). Although this principle is not currently accepted, the parallelism between phylogeny and ontogeny has been one of the main themes in modern biology (Gould, 1977), and it appears that most biologists now believe that ontogenesis and phylogenesis are intimately related developmental programs for each species (Sarnat and Netsky, 1981). How then do the evolution and ontogeny of the cholinergic BF system compare with each other? What do we learn from this comparison?

There are both similarities and differences. One similarity concerns the relative time of appearance. The cholinergic BF system is a common feature of tetrapods and bony fishes, but not of 'lower' fishes. This relative time of appearance might be reflected in the timing of neurogenesis of BF cholinergic neurons. Specifically, these neurons are generated from E12–16 in the rat. In contrast, noradrenergic, serotonergic and cholinergic neurons in the pons, which also innervate the cortex, are generated earlier, at E11–12, E12–13 and E12–13, respectively (K. Semba and H. C. Fibiger, unpublished observations; also Phelps et al., 1990). More rostrally, dopaminergic neurons in the midbrain and histaminergic neurons in the posterior hypothalamus are generated at E13–14 (K. Semba and H. C. Fibiger, unpublished observations) and E16 (Reiner et al., 1988), respectively. Thus, as a group, the cholinergic BF neurons are generated later than these aminergic and cholinergic neurons. They are also located most rostrally of all these projection groups. The relatively late appearance of cholinergic neurons, therefore, might reflect encephalization of the brain both during phylogeny and ontogeny.

Another similarity regards the timing of morphological maturation of perikarya at a relatively early stage. Morphologically, the cholinergic neurons in the ventral telencephalon in 'lower' species generally display considerable maturity, in terms of the shape of cell body, the number of dendrites and dendritic ramification, although they are still small. The degree of maturity in perikaryal shape is indeed quite comparable to that seen in adults of higher species, with the exception of primates, which tend to show even greater morphological differentiation. This relative maturity in cellular morphology in lower species parallels a relatively mature appearance of BF cholinergic perikarya at birth.

There is also a difference between phylogenetic and ontogenetic development. The cholinergic neurons in the ventral telencephalon appear to shift their positions somewhat during evolution, apparently in association with general evolution of the brain. There are indications that in 'lower' species, these cholinergic neurons are more closely associated with olfactory structures that are located rostrally, whereas in higher species, this association is weaker as the major part of cholinergic neurons are increasingly located more caudally, away from olfactory structures. In particular, remarkable growth is seen in the human brain for the magnocellular basal nucleus, which is the caudalmost part of the cholinergic BF complex. This shift might be related to the gradual decrease in the relative significance of olfaction, along with the development of lateral hemisphere and cerebral cortex, during evolution. This rostral to caudal shift, however, is in stark contrast to the characteristic caudal to rostral gradient of neurogenesis of these cholinergic neurons during ontogeny. That is, phylogenetically, the rostral cholinergic BF neurons appear to be the oldest, whereas ontogenetically, they are the newest. Nevertheless, it is currently not clear whether or not the onset of ChAT expression follows the neurogenetic gradient, and there is also a question of separate progenitor cells for the rostral, septal cholinergic BF neurons. These ontogenetic issues might be related to this discrepancy between evolutionary and ontogenetic trends.

In conclusion, the comparative and ontogenetic data on the cholinergic BF system show both differences and similarities. However, the data are yet incomplete. More species need to be examined for each class, and embryonic development needs to be studied using more sensitive markers. Nevertheless, it is tempting to speculate what might be the driving force for the evolution of the cholinergic BF system. One possibility is that if the cortex requires a cholinergic input for its function, an increasingly larger cortex will require an increasingly larger cholinergic BF system. ACh is indeed implicated in various 'higher' functions, such as attention, learning and memory. As the cerebral cortex acquires 'higher'

functions both during evolution and during ontogenesis, there will be a need for a larger and more complex cholinergic BF system. Future investigations using neurochemical, connectional and molecular markers are likely to shed more light on these intriguing questions about this important group of brain cells.

Acknowledgments

I thank Dr. Richard Wassersug for constructive discussions on an early version of the phylogenetic aspects section, and all the original authors of Figs. 1, 2A, 2B, 3, 4, 6, 9–11 and 13 for their kind permission to reproduce these figures in this chapter. I also thank Dr. Doug Rasmusson and the editors of this volume for helpful comments and suggestions on an early version of the manuscript. Ms. Joan Burns assisted in the preparation of figures. Supported by grants from Canadian Institutes of Health Research (MOP14451) and Natural Sciences and Engineering Research Council (217301-99).

Abbreviations

ACh	acetylcholine
AChE	acetylcholinesterase
BDNF	brain derived neurotrophic factor
BF	basal forebrain
ChAT	choline acetyltransferase
E	embryonic day
ir	immunoreactive
NGFR	nerve growth factor receptor
NT	neurotrophin
P	postnatal day
Vc	area ventralis telencephali pars centralis
Vd	area ventralis telencephali pars dorsalis
Vl	area ventralis telencephali pars lateralis
Vv	area ventralis telencephali pars ventralis

References

Adrio, F., Anadón, R. and Rodríguez-Modes, I. (2000) Distribution of choline acetyltransferase (ChAT) immunoreactivity in the central nervous system of a chondrostean, the Siberian sturgeon (*Acipenser baeri*). J. Comp. Neurol., 426: 602–621.

Albanese, A., Gozzo, S., Iacopino, C. and Concetta Altavista, M. (1985) Strain-dependent variations in the number of forebrain cholinergic neurons. Brain Res., 334: 380–384.

Alheid, G.F. and Heimer, L. (1988) New perspectives in basal forebrain organization of special relevance for neuropsychiatric disorders: the striatopallidal, amygdaloid, and corticopetal components of substantia innominata. Neuroscience, 27: 1–39.

Anadón, R., Molist, P., Rodríguez-Modes, I., López, J.M., Quintela, I., Cerviñ, M.C., Barja, P. and González, A. (2000) Distribution of choline acetyltransferase immunoreactivity in the brain of an elasmobranch, the lesser spotted dogfish (Scyliorhynus canucula). J. Comp. Neuro., 420: 139–170.

Aramakis, V.B., Hsieh, C.Y., Leslie, F.M. and Metherate, R. (2000) A critical period for nicotine-induced disruption of synaptic development in rat auditory cortex. J. Neurosci., 20: 6106–6116.

Arendt, T., Bigl, V., Tennestedt, A. and Arendt, A. (1985) Neuronal loss in different parts of the nucleus basalis is related to neuritic plaque formation in cortical target areas in Alzheimer's disease. Neuroscience, 14: 1–14.

Armstrong, D.M., Bruce, G., Hersh, L.B. and Gage, F.H. (1987) Development of cholinergic neurons in the septal/diagonal band complex of the rat. Dev. Brain Res., 36: 249–256.

Aubert, I., Cécyre, D., Gauthier, S. and Quirion, R. (1996) Comparative ontogenetic profile of cholinergic markers, including nicotinic and muscarinic receptors, in the rat brain. J. Comp. Neurol., 369: 31–55.

Avendaño, C., Umbriaco, D., Dykes, R.W. and Descarries, L. (1996) Acetylcholine innervation of sensory and motor neocortical areas in adult cat: a choline acetyltransferase immunohistochemical study. J. Chem. Neuroanat., 11: 113–130.

Bartus, R.T., Dean, R.L., Beer, B. and Lippa, A.S. (1982) The cholinergic hypothesis of geriatric memory dysfunction. Science, 217: 408–417.

Bayer, S. (1979) The development of the septal region in the rat. I. Neurogenesis examined with ^{3}H-thymidine autoradiography. J. Comp. Neurol., 183: 89–106.

Bayer, S. (1985) Neurogenesis of the magnocellular basal telencephalic nuclei in the rat. Int. J. Dev. Neurosci., 3: 229–243.

Bear, M.F. and Singer, W. (1986) Modulation of visual cortical plasticity by acetylcholine and noradrenaline. Nature, 320: 172–176.

Bear, M.F., Carnes, D.M. and Ebner, F.F. (1985) Postnatal changes in the distribution of acetylcholinesterase in kitten striate cortex. J. Comp. Neurol., 237: 519–532.

Bender, R., Plaschke, M., Naumann, T., Wahle, P. and Frotscher, M. (1996) Development of cholinergic and GABAergic neurons in the rat medial septum: Different onset of choline acetyltransferase and glutamate decarboxylase mRNA expression. J. Comp. Neurol., 372: 204–214.

Brady, D.R., Phelps, P.E. and Vaughn, J.E. (1989) Neurogenesis of basal forebrain cholinergic neurons in rat. Dev. Brain Res., 47: 81–92.

Brantley, R.K. and Bass, A.H. (1988) Cholinergic neurons in the brain of a teleost fish (Porichthys notatus) located with a monoclonal antibody to choline acetyltransferase. J. Comp. Neurol., 275: 87–105.

Brauer, K., Holzer, M., Brückner, G., Tremere, L., Rasmusson, D.D., Poetheke, R., Arendt, T. and Härtig, W. (1999) Two distinct populations of cholinergic neurons in the septum of raccoon (Procyon lotor): evidence for a separate subset in the lateral septum. J. Comp. Neurol., 412: 112–122.

Brauth, S.E., Reiner, A., Kitt, C.A. and Karten, H.J. (1983) The substance P-containing striatotegmental path in reptiles: an immunohistochemical study. J. Comp. Neurol., 219: 305–327.

Brauth, S.E., Kitt, C.A., Price, D.L. and Wainer, B.H. (1985) Cholinergic neurons in the telencephalon of the reptile Caiman crocodilus. Neurosci. Lett., 58: 235–240.

Brown, R. and Brooksbank, B.W.L. (1979) Developmental changes in choline acetyltransferase and glutamate decarboxylase activity in various regions of the brain of the male, female, and neonatally androgenized female rat. Neurochem. Res, 4: 127–136.

Brückner, G., Schober, W., Härtig, W., Ostermann-Latif, C., Webster, H.H., Dykes, R.W., Rasmusson, D.D. and Bigl, V. (1992) The basal forebrain cholinergic system in the raccoon. J. Chem. Neuroanat., 5: 441–452.

Butcher, L.L. and Semba, K. (1989) Reassessing the cholinergic basal forebrain: Nomenclature schemata and concepts. Trends Neurosci., 12: 483–485.

Butcher, L.L., Oh, J.D., Woolf, N.J., Edwards, R.H. and Roghani, A. (1992) Organization of central cholinergic neurons revealed by combined in situ hybridization histochemistry and choline-O-acetyltransferase immunocytochemistry. Neurochem. Int., 21: 429–445.

Calarco, C.A. and Robertson, R.T. (1995) Development of basal forebrain projections to visual cortex: DiI studies in rat. J. Comp. Neurol., 354: 608–626.

Candy, J.M., Perry, E.K., Perry, R.H., Bloxham, C.A., Thompson, J., Johnson, M., Oakley, A.E. and Edwardson, J.A. (1985) Evidence for the early prenatal development of cortical cholinergic afferents from the nucleus of Meynert in the human foetus. Neurosci. Lett., 61: 91–95.

Carden, W.B., Datskovskaia, A., Guido, W., Godwin, D.W. and Bickford, M.E. (2000) Development of the cholinergic nitrergic and GABAergic innervation of the cat dorsal lateral geniculate nucleus. J. Comp. Neurol., 418: 65–80.

Changeux, J.P., Devillers-Thiery, A. and Chemouilli, P. (1984) Acetylcholine receptor: an allosteric protein. Science, 225: 1335–1345.

Conner, J.M. and Varon, S. (1997) Developmental profile of NGF immunoreactivity in the rat brain: a possible role of NGF in the establishment of cholinergic terminal fields in the hippocampus and cortex. Dev. Brain Res., 101: 67–79.

Coyle, J.T. and Yamamura, H.I. (1976) Neurochemical aspects of the ontogenesis of cholinergic neurons in the rat brain. Brain Res., 118: 429–440.

Crawford, G.D., Correa, L. and Salvaterra, P.M. (1982) Interaction of monoclonal antibodies with mammalian choline acetyltransferase. Proc. Natl. Acad. Sci. USA, 79: 7031–7035.

Creps, E.S. (1974) Time of neuron origin in preoptic and septal areas of the mouse: an autoradiographic study. J. Comp. Neurol., 157: 161–243.

De Carlos, J.A. and O'Leary, D.D.M. (1992) Growth and targeting of subplate axons and establishment of major cortical pathways. J. Neurosci., 12: 1194–1211.

De Carlos, J.A., Schlaggar, B.L. and O'Leary, D.D.M. (1995) Development of acetylcholinesterase-positive thalamic and basal forebrain afferents to embryonic rat neocortex. Exp. Brain Res., 194: 385–401.

Dinopoulos, A., Eadie, L.A., Dori, I. and Parnavelas, J.G. (1989) The development of basal forebrain projections to the rat visual cortex. Exp. Brain Res., 76: 563–571.

Dinopoulos, A., Michaloudi, H., Daramanlidis, A.N., Antonopoulos, J. and Parnavelas, J.G. (1988) Basal forebrain neurons project to the cortical mantle of the European hedgehog (*Erinaceus europaeus*). Neurosci. Lett., 86: 127–132.

Dinopoulos, A., Uylings, H.B.M. and Parnavelas, J.G. (1992) The development of neurons in the nuclei of the horizontal and vertical limbs of the diagonal band of Broca of the rat: a qualitative and quantitative analysis of Golgi preparations. Dev. Brain Res, 65: 65–74.

Dori, I. and Parnavelas, J.G. (1989) The cholinergic innervation of the rat cerebral cortex shows two distinct phases in development. Exp. Brain Res., 76: 417–423.

Eckenstein, F. and Thoenen, H. (1983) Cholinergic neurons in the rat cerebral cortex demonstrated by immunohistochemical localization of choline acetyltransferase. Neurosci. Lett., 36: 211–215.

Ekström, P. (1987) Distribution of choline acetyltransferease-immunoreactive neurons in the brain of a Cyprinid teleost (*Phoxinus phoxinus L.*). J. Comp. Neurol., 256: 494–515.

Esclapez, M., Tillakaratne, N.J., Kaufman, D.L., Tobin, A.J. and Houser, C.R. (1994) Comparative localization of two forms of glutamic acid decarboxylase and their mRNAs in rat brain supports the concept of functional differences between the forms. J. Neurosci., 13: 1834–1855.

Everitt, B.J., Sirkia, T.E., Roberts, A.C., Jones, G.H. and Robbins, T.W. (1988) Distribution and some projections of cholinergic neurons in the brain of the common marmoset, Callithrix jacchus. J. Comp. Neurol., 271: 533–558.

Fagan, A.M., Garber, M., Barbacid, M., Silos-Santiago, I. and Holtzman, D.M. (1997) A role for TrkA during maturation of striatal and basal forebrain cholinergic neurons *in vivo*. J. Neurosci., 17: 7644–7654.

Fine, A. (1987) Cortical acetylcholine: studies on the origin, function and plasticity of the cholinergic input to the neocortex. Ann Arbor, Michigan: University Microfilm International.

Fosse, V.M., Heggelund, P. and Fonnum, F. (1989) Postnatal development of glutamatergic, GABAergic, and cholinergic neurotransmitter phenotypes in the visual cortex, lateral geniculate nucleus, pulvinar and superior colliculus in cats. J. Neurosci., 9: 426–435.

Freund, T.F. (1989) GABAergic septohippocampal neurons contain parvalbumin. Brain Res., 478: 375–381.

Goldbach, R., Allgaier, C., Heimrich, B. and Jackisch, R. (1998) Postnatal development of muscarinic autoreceptors modulating acetylcholine release in the septohippocampal cholinergic system. I. Axon terminal region: hippocampus. Dev. Brain Res., 108: 23–30.

Gorenstein, C., Gallardo, K. and Robertson, R. (1991) Molecular forms of acetylcholinesterase in cerebral cortex and dorsal thalamus of developing rats. Dev. Brain Res., 61: 271–276.

Gorry, J.D. (1963) Studies on the comparative anatomy of the ganglion basale of Meynert. Acta Anat., 55: 51–104.

Gould, E. and Butcher, L.L. (1989) Developing cholinergic basal forebrain neurons are sensitive to thyroid hormone. J. Neurosci., 9: 3347–3358.

Gould, E., Farris, T.W. and Butcher, L.L. (1989) Basal forebrain neurons undergo somatal and dendritic remodeling during postnatal development: a single-section Golgi and choline acetyltransferase analysis. Dev. Brain Res., 46: 297–302.

Gould, E., Woolf, N.J. and Butcher, L.L. (1991) Postnatal development of cholinergic neurons in the rat: I. Forebrain. Brain Res. Bull., 27: 767–789.

Gould, S.J. (1977) Ontogeny and Phylogeny. Harvard University Press, Cambridge, MA.

Gritti, I., Mainville, L., Mancia, M. and Jones, B.E. (1997) GABAergic and other noncholinergic basal forebrain neurons, together with cholinergic neurons, project to the mesocortex and isocortex in the rat. J. Comp. Neurol., 383: 163–177.

Gu, Q. and Singer, W. (1993) Effects of intracortical infusion of anticholinergic drugs on neuronal plasticity in kitten striate cortex. Eur. J. Neurosci., 5: 475–485.

Hebb, C. and Ratkovic, D. (1964) Choline acetylase in the evolution of the brain in vertebrates. In: D. Richter (Ed.), Comparative Neurochemistry. Macmillan, New York.

Hedreen, J.C., Struble, R.G., Whitehouse, P.J. and Price, D.L. (1984) Topography of the magnocellular basal forebrain system in human brain. J. Neuropathol. Exp. Neurol., 43: 1–19.

Henderson, Z. (1987) A small proportion of cholinergic neurones in the nucleus basalis magnocellularis of ferret appear to stain positively for tyrosine hydroxylase. Brain Res., 412: 363–369.

Henderson, Z. (1991) Early development of the nucleus basalis-cortical projection but late expression of its cholinergic function. Neuroscience, 44: 311–324.

Hendry, S.H.C., Jones, E.G., Killackey, H.P. and Chalupa, L.M. (1987) Choline acetyltransferase-immunoreactive neurons in fetal monkey cerebral cortex. Brain Res., 465: 313–317.

Höhmann, C.F. and Ebner, F.F. (1985) Development of cholinergic markers in mouse forebrain. I. Choline acetyltransferase enzyme activity and acetylcholinesterase histochemistry. Dev. Brain Res., 23: 225–241.

Höhmann, C.F. and Berger-Sweeney, J. (1998) Cholinergic regulation of cortical development and plasticity. Perspectives Dev. Neurobiol., 5: 401–425.

Höhmann, C.F., Brooks, A.R. and Coyle, J.T. (1988) Neonatal lesions of the basal forebrain cholinergic neurons result in abnormal cortical development. Dev. Brain Res., 42: 253–264.

Höhmann, C.F., Potter, E.D. and Levey, A.I. (1995) Development of muscarinic receptor subtypes in the forebrain of the mouse. J. Comp. Neurol., 358: 88–101.

Hoogland, P.V. and Vermeulen-VanderZee, E. (1990) Distribution of choline acetyltransferase immunoreactivity in the telencephalon of the lizard Gekko gecko. Brain Behav. Evol., 36: 378–390.

Horder, T.J. and Martin, K.A.C. (1978) Morphogenetics as an alternative to chemospecificity in the formation of nerve connections. In: A.S.G. Curtis (Ed.), Cell–Cell Recognition. Cambridge University Press, Cambridge, pp. 275–358.

Houser, C.R., Crawford, G.D., Barber, R.P., Salvaterra, P.M. and Vaughn, J.E. (1983) Organization and morphological characteristics of cholinergic neurons: an immunocytochemical study with a monoclonal antibody to choline acetyltransferase. Brain Res., 266: 97–119.

Hu, Z., Yuri, K., Ichikawa, T. and Kawata, M. (1996) Exposure of postnatal rats to glucocorticoids suppresses the development of choline acetyltransferase-immunoreactive neurons: role of adrenal streoids in the development of forebrain cholinergic neurons. J. Chem. Neuroanat., 10: 1–10.

Hubel, D.H. and Wiesel, T.N. (1970) The period of susceptibility to the physiological effects of unilateral eye closure in kittens. J. Physiol. (Lond.), 206: 419–436.

Ichikawa, T., Ajiki, K., Matsuura, J. and Masawa, H. (1997) Localization of two cholinergic markers, choline acetyltransferase and vesicular acetylcholine transporter in the central nervous system of the rat: in situ hybridization histochemistry and immunohistochemistry. J. Chem. Neuroanat., 13: 23–39.

Ishii, Y. (1957) The histochemical studies of cholinesterase in the central nervous system. II. Histochemical alteration of cholinesterase of the brain of rats from late fetal life to adults (In Japanese). Arch. Histol. Japon, 12: 613–637.

Kalsbeek, A., Voorn, P., Buijs, R.M., Pool, C.W. and Uylings, H.B. (1988) Development of the dopaminergic innervation in the prefrontal cortex of the rat. J. Comp. Neurol., 269: 58–72.

Kasashima, S., Muroishi, Y., Futakuchi, H., Nakanishi, I. and Oda, Y. (1998) In situ mRNA hybridization study of the distribution of choline acetyltransferase in the human brain. Brain Res., 806: 8–15.

Kasashima, S., Kawashima, A., Muroishi, Y., Futakuchi, H., Nakanishi, I. and Oda, Y. (1999) Neurons with choline acetyltransferase immunoreactivity and mRNA are present in the human cerebral cortex. Histochem. Cell Biol., 111: 197–207.

Kimura, H., McGeer, P.L., Peng, J.H. and McGeer, E.G. (1981) The central cholinergic system studied by choine

acetyltransferase immunohistochemistry in the cat. J. Comp. Neurol., 200: 151–201.

Kimura, H., Akiyama, H., Tago, H., Hersh, L.B., Tooyama, I. and McGeer, P.L. (1990) Choline acetyltransferase immunopositive neurons in the lateral septum. Brain Res., 533: 165–170.

Kiss, J. and Patel, A.J. (1992) Development of the cholinergic fibres innervating the cerebral cortex of the rat. Int. J. Dev. Neurosci., 10: 153–170.

Kitt, C.A., Höhmann, C., Coyle, J.T. and Price, D.L. (1994) Cholinergic innervation of mouse forebrain structures. J. Comp. Neurol., 341: 117–129.

Koh, S. and Loy, R. (1989) Localization and development of nerve growth factor-sensitive rat basal forebrain neurons and their afferent projections to hippocampus and neocortex. J. Neurosci., 9: 2999–3018.

Kordower, J.H. and Rakic, P. (1990) Neurogenesis of the magnocellular basal forebrain nuclei in the rhesus monkey. J. Comp. Neurol., 291: 637–653.

Kordower, J.H., Bartus, R.T., Marciano, F.F. and Gash, D.M. (1989) Telencephalic cholinergic system of the New World monkey (Cebus apella): morphological and cytoarchitectonic assessment and analysis of the projection to the amygdala. J. Comp. Neurol., 279: 528.

Korsching, S. (1993) The neurotrophic factor concept: a reexamination. J. Neurosci., 13: 2739–2748.

Kostović, I. (1986) Prenatal development of nucleus basalis complex and related fiber systems in man: a histochemical study. Neuroscience, 17: 1047–1077.

Kracun, I. and Rösner, H. (1986) Early cytoarchitectonic development of the anlage of the basal nucleus of Meynert in the human fetus. Int. J. Dev. Neurosci., 4: 143–149.

Krnjević, K. and Silver, A. (1966) Acetylcholinesterase in the developing forebrain. J. Anat., 100: 63–89.

Kvale, I., Fosse, V.M. and Fonnum, F. (1983) Development of neurotransmitter prameters in lateral geniculate body, superior colliculus, and visual cortex of the albino rat. Dev. Brain Res., 7: 137–145.

Ladinsky, H., Consolo, S., Peri, G. and Garattini, S. (1972) Acetylcholine, choline and choline acetyltransferase activity in the developing brain of normal and hypothyroid rats. J. Neurochem., 19: 1947–1952.

Large, T.H., Bodary, S.C., Clegg, D.O., Weskamp, G., Otten, U. and Reichardt, L.F. (1986) Nerve growth factor gene expression in the developing rat brain. Science, 234: 352–355.

Latsari, M., Dori, I., Antonopoulos, J., Chiotelli, M. and Dinopoulos, A. (2002) Noradrenergic innervation of the developing and mature visual and motor cortex of the rat brain: A light and electron microscopic immunocytochemical analysis. J. Comp. Neurol., 445: 145–158.

Lauder, J.M. and Schambra, U.B. (1999) Morphogenetic roles of acetylcholine. Environment. Health Perspectives, 107: 65–69.

Lauder, J.M., Han, V.K., Henderson, P., Verdoorn, T. and Towle, A.C. (1986) Prenatal ontogeny of the GABAergic system in the rat brain: an immunocytochemical study. Neuroscience, 19: 465–493.

Lauterborn, J.C., Isackson, P.J., Montalvo, R. and Gall, C.M. (1993) In situ hybridization localization of choline acetyltransferase mRNA in adult rat brain and spinal cord. Mol. Brain Res., 17: 59–69.

Le Jeune, H. and Fourdan, F. (1991) Postnatal development of cholinergic markers in the rat olfactory bulb: a histochemical and immunocytochemical study. J. Comp. Neurol., 314: 383–395.

Levey, A.I., Wainer, B.H., Rye, D.B., Mufson, E.J. and Mesulam, M.-M. (1984) Choline acetyltransferase-immunoreactive neurons intrinsic to rodent cortex and distinction from acetylcholinesterase-positive neurons. Neuroscience, 13: 341–353.

Lewis, P.R. and Shute, C.C.D. (1967) The cholinergic limbic system: projections to hippocampal formation, medial cortex, nuclei of the ascending cholinergic reticular system, and the subfornical organ and supra-optic crest. Brain, 90: 521–540.

Li, Y., Holtzman, D.M., Kromer, L.F., Kaplan, D.R., Chua-Couzens, J., Clary, D.O., Knusel, B. and Mobley, W.C. (1995) Regulation of TrkA and ChAT expression in developing rat basal forebrain: Evidence that both exogenous and endogenous NGF regulate differentiation of cholinergic neurons. J. Neurosci., 15: 2888–2905.

Lidov, H.W. and Molliver, M.E. (1982) An immunohistochemical study of serotonin neuron development in the rat: ascending pathways and terminal fields. Brain Res. Bull., 8: 389–430.

Linke, R. and Frotscher, M. (1993) Development of the rat septohippocampal projection: Tracing with DiI and electron microscopy of identified growth cones. J. Comp. Neurol., 332: 69–88.

Luskin, M.B. and Shatz, C.J. (1985) Neurogenesis of the cat's primary visual cortex. J. Comp. Neurol., 242: 611–631.

Lysakowski, A., Wainer, B.H., Rye, D.B., Bruce, G. and Hersh, L.B. (1986) Cholinergic innervation displays strikingly different laminar preferences in several cortical areas. Neurosci. Lett., 64: 102–108.

Makuch, R., Baratta, J., Karaelias, L.D., Lauterborn, J.C., Gall, C.M., Yu, J. and Robertson, R.T. (2001) Arrival of afferents and the differentiation of target neurons: Studies of developing cholinergic projections to the dentate gyrus. Neuroscience, 104: 81–91.

Maley, B.E., Frick, M.L., Levey, A.I., Wainer, B.H. and Elde, R.P. (1988) Immunohistochemistry of choline acetyltransferase in the guinea pig brain. Neurosci. Lett., 84: 137–142.

Manns, I.D., Mainville, L. and Jones, B.E. (2001) Evidence for glutamate, in addition to acetylcholine and GABA, neurotransmitter synthesis in basal forebrain neurons projecting to the entorhinal cortex. Neuroscience, 107: 249–263.

Marín, O., Smeets, W.J.A.J. and González, A. (1997) Distribution of choline acetyltransferase immunoreactivity in the brain of anuran (Rana perezi, Xenopus laevis) and urodele (Pleurodeles walti) amphibians. J. Comp. Neurol., 382: 499–534.

Matthews, D.A., Nadler, J.V., Lynch, G.S. and Cotman, C.W. (1974) Development of cholinergic innervation in the hippocampal formation of the rat. I. Histochemical demonstration of acetylcholinesterase activity. Dev. Biol., 36: 130–141.

Mechawar, N. and Descarries, L. (2001) The cholinergic innervation develops early and rapidly in the rat cerebral cortex: A quantitative immunohistochemical study. Neuroscience, 108: 555–567.

Mechawar, N., Watkins, K.C. and Descarries, L. (2002) Ultrastructural features of the acetylcholine innervation in the developing parietal cortex of rat. J. Comp. Neurol., 443: 250–258.

Medina, L. and Reiner, A. (1994) Distribution of choline acetyltransferase immunoreactivity in the pigeon brain. J. Comp. Neurol., 342: 497–537.

Medina, L., Smeets, W.J.A.J., Hoogland, P.V. and Puelles, L. (1993) Distribution of choline acetyltransferase immunoreactivity in the brain of the lizard Gallotia galloti. J. Comp. Neurol., 331: 261–285.

Meisami, E. and Firoozi, M. (1985) Acetylcholinesterase activity in the developing olfactory bulb: a biochemical study on normal maturation and the influence of peripheral and central connections. Dev. Brain Res., 21: 115–124.

Mellgren, S.I. (1973) Distribution of acetylcholinesterase in the hippocampal region of the rat during postnatal development. Z. Zellforsch. Mikrosk. Anat., 141: 375–400.

Mesulam, M.-M. and Geula, C. (1988) Nucleus basalis (Ch4) and cortical cholinergic innervation in the human brain: observations based on the distribution of acetylcholinesterase and choline acetyltransferase. J. Comp. Neurol., 275: 216–240.

Mesulam, M.-M. and Geula, C. (1991) Acetylcholinesterase-rich neurons of the human cerebral cortex: Cytoarchitectonic and ontogenetic patterns of distribution. J. Comp. Neurol., 306: 193–220.

Mesulam, M.-M., Mufson, E.J., Levey, A.I. and Wainer, B.H. (1983) Cholinergic innervation of cortex by the basal forebrain: cytochemistry and cortical connections of the septal area, diagonal band nuclei, nucleus basalis (substantia innominata), and hypothalamus in the Rhesus monkey. J. Comp. Neurol., 214: 170–197.

Mesulam, M.-M., Mufson, E.J., Levey, A.I. and Wainer, B.H. (1984) Atlas of cholinergic neurons in the forebrain and upper brainstem of the macaque based on monoclonal choline acetyltransferase immunohistochemistry and acetylcholinesterase histochemistry. Neuroscience, 12: 669–686.

Milner, T.A., Loy, R. and Amaral, D.G. (1983) An anatomical study of the development of the septo-hippocampal projection in the rat. Dev. Brain Res., 59: 133–142.

Mufson, E.J. and Cunningham, M.G. (1988) Observations on choline acetyltransferase containing structures in the CD-1 mouse brain. Neurosci. Lett., 84: 7–12.

Mufson, E.J., Desan, P.H., Mesulam, M.M., Wainer, B.H. and Levey, A.I. (1984) Choline acetyltransferase-like immunoreactivity in the forebrain of the red-earled pond turtle (Pseudemys scripta elegans). Brain Res., 323: 103–108.

Murakami, T., Morita, Y. and Ito, H. (1983) Extrinsic and intrinsic fiber connections of the telencephalon in a teleost, Sebastiscus marmoratus. J. Comp. Neurol., 216: 115–131.

Nadler, J.V., Matthews, D.A., Cotman, C.W. and Lynch, G.S. (1974) Development of cholinergic innervation in the hippocampal formation of the rat. II. Quantitative changes in choline acetyltransferase and acetylcholinesterase activities. Dev. Biol., 36: 142–154.

Neary, T.J. (1990) The pallium of anuran amphibians. In: Jones E.G. and Peters A. (Eds.), Comparative Structure and Evolution of Cerebral Cortex, Vol. 8A, Part 1, Plenum, New York, pp. 107–138.

Nieuwenhuys, R. (1965) The interpretation of the cell masses in the teleostean forebrain. In: Hassler R. and Stephan H. (Eds.), Evolution of the Forebrain. Phylogenesis and Ontogenesis of the Forebrain. Plenum, New York, pp. 32–39.

Northcutt, R.G. (1981) Evolution of the telencephalon in nonmammals. Ann. Rev. Neurosci., 4: 301–350.

Northcutt, R.G. and Braford, M. (1980) New observations on the organization and evolution of the telencephalon of actinopterygian fishes. In: Ebbeson S.O.E. (Ed.), Comparative Neurology of the Telencephalon. Plenum Press, New York, pp. 41–98.

Northcutt, R.G. and Davis, R.E. (1983) Telencephalic organization in ray-finned fishes. In: Davis R.E. and Northcutt R.G. (Eds.), Fish neurobiology, Vol. 2, University of Michigan Press, Ann Arbor, pp. 203–236.

Nyakas, C., Buwalda, B., Kramers, R.J.K., Traber, J. and Luitten, P.G.M. (1994) Postnatal development of hippocampal and neocortical cholinergic and serotonergic innervation in rat: effects of nitrite-induced prenatal hypoxia and nimodipine treatment. Neuroscience, 59: 541–559.

Oda, Y. (1999) Choline acetyltransferase: the structure, distribution and pathological changes in the central nervous system. Pathol. Int., 49: 921–937.

Oh, J.D., Butcher, L.L. and Woolf, N.J. (1991) Thyroid hormone modulates the development of cholinergic terminal fields in the rat forebrain: relation to nerve growth factor receptor. Dev. Brain Res., 59: 133–142.

Papadopoulos, G.C., Parnavelas, J.G. and Buijs, R.M. (1989) Light and electron microscopic immunocytochemical analysis of the noradrenaline innervation of the rat visual cortex. J. Neurocytol., 18: 1–10.

Parent, A. (1986) Comparative Neurobiology of the Basal Ganglia. John Wiley & Sons, New York.

Pérez, S.E., Yáñez, J., Marín, O., Anadón, R., González, A. and Rodríguez-Moldes, I. (2000) Distribution of choline acetyltransferase (ChAT) immunoreactivity in the brain of the adult trout and tract-tracing observations on the connections of the nuclei of the isthmus. J. Comp. Neurol., 428: 450–474.

Perry, E.K., Smith, C.J., Atack, J.R. and Candy, J.M. (1986) Neocortical cholinergic enzyme and receptor activities in the human fetal brain. J. Neurochem., 47: 1262–1269.

Phelps, P.E., Brennan, L.A. and Vaughn, J.E. (1990) Generation patterns of immunocytochemically identified cholinergic neurons in rat brainstem. Dev. Brain Res., 56: 63–74.

Plaschke, M., Naumann, T., Kasper, E., Bender, R. and Frotscher, M. (1997) Development of cholinergic and GABAergic neurons in the rat medial septum: Effect of target removal in early postnatal development. J. Comp. Neurol., 379: 467–481.

Pombal, M.A., Marín, O. and González, A. (2001) Distribution of choline acetyltransferase-immunoreactive structures in the lamprey brain. J. Comp. Neurol., 431: 105–126.

Potempska, A., Skangiel-Kramska, L. and Kossut, M. (1979) Development of cholinergic enzymes and adenosinetriphosphatase activity of optic system of cats in normal and restricted visual input conditions. Dev. Neurosci, 2: 38–45.

Powers, A.S. and Reiner, A. (1993) The distribution of cholinergic neurons in the central nervous system of turtles. Brain Behav. Evol., 41: 326–345.

Rasmusson, D.D. (2000) The role of acetylcholine in cortical synaptic plasticity. Behav. Brain Res., 115: 205–218.

Rea, M.A. and Nurnberger, J.I., Jr. (1986) Evidence for developmental synaptic regression of cholinergic afferents to the rat main olfactory bulb. Brain Res., 389: 233–237.

Reiner, A., Brauth, S.E. and Karten, H.J. (1984) Evolution of the amniote basal ganglia. Trends Neurosci., 7: 320–325.

Reiner, A., Medina, L. and Veenman, C.L. (1998) Structural and functional evolution of the basal ganglia in vertebrates. Brain Res. Rev., 28: 235–285.

Reiner, P.B., Semba, K. and Fibiger, H.C. (1988) Ontogeny of histidine-decarboxylase-immunoreactive neurons in the tuberomammillary nucleus of the rat hypothalamus: time of origin and development of transmitter phenotype. J. Comp. Neurol., 276: 304–311.

Robertson, R., Mostamand, F., Kageyama, G., Gallardo, K. and Yu, J. (1991) Primary auditory cortex in the rat: transient expression of acetylcholinesterase activity in developing geniculocortical projections. Dev. Brain. Res., 58: 81–95.

Rodríguez-Moldes, I., Molist, P., Adrio, F., Pombal, M.A., Pérez, S.E., Yáñez, J., Mandado, M., Marín, O., López, J.M., González, A., Anadón, R. (2002) Organization of cholinergic system in the brain of different fish groups: a comparative analysis. Brain Res. Bull., 57: 331–334.

Rothman, T.P., Spect, L.A., Gershon, M.D., Joh, T.H., Teitelman, G., Pickel, V.M. and Reis, D.J. (1980) Catecholamine biosynthetic enzymes are expressed in replicating cells of the peripheral but not in the central nervous system. Proc. Natl. Acad. Sci. USA, 77: 6221–6225.

Sakmann, B., Methfessel, C., Mishina, M., Takahashi, T., Takai, T., Kurasaki, M., Fukuda, K. and Numa, S. (1985) Role of acetylcholine receptor subunits in gating of the channel. Nature, 318: 538–543.

Sanes, J.R. and Yamagata, M. (1999) Formation of lamina-specific synaptic connections. Curr. Opin. Neurobiol., 9: 79–87.

Saper, C.B. (1984) Organization of cerebral cortical afferent systems in the rat. I. Magnocellular basal nucleus. J. Comp. Neurol., 222: 313–342.

Saper, C.B. and Chelimsky, T.C. (1984) A cytoarchitectonic and histochemical study of nucleus basalis and associated cell

42

groups in the normal human brain. Neuroscience, 13: 1023–1037.

Sarnat, H.B. and Netsky, M.G. (1981) Evolution of the Nervous System. Oxford University Press, New York.

Sassoè-Pognetto, M., Pairault, C., Clairambault, P. and Fasolo, A. (1991) The connections of the anterior pallium in Pleurodeles vaity and Triturus carnifex: An HRP study. J. Hirnforsch., 32: 397–407.

Satoh, K. and Fibiger, H.C. (1985) Distribution of central cholinergic neurons in the baboon (Papio papio). I. General morphology. J. Comp. Neurol., 236: 197–214.

Schambra, U.B., Sulik, K.K., Petrusz, P. and Lauder, J.M. (1989) Ontogeny of cholinergic neurons in the mouse forebrain. J. Comp. Neurol., 288: 101–122.

Schwab, C., Bruckner, G, Mares, V and Biesold, D (1988) A combined method of acetylcholinesterase histochemistry and [^{3}H]-thymidine autoradiography: application to neurogenesis of the rat basal forebrain cholinergic system. Neurosci. Lett., 90: 69–74.

Schwaber, J.S., Kapp, B.S., Higgins, G.A. and Rapp, P.R. (1982) Amygdaloid and basal forebrain direct connections with the nucleus of the solitary tract and the dorsal motor nucleus. J. Neurosci., 2: 1424–1438.

Schwaber, J.S., Rogers, W.T., Satoh, K. and Fibiger, H.C. (1987) Distribution and organization of cholinergic neurons in the rat forebrain demonstrated by computer-aided data acquisition and three-dimensional reconstruction. J. Comp. Neurol., 263: 309–325.

Séguéla, P., Watkins, K.C., Geffard, M. and Descarries, L. (1990) Noradrenaline axon terminals in adult rat neocortex: an immunocytochemical analysis in serial thin sections. Neuroscience, 35: 249–264.

Seiger, A. and Olson, L. (1973) Late prenatal ontogeny of central monoamine neurons in the rat: Fluorescence histochemical observations. Z. Anat. Entwicklungsgesch., 140: 281–318.

Semba, K. (1992) Development of central cholinergic neurons. In: A. Björklund, T. Hökfelt and M. Tohyama (Eds.), Handbook of Chemical Neuroanatomy. Vol. 10, Ontogeny of Transmitters and Peptides in the CNS, Elsevier, Amsterdam, pp. 33–62.

Semba, K. (1999) The mesopontine cholinergic system: A dual role in REM sleep and wakefulness. In: R. Lydic and H.A. Baghdoyan (Eds.), Handbook of Behavioral State Control. Cellular and Molecular Mechanisms. CRC Press, Boca Raton, FL, pp. 161–180.

Semba, K. (2000) Multiple output pathways of the basal forebrain: Organization, chemical heterogeneity, and roles in vigilance. Behav. Brain Res., 115: 117–141.

Semba, K. and Fibiger, H.C. (1988) Time of origin of cholinergic neurons in the rat basal forebrain. J. Comp. Neurol., 269: 87–95.

Semba, K. and Fibiger, H.C. (1989) Organization of central cholinergic systems. In: A. Nordberg, K. Fuxe, B. Holmstedt and A. Sundwall (Eds.), Nicotinic Receptors in the CNS. Their Roles in Synaptic Transmission. Prog. Brain Res. Vol. 79, Elsevier, pp. 37–63.

Semba, K., Vincent, S.R. and Fibiger, H.C. (1988) Different times of origin of choline acetyltransferase- and somatostatin-immunoreactive neurons in the rat striatum. J. Neurosci., 8: 3937–3944.

Shaw, C., Needler, M.C. and Cynader, M. (1984) Ontogenesis of muscimol binding sites in cat visual cortex. Brain Res. Bull., 13: 331–334.

Shelton, D. L., Nadler, J. V. and Cotman, C. W. (1979) Development of high affinity choline uptake and associated acetylcholine synthesis in the rat fascia dentata. Brain Res., 163: 263–275.

Shute, C.C.D. and Lewis, P.R. (1967) The ascending cholinergic reticular system: neocortical, olfactory and subcortical projections. Brain, 90: 497–522.

Sofroniew, M.V., Pearson, R.C.A. and Powell, T.P.S. (1987) The cholinergic nuclei of the basal forebrain of the rat: normal structure, development and experimentally induced degeneration. Brain Res., 411: 310–331.

Solbach, S. and Celio, M.R. (1991) Ontogeny of the calcium binding protein parvalbumin in the rat nervous system. Anat. Embryol., 184: 103–124.

Sperry, R.W. (1963) Chemoaffinity in the orderly growth of nerve fiber patterns and connections. Proc. Natl. Acad. Sci. USA, 50: 703–710.

St.-Jacques, R., Gorczyca, W., Mohr, G. and Schipper, H.M. (1996) Mapping of the basal forebrain cholinergic system of the dog: a choline acetyltransferase immunohistochemical study. J. Comp. Neurol., 366: 717–725.

Steriade, M., Jones, E.G. and McCormick, D.A. (1997) Thalamus. Volume I. Organisation and Function. Elsevier, New York.

Stichel, C.C. and Singer, W. (1987) Quantitative analysis of the choline acetyltransferase-immunoreactive axonal network in the cat primary visual cortex: II. Pre- and postnatal development. J. Comp. Neurol., 258: 99–111.

Sweeney, J.E., Hohmann, C.F., Oster-Granite, M.L. and Coyle, J.T. (1989) Neurogenesis of the basal forebrain in euploid and trisomy 16 mice: an animal model for developmental disorders in Down syndrome. Neuroscience, 31: 413–425.

Székely, G. (1990) Problems of the neuronal specificity concept in the development of neural organization. Concepts Neurosci., 1: 165–197.

ten Donkelaar, H.J. and Dederen, P.J. (1979) Neurogenesis in the basal forebrain of the Chinese hamster (Cricetulus griseus). I. Time of neuron origin. Anat. Embryol. (Berl.), 156: 331–348.

Thal, L.J., Gilbertson, E., Armstrong, D.M. and Gage, F.H. (1991) Development of the basal forebrain cholinergic system: Phenotype expression prior to target innervation. Neurobiol. Aging, 13: 67–72.

Umbriaco, D., Watkins, K.C., Descarries, L., Cozzari, C. and Hartman, B.K. (1994) Ultrastructural and morphometric features of the acetylcholine innervation in adult rat parietal cortex: an electron microscopic study in serial sections. J. Comp. Neurol., 348: 351–373.

Verney, C., Berger, B., Bauluc, M., Helle, K.B. and Arvarez, C. (1984) Dopamine-β-hydroxylase-like immunoreactivity in the fetal cerebral cortex of the rat: noradrenergic ascending pathways and terminal fields. Int. J. Dev. Neurosci., 2: 491–503.

Villalobos, J., Rios, O. and Barbosa, M. (2000) Postnatal development of the basal forebrain cholinergic projections to the medial prefrontal cortex in mice. Dev. Brain Res., 120: 99–103.

Villalobos, J., Rios, O. and Barbosa, M. (2001) Postnatal development of cholinergic system in mouse basal forebrain: acetylcholinesterase histochemistry and choline-acetyltransferase immunoreactivity. Int. J. Dev. Neurosci., 19: 495–502.

Vincent, S.R. and Reiner, P.B. (1987) The immunohistochemical localization of choline acetyltransferase in the cat brain. Brain Res. Bull., 18: 371–415.

Wachtler, K. (1981) The regional distribution of acetylcholine, choline acetyltransferase, and acetylcholinesterase in vertebrate brains of different phylogenetic levels. In: Pepeu G. and Ladinsky H. (Eds.), Cholinergic Mechanisms: Phylogenetic Aspects, Central and Peripheral Synapses, and Clinical Significance, Vol. 25, Plenum, New York, pp. 59–72.

Wainer, B.H., Levey, A.I., Mufson, E.J. and Mesulam, M.-M. (1984) Cholinergic systems in mammalian brain identified with antibodies against choline acetyltransferase. Neurochem. Int., 6: 163–182.

Ward, N.L. and Hagg, T. (1999) p75(NGFR) and cholinergic neurons in the developing forebrain: a re-examination. Dev. Brain Res., 118: 79–91.

Wullimann, M.F. and Rink, E. (2002) The teleostean forebrain: a comparative and developmental view based on early proliferation, Pax6, activity and catecholaminergic organization. Brain Res. Bull., 57: 363–370.

Progress in Brain Research, Vol. 145
ISSN 0079-6123

CHAPTER 2

Structural determinants of the roles of acetylcholine in cerebral cortex

Laurent Descarries*, Naguib Mechawar, Nicolas Aznavour and Kenneth C. Watkins

Departments of Pathology and Cell Biology and of Physiology, and Centre de Recherche en Sciences Neurologiques, Faculté de Médecine, Université de Montréal, Montreal, QC H3C 3J7, Canada

Introduction

The various roles attributed to acetylcholine (ACh) in mammalian cerebral cortex span across all levels of observation and knowledge at which the CNS may be studied. As illustrated by the different chapters in this book, cortical ACh has been implicated in elementary processes such as corticogenesis, neuronal excitability, shaping of receptive fields and the control of microcirculation, but also in more global functions, such as waking and sleep, attention, memory, learning, and even conscious experience. Accordingly, in humans, dysregulations of cholinergic neurons and properties have been identified in numerous neurological and psychiatric diseases, and sometimes held responsible for their occurrence and/or manifestations, raising hopes of therapies that might be better targetted and more efficient in a near future.

In general terms, it is therefore a real challenge for contemporary neuroscience to elucidate the different mechanisms underlying the many actions of ACh in cerebral cortex; notably how this neuromodulator participates in the processing of information and in the development of this part of the brain. This endeavor obviously requires a detailed knowledge of the structural and ultrastructural features of the cortical ACh innervation. For this reason, and to set the stage for current and future research, this chapter reviews the results of light and electron microscopic

immunocytochemical studies, carried over the last ten years in our laboratory, which have provided basic information on the regional distribution and relational features of the cortical ACh innervation in both the adult and the postnatal rat neocortex. Similar data are currently being acquired in the hippocampus. Our most recent observations indicate that, at least in neocortex, the ACh innervation develops early, and rapidly reaches the relatively dense laminar and regional densities measured in the adult. Moreover, we have established that this ACh innervation is largely nonsynaptic, not only in the adult but also throughout development.

ACh innervation in cerebral cortex and hippocampus: general organization

As also described and illustrated in Chapter 1 (see also Woolf, 1991), the intricate network of varicose ACh axons pervading the mammalian neocortex and hippocampus arises mainly from magnocellular neurons in the basal forebrain (Rye et al., 1984; Saper, 1984). All areas and layers of the neocortex receive an ACh projection from a subpopulation of neurons in the substantia innominata-nucleus basalis complex, named Ch4 by Mesulam (Mesulam et al., 1983). The hippocampal ACh innervation originates mostly from cell bodies located in the septal nucleus and vertical limb of the diagonal band of Broca, i.e., Ch1 and Ch2 (McKinney et al., 1983; Rye et al., 1984; Amaral and Kurz, 1985; Nyakas et al., 1987; Woolf, 1991). It is currently estimated that there are

*Corresponding author: Tel.: (514) 343-7070;
Fax: (514) 343-5755; E-mail: laurent.descarries@umontreal.ca

DOI: 10.1016/S0079-6123(03)45002-4

some 7000–9000 Ch4 neurons on each side of the rat brain (Rye et al., 1984; Gritti et al., 1993), versus some 400 in Ch1/Ch2 (Schwegler et al., 1996). In rodents at least, both regions also receive an intrinsic ACh innervation from scattered bipolar interneurons, which has been estimated to account for about 20% of ACh axon terminals in neocortex (Eckenstein and Baughman, 1987), and 5–10% in hippocampus (Gage et al., 1983; Eckenstein and Baughman, 1987). In rat neocortex, these choline acetyltransferase (ChAT)-expressing interneurons are also known to synthesize GABA and VIP (Chédotal et al., 1994; Bayraktar et al., 1997).

The ACh cortical input may therefore be considered as having a dual origin, from both extrinsic (projection) and intrinsic neurons. This input can also be viewed as a system exerting its influence indirectly, through other cortical afferent systems, as exemplified by the dense ACh projections from the midbrain and pontine tegmentum to thalamocortical relay nuclei (Steriade et al., 1988; see Woolf, 1991, and Chapters 4 and 11). Moreover, the ACh projections from the midbrain to Ch4, and probable influence of the latter on cortical ACh interneurons (Porter et al., 1999) suggest important interactions between the different ACh neuron groups. We will now review some of the basic structural determinants of ACh function in both neocortex and hippocampus.

Regional and laminar distribution of the ACh innervation in the mature cortex and hippocampus

In their light microscopic description of the distribution of ChAT-immunolabeled axons in the different cortical areas of the rat cerebral cortex, Lysakowski et al. (1989) reported a total of 13 different patterns of ACh innervation that were generally correlated to functionally similar cortical areas. However, none of the different ChAT antibodies available at that time allowed for a complete detection of the ACh axon network, prerequisite to a quantitative analysis of its distributional features. It is only in 1990 that an antibody against whole rat brain ChAT became available, that was sensitive enough to achieve a full visualization of the fine ACh varicose axons throughout the CNS of different mammalian species (Cozzari et al., 1990). As labeling of ACh axons with this

antibody could be shown to be maximal in vibratome sections of perfusion-fixed brain, it became possible to quantitate their regional and laminar distribution in adult rat cerebral cortex (Mechawar et al., 2000).

In transverse sections like those in Fig. 1, we used a semi-computerized light microscopic method to measure the actual length of the ChAT-immunoreactive axon network contained within the full thickness of transverse sections from the frontal (Fr1), parietal (Par1) and occipital (Oc1) neocortex (for technical details, see Mechawar et al., 2000). The number of varicosities (i.e., dilations > 0.5 microns in diameter) on these axons was counted directly under the light microscope on axon segments randomly selected from every layer of each cortical region. Given that a constant ratio of 4 varicosities per 10 μm of axon was found throughout neocortex, the actual number of varicosities on the axon network could be directly derived from measurements of its length.

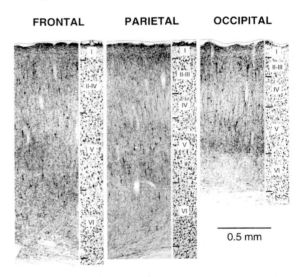

Fig. 1. Low power micrographs illustrating the laminar distribution of ACh (ChAT-immunopositive) fibers in transverse sections from three neocortical areas: frontal, primary motor, Fr1; parietal, primary somatosensory, Par 1; occipital, primary visual, Oc1. Narrow strips from adjacent Nissl-stained sections, which helped to delineate the cortical layers, are also shown. In each area, small, fusiform, bipolar cholinergic interneurons, with their vertically oriented dendrites, are scattered in layers II–VI, amidst the intricate network of fine varicose fibers pervading the entire cortical thickness. In the lower part of layer VI, immediately above the callosal radiations, smooth and thicker, transversely oriented fibers are also visible. (Reproduced with permission from Mechawar et al., 2000.)

Thus, the data could be expressed in two ways: (1) densities, i.e., axon length (in meters) and number of varicosities (in millions) per mm^3 of tissue; (2) axon length (in meters) and number of varicosities (in millions) under an arbitrary surface of 1 mm^2, to take into account the varying thickness of the layers and areas examined.

It was immediately apparent that the frontal cortex had the densest ACh innervation, followed by the occipital and the parietal cortex, with respective values of 5.4, 4.6 and 3.8×10^6 varicosities/mm^3, as shown in Table 1 and Fig. 2. Because of the lesser thickness of the occipital cortex, the order changed between parietal and occipital for the values under 1 mm^2 of cortical surface. In the three areas, layers I and V were the most densely innervated, with respective inter-areal means of 5.3 and 5.0×10^6 varicosities/mm^3. The least densely innervated layers were IV and VI of the primary sensory areas, with inter-areal means (Par1 and Oc1) of 3.4 and 3.8×10^6 varicosities/mm^3, respectively. The laminar distributions were area specific, and characterized by uniformly high densities throughout frontal cortex, and lower densities in layers II/III, IV and VI of the parietal cortex, as well as in layers IV and VI of the occipital cortex (Fig. 2).

In hippocampus, the distribution of the ACh innervation was quantified in both mouse and rat (Aznavour et al., 2002). Values of regional (Table 1) and laminar densities (Fig. 2) were obtained from three sectors of the dorsal hippocampus, CA1, CA3 and the dentate gyrus (DG). Again, the number of varicosities per unit length of ACh axon was fixed at 4 per 10 mm, allowing to express the data as both densities of axons and densities of varicosities. The

salient features were regional densities even higher than in neocortex, ranging from 4.9 million in CA1 to 6.2 million in CA3 of rat, for a regional average of 5.9 million, compared to 4.6 million in the neocortex. There were few differences between rat and mouse hippocampal ACh innervations, the only significant one being a slightly lower interareal density in CA1 of rat, accounted for by relatively low values in its stratum radiatum and lacunosum moleculare.

Ultrastructural features of the ACh innervation in adult neocortex and hippocampus

We have described the ultrastructural features of the ACh innervation in both the primary somatosensory cortex (Par 1) (Umbriaco et al., 1994) and CA1 stratum radiatum of adult rat (Umbriaco et al., 1995). In the parietal cortex, an extensive examination of the intrinsic and relational features of the ACh (ChAT-immunostained) axon varicosities was carried out in serial as well as single thin sections. Among other findings, this thorough analysis revealed that, in all layers of the parietal cortex, only a small fraction of all ACh varicosities were endowed with a junctional complex, i.e., a straightening of apposed membranes with or without postsynaptic thickening on either side of a slightly widened extracellular space. In general, the cortical ACh varicosities were relatively small, averaging 0.57 μm in diameter. Those bearing a synaptic junction were slightly but significantly larger than their nonsynaptic counterparts (0.67 μm in diameter). The junctional complexes formed by these terminals were single, occupied a small fraction of the total

Table 1. Regional distribution of the ACh innervation in adult rat cerebral cortex and hippocampus

	Neocortex			Dorsal hippocampus		
	Frontal	Parietal	Occipital	CA1	CA3	DG
Axons						
density (m/mm^3)	13.0	9.9	11.0	12.2	16.2	15.6
length under 1 mm^2 (m)	26.7	19.7	15.3	–	–	–
Varicosities						
density (10^6/mm^3)	5.4	3.8	4.6	4.9	6.6	6.2
number under 1 mm^2 (10^6)	11.1	7.7	6.4	–	–	–

Data from Mechawar et al. (2000) and Aznavour et al. (2002), as explained in the text.

48

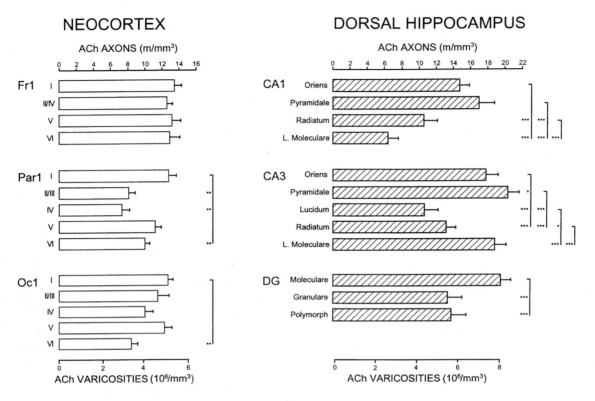

Fig. 2. Densities of ACh axons and axon varicosities in the different layers (I–VI) of the frontal (Fr1), parietal (Par1) and occipital (Oc1) neocortex and the dorsal hippocampus (CA1, CA3 and DG) of adult rat. Mean values (± s.e.m) from five and six rats in cortex and hippocampus, respectively. The data are expressed in density of ACh axons (upper scale) and density of axon varicosities (lower scale), as the number of axon varicosities per unit length of axons was stable in all regions at 4 per 10 μm. The layers showing statistically significant differences are linked by hooks, with asterisks in front of the differing layer(s). *$p < 0.005$ (cortex); ** ($p < 0.01$) and *** ($p < 0.001$) (hippocampus) indicate significant differences between layers by Student t-test. (Modified from Mechawar et al., 2000 and Aznavour et al., 2002.)

surface of varicosities (3%), and were almost always symmetrical (99%).

As shown in Table 2, the percentages of synaptic ACh varicosities that were found in the various layers of Par1 were 10%, 14%, 11%, 21% and 14%, respectively, for an interlayer mean of 14%, and a true average of 16%, when taking into account the various thickness of the different layers. In hippocampus, ACh varicosities displayed similar ultrastructural features, with an average diameter of 0.6 μm and a synaptic incidence of 7%, as extrapolated for whole varicosities from the very low proportion (3%) observed to form synaptic contact in single thin sections (Umbriaco et al., 1995). In both neocortex and hippocampus, ACh synapses were always made with dendrites, either branches or spines. In cerebral cortex, enough synaptic ACh varicosities were seen to conclude that 75% were in contact with dendritic branches and only 25% with spines (Umbriaco et al., 1994). Combined with the above-mentioned quantified data, these average synaptic proportions gave us the actual number of synaptic and nonsynaptic ACh varicosities, as well as the number of synaptic varicosities in contact with dendritic branches or spines in cerebral cortex (Table 2). It is clear from these values that the synaptic ACh varicosities represent a very small minority of all synapses in cortex or hippocampus, something in the order of 1 per 1500 synaptic terminals. This relative paucity of cortical ACh synapses should however not be taken to represent a lesser influence of ACh in cortical tissue, as discussed below.

Table 2. Synaptic and Asynaptic ACh axon varicosities in adult rat parietal cortex and dorsal hippocampus

| Layer | Parietal cortex | | | | | | Dorsal hippocampus |
	I	II–III	IV	V	VI	*I–VI*	CA1: stratum radiatum
Synaptic incidence	10%	14%	11%	21%	14%	*14%*	7%
Varicosities under 1 mm^2 (10^6)	0.69	1.05	0.70	2.82	2.44	*7.70*	0.93
asynaptic	0.62	0.90	0.62	2.23	2.10	*6.47*	0.87
synaptic	0.07	0.15	0.08	0.59	0.34	*1.23*	0.06

Synaptic incidence as estimated in Umbriaco et al. (1994) for cortex, and Umbriaco et al. (1995) for hippocampus. Data on the number of varicosities under 1 mm^2 from Mechawar et al. (2000) for cortex, and Aznavour et al. (2002) for hippocampus.

Subsequent investigations in other laboratories have confirmed that a vast majority of cortical ACh varicosities in rat cortex are asynaptic, with reported estimates of 14% and 9% for the frontoparietal and entorhinal region, respectively (Chédotal et al., 1994; Vaucher and Hamel, 1995). The value of 66% recently reported by Turrini et al. (2001) for layer V of rat parietal cortex after labeling with the vesicular ACh transporter was presumably the result of a sampling bias, as suggested by a significantly larger size of the profiles examined in that particular study.

It is interesting to note that a thorough ChAT-immunocytochemical study of the cortical ACh innervation has also been performed in the prefrontal cortex of adult rhesus monkey (Mrzljak et al., 1995). These authors found that among 100 serially sectioned ChAT-immunoreactive varicosities at the border of layers II and III, only 44% made synaptic contact. Again, small dendritic shafts represented the most frequent synaptic target (70%). Fifty-six percent of the cortical ACh varicosities were without any visible junctional specialization, even if frequently juxtaposed to dendrites or spines receiving asymmetrical synapses. More recently, Smiley et al. (1997) carried out a similar study on two samples of human anterior temporal lobe removed at surgery. Sixty-seven percent of 42 varicosities sampled in layers I and II were then observed to be endowed with small but identifiable synaptic specializations. It remains to be determined whether such variations of synaptic incidence in cortex reflect sampling biases, regional differences or species differences. Unfortunately, there are at present no such data on the proportion of synaptic ACh varicosities in primate or human hippocampus.

Regional and laminar distribution of the ACh innervation in the developing neocortex

The development of the cortical ACh innervation was described between birth (P0) and postnatal day 32 (P32) in the frontal, parietal and occipital areas of rat, using the same approach as in the adult (> P60) (Mechawar and Descarries, 2001). The main findings of this study were that this innervation develops rapidly and much earlier than previously suspected. This was apparent even without quantification. As illustrated in the original article, immunostained ACh fibers, capped with growth cones, were already visible in the cortical subplate of the three regions at birth. These fibers, initially few in number, had invaded the cortical plate and marginal zone by P4, the age at which faintly immunoreactive interneurons were first detected. Rapid ingrowth and proliferation generated an adult-like distribution of axons that was already apparent at P8. Adult densities of branched varicose axons were apparently reached by postnatal day 16 in the parietal area, while development continued until the end of the first month for the frontal area and even later for the occipital area.

Another basic finding was that, in addition to the lenghtening and branching of ACh axons, a third parameter characterized this growing innervation. Between P4 and P16, the number of axon varicosities per unit length of axon increased steadily in all layers and areas, from 2 per 10 μm at P4 to 4 at P16, the adult ratio.

Figure 3 depicts the growth of the ACh innervation in the three neocortical regions at the different postnatal ages examined. Again, both the density of varicosities (left) and their number under 1 mm^2 of cortical surface (right) are presented. The curves

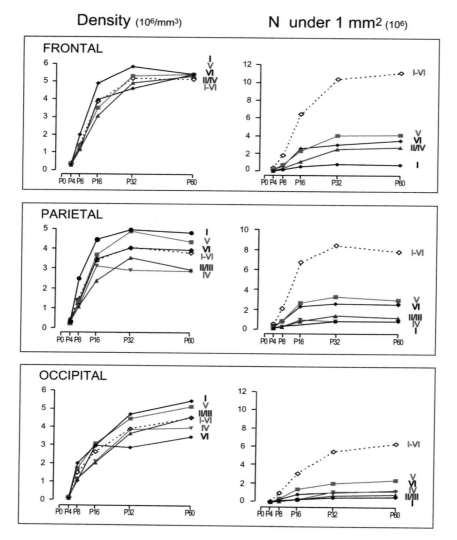

Density (10⁶/mm³)   **N under 1 mm²** (10⁶)

Fig. 3. Temporal profile of laminar and regional densities of ACh innervation in the frontal, parietal and occipital neocortex, at different postnatal ages (P4–P32) and in the adult. Means from 3 rats per age, except at P60 ($n = 5$). Data for the different layers (I, II/III or II/IV, V and VI) and the whole cortex (I–VI) are expressed as densities (left graphs) and number of ACh axon varicosities under 1 mm² of cortical surface (right graphs), as explained in the text. (Modified from Mechawar and Descarries, 2001.)

describe the time course of the growth in each layer and in the full thickness of each cortex. As may be observed for the three regions, the first two weeks are the period of fastest growth for each layer, and the temporal profile of growth is similar in every layer. There is no overproduction of fibers or varicosities at any age, only a steady increase until adult values are reached by one month. The curves for the frontal, parietal and occipital region are similar, except for a slightly slower, more protracted development in the occipital cortex.

Ultrastructural features of the ACh innervation in the developing cortex

The intrinsic and relational features of ACh varicosities in the developing cortex were examined in single sections from the parietal (Par1) region at postnatal ages P8, P16 and P32 (Mechawar et al., 2002). Figure 4 shows some examples of ChAT-immunostained varicosities, asynaptic or synaptic, at the three postnatal ages. On an average, these varicosities were of similar size throughout development, with an average diameter of 0.48 μm, and area of 0.16 μm^2, significantly smaller than in the adult (0.52 μm and 0.23 μm^2; Umbriaco et al., 1994). Interestingly, the proportion of ACh varicosity profiles containing mitochondria increased gradually with age, from 21% at 8 days to 61% in the adult. As in adult, the vast majority of synaptic varicosities displayed symmetrical junctional complexes, which averaged 0.29 μm in length and were exclusively made with dendritic branches (71%) and spines (29%). Among 300 single sectional profiles examined at each postnatal age, only a small fraction displayed a synaptic junction, yielding respective synaptic incidences of 13.2% (P8), 15.6% (P16) and 22.2% (P32) for whole varicosities (Table 3), a proportion not significantly different from the adult (14%) (Umbriaco et al., 1994). It was clear that the low frequency of synapses was characteristic of ACh varicosities as soon as they formed, as also reported recently for the noradrenaline innervation in the developing frontal and occipital cortex of rat (Latsari et al., 2002).

Since the proportion of synaptic ACh varicosities was stable during development, it could be inferred that the number of cortical ACh synapses had already reached its adult value by the end of the second week, at least in the parietal cortex. It could then be estimated at about 0.55 million per mm^3, and 1.0 million under 1 mm^2 of cortical surface (P16), compared to 0.53 and 1.23 million in the adult (>P60).

Functional properties of the ACh innervation in the mature and the developing cortex

These studies reveal major organizational features of the ACh innervations in cerebral cortex and hippocampus that should be determining their role(s) both in the adult and during development. First is their widespread distribution and abundance. As documented in Table 4, our quantitative data indicate that the ACh innervation is the densest neuromodulatory input to both neocortex and hippocampus. In neocortex, its average density of axon varicosities is more than four times higher than that of the noradrenaline (Audet et al., 1988) and a little higher than that of the serotonin innervation (Audet et al., 1989). In hippocampus, the average density of ACh varicosities is almost three-fold that of the noradrenaline (Oleskevich et al., 1989) and more than two-fold that of the serotonin innervation (Oleskevich and Descarries, 1990).

The ubiquity and density of the ACh input to cerebral cortex and hippocampus is consistent with the well documented involvement of ACh in numerous functional processes, such as the modulation of sensory information (Sillito and Kemp, 1983; Donoghue and Carroll, 1987; Lamour et al., 1988; McKenna et al., 1988; Rasmusson and Dykes, 1988), plasticity of sensory maps (Metherate et al., 1988; Tremblay et al., 1990; Juliano et al., 1991; Kilgard and Merzenich, 1998; see Chapter 10), control of microcirculation (Armstrong, 1986; Elhusseiny and Hamel, 2000; see Chapter 12), learning (Rigdon and Pirch, 1986; Pirch et al., 1992; Fine et al., 1997; Miranda and Bermúdez-Rattoni, 1999) and memory (Hasselmo et al., 1992; see also Chapter 15). The richness of the ACh input to cortex can also be related to the more global effects attributed to ACh in wakefulness and sleep (Richter and Crossland, 1949; Jasper and Tessier, 1971; Jiménez-Capdeville and Dykes, 1996; see Chapter 11), cortical activation (Krnjević and Phillis, 1963a,b; Mitchell, 1963; Celesia and Jasper, 1966; Buzsàki et al., 1988; Metherate et al., 1992; see Chapter 13), attention (Voytko et al., 1994; Gill et al., 2000), and even conscious awareness (Perry et al., 1999; see also Chapter 20).

The mechanisms underlying these numerous functions are only beginning to be unraveled. For instance, the induction of low-amplitude, fast (20–40 Hz) oscillations in neocortex by in vivo nucleus basalis stimulation is thought to underlie the cortical activating properties of ACh (Metherate et al., 1992; Cox et al., 1994; Steriade et al., 1996; see also Chapters 10 and 13). Such oscillations are viewed by some as reflecting the synchronous activity of widely distributed neuronal ensembles (Steriade and Amzica, 1996), which in turn would bind the separately processed features of a stimulus (for review, see Engel and Singer, 2001). The recent finding that nucleus basalis stimulation also elicits 40 Hz oscillations in the visual cortex has further prompted the hypothesis

52

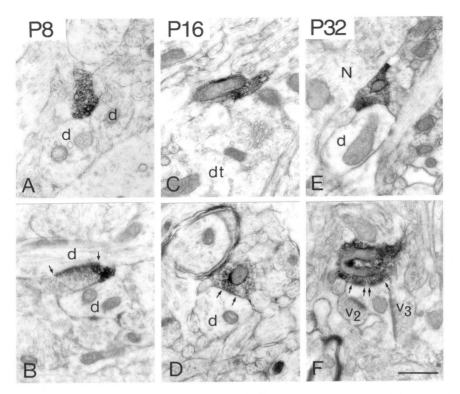

Fig. 4. (A–F) Electron micrographs of asynaptic (A, C, E) and synaptic (B, D, F) ACh axon varicosities from rat parietal cortex, at postnatal ages P8 (A–B), P16 (C–D) and P32 (E–F). These ChAT-immunostained profiles are identified as axon varicosities by their content in aggregated synaptic vesicles, often associated with a mitochondrion. The two ACh varicosities at P8 (A, B) are from layers VI and V, respectively. The one in B makes a symmetrical synaptic junction (between thin arrows) on a dendritic branch (upper d). Both the ACh varicosities at P16 are from layer VI. In C, the nonsynaptic ACh profile is juxtaposed to the base of a dendritic trunk (dt). The synapse made by the ACh varicosity in D (between thin arrows) is again symmetrical and made with a dendritic branch (d). The two ACh varicosities at P32 are from layer V. The one in E is juxtaposed to a dendritic branch (d), below, and a neuronal cell body (N), above. The synapse made by the varicosity in F is of the perforated type, asymmetrical, and made with a dendritic spine which also receives two other synaptic varicosities (V2 and V3), unlabeled. Scale bar (in F): 0.5 μm. (Reproduced with permission from Mechawar et al., 2002.)

Table 3. Synaptic and Asynaptic ACh axon varicosities in the developing parietal cortex of rat

Age	P4	P8	P16	P32
Synaptic incidence	–	13.2%	15.6%	22.2%
Varicosities under 1 mm^2 (10^6)	0.40	2.0	6.6	8.3
asynaptic	–	1.7	5.6	6.5
synaptic	–	0.3	1.0	1.8

Synaptic incidence as estimated in Mechawar et al. (2002). Data on the number of varicosities under 1 mm^2 from Mechawar and Descarries (2001).

that ACh is crucial for the temporal binding of visual stimuli (W. Singer, personal communication). Of course, this does not preclude that other modulatory systems,

differentially afferented and ubiquitously distributed in the cortex, might operate in concert to achieve the binding of information from various sources.

It has also been proposed that the involvement of ACh in learning and memory formation is rooted in its ability to differentially modulate the processing of afferent and efferent information (Hasselmo and Schnell, 1994; Kimura et al., 1999; Kimura, 2000; see also Chapter 15). Interestingly, the respective laminar distributions of the ACh innervation in primary motor and primary sensory neocortex (Mechawar et al., 2000), as well as in dorsal hippocampus (Aznavour et al., 2002), are consistent with this hypothesis.

A second feature of obvious functional consequences is the predominantly asynaptic nature of the

Table 4. Density of modulatory inputs to adult rat cerebral cortex and hippocampus

	Neocortex			Dorsal hippocampus		
	ACh	NA	5-HT	ACh	NA	5-HT
Synaptic incidence	14% [a]	17% [b]	38% [c]	7% [d]	16% [e]	22% [f]
Number per mm^3 (10^6)	4.6 [g]	1.05 [h]	4.55 [i]	5.9 [j]	2.0 [k]	2.65 [l]
asynaptic	4.0	0.87	2.82	5.5	1.28	2.07
synaptic	0.6	0.18	1.73	0.4	0.32	0.58

Values of synaptic incidence are from:
[a]The parietal cortex (Umbriaco et al., 1994).
[b,c]The frontal (Fr1), parietal (Par1) and occipital (Oc1) cortex (Séguéla et al., 1989, 1990).
[d,e]The stratum radiatum of CA1 (Umbriaco et al., 1995).
[f]CA1, CA3, and DG (Oleskevich et al., 1991; Umbriaco et al., 1995).
[g]Numbers of varicosities per mm^3 are from the frontal, parietal, and occipital cortex (Mechawar et al., 2000).
[h,i]The frontal and parietal cortex (Audet et al., 1988, 1989).
[j,k,l]CA1, CA3, and DG (Oleskevich et al., 1989; Oleskevich and Descarries, 1990; Aznavour et al., 2002).

cortical and hippocampal ACh innervations, as first demonstrated in adult rat parietal cortex (Umbriaco et al., 1994) and stratum radiatum of CA1 (Umbriaco et al., 1995). It is generally assumed that such innervations exert many of their effects by diffuse transmission, in addition to synaptic transmission (Descarries et al., 1997; Descarries and Mechawar, 2000). Considering the density of ACh innervation in both neocortex and hippocampus, it is likely that much of the ACh released from their predominantly asynaptic ACh varicosities reaches muscarinic and nicotinic receptors by diffusion in the extracellular space. To date, several immunocytochemical reports on the subcellular localization of muscarinic or nicotinic receptor subunits in cerebral cortex or hippocampus have underlined the fact that both receptor subtypes are mainly found at extrasynaptic locations (e.g., Hill et al., 1993; Mrzljak et al., 1998; Lubin et al., 1999; Rouse et al., 2000; see also Chapters 3 and 7).

Because of the high density and predominantly asynaptic nature of the ACh axon varicosities, we have also proposed that in richly ACh-innervated regions such as cortex and hippocampus, as well as in neostriatum, a low ambient level of ACh might be permanently maintained in the extracellular space, and capable of influencing nicotinic and muscarinic receptors distributed on a variety of neighboring elements, including endothelial cells (Fig. 5; Descarries et al., 1997; Descarries, 1998). This hypothesis is consistent with the basal levels of extracellular ACh measured in cerebral cortex by microdialysis (e.g.,

Nilsson et al., 1990; Kosasa et al., 1999). It is also consistent with the fact that most of the acetylcholinesterase in the CNS is of the tetrameric globular form (G4). This isoform, which is found extrasynaptically at the neuromuscular junction (Gisiger and Stephens, 1988), is secreted by nerve terminals in the CNS and thus thought to regulate levels of ACh diffusing in the extracellular space (see Descarries et al., 1997). The existence of an ambient level of ACh and its functional impact in cortex and hippocampus could also account for the fact that heavy losses of cholinergic neurons (and hence of ACh innervation) must be occuring in Alzheimer's disease before it becomes clinically manifested (e.g., Whitehouse et al., 1981; see also Chapter 22).

A third, previously unsuspected feature of the cortical ACh innervation, is its early and rapid development. Early, because we could observe ingrowing ChAT-immunoreactive axons in the cortical subplate on the very day of birth, almost two weeks sooner than reported in earlier ChAT-immunocytochemical investigations (Dori and Parnavelas, 1989; Gould et al., 1991; see also Chapter 1). The precocity of this innervation is consistent with the expression of several parameters of ACh function in the embryonic and early postnatal neocortex (e.g., Aubert et al., 1996), and the various roles assigned to ACh in corticogenesis. However, the relatively high concentrations of ACh measured in rat neocortex at birth (Coyle and Yamamura, 1976) remain to be explained. They are difficult to reconcile with the existence of only few, scarcely

54

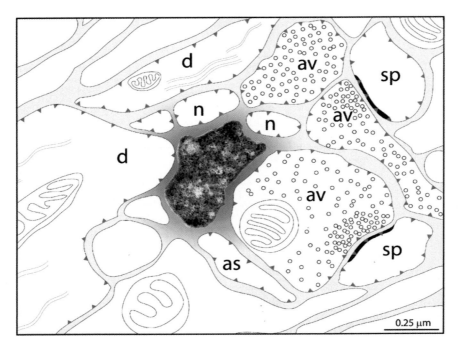

Fig. 5. Schematic representation of a nonsynaptic ACh axon terminal (varicosity) in the cerebral cortex (see Fig. 1A in Descarries et al., 1997), from which ACh has been released to diffuse in the extracellular space and contribute to a low ambient level of ACh around a variety of neighboring cellular elements. The extracellular space has been widened for illustrative shading purposes. ACh receptors, many of which are located extrasynaptically on diverse potential targets, are represented by small red triangles. as, astrocyte; av, axon varicosity; d, dendrite; n, neurite; sp, dendritic spine.

distributed, unbranched ChAT-immunoreactive fibers observed in this brain region at birth (Mechawar and Descarries, 2001), even if these fibers are already capable of ACh release (Pedata et al., 1983). It is unlikely that this discrepancy reflects a transient synthesis of ACh by other cortical elements (e.g., glial cells), since ChAT protein or mRNA have never been detected at 'ectopic' locations in developing cortex (Gould et al., 1991; Ibanez et al., 1991; Mechawar and Descarries, 2001). Perhaps the absence of a fully mature (and impermeable) blood-brain barrier until the end of the first postnatal week in rodents allows for the penetration and diffusion in cortex of nonneuronal ACh synthesized in the placenta or in peripheral organs (for review on nonneuronal ACh, see Wessler et al., 1998).

In addition to demonstrating the speed with which the ACh innervation is formed in the developing neocortex, our quantitative study revealed a new parameter of ACh axon growth, i.e., the increase in number of varicosities per unit length of axon, which

takes place in the first two weeks. During this period, if broken down to a single basalo-cortical neuron, the speed of axon growth may then be estimated at 2 cm of axon and 9000 varicosities per day, i.e., almost 1 mm of axon and 400 varicosities per hour.

Our ultrastructural study in postnatal rat indicated that diffuse transmission by ACh might apply to the developing as well as the adult cortex. Indeed, the intrinsic and relational features of ACh varicosities in the developing somatosensory area were adult-like throughout its postnatal development (Mechawar et al., 2002). The low proportion of synapses made by cortical ACh neurons may be considered as an innate feature, since it does not show significant age- or area-related variations, as also being observed in an ongoing study of the cholinergic innervation in the developing neostriatum (Aznavour et al., 2003). Also noteworthy is the fact that this proportion remains stable at a time when the varicosities are forming and greatly increasing in number along axons. It makes it unlikely that transient, 'instructive' synapses are being established by the growing ACh neurons.

Even though the subcellular distribution of ACh receptors in the developing cortex has not yet been determined, functional receptors of both the nicotinic and muscarinic families are known to be expressed by cortical cells in the late embryonic and early postnatal periods (Ostermann et al., 1995; Zoli et al., 1995; Aubert et al., 1996; Ma et al., 2000; see also Chapter 1). Thus, diffuse transmission and an ambient level of ACh, as discussed above, could participate in the many roles attributed to ACh during corticogenesis (Bear and Singer, 1986; Broide et al., 1996; Zhu and Waite, 1998; Aramakis et al., 2000; Peinado, 2000). This again raises the question as to the source(s) of the high concentrations of ACh measured in the cortex during the perinatal period, and how this ACh might be involved in corticogenesis? One might also ask if distinct or complementary functions are then being subserved by the synaptic versus asynaptic ACh varicosities? Which trophic factors/molecular cues guide the early and rapid growth of the cortical ACh innervation and the establishment and maintenance of its laminar- and area-specific distribution during the postnatal period and adult life? Answering these questions would undoubtedly provide useful information from both the fundamental and clinical standpoints.

The studies altogether convey a new image of the ACh neurons innervating the cerebral cortex or hippocampus. Early tracing experiments had shown the widespread (and probably overlapping) distribution of the axonal arborization of individual nucleus basalis neurons innervating the cerebral cortex (McKinney et al., 1983). On the basis of the presen data, unitary ACh neurons may be calculated to project 0.5 to 1 meter of axon endowed with 200,000 and 500,000 varicosities in cerebral cortex and hippocampus, respectively (Mechawar et al., 2000; Aznavour et al., 2002). In adult rat cortex alone, this adds up to a staggering total of 7.6 km of ACh axons bearing more than 3 billion varicosities. Such numbers together with the largely asynaptic nature of ACh axon varicosities suggest that these neurons are ideally shaped to exert diverse and widespread influences in vast expanses of cortex.

Acknowledgments

These studies were supported by grant MT-3544 to L.D., from the Medical Research Council of Canada (MRC), now grant NRF 3544 from the Canadian Institutes for Health Research. N.M. held a Research Studentship from MRC, and N.A. is recipient of a Ph.D. studentship from the Groupe de recherche sur le système nerveux central (FCAR) at the Université de Montréal. The authors are grateful to Costantino Cozzari and Boyd K. Hartman for their generous gift of ChAT antibody. They also thank Jean Léveillé and Gaston Lambert for photographic work.

References

Amaral, D.G. and Kurz, J. (1985) An analysis of the origins of the cholinergic and noncholinergic septal projections to the hippocampal formation of the rat. J. Comp. Neurol., 240: 37–59.

Aramakis, V.B., Hsieh, C.Y., Leslie, F.M. and Metherate, R. (2000) A critical period for nicotine-induced disruption of synaptic development in rat auditory cortex. J. Neurosci., 20: 6106–6116.

Armstrong, D.M. (1986) Ultrastructural characterization of choline acetyltransferase-containing neurons in the basal forebrain of rat: Evidence for a cholinergic innervation of intracerebral blood vessels. J. Comp. Neurol., 250: 81–92.

Aubert, I., Cécyre, D., Gauthier, S. and Quirion, R. (1996) Comparative ontogenic profile of cholinergic markers, including nicotinic and muscarinic receptors, in the rat brain. J. Comp. Neurol., 369: 31–55.

Audet, M.A., Doucet, G., Oleskevich, S. and Descarries, L. (1988) Quantified regional and laminar distribution of the noradrenaline innervation in the anterior half of the adult rat cerebral cortex. J. Comp. Neurol., 274: 307–318.

Audet, M.A., Descarries, L. and Doucet, G. (1989) Quantified regional and laminar distribution of the serotonin innervation in the anterior half of adult rat cerebral cortex. J. Chem. Neuroanat., 2: 29–44.

Aznavour, N., Mechawar, N. and Descarries, L. (2002) Comparative analysis of cholinergic innervation in the dorsal hippocampus of adult mouse and rat: a quantitative immunocytochemical study. Hippocampus, 12: 206–217.

Aznavour, N., Mechawar, N., Watkins, K.C. and Descarries, L. (2003) Fine structural features of the acetylcholine innervation in the developing neostriatum of rat. J. Comp. Neurol., 467: 280–291.

Bayraktar, T., Staiger, J.F., Acsady, L., Cozzari, C., Freund, T.F. and Zilles, K. (1997) Co-localization of vasoactive intestinal polypeptide, γ-aminobutyric acid and choline acetyltransferase in neocortical interneurons of the adult rat. Brain Res., 757: 209–217.

Bear, M.F. and Singer, W. (1986) Modulation of visual cortical plasticity by acetylcholine and noradrenaline. Nature, 320: 172–176.

Broide, R.S., Robertson, R.T. and Leslie, F.M. (1996) Regulation of α7 nicotinic acetylcholine receptors in the developing rat somatosensory cortex by thalamocortical afferents. J. Neurosci., 16: 2956–2971.

Buzsàki, G., Bickford, R.G., Ponomareff, G., Thal, L.J., Mandel, R. and Gage, F.H. (1988) Nucleus basalis and thalamic control of neocortical activity in the freely moving rat. J. Neurosci., 8: 4007–4026.

Celesia, G.G. and Jasper, H.H. (1966) Acetylcholine released from cerebral cortex in relation to state of activation. Neurology, 16: 1053–1063.

Chédotal, A., Cozzari, C., Faure, M.P., Hartman, B.K. and Hamel, E. (1994) Distinct choline acetyltransferase (ChAT) and vasoactive intestinal polypeptide (VIP) bipolar neurons project to local blood vessels in the rat cerebral cortex. Brain Res., 646: 181–193.

Cox, C.L., Metherate, R. and Ashe, J.H. (1994) Modulation of cellular excitability in neocortex: muscarinic receptor and second messenger-mediated actions of acetylcholine. Synapse, 16: 123–136.

Coyle, J.T. and Yamamura, H.I. (1976) Neurochemical aspects of the ontogenesis of cholinergic neurons in the rat brain. Brain Res., 118: 429–440.

Cozzari, C., Howard, J. and Hartman, B. (1990) Analysis of epitopes on choline acetyltransferase (ChAT) using monoclonal antibodies (Mabs). Soc. Neurosci. Abstr., 16: 200.

Descarries, L. (1998) The hypothesis of an ambient level of acetylcholine in the central nervous system. J. Physiol. (Paris), 92: 215–220.

Descarries, L. and Mechawar, N. (2000) Ultrastructural evidence for diffuse transmission by monoamine and acetylcholine neurons of the central nervous system. Progr. Brain Res., 125: 27–47.

Descarries, L., Gisiger, V. and Steriade, M. (1997) Diffuse transmission by acetylcholine in the CNS. Progr. Neurobiol., 53: 603–625.

Donoghue, J.P. and Carroll, K.L. (1987) Cholinergic modulation of sensory responses in rat primary somatic sensory cortex. Brain Res., 408: 367–371.

Dori, I. and Parnavelas, J.G. (1989) The cholinergic innervation of the rat cerebral cortex shows two distinct phases in development. Exp. Brain Res., 76: 417–423.

Eckenstein, F. and Baughman, R.W. (1987) Cholinergic innervation in cerebral cortex. In: E.G. Jones and A. Peters (Eds.), Cerebral Cortex. Further Aspects of Cortical Function, Including Hippocampus, Vol. 6, Plenum Press, New York, pp. 129–160.

Elhusseiny, A. and Hamel, E. (2000) Muscarinic, but not nicotinic, acetylcholine receptors mediate a nitric oxide-dependent dilation in brain cortical arterioles: a possible role for the M5 receptor subtype. J. Cereb. Blood Flow Metab., 20: 298–305.

Engel, A.K. and Singer, W. (2001) Temporal binding and the neural correlates of sensory awareness. Trends Cog. Sci., 5: 16–25.

Fine, A., Hoyle, C., Maclean, C.J., Levatte, T.L., Baker, H.F. and Ridley, R.M. (1997) Learning impairments following injection of a selective cholinergic immunotoxin, ME20.4 IgG-saporin, into the basal nucleus of Meynert in monkeys. Neuroscience, 81: 331–343.

Gage, F.H., Björklund, A. and Stenevi, U. (1983) Reinnervation of the partially deafferented hippocampus by compensatory collateral sprouting from spared cholinergic and noradrenergic afferents. Brain Res., 268: 27–37.

Gill, T.M., Sarter, M. and Givens, B. (2000) Sustained visual attention performance-associated prefrontal neuronal activity: evidence for cholinergic modulation. J. Neurosci., 20: 4745–4757.

Gisiger, V. and Stephens, H.R. (1988) Localization of the pool of G4 acetylcholinesterase characterizing fast muscles and its alteration in murine muscular dystrophy. J. Neurosci. Res., 19: 62–78.

Gould, E., Woolf, N.J. and Butcher, L.L. (1991) Postnatal development of cholinergic neurons in the rat: I. Forebrain. Brain Res. Bull., 27: 767–789.

Gritti, I., Mainville, L. and Jones, B.E. (1993) Codistribution of GABA- with acetylcholine-synthesizing neurons in the basal forebrain of the rat. J. Comp. Neurol., 329: 438–457.

Hasselmo, M.E. and Schnell, E. (1994) Laminar selectivity of the cholinergic suppression of synaptic transmission in rat hippocampal region CA1: computational modeling and brain slice physiology. J. Neurosci., 14: 3898–3914.

Hasselmo, M.E., Anderson, B.P. and Bower, J.M. (1992) Cholinergic modulation of cortical associative memory function. J. Neurophysiol., 67: 1230–1246.

Hill, J.A., Zoli, M., Bourgeois, J.-P. and Changeux, J.-P. (1993) Immunocytochemical localization of a neuronal nicotinic receptor: the β2-subunit. J. Neurosci., 13: 1551–1568.

Ibanez, C.F., Ernfors, P. and Persson, H. (1991) Developmental and regional expression of choline acetyltransferase mRNA in the rat central nervous system. J. Neurosci. Res., 29: 163–171.

Jasper, H.H. and Tessier, J. (1971) Acetylcholine liberation from cerebral cortex during paradoxical (REM) sleep. Science, 172: 601–602.

Jiménez-Capdeville, M.E. and Dykes, R.W. (1996) Changes in cortical acetylcholine release in the rat during day and night: differences between motor and sensory areas. Neuroscience, 71: 567–579.

Juliano, S.L., Ma, W. and Eslin, D. (1991) Cholinergic depletion prevents expansion of topographic maps in somatosensory cortex. Proc. Natl Acad. Sci. USA, 88: 780–784.

Kilgard, M.P. and Merzenich, M.M. (1998) Cortical map reorganization enabled by nucleus basalis activity. Science, 279: 1714–1718.

Kimura, F. (2000) Cholinergic modulation of cortical function: a hypothetical role in shifting the dynamics in cortical network. Neurosci. Res., 38: 19–26.

Kimura, F., Fukuda, M. and Tsumoto, T. (1999) Acetylcholine suppresses the spread of excitation in the visual cortex revealed by optical recording: possible differential effect depending on the source of input. Eur. J. Neurosci., 11: 3597–3609.

Kosasa, T., Kuriya, Y. and Yamanishi, Y. (1999) Effect of donepezil hydrochloride (E2020) on extracellular acetylcholine concentration in the cerebral cortex of rats. Jpn. J. Pharmacol., 81: 216–222.

Krnjević, K. and Phillis, J.W. (1963a) Acetylcholine-sensitive cells in the cerebral cortex. J. Physiol. (London), 166: 296–327.

Krnjević, K. and Phillis, J.W. (1963b) Pharmacological properties of acetylcholine-sensitive cells in the cerebral cortex. J. Physiol. (London), 166: 328–350.

Lamour, Y., Dutar, P., Jobert, A. and Dykes, R.W. (1988) An iontophoretic study of single somatosensory neurons in rat granular cortex serving the limbs: a laminar analysis of glutamate and acetylcholine effects on receptive-field properties. J. Neurophysiol., 60: 725–750.

Latsari, M., Dori, I., Antonopoulos, J., Chiotelli, M. and Dinopoulos, A. (2002) Noradrenergic innervation of the developing and mature visual and motor cortex of the rat brain: a light and electron microscopic immunocytochemical analysis. J. Comp. Neurol., 445: 145–158.

Lubin, M., Erisir, A. and Aoki, C. (1999) Ultrastructural immunolocalization of the α7 nAChR subunit in guinea pig medial prefrontal cortex. Ann. N.Y. Acad. Sci., 868: 628–632.

Lysakowski, A., Wainer, B.H., Bruce, G. and Hersh, L.B. (1989) An atlas of the regional and laminar distribution of choline acetyltransferase immunoreactivity in rat cerebral cortex. Neuroscience, 28: 291–336.

Ma, W., Maric, D., Li, B.-s., Hu, Q., Andreadis, J.D., Grant, G.M., Liu, Q.-Y., Shaffer, K.M., Chang, Y.H., Zhang, L., Pancrazio, J.J., Pant, H.C., Stenger, D.A. and Barker, J.L. (2000) Acetylcholine stimulates cortical precursor cell proliferation in vitro via muscarinic receptor activation and MAP kinase phosphorylation. Eur. J. Neurosci., 12: 1227–1240.

McKenna, T.M., Ashe, J.H., Hui, G.K. and Weinberger, N.M. (1988) Muscarinic agonists modulate spontaneous and evoked unit discharge in auditory cortex of cat. Synapse, 2: 54–68.

McKinney, M., Coyle, J.T. and Hedreen, J.C. (1983) Topographic analysis of the innervation of the rat neocortex and hippocampus by the basal forebrain cholinergic system. J. Comp. Neurol., 217: 103–121.

Mechawar, N. and Descarries, L. (2001) The cholinergic innervation develops early and rapidly in the rat cerebral cortex: a quantitative immunocytochemical study. Neuroscience, 108: 555–567.

Mechawar, N., Cozzari, C. and Descarries, L. (2000) Cholinergic innervation in adult rat cerebral cortex: a quantitative immunocytochemical description. J. Comp. Neurol., 428: 305–318.

Mechawar, N., Watkins, K.C. and Descarries, L. (2002) Ultrastructural features of the acetylcholine innervation in the developing parietal cortex of rat. J. Comp. Neurol., 443: 250–258.

Mesulam, M.M., Mufson, E.J., Wainer, B.H. and Levey, A.I. (1983) Central cholinergic pathways in the rat: an overview based on an alternative nomenclature (Ch1-Ch6). Neuroscience, 10: 1185–1201.

Metherate, R., Tremblay, N. and Dykes, R.W. (1988) Transient and prolonged effects of acetylcholine on responsiveness of cat somatosensory cortical neurons. J. Neurophysiol., 59: 1253–1276.

Metherate, R., Cox, C.L. and Ashe, J.H. (1992) Cellular bases of neocortical activation: modulation of neural oscillations by the nucleus basalis and endogenous acetylcholine. J. Neurosci., 12: 4701–4711.

Miranda, M.I. and Bermúdez-Rattoni, F. (1999) Reversible inactivation of the nucleus basalis magnocellularis induces disruption of cortical acetylcholine release and acquisition, but not retrieval, of aversive memories. Proc. Natl Acad. Sci. USA, 96: 6478–6482.

Mitchell, J.F. (1963) The spontaneous and evoked release of acetylcholine from the cerebral cortex. J. Physiol. (London), 165: 98–116.

Mrzljak, L., Pappy, M., Leranth, C. and Goldman-Rakic, P.S. (1995) Cholinergic synaptic circuitry in the macaque prefrontal cortex. J. Comp. Neurol., 357: 603–617.

Mrzljak, L., Levey, A.I., Belcher, S. and Goldman-Rakic, P.S. (1998) Localization of the m2 muscarinic acetylcholine receptor protein and mRNA in cortical neurons of the normal and cholinergically deafferented rhesus monkey. J. Comp. Neurol., 390: 112–132.

Nilsson, O.G., Kalén, P., Rosengren, E. and Björklund, A. (1990) Acetylcholine release in rat hippocampus as studied by microdialysis is dependent on axonal impulse flow and increases during behavioral activation. Neuroscience, 36: 325–328.

Nyakas, C., Luiten, P.G., Spencer, D.G. and Traber, J. (1987) Detailed projection patterns of septal and diagonal band efferents to the hippocampus in the rat with emphasis on innervation of CA1 and dentate gyrus. Brain Res. Bull., 18: 533–545.

Oleskevich, S. and Descarries, L. (1990) Quantified distribution of the serotonin innervation in adult rat hippocampus. Neuroscience, 34: 19–33.

Oleskevich, S., Descarries, L. and Lacaille, J.C. (1989) Quantified distribution of the noradrenaline innervation in the hippocampus of adult rat. J. Neurosci., 9: 3803–3815.

Oleskevich, S., Descarries, L., Watkins, K.C., Séguéla, P. and Daszuta, A. (1991) Ultrastructural features of the serotonin innervation in the adult rat hippocampus. An immunocytochemical description in single and serial thin sections. Neuroscience, 42: 777–791.

Ostermann, C.-H., Grunwald, J., Wevers, A., Lorke, D.E., Reinhardt, S., Maelicke, A. and Schröder, H. (1995) Cellular expression of α4 subunit mRNA of the nicotinic acetylcholine receptor in the developing rat telencephalon. Neurosci. Lett., 192: 21–24.

Pedata, F., Slavikova, J., Kotas, A. and Pepeu, G. (1983) Acetylcholine release from rat cortical slices during postnatal development and aging. Neurobiol. Aging, 4: 31–35.

58

Peinado, A. (2000) Traveling slow waves of neural activity: a novel form of network activity in developing neocortex. J. Neurosci., RC54: 1–6.

Perry, E., Walker, M., Grace, J. and Perry, R. (1999) Acetylcholine in mind: A neurotransmitter correlate of consciousness. Trends Neurosci., 22: 273–280.

Pirch, J.H., Turco, K. and Rucker, H.K. (1992) A role for acetylcholine in conditioning-related responses of rat frontal cortex neurons: microiontophoretic evidence. Brain Res., 586: 19–26.

Porter, J.T., Cauli, B., Tsuzuki, K., Lambolez, B., Rossier, J. and Audinat, E. (1999) Selective excitation of subtypes of neocortical interneurons by nicotinic receptors. J. Neurosci., 19: 5228–5235.

Rasmusson, D.D. and Dykes, R.W. (1988) Long-term enhancement of evoked potentials in cat somatosensory cortex by coactivation of the basal forebrain and cutaneous receptors. Exp. Brain Res., 70: 276–286.

Richter, D. and Crossland, J. (1949) Variation in acetylcholine content of the brain with physiological state. Am. J. Physiol., 159: 247–255.

Rigdon, G.C. and Pirch, J.H. (1986) Nucleus basalis involvement in conditioned neuronal responses in the rat frontal cortex. J. Neurosci., 6: 2535–2542.

Rouse, S.T., Edmunds, S.M., Yi, H., Gilmor, M.L. and Levey, A.I. (2000) Localization of M2 muscarinic acetylcholine receptor protein in cholinergic and non-cholinergic terminals in rat hippocampus. Neurosci. Lett., 284: 182–186.

Rye, D.B., Wainer, B.H., Mesulam, M.-M., Mufson, E.J. and Saper, C.B. (1984) Cortical projections arising from the basal forebrain: a study of cholinergic and noncholinergic components employing combined retrograde tracing and immunohistochemical localization of choline acetyltransferase. Neuroscience, 13: 627–643.

Saper, C.B. (1984) Organization of the cerebral cortical afferent systems in the rat. II. Magnocellular basal nucleus. J. Comp. Neurol., 222: 313–342.

Schwegler, H., Boldyreva, M., Linke, R., Wu, J., Zilles, K. and Crusio, W.E. (1996) Genetic variation in the morphology of the septo-hippocampal cholinergic and GABAergic systems in mice: II. Morpho-behavioral correlations. Hippocampus, 6: 535–545.

Séguéla, P., Watkins, K.C. and Descarries, L. (1989) Ultrastructural relationships of serotonin axon terminals in the crebral cortex of the adult rat. *J. Comp. Neurol.*, 289: 129–142.

Séguéla, P., Watkins, K.C., Geffard, M. and Descarries, L. (1990) Nonadrenaline axon terminals in adult rat neocortex: an immunocytochemical analysis in Serial thin sections. *Neuroscience*, 35: 249–264.

Sillito, A.M. and Kemp, J.A. (1983) Cholinergic modulation of the functional organization of the cat visual cortex. Brain Res., 289: 143–155.

Smiley, J.F., Morrell, F. and Mesulam, M.-M. (1997) Cholinergic synapses in human cerebral cortex: an ultrastructural study in serial sections. Exp. Neurol., 144: 361–368.

Steriade, M. and Amzica, F. (1996) Intracortical and corticothalamic coherency of fast spontaneous oscillations. Proc. Natl. Acad. Sci. USA, 93: 2533–2538.

Steriade, M., Paré, D., Parent, A. and Smith, Y. (1988) Projections of cholinergic and non-cholinergic neurons of the brainstem core to relay and associational thalamic nuclei in the cat and macaque monkey. Neuroscience, 25: 47–67.

Steriade, M., Amzica, F. and Contreras, D. (1996) Synchronization of fast (30–40 Hz) spontaneous cortical rhythms during brain activation. J. Neurosci., 16: 392–417.

Tremblay, N., Warren, R.A. and Dykes, R.W. (1990) Electrophysiological studies of acetylcholine and the role of the basal forebrain in the somatosensory cortex of the cat. II. Cortical neurons excited by somatic stimuli. J. Neurophysiol., 64: 1212–1222.

Turrini, P., Casu, M.A., Wong, T.P., De Koninck, Y., Ribeiro-Da-Silva, A. and Cuello, A.C. (2001) Cholinergic nerve terminals establish classical synapses in the rat cerebral cortex: synaptic pattern and age-related atrophy. Neuroscience., 105: 277–285.

Umbriaco, D., Watkins, K.C., Descarries, L., Cozzari, C. and Hartman, B.K. (1994) Ultrastructural and morphometric features of the acetylcholine innervation in adult rat parietal cortex. An electron microscopic study in serial sections. J. Comp. Neurol., 348: 351–373.

Umbriaco, D., Garcia, S., Beaulieu, C. and Descarries, L. (1995) Relational features of acetylcholine, noradrenaline, serotonin and GABA axon terminals in the stratum radiatum of adult rat hippocampus (CA1). Hippocampus, 5: 605–620.

Vaucher, E. and Hamel, E. (1995) Cholinergic basal forebrain neurons project to cortical microvessels in the rat: electron microscopic study with anterogradely transported *Phaseolus vulgaris* leucoagglutinin and choline acetyltransferase immunocytochemistry. J. Neurosci., 15: 7427–7441.

Voytko, M.L., Olton, D.S., Richardson, R.T., Gorman, L.K., Tobin, J.R. and Price, D.L. (1994) Basal forebrain lesions in monkeys disrupt attention but not learning and memory. J. Neurosci., 14: 167–186.

Wessler, I., Kirkpatrick, C.J. and Racké, K. (1998) Non-neuronal acetylcholine, a locally acting molecule, widely distributed in biological systems: expression and function in humans. Pharmacol. Ther., 77: 59–79.

Whitehouse, P.J., Price, D.L., Clark, A.W., Coyle, J.T. and DeLong, M.R. (1981) Alzheimer disease: evidence for selective loss of cholinergic neurons in the nucleus basalis. Ann. Neurol., 10: 122–126.

Woolf, N.J. (1991) Cholinergic systems in mammalian brain and spinal cord. Progr. Neurobiol., 37: 475–524.

Zhu, X.O. and Waite, P.M.E. (1998) Cholinergic depletion reduces plasticity of barrel field cortex. Cereb. Cortex, 8: 63–72.

Zoli, M., Le Novère, Hill, J.A. Jr. and Changeux, J.-P. (1995) Developmental regulation of nicotinic ACh receptor subunit mRNAs in the central and peripheral nervous system. *J. Neurosci.*, 15: 1912–1939.

Progress in Brain Research, Vol. 145
ISSN 0079-6123

CHAPTER 3

Muscarinic acetylcholine receptor subtypes in cerebral cortex and hippocampus

Laura A. Volpicelli and Allan I. Levey*

*Department of Neurology and Center for Neurodegenerative Disease, Emory University
School of Medicine, Whitehead Biomedical Research Building Suite 505, 615
Michael St., Atlanta, GA 30322, USA*

In the central nervous system, acetylcholine (ACh) facilitates many functions, such as learning, memory, attention and motor control. Two classes of receptors play a role in cholinergic transmission: the nicotinic ion channels which mediate fast postsynaptic transmission and the muscarinic G protein-coupled receptors (GPCRs) which play a modulatory role. This chapter focuses on the muscarinic acetylcholine receptor (mAChR) family, whose five members participate in critical cholinergic functions including learning, memory and attention.

Five subtypes of mAChRs, M_1, M_2, M_3, M_4 and M_5, have been revealed by molecular cloning (Kubo et al., 1986; Bonner et al., 1987; Peralta et al., 1987; Buckley et al., 1988). These receptors have seven transmembrane regions that are highly conserved among GPCR subtypes. The large cytoplasmic mAChR third intracellular loop between transmembrane domains 5 and 6 interacts with the G protein and is the most highly divergent domain among the 5 mAChR subtypes. The signaling pathways activated by mAChRs include: activation of phospholipase C, by M_1, M_3 and M_5, and inhibition of adenylyl cyclase activity, by M_2 and M_4. mAChRs also modulate several types of ion channels (Marino et al., 1998; Shapiro et al., 2001; Fisahn et al., 2002), thus regulating neuronal activity. In addition to these 'traditional' signaling pathways, mAChRs activate mitogen-activated protein kinases

(MAPKs), which are regulators of cell survival, differentiation and synaptic plasticity (Crespo et al., 1994; Wotta et al., 1998; Berkeley and Levey, 2000; Berkeley et al., 2001; Hamilton and Nathanson, 2001). In addition, small GTPases such as rho and rac, which are involved in cytoskeletal morphology and neurite outgrowth, are also activated by mAChRs (Kozma et al., 1997; Kiyono et al., 1999; Linseman et al., 2000). Thus, mAChRs can activate a multitude of signaling pathways involved in modulating neuronal function.

Muscarinic receptor expression in the brain

The diversity of mAChR subtypes and signaling pathways indicates that the localization of each receptor dictates the potential muscarinic responses to ACh in various cell types. Thus, considerable effort has been made to map the distribution of the mAChRs using a variety of approaches to detect each receptor. Many ligand autoradiographic binding studies have yielded insights into the relative abundance of the mAChR subtypes in brain (Cortes and Palacios, 1986; Mash and Potter, 1986; Spencer et al., 1986). The cortex and hippocampus express ~ 1000 fmol receptor per mg of protein in rodents. In comparison, the striatum expresses 1500 fmol receptor per mg of protein and thus expresses much higher levels (Cortes and Palacios, 1986). The majority of these binding sites are pharmacologically classified as M_1 and M_2 in cortex, and M_1 and M_4 in striatum.

*Corresponding author: Tel.: 404-727-3727; Fax: 404-727-3999;
E-mail: alevey@emory.edu

DOI: 10.1016/S0079-6123(03)45003-6

60

However, limitations in the binding methods have made it difficult to establish the precise identity of the receptor subtypes (i.e., the ligands do not discriminate among all of the subtypes), and more critically, to establish the cellular localization of the receptors. For this reason, in situ hybridization and immunohistochemical studies have been performed to delineate the cellular distribution of the subtype mRNA and proteins, respectively. In general, there has been remarkable agreement in these approaches (Buckley et al., 1988; Weiner et al., 1990; Levey et al., 1991; Vilaro et al., 1991; Levey, 1996). Here, we review the distribution of the mAChR proteins, as the immunological methods used to study the proteins have provided the critical details about the actual levels of the proteins and their cellular and subcellular distributions. This information provides a basis for understanding the functional relevance of the receptor subtypes in cholinergic transmission in cerebral cortex, hippocampus, and other brain regions.

M_1, M_2 and M_4 distribution in the cortex

Quantitative analysis of mAChR protein distribution performed in the rat brain using immunoprecipitations with subtype selective antibodies reveals that the M_1, M_2 and M_4 mAChRs are the predominate mAChR subtypes expressed in the brain (Levey et al., 1991). M_3 and M_5 are also expressed throughout the brain, but in low abundance. Approximately 40% of total mAChRs in the cortex of the rodent brain represent M_1, 37% M_2 and approximately 15% M_4 (Levey et al., 1991). In the human brain, M_1 is the primary receptor in the frontal, temporal, parietal and occipital cortical areas, representing 35–60% of total mAChRs (Flynn et al., 1995). M_2 is more abundant in the occipital cortex of human brain (36%) than in the frontal, temporal or parietal cortex (approximately 20%). M_4 represents only approximately 20% of total mAChRs in the human brain cortex.

Immunohistochemistry reveals distinct cellular localizations of mAChR subtypes in various brain regions. As shown in Fig. 1, M_1, M_2 and M_4 mAChRs are expressed throughout the primate cortex, with different laminar distributions. The cortical receptors appear much less abundant than in the neostriatum. At the cellular level, as visualized by immunocytochemistry in the rat brain (Levey et al., 1991), cortical M_1 localizes to pyramidal cells in all layers, and appears enriched in layers II/III and VI. Cortical M_2 localizes to rare interneuron cell bodies and terminals that are particularly dense in layer IV and the border of layer V/VI. M_4 localizes to perikarya of layers II/III and IV.

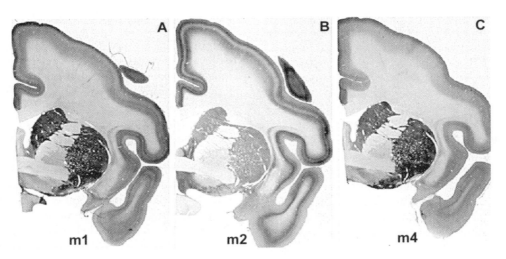

Fig. 1. M_1, M_2 and M_4 distribution in the cortex and striatum of the primate brain. Immunohistochemistry was performed using subtype selective mAChR antibodies in coronal sections from the forebrain of rhesus monkeys. Note the distinct laminar distributions of the three major mAChR in cortex, M_1 (A), M_2 (B) and M_4 (C). In striatum, the receptors are more abundant, and M_4 predominates.

Electron microscopic studies in primate brain demonstrate that M_1 localizes to postsynaptic dendrites and spines that associate with both asymmetric synapses and symmetric cholinergic synapses, indicating that M_1 can modulate excitatory as well as cholinergic transmission (Mrzljak et al., 1993). Presynaptically localized M_2 in the primate visual cortex associates primarily with asymmetric synapses, suggesting that M_2 can act as a hetero-receptor regulating excitatory transmission (Mrzljak et al., 1996). M_2 also localizes postsynaptically to pyramidal cells in layers 3 and 5 and nonpyramidal cells distributed throughout all cortical layers (Mrzljak et al., 1998). In both cell types, M_2 localizes to the periphery of the synaptic specialization suggesting that activation of M_2 requires diffusion of ACh. The majority of M_2 immunoreactivity is spared in the cortex following lesions of cholinergic neurons that project from the nucleus basalis to the cortex. These data thus suggest that M_2 predominantly localizes to neurons intrinsic to the primate cortex and noncholinergic terminals.

M_1, M_2 and M_4 distribution in the striatum

As in the rat caudate-putamen (Hersch et al., 1994) M_4 is the predominant subtype in the putamen of human, representing 50% of total mAChR protein levels (Flynn et al. 1995), with lower levels of M_1 (37%) and M_2 (16%). Immunohistochemistry in the primate brain (Fig. 1) shows correspondingly dense immunoreactivity for M_1 and M_4 in the caudate and putamen, while M_2 immunoreactivity is less intense. Immunohistochemistry and electron microscopic studies in the rat brain show that at the cellular level, M_1 primarily localizes postsynaptically to cell bodies, dendrites, and spines of medium spiny neurons. But, immuno-electron microscopy also shows localization of the receptor to presynaptic terminals that make asymmetrical synapses, suggesting that these terminals are excitatory (Hersch et al., 1994). Electron microscopic studies of the primate cortex demonstrate that the majority of postsynaptic M_1 localizes to spines that associate with asymmetric excitatory synapses originating from the cortex, but also shows association with symmetric synapses (Alcantara et al., 2001). M_2 is expressed in cell bodies, aspiny dendrites and axon terminals of cholinergic interneurons as demonstrated by electron microscopic studies of the rodent brain (Hersch et al., 1994). The majority of striatal M_4 localizes post-synaptically in cell bodies, dendrites and spines of medium spiny neurons. However, some M_4 also localizes to cholinergic interneurons (Bernard et al., 1999), cholinergic synapses (Zhang et al., 2002), and presynaptically in asymmetrical excitatory synapses (Hersch et al., 1994).

M_1, M_2 and M_4 in the hippocampus

Quantitative immunoprecipitations in the rat brain demonstrate 36% M_1, 33% M_2 and 27% M_4 expression in the hippocampus (Levey et al., 1991). In the human hippocampus, M_1 is the predominant receptor (approximately 60%) while M_2 and M_4 are less abundant (approximately 20%) (Flynn et al., 1995). As seen in Fig. 2, M_1, M_2 and M_4 also have distinct localizations in the primate hippocampus. Similar regional and cellular distributions are seen in rodents, for which more detailed neuroanatomical and physiological data are available. Throughout the rat hippocampus, M_1 is expressed in the pyramidal cell bodies and apical and basal dendrites of the stratum radiatum and stratum oriens (Levey et al., 1991, 1995b). M_2 localizes to cholinergic and noncholinergic presynaptic terminals which are concentrated in the pyramidal cell (Levey et al., 1991, 1995b). In addition, M_2 is found in cell bodies and processes of GABAergic interneurons in the oriens/alveus (Hajos et al., 1998). M_4 shows mostly fine punctate immunoreactivity in the neuropil, enriched in the inner third of the dentate gyrus molecular layer. Immunohistochemistry and lesion studies have demonstrated that M_4 localizes to presynaptic terminals of the perforant, associational and commissural pathways, and to the corresponding cell bodies of origin (Rouse et al., 1997, 1999).

Other brain regions that express M_1, M_2 and M_4

Additional brain regions that express mAChRs include: the basal forebrain, in which M_2 predominates; the amygdala which primarily expresses M_1, but also lesser amounts of M_2 and M_4; the thalamus,

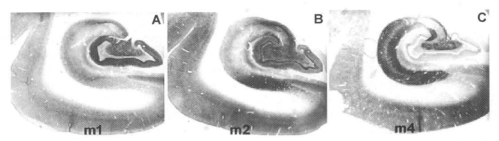

Fig. 2. M_1, M_2 and M_4 distribution in the hippocampus of the primate brain. Immunohistochemistry was performed using subtype selective mAChR antibodies in hippocampal sections from the brain of rhesus monkeys. The subtypes have distinct and complementary regional and laminar distributions in the dentate gyrus, CA1, CA3, and subiculum.

in which M_1, M_2 and M_4 are present and M_2 predominates; the substantia nigra which expresses low levels of M_2 and M_4; the brainstem in which M_2 predominates, and motor neurons which primarily express M_2 (Levey et al., 1991, 1995a).

Postsynaptic and presynaptic mAChR function

M_1 is the major postsynaptic mAChR in the hippocampus

mAChRs expressed in the hippocampus, play a role in learning and memory (Jerusalinsky et al., 1997), and degeneration of cholinergic projections to the hippocampus has been implicated in Alzheimer's disease (Bartus et al., 1982; Gallagher and Colombo, 1995). Therefore, much research has focused on identifying the cellular mechanisms by which mAChR activation might contribute to learning and memory. One of the effects of mAChR activation in the hippocampus is potentiation of current through the NMDA glutamate ion channel, a major component of synaptic plasticity. Identifying the mAChR subtype responsible for potentiation of NMDA current has previously been difficult because of the lack of subtype selective mAChR antagonists. The highly selective M_1 antagonist, M_1-toxin (Max et al., 1993a,b), allows the investigation of the role the M_1 mAChR plays in regulating hippocampal excitability. Application of NMDA causes an increase in inward current in CA1 pyramidal cells. Treatment with the mAChR agonist, carbachol (CCh), causes a two-fold increase in NMDA-induced current. Pretreatment of hippocampal slices with M_1-toxin blocks the CCh-induced potentiation of NMDA current (Marino et al., 1998).

Furthermore, at the electron microscopic level, M_1 colocalizes with the NMDA receptor in CA1 pyramidal cells bodies and dendrites. Therefore, selective activation of M_1 can potentiate excitatory transmission at pyramidal cells in the hippocampus.

The serine/threonine kinase, MAPK also plays a role in synaptic plasticity. CCh causes a dose-dependent and atropine sensitive activation of MAPK in pyramidal cell bodies and dendrites of the hippocampus (Berkeley et al., 2001). MAPK activation is inhibited by pretreatment of slices with M_1-toxin. In addition, mutant mice lacking each mAChR gene were utilized to further evaluate the role of each mAChR subtype in the activation of MAPK. In the absence of specific ligands for individual mAChR subtypes, mAChR knockout mice provide a useful tool for identifying the subtype responsible for activation of signaling cascades. In the M_1 knockout mouse, the CCh induced increase in phospho-MAPK is abolished. In M_2, M_3 and M_4 knockout mice, MAPK is activated to the same degree in the wild type mice. Therefore, by potentiating NMDA current and activating MAPK, M_1 plays an important role in synaptic plasticity in the hippocampus, implicating M_1 as an important receptor in learning and memory.

Distinct presynaptic mAChR subtypes regulate ACh release in different brain regions

In addition to modulating cell excitability and plasticity, mAChRs localized on presynaptic cholinergic terminals can regulate the release of ACh by feedback inhibition. However, pharmacological studies on the mAChR subtype responsible for regulating ACh release have produced conflicting results. The use of mAChR knockout mice has allowed direct determination of the mAChR subtype responsible for regulating ACh release in several brain regions (Zhang et al., 2002). In wild type control mice, the mAChR agonist, oxotremorine, inhibits potassium stimulated ACh release in slices from the cortex, hippocampus and striatum. This effect is diminished in hippocampal and cortical slices from M_2 knockout mice. In contrast, in striatal slices, the agonist-induced inhibition of ACh release is impaired in slices from M_4 but not from M_2 knockout mice. These results are consistent with localization studies demonstrating M_2 and M_4 colocalization with cholinergic markers in the hippocampus and striatum, respectively (Rouse et al., 2000; Zhang et al., 2002). Therefore, M_2 is the primary autoreceptor in the hippocampus and cortex and M_4 is the primary autoreceptor in the striatum. It is also clear from these data that distinct mAChR subtypes are responsible for regulating ACh release in different brain regions.

Muscarinic expression and subcellular distribution is activity-dependent

mAChR trafficking following acute stimulation

Following agonist stimulation, GPCRs undergo desensitization mediated by phosphorylation and binding of arrestin (Lefkowitz, 1998). In addition to preventing receptor interaction with the G protein, arrestin facilitates the binding of clathrin which induces receptor internalization via clathrin coated vesicles. The internalized receptors then traffic to early endosomes in which an acidic pH is thought to mediate a conformational change in the receptor, allowing interaction with phosphatases and consequent receptor resensitization (Krueger et al., 1997). The receptor can then return to the cell surface where it can again be activated by ligand. Following prolonged agonist stimulation, the receptor can be targeted to lysosomes for degradation.

Following cholinergic stimulation, the M_4 subtype of mAChR has been shown to internalize from the cell surface to early endosomes and multivesicular bodies, organelles involved in targeting receptors to lysosomes (Bernard et al., 1999; Volpicelli et al., 2001). Figure 3 demonstrates that, in saline treated rats, M_4 localizes to the cell surface and neuropil of medium spiny neurons in the striatum. Forty minutes following intraperitoneal injection of the mAChR agonist, oxotremorine, M_4 shows a predominantly intracellular distribution and decreased neuropil immunoreactivity. Electron microscopic studies of M_4 in medium spiny neurons of rat neostriatum demonstrate that the proportion of intracellular M_4 depends on the extent of cholinergic activity (Bernard et al., 1999). In control, untreated rats, the amount of

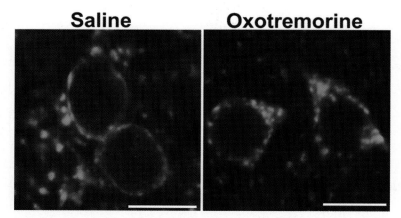

Fig. 3. M_4 internalization following agonist treatment in medium spiny neurons of the striatum. Rats received an intraperitoneal injection of saline or 0.5 mg/kg oxotremorine and were perfused 40 min later. Striatal sections were incubated with a primary rabbit antibody that recognizes M_4, and developed using a tyramide-fluorescein reagent. In saline-treated animals, M_4 shows a continuous, linear distribution along the cell surface of medium spiny neuron cell bodies. Punctate, neuropil immunoreactivity is also visible. Following oxotremorine treatment, M_4 redistributes from the cell surface to intracellular puncta. M_4 immunoreactivity in the neuropil is also reduced. Scale bars = 10 μm.

64

cell surface M_4 is higher in striatal regions with low levels of ACh (patches) than in striatal regions with high ACh levels (matrix). This suggests that the degree of internalization in other regions of the brain, such as cortex and hippocampus, may also reflect the extent of ACh released.

M_2 mAChR trafficking has been examined by electron microscopy in cholinergic interneurons of the striatum (Bernard et al., 1998). Following oxotremorine treatment, M_2 shows decreased cell surface localization and enhanced localization to endosomes and multivesicular bodies. M_2 expressed in cell bodies of the medial septum also shows internalization following cholinergic stimulation. Immunofluorescence reveals that M_2 shows a linear, continuous distribution along the cell surface in saline treated rats (Fig. 4). Forty minutes after intraperitoneal injection of oxotremorine, its distribution is mostly intracellular and punctate. Overall, these experiments suggest that mAChR internalization and intracellular trafficking is an important component of receptor regulation in the brain. The effects of chronic cholinergic stimulation on muscarinic receptor subtype regulation will be critical, given the widespread use of cholinesterase inhibitors for the treatment of Alzheimer's disease. Recently developed acetylcholinesterase knockout mice will provide an excellent model to study the effects of chronic hypercholinergic tone on mAChR trafficking and expression (Xie et al., 2000; Mesulam et al., 2002).

Summary

The M_1, M_2 and M_4 subtypes of mAChRs are the predominant receptors in the CNS. These receptors activate a multitude of signaling pathways important for modulating neuronal excitability, synaptic plasticity and feedback regulation of ACh release. In addition, novel functions mediated by mAChRs are currently being discovered. These studies are greatly facilitated by the recent development of subtype selective toxins and mice lacking individual mAChR genes.

Studies in cell culture and the rodent brain demonstrate that mAChR internalization and intracellular trafficking is an important component of mAChR regulation. Characterizing mAChR intracellular trafficking could help facilitate the development of selective mAChR ligands. For example, a selective M_1 agonist would cause a shift in the distribution of M_1 from the cell surface to an intracellular distribution, while M_2 and M_4 would remain on the cell surface. Characterizing mAChR intracellular trafficking is also important for understanding the cellular mechanisms that regulate mAChR cell surface expression and signaling. Furthermore, intracellular trafficking has recently been demonstrated to play a role in the development of tolerance to drugs (Whistler et al., 1999; He et al., 2002). Because individual mAChR subtypes are novel targets for treatments of diseases such as Alzheimer's disease

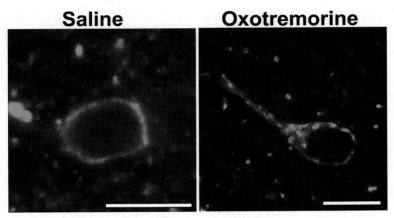

Fig. 4. M_2 internalization following agonist treatment in cell bodies of the medial septum. Rats received intraperitoneal injections of saline or 0.5 mg/kg oxotremorine and were perfused 40 min after injection. Sections including medial septum were incubated with a primary rabbit antibody that recognizes M_2 and developed using a tyramide-fluorescein reagent. In saline treated rats, M_2 shows a linear distribution along the cell surface. Oxotremorine treatment causes a redistribution of M_2 from the cell surface to intracellular puncta.

and schizophrenia, understanding the mechanisms that regulate mAChR signaling and intracellular trafficking following acute and chronic stimulation might lead to the development of rational strategies.

References

Alcantara, A.A., Mrzljak, L., Jakab, R.L., Levey, A.I., Hersch, S.M. and Goldman-Rakic, P.S. (2001) Muscarinic m1 and m2 receptor proteins in local circuit and projection neurons of the primate striatum: anatomical evidence for cholinergic modulation of glutamatergic prefronto-striatal pathways. J. Comp. Neurol., 434: 445–460.

Bartus, R.T., Dean, R.L., 3rd, Beer, B. and Lippa, A.S. (1982) The cholinergic hypothesis of geriatric memory dysfunction. Science, 217: 408–414.

Berkeley, J.L. and Levey, A.I. (2000) Muscarinic activation of mitogen-activated protein kinase in PC12 cells. J. Neurochem., 75: 487–493.

Berkeley, J.L., Gomeza, J., Wess, J., Hamilton, S.E., Nathanson, N.M. and Levey, A.I. (2001) M1 muscarinic acetylcholine receptors activate extracellular signal-regulated kinase in CA1 pyramidal neurons in mouse hippocampal slices. Mol. Cell. Neurosci., 18: 512–524.

Bernard, V., Laribi, O., Levey, A.I. and Bloch, B. (1998) Subcellular redistribution of m2 muscarinic acetylcholine receptors in striatal interneurons in vivo after acute cholinergic stimulation. J. Neurosci., 18: 10207–10218.

Bernard, V., Levey, A.I. and Bloch, B. (1999) Regulation of the subcellular distribution of m4 muscarinic acetylcholine receptors in striatal neurons in vivo by the cholinergic environment: evidence for regulation of cell surface receptors by endogenous and exogenous stimulation. J. Neurosci., 19: 10237–10249.

Bonner, T.I., Buckley, N.J., Young, A.C. and Brann, M.R. (1987) Identification of a family of muscarinic acetylcholine receptor genes [published erratum appears in Science 1987 Sep 25; 237(4822): 237]. Science 237: 527–532.

Buckley, N.J., Bonner, T.I. and Brann, M.R. (1988) Localization of a family of muscarinic receptor mRNAs in rat brain. J. Neurosci., 8: 4646–4652.

Cortes, R. and Palacios, J.M. (1986) Muscarinic cholinergic receptor subtypes in the rat brain. I. Quantitative autoradiographic studies. Brain Res., 362: 227–238.

Crespo, P., Xu, N., Simonds, W.F. and Gutkind, J.S. (1994) Ras-dependent activation of MAP kinase pathway mediated by G-protein beta gamma subunits. Nature, 369: 418–420.

Fisahn, A., Yamada, M., Duttaroy, A., Gan, J.W., Deng, C.X., McBain, C.J. and Wess, J. (2002) Muscarinic induction of hippocampal gamma oscillations requires coupling of the M1 receptor to two mixed cation currents. Neuron, 33: 615–624.

Flynn, D.D., Ferrari-DiLeo, G., Mash, D.C. and Levey, A.I. (1995) Differential regulation of molecular subtypes of muscarinic receptors in Alzheimer's disease. J. Neurochem., 64: 1888–1891.

Gallagher, M. and Colombo, P.J. (1995) Ageing: the cholinergic hypothesis of cognitive decline. Curr. Opin. Neurobiol., 5: 161–168.

Hajos, N., Papp, E.C., Acsady, L., Levey, A.I. and Freund, T.F. (1998) Distinct interneuron types express m2 muscarinic receptor immunoreactivity on their dendrites or axon terminals in the hippocampus. Neuroscience, 82: 355–376.

Hamilton, S.E. and Nathanson, N.M. (2001) The M1 receptor is required for muscarinic activation of mitogen-activated protein (MAP) kinase in murine cerebral cortical neurons. J. Biol. Chem., 276: 15850–15853.

He, L., Fong, J., von Zastrow, M. and Whistler, J.L. (2002) Regulation of opioid receptor trafficking and morphine tolerance by receptor oligomerization. Cell, 108: 271–282.

Hersch, S.M., Gutekunst, C.A., Rees, H.D., Heilman, C.J. and Levey, A.I. (1994) Distribution of m1-m4 muscarinic receptor proteins in the rat striatum: light and electron microscopic immunocytochemistry using subtype-specific antibodies. J. Neurosci., 14: 3351–3363.

Jerusalinsky, D., Kornisiuk, E. and Izquierdo, I. (1997) Cholinergic neurotransmission and synaptic plasticity concerning memory processing. Neurochem. Res., 22: 507–515.

Kiyono, M., Satoh, T. and Kaziro, Y. (1999) G protein beta gamma subunit-dependent Rac-guanine nucleotide exchange activity of Ras-GRF1/CDC25(Mm). Proc. Natl. Acad. Sci. USA, 96: 4826–4831.

Kozma, R., Sarner, S., Ahmed, S. and Lim, L. (1997) Rho family GTPases and neuronal growth cone remodelling: relationship between increased complexity induced by Cdc42Hs, Rac1, and acetylcholine and collapse induced by RhoA and lysophosphatidic acid. Mol. Cell. Biol., 17: 1201–1211.

Krueger, K.M., Daaka, Y., Pitcher, J.A. and Lefkowitz, R.J. (1997) The role of sequestration in G protein-coupled receptor resensitization. Regulation of beta2-adrenergic receptor dephosphorylation by vesicular acidification. J. Biol. Chem., 272: 5–8.

Kubo, T., Fukuda, K., Mikami, A., Maeda, A., Takahashi, H., Mishina, M., Haga, T., Haga, K., Ichiyama, A., Kangawa, K., et al. (1986) Cloning, sequencing and expression of complementary DNA encoding the muscarinic acetylcholine receptor. Nature, 323: 411–416.

Lefkowitz, R.J. (1998) G protein-coupled receptors. III. New roles for receptor kinases and beta-arrestins in receptor signaling and desensitization. J. Biol. Chem., 273: 18677–18680.

Levey, A.I. (1996) Muscarinic acetylcholine receptor expression in memory circuits: implications for treatment of Alzheimer disease. Proc. Natl. Acad. Sci. USA, 93: 13541–13546.

Levey, A.I., Kitt, C.A., Simonds, W.F., Price, D.L. and Brann, M.R. (1991) Identification and localization of muscarinic acetylcholine receptor proteins in brain with subtype-specific antibodies. J. Neurosci., 11: 3218–3226.

Levey, A.I., Edmunds, S.M., Hersch, S.M., Wiley, R.G. and Heilman, C.J. (1995a) Light and electron microscopic study of m2 muscarinic acetylcholine receptor in the basal forebrain of the rat. J. Comp. Neurol., 351: 339–356.

Levey, A.I., Edmunds, S.M., Koliatsos, V., Wiley, R.G. and Heilman, C.J. (1995b) Expression of m1-m4 muscarinic acetylcholine receptor proteins in rat hippocampus and regulation by cholinergic innervation. J. Neurosci., 15: 4077–4092.

Linseman, D.A., Hofmann, F. and Fisher, S.K. (2000) A role for the small molecular weight GTPases, Rho and Cdc42, in muscarinic receptor signaling to focal adhesion kinase. J. Neurochem., 74: 2010–2020.

Marino, M.J., Rouse, S.T., Levey, A.I., Potter, L.T. and Conn, P.J. (1998) Activation of the genetically defined m1 muscarinic receptor potentiates N-methyl-D-aspartate (NMDA) receptor currents in hippocampal pyramidal cells. Proc. Natl. Acad. Sci. USA, 95: 11465–11470.

Mash, D.C. and Potter, L.T. (1986) Autoradiographic localization of M1 and M2 muscarine receptors in the rat brain. Neuroscience, 19: 551–564.

Max, S.I., Liang, J.S. and Potter, L.T. (1993a) Purification and properties of m1-toxin, a specific antagonist of m1 muscarinic receptors. J. Neurosci., 13: 4293–4300.

Max, S.I., Liang, J.S., Valentine, H.H. and Potter, L.T. (1993b) Use of m1-toxin as a selective antagonist of m1 muscarinic receptors. J. Pharmacol. Exp. Ther., 267: 480–485.

Mesulam, M.M., Guillozet, A., Shaw, P., Levey, A., Duysen, E.G. and Lockridge, O. (2002) Acetylcholinesterase knockouts establish central cholinergic pathways and can use butyrylcholinesterase to hydrolyze acetylcholine. Neuroscience, 110: 627–639.

Mrzljak, L., Levey, A.I. and Goldman-Rakic, P.S. (1993) Association of m1 and m2 muscarinic receptor proteins with asymmetric synapses in the primate cerebral cortex: morphological evidence for cholinergic modulation of excitatory neurotransmission. Proc. Natl. Acad. Sci. USA, 90: 5194–5198.

Mrzljak, L., Levey, A.I. and Rakic, P. (1996) Selective expression of m2 muscarinic receptor in the parvocellular channel of the primate visual cortex. Proc. Natl. Acad. Sci. USA, 93: 7337–7340.

Mrzljak, L., Levey, A.I., Belcher, S. and Goldman-Rakic, P.S. (1998) Localization of the m2 muscarinic acetylcholine receptor protein and mRNA in cortical neurons of the normal and cholinergically deafferented rhesus monkey. J. Comp. Neurol., 390: 112–132.

Peralta, E.G., Winslow, J.W., Peterson, G.L., Smith, D.H., Ashkenazi, A., Ramachandran, J., Schimerlik, M.I. and Capon, D.J. (1987) Primary structure and biochemical properties of an M2 muscarinic receptor. Science, 236: 600–605.

Rouse, S.T., Thomas, T.M. and Levey, A.I. (1997) Muscarinic acetylcholine receptor subtype, m2: diverse functional implications of differential synaptic localization. Life Sci., 60: 1031–1038.

Rouse, S.T., Marino, M.J., Potter, L.T., Conn, P.J. and Levey, A.I. (1999) Muscarinic receptor subtypes involved in hippocampal circuits. Life Sci., 64: 501–519.

Rouse, S.T., Edmunds, S.M., Yi, H., Gilmor, M.L. and Levey, A.I. (2000) Localization of M(2) muscarinic acetylcholine receptor protein in cholinergic and non-cholinergic terminals in rat hippocampus. Neurosci. Lett., 284: 182–186.

Shapiro, M.S., Gomeza, J., Hamilton, S.E., Hille, B., Loose, M.D., Nathanson, N.M., Roche, J.P. and Wess, J. (2001) Identification of subtypes of muscarinic receptors that regulate $Ca2+$ and $K+$ channel activity in sympathetic neurons. Life Sci., 68: 2481–2487.

Spencer, D.G., Jr., Horvath, E. and Traber, J. (1986) Direct autoradiographic determination of M1 and M2 muscarinic acetylcholine receptor distribution in the rat brain: relation to cholinergic nuclei and projections. Brain Res., 380: 59–68.

Vilaro, M.T., Wiederhold, K.H., Palacios, J.M. and Mengod, G. (1991) Muscarinic cholinergic receptors in the rat caudate-putamen and olfactory tubercle belong predominantly to the m4 class: in situ hybridization and receptor autoradiography evidence. Neuroscience, 40: 159–167.

Volpicelli, L.A., Lah, J.J. and Levey, A.I. (2001) Rab5-dependent trafficking of the m4 muscarinic acetylcholine receptor to the plasma membrane, early endosomes, and multivesicular bodies. J. Biol. Chem., 276: 47590–47598.

Weiner, D.M., Levey, A.I. and Brann, M.R. (1990) Expression of muscarinic acetylcholine and dopamine receptor mRNAs in rat basal ganglia. Proc. Natl. Acad. Sci. USA, 87: 7050–7054.

Whistler, J.L., Chuang, H.H., Chu, P., Jan, L.Y. and von Zastrow, M. (1999) Functional dissociation of mu opioid receptor signaling and endocytosis: implications for the biology of opiate tolerance and addiction. Neuron, 23: 737–746.

Wotta, D.R., Wattenberg, E.V., Langason, R.B. and el-Fakahany, E.E. (1998) M1, M3 and M5 muscarinic receptors stimulate mitogen-activated protein kinase. Pharmacology, 56: 175–186.

Xie, W., Stribley, J.A., Chatonnet, A., Wilder, P.J., Rizzino, A., McComb, R.D., Taylor, P., Hinrichs, S.H. and Lockridge, O. (2000) Postnatal developmental delay and supersensitivity to organophosphate in gene-targeted mice lacking acetylcholinesterase. J. Pharmacol. Exp. Ther., 293: 896–902.

Zhang, W., Basile, A.S., Gomeza, J., Volpicelli, L.A., Levey, A.I. and Wess, J. (2002) Characterization of central inhibitory muscarinic autoreceptors by the use of muscarinic acetylcholine receptor knock-out mice. J. Neurosci., 22: 1709–1717.

Progress in Brain Research, Vol. 145
ISSN 0079-6123

CHAPTER 4

The cholinergic innervation of the human cerebral cortex

M.-Marsel Mesulam*

Cognitive Neurology and Alzheimer's Disease Center, Departments of Neurology and Psychiatry, Feinberg Medical School, Northwestern University, 320 East Superior Street, Chicago, IL 60611, USA

Introduction

In 1920s, Otto Loewi identified acetylcholine (ACh) as the cardioactive substance released by the vagus nerve (Loewi, 1921). It took years of additional work to establish that ACh was also a neurotransmitter of the central nervous system. The past 20 years have witnessed the emergence of powerful new methods for exploring the organization of central cholinergic pathways and their relevance to the chemical neuroanatomy of cognition, age-related memory impairments, and Alzheimer's disease. This chapter aims to summarize current knowledge on the neuroanatomy of cortical cholinergic innervation in the human brain and its implications for behavior.

The source of cortical cholinergic innervation—the nucleus basalis of Meynert

The 'Ch' nomenclature

The basal forebrain contains four partially overlapping cell groups where cholinergic and noncholinergic neurons are intermingled with each other. The Ch1-Ch4 nomenclature was introduced to designate the cholinergic (i.e., choline acetyltransferase (ChAT)-containing) neurons within these 4 cell groups (Mesulam et al., 1983a,b; Mesulam and Geula, 1988b). According to this nomenclature, Ch1

*Tel.: (312) 908-9339; E-mail: mmesulam@northwestern.edu

DOI: 10.1016/S0079-6123(03)45004-8

designates the cholinergic cells associated predominantly with the medial septal nucleus; Ch2 those associated with the vertical nucleus of the diagonal band; Ch3 those associated with the horizontal limb of the diagonal band nucleus; and Ch4 those associated with the nucleus basalis of Meynert. Tracer experiments in a number of animal species have shown that Ch1 and Ch2 provide the major cholinergic innervation for the hippocampal complex, Ch3 for the olfactory bulb, and Ch4 for the rest of the cerebral cortex and the amygdala (Mesulam et al., 1983a). In the primate brain, the Ch4 group contains a compact component in the nucleus basalis of Meynert and interstitial components embedded within the internal capsule, medullary laminae of the globus pallidus, ansa peduncularis and ansa lenticularis (Mesulam et al., 1983a; Satoh and Fibiger, 1985; Everitt et al., 1988; Mesulam and Geula, 1988b). The term 'nucleus basalis' can be used to designate the cholinergic as well as noncholinergic components in this nucleus whereas the more restrictive Ch4 designation is reserved for its cholinergic neurons.

Anatomy of the nucleus basalis

The nucleus basalis is a phylogenetically progressive nucleus which displays its greatest differentiation in the cetacean and human brains (Gorry, 1963). The human nucleus basalis extends from the level of the olfactory tubercle to that of the posterior amygdala, spanning a distance of 13–14 mm in the antero-posterior axis and attaining a medio-lateral width of 18 mm within the substantia innominata (subcommissural gray). Arendt et al. (1985) have estimated that the human nucleus basalis contains 200,000 neurons in each hemisphere. On topographical grounds, the human nucleus basalis can be subdivided into sectors that occupy its anteromedial (nb-Ch4am), anterolateral (nb-Ch4al), anterointermediate (nb-Ch4ai), intermediate (nb-Ch4i), and posterior (nb-Ch4p) regions (Mesulam and Geula, 1988b).

The cell bodies of Ch4 neurons display large amounts of cytoplasm, abundant organelles, well developed stacks of short parallel cisternae of rough endoplasmic reticulum, indented nuclei, and prominent nucleoli (Walker et al., 1983; Ingham et al., 1985; Palacios, 1990). There are no strict boundaries delineating the nucleus basalis from the adjacent cell groups of the olfactory tubercle, preoptic area, hypothalamic nuclei, diagonal band nuclei, amygdaloid complex and globus pallidus. In addition to this 'open' nuclear structure, the neurons of nucleus basalis display physiological and morphological heterogeneity. The majority of the Ch4 neurons have an isodendritic morphology with overlapping dendritic fields, many of which extend into fiber tracts traversing the basal forebrain. These characteristics are also present in the nuclei of the brainstem reticular formation and have led to the suggestion that the nb-Ch4 complex could be conceptualized as a telencephalic extension of the brainstem reticular core (Ramon-Moliner and Nauta, 1966).

Cytochemical signature of the nucleus basalis

In the human brain, Ch4 neurons express ChAT, the vesicular acetylcholine transporter, acetyl cholinesterase (AChE), calbindin-d28k, the high affinity nerve growth factor receptor trkA, and the low affinity p75 nerve growth factor receptor (NGFr) (Geula et al., 1993b; Kordower et al., 1994; Gilmor et al., 1999). A minority of Ch4 neurons are NGFr-negative and, at least in the rat, project preferentially to the amygdala (Mufson et al., 1989; Henderson and Evans, 1991; Heckers et al., 1994). The nucleus

basalis also contains a complex mosaic of non-cholinergic neurons that are NADPHd-positive, GABAergic, peptidergic, and tyrosine hydroxylase (TH)-positive (Henderson, 1987; Mesulam et al., 1989; Walker et al., 1989; Gouras et al., 1992; Wisniowski et al., 1992; Gritti et al., 1993). The GABAergic neurons of the nucleus basalis may be as numerous as the cholinergic neurons with which they are intermingled. In the rat, some of these GABAergic neurons project to the cerebral cortex and innervate cortical inhibitory interneurons (Freund and Meskenaite, 1992; Gritti et al., 1993). The importance of the non-cholinergic neurons in the nucleus basalis has been highlighted by recent evidence showing that selective immunotoxic lesions of cholinergic neurons in the nucleus basalis yield behavioral and physiological impairments which are distinctly less severe than those obtained by excitotoxic lesions which destroy all cell types (Berger-Sweeney et al., 1994; Wenk et al., 1994).

Inputs and neurotransmitter circuitry of the nucleus basalis

Although the primate nucleus basalis projects to the entire cerebral cortex, it receives cortical projections only from limbic and paralimbic regions of the brain, including the amygdala (Mesulam and Mufson, 1984). The cortical inputs are mostly glutamatergic, but can also be GABAergic. At least in the rat, the glutamatergic inputs from the cerebral cortex synapse almost exclusively on GABAergic neurons in the nucleus basalis (Záborszky et al., 1997).

The nucleus basalis contains receptor sites for 5-HT (serotonin), dopamine and norepinephrine (Zilles et al., 1991). In the rat, projections from dopaminergic ventral tegmental neurons, serotonergic raphe neurons, and noradrenergic locus coeruleus neurons have been identified (Jones and Cuello, 1989). In the human, TH and DBH immunoreactivity has been described in the septal area (Gaspar et al., 1985). Our observations with the light microscope revealed intense, preterminal-like TH and DBH immunoreactive axonal profiles in the human nucleus basalis (Smiley et al., 1999b). In the monkey, electron microscopic analyzes of tissue prepared for the concurrent visualization of ChAT and TH showed the presence of TH-positive (presumably dopaminergic) asymmetric synapses onto ChAT-positive cholinergic Ch4 neurons (Smiley and Mesulam, 1999). Reliable serotonin immunoreactivity has been difficult to obtain in the human brain. In the monkey brain, we detected a dense plexus of serotonin-immunoreactive axons within the nucleus basalis (Smiley et al., 1999b). The primate nucleus basalis therefore provides a site for extensive cholinergic-monoaminergic interactions.

Electron microscopic investigations of ChAT-immunolabeled nucleus basalis tissue reveals the existence of cholinergic terminals. In the rat, the vast majority of these terminals make contact with non-cholinergic neurons (Martinez-Murillo et al., 1990). In the monkey, cholinergic synapses onto cholinergic Ch4 neurons are quite frequent and take the form of large asymmetrical synapses (Smiley and Mesulam, 1999). The precise source of the cholinergic input to the nucleus basalis is unknown but could include collaterals from cholinergic Ch1-Ch4 neurons of the basal forebrain or ascending projections from the Ch5-Ch6 group of pontomesencephalic cholinergic nuclei (Jones and Cuello, 1989). In the monkey, electron microscopic analyzes also revealed GABAergic symmetrical synapses on the dendrites of both cholinergic and non-cholinergic nucleus basalis neurons (Smiley and Mesulam, 1999).

Dissociated cell cultures of nucleus basalis neurons show that they are responsive to acetylcholine, neurotensin, substance P, L-glutamate, and orexin (Nakajima et al., 1985; Farkas et al., 1994; Eggerzmann et al., 2001). Patch clamp techniques indicate the presence of currents associated with nicotinic as well as m2 muscarinic receptors (Harata et al., 1991). Nucleus basalis neurons in the rat express m2 receptor mRNA and contain m2 and m3 receptors. The m2 receptor appears to be the dominant species of cholinergic receptor in the nucleus basalis (Levey et al., 1991; Vilaró et al., 1992; Levey et al., 1994). In the rat and monkey, this receptor subtype is expressed by approximately a third of Ch4 neurons (Smiley et al., 1999a). Therefore, the often cited assumptions that all Ch4 neurons express m2, and that m2 is a universal presynaptic marker of cortical cholinergic innervation need to be modified.

Cholinergic fibers and synapses of the cerebral cortex

Topography and distribution of cortical cholinergic projections

In the monkey brain, individual cortical areas receive their major cholinergic input from different sectors of the nucleus basalis-Ch4 complex. Thus, Ch4am provides the major source of cholinergic input to medial cortical areas including the cingulate gyrus; Ch4al to frontoparietal cortex, opercular regions, and the amygdaloid nuclei; Ch4i to laterodorsal frontoparietal, peristriate and mid-temporal regions; and Ch4p to the superior temporal and temporopolar areas (Mesulam et al., 1983a). The experimental methods that are needed to reveal this topographic arrangement cannot be used in the human brain. However, indirect evidence for the existence of a similar topographical arrangement can be gathered from patients with Alzheimer's disease. We described two patients in whom extensive loss of cholinergic fibers in temporopolar but not frontal opercular cortex was associated with marked cell loss in the posterior (Ch4p) but not the anterior (Ch4am + Ch4al) sectors of Ch4 (Mesulam and Geula, 1988b). This relationship is consistent with the topography of the projections in the monkey brain.

The density of cholinergic axons is higher in the more superficial layers of the cerebral cortex, suggesting that the axons which enter the cortex from the underlying gray matter undergo branching as they course towards the pial surface, an interpretation which is supported by physiological evidence (Aston-Jones et al., 1985; Mesulam et al., 1992). The density of cholinergic innervation is lower within unimodal and heteromodal association areas than in paralimbic areas of the brain (Mesulam and Geula, 1993). Core limbic areas such as the amygdala and hippocampus contain the highest densities of cholinergic innervation.

Trajectory of cholinergic pathways from the nucleus basalis to the cerebral cortex

Two highly organized and discrete bundles of cholinergic fibers extend from the nucleus basalis to the cerebral cortex and amygdala in the human brain. A *medial pathway* joins the white matter of the gyrus rectus, curves around the rostrum of the corpus callosum to enter the cingulum, and merges with fibers of the lateral pathway within the occipital lobe. It supplies the parolfactory, cingulate, peri-cingulate and retrosplenial cortices. The *lateral pathway* is subdivided into a capsular division traveling in the white matter of the external capsule and a peri-Sylvian division traveling within the claustrum. Branches of the *peri-Sylvian division* supply the fronto-parietal operculum, insula, and superior temporal gyrus. Branches of the *capsular division* innervate the remaining parts of the frontal, parietal and temporal neocortex. These cholinergic pathways have been represented within a three-dimensional MRI volume in a manner that helps to determine the location of subcortical cerebrovascular lesions that are likely to interrupt the cholinergic projections to the cerebral cortex (Selden et al., 1998).

Cholinergic synapses in the cerebral cortex

In the rat and cat cerebral cortex, ChAT-immunoreactive fibers are almost exclusively unmyelinated, display numerous varicosities, and make mostly symmetrical synapses on the perikaryon, dendritic shaft and spines of pyramidal as well as non-pyramidal neurons (Wainer et al., 1984; Frotscher and Leranth, 1985; DeLima and Singer, 1986). The cerebral cortex of the rat also contains cholinergic varicosities not associated with identifiable synapses, leading to the inference that cholinergic neurotransmission may partially rely on volume transmission (Umbriaco et al., 1994). In the temporal cortex of the human brain, almost two thirds of ChAT-positive synapses formed identifiable synaptic specializations. These were usually quite small and symmetric and were located predominantly on the dendritic shafts and spines of pyramidal neurons (Smiley et al., 1997).

Cholinergic receptors in the cerebral cortex

The acetylcholine (ACh) which is released by cholinergic axons exerts its influence upon the cerebral cortex by interacting with neurons expressing muscarinic and nicotinic receptors (Schröder

et al., 1989, 1990). In mammals, the m1 subtype of muscarinic receptor is the most common species of cholinergic receptor in the cerebral cortex. Methods based on receptor autoradiography have detected regional and laminar variations in the distribution of muscarinic receptor subtypes in both the human and monkey brain (Mash et al., 1988; Lidow et al., 1989; Zilles et al., 1995).

The investigation of cholinergic pathways has entered a new and very productive phase with the molecular identification and cloning of five subtypes (m1-m5) of muscarinic receptors (Kubo et al., 1986; Bonner et al., 1987; Peralta et al., 1987; Bonner et al., 1988). All five of these receptors are G protein coupled: m1, m3, and m5 preferentially activate phospholipase C, whereas m2 and m4 inhibit adenylyl cyclase activity (Bonner et al., 1988). Immunocytochemical experiments and in situ hybridization studies in the rat show that the m1 receptor subtype is found in the majority of cortical neurons, a distribution which is consistent with its role as the major postsynaptic cholinergic receptor of the cerebral cortex (Buckley et al., 1988; Levey et al., 1991). The m1-like immunoreactivity is seen not only in association with symmetrical synapses characteristic of cortical cholinergic pathways, but also in association with asymmetrical synapses characteristic of excitatory amino acid pathways, raising the possibility that ACh may modulate excitatory neurotransmission in cortical neurons via an m1 receptor site (Mrzljak et al., 1993).

Cortical immunostaining related to the m2 receptor subtype is located mostly in the neuropil (representing, at least in part, incoming axons of m2-positive nucleus basalis neurons) but also in some perikarya, suggesting that this receptor subtype may function both as a presynaptic and postsynaptic receptor (Levey et al., 1991). The m2 subtype is also associated with non-cholinergic terminals, suggesting that it may additionally act as a presynaptic heteroceptor through which ACh may modulate the release of other transmitters (Mrzljak et al., 1993). Immunostaining and message for the m3 receptor is detected in several areas of the forebrain, including the pyramidal layer of the hippocampus, entorhinal cortex and the superficial layers of the cerebral cortex (Buckley et al., 1988). In immunocytochemical experiments, the staining is located mostly in the neuropil, but faint perikaryal staining can also be discerned, suggesting that the m3 receptor subtype can function as a pre- as well as postsynaptic receptor in the cerebral cortex (Levey et al., 1994). Cortical immunostaining for m4 protein is seen mostly in the neuropil, but at a lesser density than that associated with the m1 or m2 subtypes. Immunostaining for the m5 receptor subtype has not yet been detected reliably in the cerebral cortex (Levey et al., 1991). These observations show that the muscarinic receptor subtypes can function not only as postsynaptic receptors at traditional cholinoceptive sites but also as postsynaptic modulators of non-cholinergic transmission and as presynaptic autoreceptors and heteroceptors that influence the release of acetylcholine and other transmitters.

Cholinoceptive neurons of the cerebral cortex

Incoming cholinergic axons innervate at least three kinds of cortical neurons: glutamatergic neurons with nicotinic or muscarinic receptors, GABAergic interneurons and NADPHd-positive infracortical neurons. The one marker most closely associated with cholinoceptive neurons is acetylcholinesterase (AChE), the enzyme which terminates cholinergic neurotransmission through the rapid hydrolysis of ACh into acetate and choline. Although the vast majority of cholinoceptive neurons probably express some AChE, only a subset yields an AChE-rich histochemical staining pattern. In the cerebral cortex of the adult rat, the AChE-rich cytochemical pattern is limited to a few polymorphic neurons (Silver, 1974; Kutscher, 1991; Geula et al., 1993a). The situation is dramatically different in the human cerebral cortex which contains a dense network of AChE-rich cortical neurons, especially in layers III and V of premotor and sensory association cortex (Kostovic et al., 1988; Mesulam and Geula, 1988a, 1991). These neurons are mostly pyramidal in shape and are likely to represent a glutamatergic contingent of cholinoceptive neurons (Fig. 1).

The density and staining intensity of AChE-rich cortical neurons is higher in the human brain than in any other species that we have studied, including the macaque and baboon. These neurons also display a most unusual ontogenetic profile: their AChE-rich

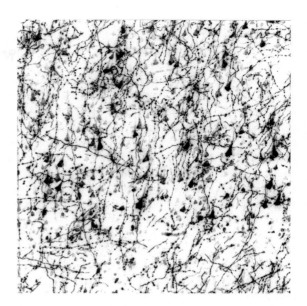

Fig. 1. Anterior orbitofrontal cortex of the human brain. AChE histochemistry. Varicose cholinergic axons (coming from Ch4) are contacting AChE-rich cholinoceptive cell bodies. Magnification 10×.

staining pattern is not detectable as late as the 10th year of life and becomes fully established during adulthood. These AChE-rich neurons are ChAT-negative and, therefore, non-cholinergic. Paralimbic and limbic areas of the human brain receive a very dense cholinergic input, but have very few AChE-rich neurons. Furthermore, AChE-rich intracortical neurons are rare during infancy, when the cerebral cortex contains a dense net of cholinergic afferents, and presumably a correspondingly large number of cholinoceptive neurons. It is therefore reasonable to assume that the AChE-rich cortical neurons constitute a special subset of cholinoceptive neurons and that their high AChE content may reflect affiliations that transcend the necessary requirements of standard cholinergic transmission.

Another set of cholinoceptive neurons, also AChE-rich, is located in the infracortical region, just under the VIth layer of the cerebral cortex. They display a pattern of ontogenetic development that is quite different in that they are AChE-rich at least as early as a few weeks after birth. They are polymorphic in shape, display abundant rough endoplasmic reticulum, a prominent nucleolus and

72

invaginations of the nuclear membrane. Their dendrites and axons extend into the overlying cerebral cortex. In the monkey and human brains, these neurons are found in all cytoarchitectonic regions, with few regional variations in density. Many of these neurons are intensely immunoreactive for the m2 subtype muscarinic receptor and express nitric oxide synthase as revealed by NADPHd histochemistry (Smiley et al., 1998). This subgroup of cholinoceptive neurons provides an m2-linked relay through which cholinergic innervation from Ch4 can regulate the release of nitric oxide in the cerebral cortex and subjacent white matter. This connection may play an important role in regulating cerebral blood-flow and also in modulating physiological and morphological plasticity in the cerebral cortex.

The cerebral cortex also contains GABAergic cholinoceptive interneurons that respond to ACh through muscarinic receptors (McCormick and Prince, 1985). These neurons may receive additional inputs from the GABAergic projection neurons of the nucleus basalis (Freund and Meskenaite, 1992).

Changes of cortical cholinergic innervation in aging and Alzheimer's disease

Age-related changes in the cholinergic innervation of the human cerebral cortex are regionally selective and relatively modest. We find that most amygdaloid nuclei and the cingulate cortex display virtually no age-related changes, whereas the inferotemporal and entorhinal cortices appear to sustain a significant but relatively modest loss of approximately 20%, when densities of cholinergic axons in specimens from 22 to 43-year-old subjects are compared to those from subjects above 68 years of age (Geula and Mesulam, 1989; Emre et al., 1993). Although modest, these changes in cortical cholinergic innervation may provide an important anatomical substrate for age-related changes in memory function.

Age-related changes in the nucleus basalis are also generally modest and do not become established until advanced senescence (Geula and Mesulam, 1994). An initial perikaryal hypertrophy around the age of 60 followed by shrinkage and cell loss appears to represent a characteristic pattern of age-related change. Aging may lead to a selective loss of calbindin in Ch4

neurons (Wu et al., 1997). The resultant impairment of intracellular calcium binding may make these neurons more susceptible to age-related involutional changes, including neurofibrillary degeneration.

A loss of cortical cholinergic innervation in Alzheimer's disease has been reported in more than 30 papers (Geula and Mesulam, 1994). We find that this depletion is *severe* (76–85%) in inferotemporal, midtemporal and entorhinal cortex and in parts of the amygdala; *modest* (40–67%) in prefrontal, posterior parietal, peristriate, orbitofrontal, insular, posterior cingulate, primary auditory and hippocampal cortex; and *light* (4–28%) in primary visual, primary somatosensory, primary motor, premotor, and anterior cingulate cortex (Emre et al., 1993; Geula and Mesulam, 1994). In general, the cholinergic depletion tends to be the most accentuated within the temporal lobe, including its limbic, paralimbic and association components.

Even in very advanced stages of Alzheimer's disease, and even in areas with a severe depletion of cholinergic innervation, the cerebral cortex still contains some cholinergic axons. The regional densities of these residual cholinergic axons appear to reflect differences in premorbid levels of innervation. Thus, severely affected regions with a relatively low basal density of cholinergic innervation, such as the inferior, middle and superior temporal association areas, appear almost completely denuded of cholinergic fibers whereas the entorhinal area, which loses an equal proportion of its cholinergic innervation but has a much higher premorbid density of innervation, retains a considerably higher residual innervation (Geula and Mesulam, 1994).

Physiological and behavioral considerations related to cortical cholinergic innervation

Cortical cholinergic innervation and attention

From the vantage point of comparative neuroanatomy, the nucleus basalis constitutes a telencephalic extension of the brainstem reticular core and a mediobasal component of the limbic system (Ramon-Moliner and Nauta, 1966; Mesulam and Geula, 1988b; Mesulam et al., 1989). In keeping with these anatomical affiliations, the two major behavioral specializations of the nucleus basalis and of its cortical projections are in the realms of attention and memory.

The major effect of ACh upon cortical neurons is mediated through muscarinic m1 receptors and causes a relatively prolonged reduction of potassium conductance so as to make cortical cholinoceptive neurons more susceptible to other excitatory inputs. In primary visual cortex, for example, cholinergic stimulation increases the likelihood that a neuron will fire in response to its preferred stimulus (Sato et al., 1987). The fact that the nucleus basalis projects to all cortical areas while receiving its cortical inputs only from components of the limbic system suggests that it can selectively enhance the release of cortical ACh throughout the cerebral cortex in response to events that are of limbic relevance. Cortical cholinergic innervation is thus in a position to preferentially promote the cortical impact of events that are of emotional and motivational significance. In keeping with this formulation, neurons of the nucleus basalis in the monkey are selectively sensitive to novel and motivationally relevant sensory events (DeLong, 1971; Wilson and Rolls, 1990). Furthermore, the novelty-related P300 potential in the human cerebral cortex is abolished upon the administration of cholinergic blockers (Hammond et al., 1987).

Stimulation of the nucleus basalis elicits cortical EEG activation via muscarinic receptors, depolarizes cortical neurons and produces a change in subthreshold membrane potential fluctuations from large amplitude slow oscillations to low amplitude fast (20–40 Hz) oscillations (Metherate et al., 1992). Inactivation of the nucleus basalis, on the other hand, suppresses low voltage fast EEG activity in the cerebral cortex (Dringenberg and Vanderwolf, 1996). In the monkey, cortical cholinergic blockade interferes with the attentive on-line maintenance of sensory information (Dias et al., 1996). These observations provide considerable support for implicating the nucleus basalis and its cortical cholinergic projections in the modulation of attention and arousal (Mesulam, 1995; Himmelheber et al., 2001). Intracellular recordings show that stimulation of the brainstem-thalamic-cortical pathways can activate the cerebral cortex even in animals with nucleus basalis lesions (Steriade et al., 1993), illustrating the synergistic effects of the basal forebrain and brainstem upon arousal states.

Cortical cholinergic innervation and memory

Various observations have implicated cortical cholinergic pathways in memory and learning. Cortical cholinergic afferents, for example, promote oscillatory and covariant states of neural activity which enable otherwise ineffective stimulation to induce experience-dependent physiological plasticity in the course of new learning (Huerta and Lisman, 1993; Cruikshank and Weinberger, 1996). During Pavlovian conditioning, approximately half of nucleus basalis neurons show a significantly greater change of activity in response to a tone that predicts the occurrence of a mildly aversive unconditioned stimulus than to a tone that does not (Whalen et al., 1994). In the human, the systemic administration of the muscarinic blocker scopolamine causes memory impairments (Drachman and Leavitt, 1974). Pharmacological experiments in animals have indicated that the cholinergic innervation of the amygdala plays an important role in memory consolidation (McGaugh et al., 1993; Power et al., 2002). Destructive lesions which include the cholinergic as well as the non-cholinergic components of the nucleus basalis lead to memory deficits in some experiments (Ridley et al., 1986), but not in others (Voytko et al., 1994). The recent availability of 192 IgG-saporin (a ribosome inactivating neurotoxin conjugated to an antibody that recognizes NGFr) has made it possible to produce a selective destruction of only the cholinergic neurons in the nucleus basalis. Such experiments show that the resultant destruction of cortical cholinergic innervation causes severe impairments of learning in spatial navigation, but not in delayed alteration or passive avoidance tasks (Berger-Sweeney et al., 1994; Wenk et al., 1994).

The mechanisms that link the cholinergic projections of the Ch1-Ch4 cell group to memory function are incompletely understood and have led to several speculations. The role of acetylcholine in hippocampal long-term potentiation (Tanaka et al., 1989; Auerbach and Segal, 1994) may provide a cellular mechanism that underlies the relationship of cholinergic pathways to memory. Brain slice experiments in piriform cortex of the rat have shown that acetylcholine can selectively suppress intrinsic synaptic transmission through a presynaptic mechanism, while leaving extrinsic afferent input unaffected.

74

This selective suppression, could prevent interference from previously stored patterns during the learning of new patterns (Hasselmo, 1992). Buzsáki (Buszáki, 1989) has proposed a different model according to which the cholinergic innervation, especially of the hippocampal complex, plays a major role in switching from on-line attentive processing, characterized by the hippocampal theta rhythm, to off-line memory consolidation, characterized by sharp wave activity.

Cortical acetylcholine and plasticity

The role of ACh in plasticity has been investigated for many years (Bear and Singer, 1986). The selective lesioning of cortical cholinergic innervation in the rat was shown to interfere with experience-dependent plasticity in the barrel fields. In one experiment, all whiskers except for D2 and D3 were trimmed. This led to a pairing between the D2 and D3 barrel fields in the cerebral cortex so that the D2 neurons started to show a greater responsivity to D3 that to the adjacent D1 which had been trimmed. This pairing, indicative of experience-induced synaptic plasticity, could not be obtained in rats with selective immunotoxic lesions of the cholinergic neurons in the nbM (Baskerville et al., 1997). In another experiment on newborn rat pups, barrels representing intact whiskers failed to show the expected expansion into the territory of barrels representing trimmed whiskers in animals with nucleus basalis lesions (Zhu and Waite, 1998). Furthermore, pairing auditory stimuli with the electrical stimulation of the nucleus basalis in adult rats caused a long-lasting reorganization of primary auditory cortex so that the area optimally responsive to the paired tone expanded substantially. This plasticity was not observed following the selective immunotoxic destruction of cholinergic nucleus basalis neurons (Kilgard and Merzenich, 1998).

Conclusions

The cholinergic innervation of the human cerebral cortex is remarkably dense and widespread. It arises within the basal forebrain and reaches its highest density within components of the limbic system. The cortically projecting cholinergic neurons of the nucleus basalis (the Ch4 cell group) belong to the limbic system as well as to the ascending reticular activating system. In keeping with these relationships, the behavioral affiliations of cortical cholinergic pathways encompass the realms of attention and memory. Cortical cholinergic pathways may also influence cerebral blood flow (through the m2-mediated release of nitric oxide) and the potential for experience-induced neuroplasticity in the cerebral cortex. Aging is associated with a modest decline of cortical cholinergic axons that may provide an anatomical substrate for age-related memory changes. The substantially more severe decline seen in Alzheimer's disease may contribute to the emergence of the dementia and may erode the capacity for neuroplasticity to the point where the brain becomes more susceptible to neurodegenerative changes (Mesulam, 1999).

Acknowledgments

Supported in part by a Javits Neuroscience Investigator Award (NS20285) and an Alzheimer's Disease Center Grant (AG13854).

References

Arendt, T., Bigl, V., Tennstedt, A. and Arendt, A. (1985) Neuronal loss in different parts of the nucleus basalis is related to neuritic plaque formation in cortical target areas in Alzheimer's disease. Neuroscience, 14: 1–14.

Aston-Jones, G., Shaver, R. and Dinan, T.G. (1985) Nucleus basalis neurons exhibit axonal branching with decreased impulse conduction velocity in rat cerebrocortex. Brain Res., 325: 271–285.

Auerbach, J.M. and Segal, M. (1994) A novel cholinergic induction of long-term potentiation in rat hippocampus. J. Neurophysiol., 72: 2034–2040.

Baskerville, K.A., Schweitzer, J.B. and Herron, P. (1997) Effects of cholinergic depletion on experience-dependent plasticity in the cortex of the rat. Neuroscience, 80: 1159–1169.

Bear, M.F. and Singer, W. (1986) Modulation of visual cortical plasticity by acetylcholine and noradrenaline. Nature, 320: 172–176.

Berger-Sweeney, J., Heckers, S., Mesulam, M.-M., Wiley, R.G., Lappi, D.A. and Sharma, M. (1994) Differential effects upon spatial navigation of immunotoxin-induced cholinergic lesions of the medial septal area and nucleus basalis magnocellularis. J. Neurosci., 14: 4507–4519.

Bonner, T.I., Buckley, N.J., Young, A.C. and Brann, M.R. (1987) Identification of a family of muscarinic acetylcholine receptor genes. Science, 237: 527–532.

Bonner, T.I., Young, A.C., Brann, M.R. and Buckley, N.J. (1988) Cloning and expression of the human and rat m5 muscarinic acetylcholine receptor genes. Neuron, 1: 403–410.

Buckley, N.J., Bonner, T.I. and Brann, M.R. (1988) Localization of muscarinic receptor mRNAs in rat brain. J. Neurosci., 8: 4646–4652.

Buszáki, G. (1989) Commentary: Two-stage model of memory trace formation: A role for 'noisy' brain states. Neuroscience, 31: 551–570.

Cruikshank, S.J. and Weinberger, N.M. (1996) Evidence for the Hebbian hypothesis in experience-dependent physiological plasticity of neocortex: a critical review. Brain Res. Rev., 22: 191–228.

DeLima, A.D. and Singer, W. (1986) Cholinergic innervation of the cat striate cortex: A choline acetyltransferase immunocytochemical analysis. J. Comp. Neurol., 250: 324–338.

DeLong, M.R. (1971) Activity of pallidal neurons during movement. J. Neurophysiol., 34: 414–427.

Dias, E.C., Compaan, D.M., Mesulam, M.-M. and Segraves, M.A. (1996) Selective disruption of memory-guided saccades with injecton of a cholinergic antagonist in the frontal eye field of monkey. Soc. Neurosci. Abstr., 22: 418.

Drachman, D.A. and Leavitt, J. (1974) Human memory and the cholinergic system-A relationship to aging? Arch. Neurol., 30: 113–121.

Dringenberg, H.C. and Vanderwolf, C.H. (1996) Cholinergic activation of the electrocorticogram: an amygdaloid activating system. Exp. Brain Res., 108: 285–296.

Eggermann, E., Serafin, M., Bayer, L., Machard, D., Saint-Mleux, B., Jones, B.E. and Mühlethaler, M. (2001) Orexins/hyporectins excite basal forebrain cholinergic neurons. Neuroscience, 108: 177–181.

Emre, M., Heckers, S., Mash, D.C., Geula, C. and Mesulam, M.-M. (1993) Cholinergic innervation of the amygdaloid complex in the human brain and its alterations in old age and Alzheimer's disease. J. Comp. Neurol., 336: 117–134.

Everitt, B.J., Sirkiä, T.E., Roberts, A.C., Jones, G.H. and Robbins, T.W. (1988) Distribution and some projections of cholinergic neurons in the brain of the common marmoset, Callithrix jaccus. J. Comp. Neurol., 271: 533–558.

Farkas, R.H., Nakajima, S. and Nakajima, Y. (1994) Neurotensin excites basal forebrain cholinergic neurons: ionic and signal-transduction mechanisms. Proc. Natl. Acad. Sci. (USA), 91: 2853–2857.

Freund, T.F. and Meskenaite, V. (1992) γ-Aminobutyric acid-containing basal forebrain neurons innervate inhibitory interneurons in the neocortex. Proc. Natl. Acad. Sci. (USA), 89: 738–742.

Frotscher, M. and Leranth, C. (1985) Cholinergic innervation of the rat hippocampus as revealed by choline acetyltransferase immunocytochemistry: A combined light and electron microscopic study. J. Comp. Neurol., 239: 237–246.

Gaspar, P., Berger, B., Alvarez, C., Vigny, A. and Henry, J.P. (1985) Catecholaminergic innervation of the septal area in man: immunocytochemical study using TH and DBH antibodies. J. Comp. Neurol., 241: 12–33.

Geula, C. and Mesulam, M.-M. (1989) Cortical cholinergic fibers in aging and Alzheimer's disease: a morphometric study. Neuroscience, 33: 469–481.

Geula, C. and Mesulam, M.-M. (1994) Cholinergic systems and related neuropathological predilection patterns in Alzheimer disease. In: R.D. Terry, R. Katzman and K.L. Bick (Eds.), Alzheimer Disease. Raven Press, New York, pp. 263–294.

Geula, C., Mesulam, M.-M., Tokuno, H. and Kuo, C.C. (1993a) Developmentally transient expression of acetylcholinesterase within cortical pyramidal neurons of the rat brain. Dev. Brain Res., 76: 23–31.

Geula, C., Schatz, C.R. and Mesulam, M.-M. (1993b) Differential localization of NADPH-diaphorase and calbindin-D28k within the cholinergic neurons of the basal forebrain, striatum and brainstem in the rat, monkey, baboon and human. Neuroscience, 54: 461–476.

Gilmor, M.L., Erickson, J.D., Varoqui, H., Hersh, L.B., Bennett, D.A., Cochran, E.J., Mufson, E.J. and Levey, A.I. (1999) Preservation of nucleus basalis neurons containing choline acetyltransferase and the vesicular acetylcholine transporter in the elderly with mild cognitive impairment and earlty Alzheimer's disease. J. Comp. Neurol., 411: 693–704.

Gorry, J.D. (1963) Studies on the comparative anatomy of the ganglion basale of Meynert. Acta Anat., 55: 51–104.

Gouras, G.K., Rance, N.E., Young, W.S., III and Koliatsos, V.E. (1992) Tyrosine-hydroxylase containing neurons in the primate basal forebrain magnocellular complex. Brain Res., 584: 287–293.

Gritti, I., Mainville, L. and Jones, B.E. (1993) Codistribution of GABA with acetylcholine-synthesizing neurons in the basal forebrain of the rat. J. Comp. Neurol., 329: 438–457.

Hammond, E.J., Meador, K.J., Aunq-Din, R. and Wilder, B.J. (1987) Cholinergic modulation of human P3 event-related potentials. Neurology, 37: 346–350.

Harata, N., Tateishi, N. and Akaike, N. (1991) Acetylcholine receptors in dissociated nucleus basalis of Meynert neurons of the rat. Neurosci. Lett., 130: 153–156.

Hasselmo, M.E. (1992) cholinergic modulation of cortical associative memory function. J. Neurophysiol., 67: 1230–1246.

Heckers, S., Ohtake, T., Wiley, R.G., Lappi, D.A., Geula, C. and Mesulam, M.-M. (1994) Complete and selective cholinergic denervation of rat neocortex and hippocampus but not amygdala by an immunotoxin agoinst the p75 NGF receptor. J. Neurosci., 14: 1271–1289.

Henderson, Z. (1987) A small proportion of cholinergic neurones in the nucleus basalis magnocellularis of ferret appear to stain positively for tyrosine hydroxylase. Brain Res., 412: 363–369.

Henderson, Z. and Evans, S. (1991) Presence of a cholinergic projection from ventral striatum to amygdala that is not immunoreactive for NGF receptor. Neurosci. Lett., 127: 73–76.

Himmelheber, A.M., Sarter, M. and Bruno, J.P. (2001) The effects of manipulations of attentional demand on cortical acetylcholine release. Cog. Brain Res., 12: 353–370.

Huerta, P.T. and Lisman, J.E. (1993) Heightened synaptic plasticity of hippocampal CA1 neurons during a cholinergically induced rhythmic state. Nature, 364: 723–725.

Ingham, C.A., Bolam, J.P., Wainer, B.H. and Smith, A.D. (1985) A correlated light and electron microscopic study of identified cholinergic basal forebrain neurons that project to the cortex in the rat. J. Comp. Neurol., 239: 176–192.

Jones, B.E. and Cuello, A.C. (1989) Afferents to the basal forebrain cholinergic cell area from pontomesencephalic-catecholamine, serotonin and acetylcholine-neurons. Neuroscience, 31: 37–61.

Kilgard, M.P. and Merzenich, M.M. (1998) Cortical map reorganization enabled by nucleus basalis activity. Science, 279: 1714–1718.

Kordower, J.H., Chen, E.-Y., Sladek, J.R., Jr. and Mufson, E.J. (1994) TRK-immunoreactivity in the monkey central nervous system: Forebrain. J. Comp. Neurol., 349: 20–35.

Kostovic, I., Skavic, J. and Strinovic, D. (1988) Acetylcholinesterase in the human frontal associative cortex during the period of cognitive development: Early laminar shifts and late innervation of pyramidal neurons. Neurosci. Lett., 90: 107–112.

Kubo, T., Fukuda, K., Mikami, A., Maeda, A., Takahashi, H., Mishina, M., Haga, T., Haga, K., Ichiyama, A., Kangawa, K., Kojima, M., Matsuo, H., Hirose, T. and Numa, S. (1986) Cloning, sequencing and expression of complementary DNA encoding the muscarinic acetylcholine receptor. Nature, 323: 411–416.

Kutscher, C.L. (1991) Development of transient acetylcholinesterase staining in cells and permanent staining in fibers in cortex of rat brain. Brain Res. Bull., 27: 641–649.

Levey, A.I., Kitt, C.A., Simonds, W.F., Price, D.L. and Brann, M.R. (1991) Identification and localization of muscarinic acetylcholine receptor protein in brain with subtype-specific antibodies. J. Neurosci., 11: 3218–3226.

Levey, A.I., Edmunds, S.M., Heilman, C.J., Desmond, T.J. and Frey, K.A. (1994) Localization of muscarinic m3 receptor protein and M3 receptor binding in the rat brain. Neuroscience, 63: 207–221.

Lidow, M.S., Gallager, D.W., Rakic, P. and Goldman-Rakic, P.S. (1989) Regional differences in the distribution of muscarinic cholinergic receptors in the macaque cerebral cortex. J. Comp. Neurol., 289: 247–259.

Loewi, O. (1921) Über humorale Übertragbarkeit der Herznervenwirkung. Pflügers Arch. Ges. Physiol., 189: 239–242.

Martinez-Murillo, R., Villalba, R.M. and Rodrigo, J. (1990) Immunocytochemical localization of cholinergic terminals in the region of the nucleus basalis magnocellularis of the rat: A correlated light and electron microscopic study. Neuroscience, 36: 361–376.

Mash, D.C., White, W.F. and Mesulam, M.-M. (1988) Distribution of muscarinic receptor subtypes within architectonic subregions of the primate cerebral cortex. J. Comp. Neurol., 278: 265–274.

McCormick, D.A. and Prince, D.A. (1985) Two types of muscarinic responses to acetylcholine in mammalian cortical neurons. Proc. Natl. Acad. Sci. USA, 82: 6344–6348.

McGaugh, J.L., Introini-Collison, I.B., Cahill, L.F., Castellano, C., Dalmaz, C., Parent, M.B. and Williams, C.L. (1993) Neuromodulatory systems and memory storage: role of the amygdala. Behav. Brain Res., 58: 81–90.

Mesulam, M.-M. (1995) Cholinergic pathways and the ascending reticular activating system of the human brain. Ann. N.Y. Acad. Sci., 757: 169–179.

Mesulam, M.-M. (1999) Neuroplasticity failure in Alzheimer's disease: Bridging the gap between plaques and tangles. Neuron, 24: 521–529.

Mesulam, M.-M. and Geula, C. (1988a) Acetylcholinesterase-rich pyramidal neurons in the human neocortex and hippocampus: absence at birth, development during the life span, and dissolution in Alzheimer's disease. Ann. Neurol., 24: 765–773.

Mesulam, M.-M. and Geula, C. (1988b) Nucleus basalis (Ch4) and cortical cholinergic innervation in the human brain: observations based on the distribution of acetylcholinesterase and choline acetyltransferase. J. Comp. Neurol., 275: 216–240.

Mesulam, M.-M. and Geula, C. (1991) Acetylcholinesterase-rich neurons of the human cerebral cortex: cytoarchitectonic and ontogenetic patterns of distribution. J. Comp. Neurol., 306: 193–220.

Mesulam, M.-M. and Geula, C. (1993) Chemoarchitectonics of axonal and perikaryal acetylcholinesterase along information processing systems of the human cerebral cortex. Brain Res. Bull., 33: 137–153.

Mesulam, M.-M. and Mufson, E.J. (1984) Neural inputs into the nucleus basalis of the substantia innominata (Ch4) in the rhesus monkey. Brain, 107: 253–274.

Mesulam, M.-M., Mufson, E.J., Levey, A.I. and Wainer, B.H. (1983a) Cholinergic innervation of cortex by the basal forebrain: cytochemistry and cortical connections of the septal area, diagonal band nuclei, nucleus basalis (substantia innominata), and hypothalamus in the rhesus monkey. J. Comp. Neurol., 214: 170–197.

Mesulam, M.M., Mufson, E.J., Wainer, B.H. and Levey, A.I. (1983b) Central cholinergic pathways in the rat: an overview based on an alternative nomenclature (Ch1-Ch6). Neuroscience, 10: 1185–1201.

Mesulam, M.-M., Geula, C., Bothwell, M.A. and Hersh, L.B. (1989) Human reticular formation: cholinergic neurons of the pedunculopontine and laterodorsal tegmental nuclei and some cytochemical comparisons to forebrain cholinergic neurons. J. Comp. Neurol., 283: 611–633.

Mesulam, M.-M., Hersh, L.B., Mash, D.C. and Geula, C. (1992) Differential cholinergic innervation within functional subdivisions of the human cerebral cortex: a choline acetyltransferase study. J. Comp. Neurol., 318: 316–328.

Metherate, R., Cox, C.L. and Ashe, J.H. (1992) Cellular bases of neocortical activation: Modulation of neuronal oscillations by the nucleus basalis and endogenous acetylcholine. J. Neurosci., 12: 4701–4711.

Mrzljak, L., Levey, A.I. and Goldman-Rakic, P.S. (1993) Association of m1 and m2 muscarinic receptor proteins with asymmetric synapses in the primate cerebral cortex: Morphological evidence for cholinergic modulation of excitatory neurotransmission. Proc. Nat. Acad. Sci. (USA), 90: 5194–5198.

Mufson, E.J., Bothwell, M., Hersh, L.B. and Kordower, J.H. (1989) Nerve growth factor receptor immunoreactive profiles in the normal, aged human basal forebrain: Colocalization with cholinergic neurons. J. Comp. Neurol., 285: 196–217.

Nakajima, Y., Nakajima, S., Obata, K., Carlson, C.G. and Yamaguchi, K. (1985) Dissociated cell culture of cholinergic neurons from nucleus basalis of Meynert and other basal forebrain nuclei. Proc. Nat. Acad. Sci. (USA), 82: 6325–6329.

Palacios, G. (1990) The endomembrane system of cholinergic and non-cholinergic neurons in the medial septal nucleus and vertical limb of the diagonal band of Broca: A cytochemical and immunocytochemical study. J. Histochem. Cytochem., 38: 563–571.

Peralta, E.G., Ashkenazi, A., Winslow, J.W., Smith, D.H., Ramachandran, J. and Capon, D.J. (1987) Distinct primary structures, ligand-binding properties and tissue-specific expression of four human muscarinic acetylcholine receptors. EMBO J., 6: 3923–3929.

Power, A.E., Thal, L.J. and McGaugh, J.L. (2002) Lesions of the nucleus basalis magnocellularis induced by 192 IgG-saporin block memory enhancement with posttraining norepinephrine in the basolateral amygdala. Proc. Natl. Acad. Sci. (USA), 99: 2315–2319.

Ramon-Moliner, E. and Nauta, W.J.H. (1966) The isodendritic core of the brain. J. Comp. Neurol., 126: 311–336.

Ridley, R.M., Murray, T.K., Johnson, J.A. and Baker, H.F. (1986) Learning impairment following lesion of the basal nucleus of Meynert in the marmoset: Modification by cholinergic drugs. Brain Res., 376: 108–116.

Sato, H., Hata, V., Hagihara, K. and Tsumoto, T. (1987) Effects of cholinergic depletion on neuron activities in the cat visual cortex. J. Neurophysiol., 58: 781–794.

Satoh, K. and Fibiger, H.C. (1985) Distribution of central cholinergic neurons in the baboon (Papio papio). I. General morphology. J. Comp. Neurol., 236: 197–214.

Schröder, H., Zilles, K., Maelicke, A. and Hajós, F. (1989) Immunohisto- and cytochemical localization of cortical cholinoceptors in rat and man. Brain Res., 502: 287–295.

Schröder, H., Zilles, K., Luiten, P.G.M. and Strosberg, A.D. (1990) Immunocytochemical visualization of muscarinic cholinoceptors in the human cerebral cortex. Brain Res., 514: 249–258.

Selden, N.R., Gitelman, D.R., Salamon-Murayama, N., Parrish, T.B. and Mesulam, M.-M. (1998) Trajectories of cholinergic pathways within the cerebral hemispheres of the human brain. Brain, 121: 2249–2257.

Silver, A. (1974) The Biology of Cholinesterases. American Elsevier Publishing, New York.

Smiley, J.F. and Mesulam, M.-M. (1999) Cholinergic neurons of the nucleus basalis of Meynert (Ch4) receive cholinergic, catecholaminergic, and GABAergic synapses: an electron microscopic investigation in the monkey. Neuroscience, 88: 241–255.

Smiley, J.F., Morrell, F. and Mesulam, M.-M. (1997) Cholinergic synapses in human cerebral cortex: an ultrastructural study in serial sections. Exper. Neurol., 144: 361–368.

Smiley, J.F., Levey, A.I. and Mesulam, M.-M. (1998) Infracortical interstitial cells concurrently expressing m2-muscarinic receptors, AChE, and NADPH-d in the human and monkey cerebral cortex. Neuroscience, 84: 755–769.

Smiley, J.F., Levey, A.I. and Mesulam, M.-M. (1999a) m2 muscarinic receptor immunolocalization in cholinergic cells of the monkey basal forebrain and striatum. Neuroscience, 90: 803–814.

Smiley, J.F., Subramanian, M. and Mesulam, M.-M. (1999b) Monoaminergic-cholinergic interactions in the primate basal forebrain. Neuroscience, 93: 817–829.

Steriade, M., Amzica, F. and Nunez, A. (1993) Cholinergic and noradrenergic modulation of the slow (approximately 0.3 Hz) oscillation in neocortical cells. J. Neurophysiol., 70: 1385–1400.

Tanaka, Y., Sakurai, M. and Hayashi, S. (1989) Effect of scopolamine and HP029, a cholinesterase inhibitor, on long-term potentiation in hippocampal slices of guinea pig. Neurosci. Lett., 98: 179–183.

Umbriaco, D., Watkins, K.C., Descarries, L., Cozzari, C. and Hartman, B.K. (1994) Ultrastructural and morphometric features of the acetylcholine innervation in adult rat parietal cortex: an electron microscopic study in serial sections. J. Comp. Neurol., 348: 351–373.

Vilaró, M.T., Wiederhold, K.-H., Palacios, J.M. and Mengod, G. (1992) Muscarinic M2 receptor mRNA expression and receptor binding in cholinergic and non-cholinergic cells in the rat brain: a correlative study using in situ hybridization histochemistry and receptor autoradiography. Neuroscience, 47: 367–393.

Voytko, M.L., Olton, D.S., Richardson, R.T., Gorman, L.K., Tobin, J.R. and Price, D.L. (1994) Basal forebrain lesions in monkeys disrupt attention but not learning and memory. J. Neurosci., 14: 167–186.

Wainer, B.H., Bolam, J.P., Freund, T.F. and Henderson, Z. (1984) Cholinergic synapses in the rat brain: A correlated light and electron microscopic immunohistochemical study employing a monoclonal antibody against choline acetyltransferase. Brain Res., 308: 69–76.

Walker, L.C., Tigges, M. and Tigges, J. (1983) Ultrastructure of neurons in the nucleus basalis of Meynert in squirrel monkey. J. Comp. Neurol., 217: 158–166.

Walker, L.C., Koliatsos, V.E., Kitt, C.A., Richardson, R.T., Rökaeus, Å. and Price, D.L. (1989) Peptidergic neurons in the basal forebrain magnocellular complex of the rhesus monkey. J. Comp. Neurol., 280: 272–282.

Wenk, G.L., Stoehr, J.D., Quintana, G., Mobley, S. and Wiley, R.G. (1994) Behavioral, biochemical, histological, and electrophysiological effects of 192 IgG-saporin injections into the basal forebrain of rats. J. Neurosci., 14: 5986–5995.

Whalen, P.J., Knapp, B.S. and Pascoe, J.P. (1994) Neuronal activity within the nucleus basalis and conditioned neocortical electroencephalographic activation. J. Neurosci., 14: 1623–1633.

Wilson, F.A.W. and Rolls, E.T. (1990) Neuronal responses related to novelty and familiarity of visual stimuli in the substantia innominata, diagonal band of Broca and periventricular region of the primate basal forebrain. Exp. Brain Res., 80: 104–120.

Wisniowski, L., Ridley, R.M., Baker, H.F. and Fine, A. (1992) Tyrosine hydroylase-immunoreactive neurons in the nucleus basalis of the common marmoset (Callithrix jaccus). J. Comp. Neurol., 325: 379–387.

Wu, C.-K., Mesulam, M.-M. and Geula, C. (1997) Age-related loss of calbindin from basal forebrain cholinergic neurons. NeuroReport, 8: 2209–2213.

Záborszky, L., Gaykema, R.P., Swanson, D.J. and Cullinan, W.E. (1997) Cortical input to the basal forebrain. Neuroscience, 79: 1051–1078.

Zhu, X.O. and Waite, P.M.E. (1998) Cholinergic depletion reduces plasticity of barrel field cortex. Cereb. Cortex, 8: 63–72.

Zilles, K., Werner, L., Qü, M., Schleicher, A. and Gross, G. (1991) Quantitative autoradiography of 11 different transmitter binding sites in the basal forebrain region of the rat-evidence of heterogeneity in distribution patterns. Neuroscience, 42: 473–481.

Zilles, K., Schlaug, G., Matelli, M., Luppino, G., Schleicher, A., Qu, M., Dabringhaus, A., Seitz, R. and Roland, P. (1995) Mapping of human and macaque sensorimotor areas by integrating architectonic, transmitter receptor, MRI and PET data. J. Anat., 187: 515–537.

Modes of action of acetylcholine in the cerebral cortex

Progress in Brain Research, Vol. 145
ISSN 0079-6123

CHAPTER 5

Synaptic mechanisms modulated by acetylcholine in cerebral cortex

Krešimir Krnjević*

*Anaesthesia Research Unit and Physiology Department, McGill University, Room 1215, McIntyre Building,
3655 Promenade Sir William Osler, Montreal, QC H3G 1Y6, Canada*

History: pioneers

It is almost 100 years since Hunt and Taveau (1906) 'discovered' ACh when they first demonstrated its powerful hypotensive action. Dale (1914) made the crucial distinction between the nicotinic and muscarinic actions; and within 30 years, ACh became established as an important transmitter at ganglionic and neuromuscular synapses (Dale, 1938). Major progress on a possible role of cholinergic transmission in the CNS required another two decades: arguing that intraspinal branches of motoneurons are also likely to be cholinergic, Eccles et al. (1954) were indeed able to show that these axons excite Renshaw (inhibitory) cells by a fast nicotinic action. But comparable fast cholinergic synapses were not observed elsewhere in the CNS for another 40 years. Only recently has it become clear that nicotinic synapses are characteristic of inhibitory neurons in the brain—as we shall hear about in much greater detail from other participants.

It had been known since the 1930s that slices of cerebral cortex can synthesize ACh in vitro (Quastel et al., 1936); the relevant enzyme was later identified as choline acetylase (acetyltransferase) by Nachmansohn and Machado (1943). But the first

compelling evidence that ACh is released from the intact cerebral cortex came much later (Elliott et al. 1950; MacIntosh and Oborin, 1953). Using a 'surface cup'—as illustrated in Fig. 1A—ACh was assayed in physiological saline that had been exposed to the surface of the cortex. In several such studies (MacIntosh and Oborin, 1953; Mitchell 1963; Celesia and Jasper, 1966; Collier and Mitchell, 1967), the amounts of ACh released clearly varied with the state of wakefulness (Fig. 1B). Early in the 1960's, came the first evidence that this ACh must be released from fibers that do not originate from cortical cholinergic cells, but rather from cells situated deep in the forebrain (Shute and Lewis, 1963; Krnjević and Silver, 1965, 1966; Shute and Lewis, 1967). But what did this ACh do in the cortex?

Suggestive excitatory effects of ACh had been reported earlier (e.g., after intracarotid injections, Bremer and Chatonnet, 1949); but a clearer picture emerged only with the use of the microiontophoretic technique. Working independently, Krnjević and Phillis (1963a) and Spehlmann (1963) found that ACh causes a slow and prolonged excitation of many cortical neurons (Fig. 2A). An unusual feature was that, though sometimes apparently ineffective, ACh could strongly potentiate other forms of excitation. Thus, as shown in Fig. 2B and C, even when, under conditions of deep anaesthesia, ACh did not cause overt excitation, it greatly enhanced firing evoked by a

*Tel.: (514) 398 6001; Fax: (514) 398 4376;
E-mail: Kresimir.krnjevic@cgill.ca

DOI: 10.1016/S0079-6123(03)45005-X

82

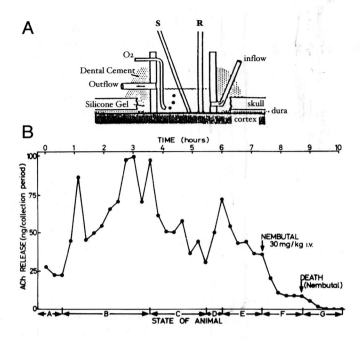

Fig. 1. ACh release from cortex increases with state of wakefulness. (A) Agents released from cortex of anaesthetized cat were collected in superfusate inside a cup applied to cortical surface; to prevent breakdown of ACh, an anticholinesterase was added; ACh was subsequently measured by bioassay (S and R are stimulating and recording electrodes) (from Celesia and Jasper, 1966). (B) ACh release observed with 'push–pull' cannula inserted into visual cortex of free-moving rabbit varied with degree of wakefulness or anaesthesia (indicated below: A, during initial anaesthesia; B, after recovery from anaesthesia, animal is moving about, exploring cage; C, now quiet; D, active; E, quiet again; F,G, increasing doses of pentobarbitone (from Collier and Mitchell, 1967).

brief application of glutamate. This muscarinic action (Krnjević and Phillis, 1963b) was the prototype of what later became known as 'synaptic modulation'.

Muscarinic actions in cortex

What could be the mechanism of such peculiar excitation?

Suppression of K conductance

In intracellular recordings (Krnjević et al., 1971), the depolarizing action on neocortical neurons was consistently associated with increased membrane resistance (Fig. 2E) and it had a reversal potential near −90 mV; it was therefore proposed that ACh increases excitability by reducing K conductance. This was confirmed by many subsequent studies, which have shown that a variety of K currents are suppressed

by ACh. Of particular interest is the low threshold, slowly activating and non-inactivating M current (I_M), originally discovered in frog sympathetic ganglia (Brown and Adams, 1980) and later in rat hippocampal slices (Halliwell and Adams, 1982) (Fig. 3). Because it is voltage-dependent and is activated by small depolarizations close to the firing threshold, I_M tends to prevent neuronal firing. Inactivation of the muscarine-sensitive I_M readily accounted for some, if by no means all, of the effects of ACh.

Indeed, further studies (Cole and Nicoll, 1984; Constanti and Sim, 1987; Madison et al., 1987) showed that low doses of ACh act predominantly on two other K currents of cortical neurons (Fig. 4): an ongoing 'leak' current, block of which accounts for the 'slow' EPSP seen when cortex is stimulated repetitively, especially in the presence of an anticholinesterase; and a Ca-dependent K current ('SK'), responsible for the slow after-hyperpolarization (sAHP) that follows bursts of action potentials. The

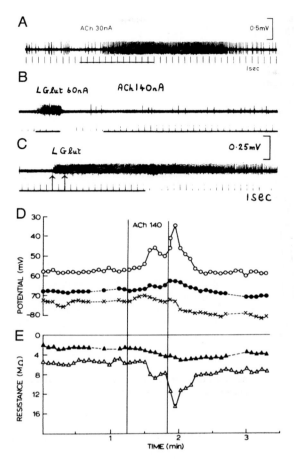

Fig. 2. Slow excitatory/facilitatory action of ACh on neocortical neurons is associated with increase in membrane resistance. (A) Firing with characteristic slow onset and prolonged after-discharge evoked by microiontophoretic application of ACh in cat cortex (from Krnjević and Phillis, 1963b). (B) In deeply anaesthetized cat, cortical unit was readily excited by glutamate but not ACh; during long application of ACh (note 5 s break between end of B and start of (C), identical brief application of glutamate elicited very prolonged firing (from Krnjević and Phillis, 1963a). Times of microiontophoretic applications are indicated by horizontal bars below traces. Plots of membrane potential (D) and input resistance (E) illustrate changes observed during intracellular recording and similar application of ACh: open symbols, resting potential or resistance (measured with brief current pulses); closed symbols, corresponding values during IPSPs; and crosses, IPSP reversal potential (from Krnjević et al., 1971).

cumulative hyperpolarization produced by successive sAHPs normally prevents sustained firing in response to a depolarizing input; the suppression of sAHPs by ACh (and other modulators such as NA) therefore promotes sustained firing and after-discharges. This effect is reinforced by the ACh-induced block of the fast inactivating A current (Nakajima et al., 1986).

Receptors and internal messengers

Like all muscarinic receptors, those involved in the depression of G_K's are blocked by atropine. Sensitivity to pirenzipine (and some other antagonists) indicates M1, M3 and M5 subtypes (Hammer et al., 1980; Levey et al., 1991); typically, these are linked to the breakdown of phosphatidyl inositol rather than changes in cyclic AMP levels (which are induced via the M2 and M4 subtypes). Further intracellular signals are triggered partly by a pertussis-sensitive G protein and by $G\alpha_q$ (Haley et al., 2000). In view of the slow and prolonged character of muscarinic action, a diffusible internal second messenger was also likely to be involved; but its identity remains controversial. There is strong evidence that Ca^{2+} can play such a role (Selyanko and Brown, 1996); but some other, as yet unknown, agent has been invoked to explain slow actions that cannot be ascribed to Ca^{2+} or breakdown products of phosphatidyl inositol (Shapiro et al., 2000).

Block of Ca^{2+} currents

Several groups (Toselli et al., 1989; Brown et al., 1990; Shapiro et al., 2000) first reported that ACh suppresses somatic Ca currents, especially high-voltage activated currents, such as N, P and Q/R types. This action is mediated by M2/M4 receptors and $G\alpha_q$ (Haley et al., 2000). A similar block of Ca currents in nerve terminals may account for presynaptic modulation of transmitter release, via muscarinic as well as nicotinic receptors (Hounsgaard, 1978; Rovira et al., 1983; Hasselmo and Schnell, 1994).

Activation of a non-selective cationic current

Studies on slices of neocortical, hippocampal and entorhinal cortex have revealed another interesting mechanism of ACh action, activation of a Ca-dependent non-selective cationic inward current, carrying both Na and Ca ions, and with a reversal potential near zero (Fraser and MacVicar, 1996;

84

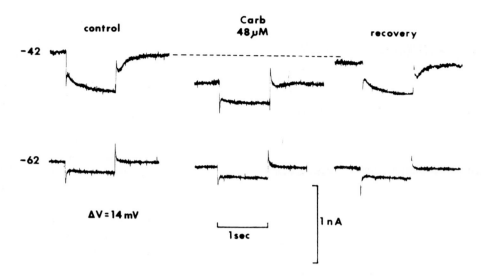

Fig. 3. Voltage-clamp traces from CA1 hippocampal neuron in slice illustrate carbachol-sensitive voltage-dependent M current. In left upper trace, at holding potential of −42 mV, hyperpolarizing step evokes slowly increasing inward current produced by closing of non-inactivating outward current through K channels. By suppressing this ongoing current, carbachol (Carb) causes inward shift in baseline current and abolishes the slow inward relaxation (middle trace). In keeping with the M current voltage-dependence, at −62 mV similar hyperpolarizing step evokes no comparable slow inward current and carbachol has no obvious effect (from Halliwell and Adams, 1982).

Haj-Dahmane and Andrade, 1996; Klink and Alonso, 1997). This current promotes firing in bursts and prolonged action potentials. Thus, in the presence of ACh, brief depolarizing pulses can evoke long 'plateau potentials' (Fig. 5). According to a recent analysis (Kuzmiski and MacVicar, 2001), these channels may be opened by cyclic GMP formed by the combined action of M1/M2 receptors and Ca^{2+} influx (Fig. 6). By facilitating paroxysmal firing, this kind of cholinergic action would predispose to epileptiform activity.

Depression of transmitter release

MacIntosh and Oborin (1953) had already noted that atropine greatly increased the liberation of ACh from the cortex. Autoregulation of release from cortical cholinergic terminals was a likely explanation, which has been confirmed by later investigators (Mitchell, 1963; Dudar, 1977). There is now compelling evidence that this action is mediated by M2 autoreceptors (Zhang et al., 2002).

Presynaptic depression of EPSPs

Perhaps more surprising was Hounsgaard (1978) finding that ACh depresses glutamatergic EPSPs in hippocampal slices, evidently by a presynaptic, not postsynaptic action. A more detailed investigation of the hippocampus in the brain in situ showed that this depression is a muscarinic effect; indeed, nicotinic agents enhanced EPSP (as they do elsewhere in the nervous system, see below), but the muscarinic depression predominated when ACh itself is applied in the region of the excitatory synapses in stratum radiatum (Rovira et al., 1983).

Presynaptic depression of IPSPs

By contrast, when ACh was applied in the pyramidal cell layer, where inhibitory synapses are most prominent, ACh markedly enhanced population spikes (Krnjević et al., 1981), probably by reducing IPSPs via a presynaptic action (Ben-Ari et al., 1981). Here, both muscarinic and nicotinic agents had a similar effect (Rovira et al., 1983). In another study, Krnjević

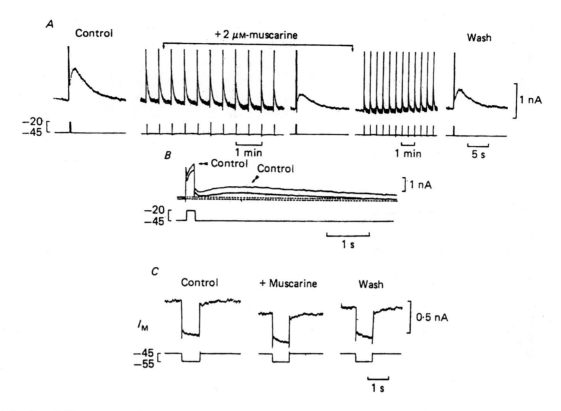

Fig. 4. In slice of olfactory cortex, low concentration of muscarine induces inward current and depresses slow outward after-current, but has little effect on M current. (A) Sequence of voltage-clamp traces recorded before, during and after application of muscarine; note different speeds of recording. (B) Superimposed expanded traces of outward after-current, before (control) and during muscarine application (at lower gain). (C) Slow inward relaxations (M channels closing) evoked by hyperpolarizing pulses are not affected by muscarine, though inward shift of middle trace indicates suppression of ongoing outward 'leak' current (from Constanti and Sim, 1987).

and Ropert (1981) found that stimulation of the medial septum/diagonal band region could produce a comparable facilitation of population spikes (Fig. 7A and B), at least in part by activation of septo-hippocampal cholinergic fibers—judging by the sharp reduction of this effect seen after intraventricular injections of hemicholinium, a selective blocker of ACh synthesis (Glavinovic et al., 1983) (Fig. 7C). This cholinergic disinhibition is probably mediated by M2 receptors located on perisomatic GABAergic terminals (Hajos et al., 1998). A major disinhibitory effect of septal stimulation is also produced by inhibition of inhibitory interneurons (Toth et al., 1997), via the major GABAergic projection from the medial septum/diagonal band (Köhler et al., 1984; Freund and Antal, 1988).

Relevance of muscarinic actions for cortical function

By enhancing and prolonging cortical responses to sensory inputs (Krnjević and Phillis, 1963a; Sillito and Kemp, 1983), ACh may play a significant role in cortical 'arousal' (Krnjević, 1969). In addition to the facilitation of neuronal firing, ACh promotes oscillations in cortical networks (MacVicar and Tse, 1989). These oscillations may be relatively slow, like the hippocampal theta rhythm—which has long been known to have a cholinergic (muscarinic) component (Stumpf, 1965; Vanderwolf and Stewart, 1986)—or at much higher (40–80 Hz or 'gamma') frequency in neocortex: such gamma waves are believed to be important for 'binding' sensory information arriving

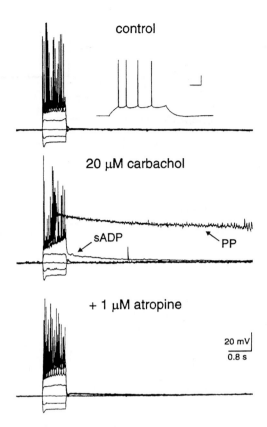

control

20 μM carbachol

sADP

PP

+ 1 μM atropine

20 mV
0.8 s

Fig. 5. Another muscarinic action in hippocampal neurons takes the form of a slow after-depolarization (sADP) or very prolonged 'plateau potential' (pp), following weak or strong depolarizing pulses, respectively. Recording was from CA1 neuron, at −65 mV. In upper control trace, note absence of such after-potentials and inset faster trace at much lower gain showing firing during depolarizing pulse (from Fraser and MacVicar, 1996). The underlying inward current proved to be a Ca-dependent non-selective inward current.

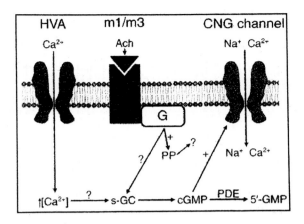

Fig. 6. Proposed scheme of internal messengers involved in generation of Ca-dependent non-selective inward current underlying after-depolarizations and plateau potentials induced by activation of muscarinic receptors (illustrated in Fig. 5). Stimulation of muscarinic receptors ($m1/m3$), coupled to G-proteins, in combination with Ca^{2+} influx through high-voltage activated Ca^{2+} channels (HVA) result in formation of cyclic GMP, opening of cyclic nucleotide-gated (CNG) non-selective cationic channels and influx of Na^+ and Ca^{2+}. Activation of protein phosphatase (PP), required for plateau potential generation, may increase sensitivity of CNG channels to cyclic nucleotides; specific phosphodiesterase (PDE) terminates inward current by metabolizing cGMP (from Kuzmiski and MacVicar, 2001).

in different cortical areas into a coherent sensory picture (Eckhorn et al., 1988; Maloney et al., 1997; Singer, 2001)—perhaps an essential cellular component of 'consciousness'.

Because it suppresses various G_K's and reduces synaptic inhibition, ACh facilitates the induction of long term potentiation (Brocher et al., 1992; Hasselmo and Barkai, 1995) and synaptic plasticity more generally (Gu and Singer, 1993; Segal and Auerbach, 1997; Rasmusson, 2000). Thus, the cholinergic input to cortex is likely to be beneficial for learning and memory, and, during development,

for the growth and organization of synaptic connections. This notion is supported by the fact that drugs that act as 'cognitive enhancers' bind to M type K channels (see below) and, of course by, the notorious progressive failure of cognitive function that is so characteristic of senile dementia (Petersen and Kanow, 2001; Rogan and Lippa, 2002).

Molecular identification of muscarine-activated K channels

In recent years, there has been a rapid expansion of studies on muscarinic mechanisms in nerve cells (Brown et al., 1997; Shapiro et al., 2001). Especially exciting is new information about two types of voltage-dependent K channels that can generate M-like currents. These channels are present in the brain and therefore may well be important targets for muscarinic actions in the cortex.

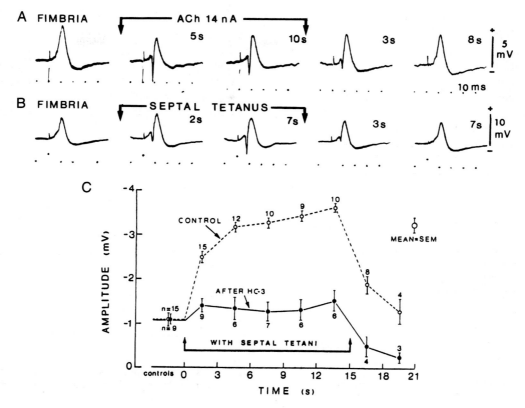

Fig. 7. In rat hippocampus in situ, local ACh application and medial septal stimulation produce a similar facilitation of CA1 population spikes; effect of septal stimulation is greatly reduced after 15 µg intraventicular injection.of hemicholinium-3 (HC-3, blocker of ACh synthesis). A, B, traces at left: in control conditions fimbrial stimulation (at 0.6 Hz) evoked only large positive field response; population spikes appeared only when ACh was applied (A) or when fimbrial stimulus was preceded by brief septal tetanus (110 Hz for 70 ms, ending 30 ms earlier) (from Fig. 1 in Krnjević and Ropert, 1981). (C) In similar experiment, facilitation of CA1 population spike by brief tetanic stimulation of medial septum (10 pulses at 70 Hz) was much reduced when identical tests were repeated ~ 30 min after HC-3 intraventricular injection. Mean data ± SEM; numbers of observations are indicated (from Glavinovic et al., 1983).

R-eag K channels

Cloned K channels of the *eag* type ('ether-a-gogo') were shown to be muscarine-sensitive by Stansfeld et al. (1996). An interesting point is that they are readily blocked by a small rise in internal Ca^{2+}. Thus, Ca^{2+} could be the second messenger between these receptors and K channels (see also Selyanko and Brown, 1996). On the other hand, albeit present in cortex, *eag* channels may account for only part of cortical M current—perhaps only a slow component (Meves et al., 1999; Selyanko et al., 1999, 2002).

KCNQ-type K channels

These channels were originally cloned because of their apparent involvement in a benign form of familial infantile epilepsy. Systematic surveys localized the genetic defect to chromosome 20q and 8q. By manipulating the appropriate locus, three groups in 1998 identified K channel subunits that were closely related to the cardiac KvLQT channel—mutations of which results in the cardiac 'long QT' syndrome (Biervert et al., 1998; Singh et al., 1998; Yang et al., 1998). These subunits were subsequently renamed KCNQ1–5. Unlike KCNQ1, which is found only in the heart, the other subunits are present in the brain. As illustrated in Fig. 8, when KCNQ2 and KCNQ3 are expressed in oocytes, depolarization of the cells initiates a slow current that is selective for K^+ (Biervert et al., 1998).

In the same year, the KCNQ2/3 current in oocytes was shown to be virtually identical to ganglionic M current (Wang et al., 1998). Crucial points in this study included the following. (i) The voltage-dependence and activation kinetics of these currents are very similar to those of M current (Fig. 9). (ii) Both M channels and KCNQ2/3 channels are blocked in a very similar manner by two cognitive enhancers (linopirdine and XE991)—which act directly, independently of any G proteins and ACh receptors (Costa and Brown, 1997; Lamas et al., 1997). (iii) Heteromeric combinations of KCNQ2 and KCNQ3 produce together much larger currents than the individual subunits. (iv) While the KCNQ2 gene is rather widely distributed in the brain—including the cerebellum (not known to have M current)—KVNQ3 is more selectively localized in areas that do express M current, such as neocortex and hippocampus: the presence of KCNQ3 is especially strong evidence of a functionally significant M current. Further reports have confirmed the major role of KCNQ2/3 channels as mediators of M current (Pan et al., 2001; Selyanko et al., 2001, 2002).

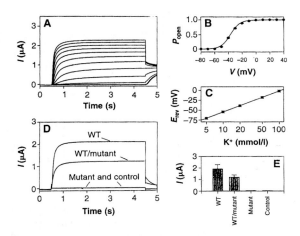

Fig. 8. Electrophysiological analysis of KCNQ2-type, non-inactivating, low-threshold K channel and its mutant in Xenopus oocytes. (A) Two-electrode voltage clamp traces of an oocyte expressing KCNQ2. From a holding potential of -80 mV, the oocyte was clamped for 4 s to values between -80 and $+40$ mV in steps of 10 mV, followed by a constant pulse to -30 mV. (B) Voltage dependence of open probability (p_{open}), as determined by tail current analysis: half-maximal p_{open} is at -37 ± 2 mV; the apparent gating charge is 3.7 ± 0.4 ($n = 12$, $\pm$SD). (C) The 53 mV per decade shift of reversal potential (E_{rev}) with changes in external K^+ ($n = 8$) indicates a channel predominantly selective for K^+. Substitution of external K^+ by other cations yields the following permeability ratios: $P_K/P_{Rb} = 1.27 \pm 0.01$; $P_K/P_{Cs} = 7.4 \pm 0.5$; and $P_K/P_{Na} = 51 \pm 4$ ($\pm$SEM, $n = 9$). (D) Current traces of WT KCNQ2, a 1:1 coinjection of wild-type and mutant KCNQ2 (WT/mutant), mutant KCNQ2 (with inactive channel), and mock-injected control oocytes; from a holding potential of -80 mV, the voltage was stepped to $+20$ mV for 4 s. Note, the traces for mutant KCNQ2 and control oocytes cannot be distinguished. Except for the last set of experiments, the same total amount of cRNA (5 ng) was injected into single oocytes. (E) Mean currents from several experiments as in (D): error bars indicate SEM with $n = 5$–10 (from Biervert et al., 1998).

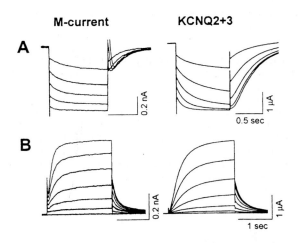

Fig. 9. Comparison of kinetic properties of native M-current in superior cervical ganglion (SCG) with currents in oocytes expressing KCNQ2 + KCNQ3 heteromultimers. (A) Current response to traditional M-current voltage-clamp protocol (see Figs. 3 and 4C): from holding potential of -30 mV, membrane was hyperpolarized in 10-mV increments, each for 1 s. Apparent differences between current waveforms are largely due to smaller leak current in oocytes. Initial phase of M-current reactivation in SCG neurons is obscured by transient A-current. (B) Activation of M-current and KCNQ2 + KCNQ3 channels from a holding potential of -60 mV in 5-mV increments. The native M-current was recorded from SCG neurons in intact, isolated ganglia and the KCNQ2 + KCNQ3 currents were recorded in Xenopus oocytes, both at room temperature (from Wang et al., 1998).

Nicotinic actions in cerebral cortex

The initial iontophoretic tests of ACh in the cortex failed to detect any consistent excitatory or inhibitory nicotinic actions (Krnjević and Phillis, 1963a,b). Nevertheless, the demonstration of numerous binding sites in the cortex of such a specific antagonist as α-bungarotoxin (Eterovic and Bennett, 1974) was strong evidence of the presence of nicotinic receptors: the most common subtypes seem to be α7 and α4β2 (Clarke, 1993). Interest in central nicotinic actions has been greatly stimulated by two major health concerns: one, the addictive, indeed lethal properties of nicotine (notably for smokers (Peto et al., 1996)); the second, senile dementias—on the one hand, because such dementias are associated with a selective loss of cortical nicotinic receptors (especially the α4 subtype (Court et al., 2001)) and, on the other, because nicotine has a pronounced neuroprotective action against NMDA receptor-induced excitotoxicity (Shimohama et al., 1996; Carlson et al., 1998; Prendergast et al., 2001).

The lack of clear actions of nicotine on most cortical neurons could be explained if the receptors did not form functional channels. But two obvious possibilities had not been excluded: as in the peripheral nervous system, many nAChRs could be situated on nerve terminals and thus influence ACh and other transmitter release; alternatively, nAChRs might be present on a population of neurons not commonly recorded by investigators. Both proved to be the case.

Presynaptic nicotinic actions

As already mentioned, activation of muscarinic receptors on cholinergic, glutamatergic and GABAergic terminals in cortex depresses the action-potential-induced release of ACh, glutamate and GABA. Pharmacological tests of nicotine agonists and antagonists revealed similar depressant nicotinic and muscarinic effects on GABAergic inhibition (Rovira et al., 1983). By contrast, nicotinic agonists potentiated glutamatergic transmission (Rovira et al., 1983), in good agreement with studies on peripheral junctions, where presynaptic nicotine receptors also enhance transmitter release (Ginsborg, 1971).

It thus seemed that nicotine exerts its main effects in the CNS via α42 receptors on nerve terminals (and perhaps the preterminal region of axons) (Gray et al., 1998; Wonnacott, 1997). But there was more to come.

Postsynaptic excitation of inhibitory interneurons

Nicotinic fast-decaying inward currents were first observed in hippocampal GABAergic neurons by

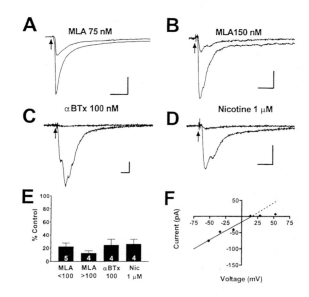

Fig. 10. Pharmacology of nicotinic responses observed in hippocampal inhibitory interneurons in presence of blockers of glutamate, GABA, serotonin and ATP. Superimposed are averaged EPSCs recorded before and during superfusion with nicotinic agonists and antagonists. Slices were stimulated electrically (arrows), and three to five successive responses were averaged before and during drug superfusion. (A) and (B) illustrate effects of two different concentrations of methyllicaconitine (MLA), whereas (C) shows effect of α-bungarotoxin (αBTx). (D) Functional antagonism induced by superfusion with nicotine. (E) Means ± SEM of EPSCs in presence of each agent as percentage of mean control EPSC amplitude; numbers of cells and drug concentrations (in nM) are indicated (< 100 represents combined results with 50 and 75 nM MLA; > 100 includes data obtained with 100 and 150 nM MLA). (F) I–V plot for EPSC: cell was clamped with Cs-gluconate/QX-314-filled electrode at indicated voltages for 10 s before synaptic stimulus to inactivate most voltage-dependent conductances; reversal potential was 15 mV. Calibrations: 25 ms/200 pA in A; 25 ms/50 pA in B–D (from Frazier et al., 1998a).

Jones and Yakel (1997) and Alkondon et al. (1997). These findings were soon confirmed by other groups (Frazier et al., 1998a,b; Ji and Dani, 2000). Frazier et al. (1998a) showed moreover that nicotinic EPSPs can be evoked by electrically stimulating the slice in the presence of blockers of other forms of synaptic transmission (Fig. 10). The relevant nAChRs, mainly of the $\alpha7$ subtype, are relatively sensitive to choline and thus may be subject to ongoing modulation by ambient choline—in keeping with an earlier suggestion by Krnjević and Reinhardt (1979). Nicotinic synapses may also be present on some neocortical pyramidal cells (Chu et al., 2000).

Nicotinic excitation of inhibitory cells perhaps should not have been surprising, in view of the long-established fact that spinal Renshaw cells (also inhibitory) are excited by nicotinic synapses (Eccles et al., 1954). Nevertheless, it seems curious that in the mammalian CNS, fast nicotinic excitations appear to be a prominent feature only of cholinergic inputs to inhibitory neurons.

Conclusions

By and large, the great abundance of information now available has tended to confirm the initial predictions that ACh's role in the cortex (and probably elsewhere in the CNS) is predominantly as a modulator rather than a fast synaptic transmitter, such as glutamate. This is obviously true of the relatively slow, G-protein mediated muscarinic actions, whether postsynaptic (on K and Ca channels, via M1 receptors) or presynaptic, regulating its own release or that of other transmitters, via M2 receptors. The general enhancement of activity produced by suppression of K conductances (as well as non-specific cationic inward currents and some striking disinhibitory actions) is probably a major element of an ascending arousal mechanism that facilitates thalamocortical interactions and is thus somehow involved in the sleep–waking cycle—though how the observed changes in cellular and synaptic activity relate to consciousness remains a matter of imaginative speculation. Equally intriguing are the 'newer' nicotinic actions: notably their beneficial effect, on the one hand on neuronal survival and, on the other on cognitive function. Can these be ascribed simply to the enhancement of

inhibitory mechanisms? No doubt, some of the answers to these questions will be unexpected, like much of our current knowledge.

References

Alkondon, M., Pereira, E.F., Barbosa, C.T. and Albuquerque, E.X. (1997) Neuronal nicotinic acetylcholine receptor activation modulates gamma-aminobutyric acid release from CA1 neurons of rat hippocampal slices. J. Pharmacol. Exp. Ther., 283: 1396–1411.

Ben-Ari, Y., Krnjević, K., Reinhart, W. and Ropert, N. (1981) Intracellular observations on the disinhibitory action of acetylcholine in the hippocampus. Neuroscience, 6: 2475–2484.

Biervert, C., Schroeder, B.C., Kubisch, C., Berkovic, S.F., Propping, P., Jentsch, T.J. and Steinlein, O.K. (1998) A potassium channel mutation in neonatal human epilepsy. Science, 279: 403–406.

Bremer, F. and Chatonnet, J. (1949) Acétylcholine et cortex cérébral. Arch. Int. Physiol., 57: 106–109.

Brocher, S., Artola, A. and Singer, W. (1992) Agonists of cholinergic and noradrenergic receptors facilitate synergistically the induction of long-term potentiation in slices of rat visual cortex. Brain Res., 573: 27–36.

Brown, D.A., Abogadie, F.C., Allen, T.G., Buckley, N.J., Caulfield, M.P., Delmas, P., Haley, J.E., Lamas, J.A. and Selyanko, A.A. (1997) Muscarinic mechanisms in nerve cells. Life Sci., 60: 1137–1144.

Brown, D.A. and Adams, P.R. (1980) Muscarinic suppression of a novel voltage-sensitive K^+ current in a vertebrate neurone. Nature, 283: 673–676.

Brown, D.A., Gähwiler, B.H., Griffith, W.H. and Halliwell, J.V. (1990) Membrane currents in hippocampal neurons. Prog. Brain Res., 83: 141–160.

Carlson, N.G., Bacchi, A., Rogers, S.W. and Gahring, L.C. (1998) Nicotine blocks TNF-alpha-mediated neuroprotection to NMDA by an alpha-bungarotoxin-sensitive pathway. J. Neurobiol., 35: 29–36.

Celesia, G.G. and Jasper, H.H. (1966) Acetylcholine released from cerebral cortex in relation to state of activation. Neurology, 16: 1053–1063.

Chu, Z.G., Zhou, F.M. and Hablitz, J.J. (2000) Nicotinic acetylcholine receptor-mediated synaptic potentials in rat neocortex. Brain Res., 887: 399–405.

Clarke, P.B. (1993) Nicotinic receptors in mammalian brain: localization and relation to cholinergic innervation. Prog. Brain Res., 98: 77–83.

Cole, A.E. and Nicoll, R.A. (1984) Characterization of a slow cholinergic post-synaptic potential recorded in vitro from rat hippocampal pyramidal. J. Physiol. (Lond.), 352: 173–188.

Collier, B. and Mitchell, J.F. (1967) The central release of acetylcholine during consciousness and after brain lesions. J. Physiol. (Lond.), 188: 83–98.

Constanti, A. and Sim, J.A. (1987) Calcium-dependent potassium conductance in guinea-pig olfactory cortex neurones *in vitro*. J. Physiol. (Lond.), 387: 173–194.

Costa, A.M. and Brown, B.S. (1997) Inhibition of M-current in cultured rat superior cervical ganglia by linopirdine: mechanism of action studies. Neuropharmacology, 36: 1747–1753.

Court, J., Martin-Ruiz, C., Piggott, M., Spurden, D., Griffiths, M. and Perry, E. (2001) Nicotinic receptor abnormalities in Alzheimer's disease. Biol. Psychiatry, 49: 175–184.

Dale, H.H. (1914) The action of certain esters and ethers of choline, and their relation to muscarine., J. Pharm. Exp. Ther., 6: 147–190.

Dale, H.H. (1938) Acetylcholine as chemical transmitter of the effects of nerve impulses. J. Mt. Sinai Hosp., 5: 401–429.

Dudar, J.D. (1977) The role of the septal nuclei in the release of acetylcholine from the rabbit cerebral cortex and dorsal hippocampus and the effect of atropine. Brain Res., 129: 237–246.

Eccles, J.C., Fatt, P. and Koketsu, K. (1954) Cholinergic and inhibitory synapses in a pathway from motor axon-collaterals to motorneurones. J. Physiol., 126: 524–562.

Eckhorn, R., Bauer, R., Jordan, W., Brosch, M., Kruse, W., Munk, M. and Reitboeck, H.J. (1988) Coherent oscillations: a mechanism of feature linking in the visual cortex? Multiple electrode and correlation analyses in the cat. Biol. Cybern., 60: 121–130.

Elliott, K.A.C., Swank, R.L. and Henderson, N. (1950) Effects of anaesthetics and convulsants on acetylcholine content of brain. Am. J. Physiol., 162: 469–474.

Eterovic, V.A. and Bennett, E.L. (1974) Nicotinic cholinergic receptor in brain detected by binding of alpha-(3 H) bungarotoxin. Biochim. Biophys. Acta, 362: 346–355.

Fraser, D.D. and MacVicar, B.A. (1996) Cholinergic-dependent plateau potential in hippocampal CA1 pyramidal neurons. J. Neurosci., 16: 4113–4128.

Frazier, C.J., Buhler, A.V., Weiner, J.L. and Dunwiddie, T.V. (1998a) Synaptic potentials mediated via alpha-bungarotoxin-sensitive nicotinic acetylcholine receptors in rat hippocampal interneurons. J. Neurosci., 18: 8228–8235.

Frazier, C.J., Rollins, Y.D., Breese, C.R., Leonard, S., Freedman, R. and Dunwiddie, T.V. (1998b) Acetylcholine activates an alpha-bungarotoxin-sensitive nicotinic current in rat hippocampal interneurons, but not pyramidal cells. J. Neurosci., 18: 1187–1195.

Freund, T.F. and Antal, M. (1988) GABA-containing neurons in the septum control inhibitory interneurons in the hippocampus. Nature, 336: 170–173.

Ginsborg, B.L. (1971) On the presynaptic acetylcholine receptors in sympathetic ganglia of the frog. J. Physiol., 216: 237–246.

Glavinovic, M., Ropert, N., Krnjević, K. and Collier, B. (1983) Hemicholinium impairs septo-hippocampal facilitatory action. Neuroscience, 9: 319–330.

Gray, R., Rajan, A.S., Radcliffe, K.A., Yakehiro, M. and Dani, J.A. (1996) Hippocampal synaptic transmission enhanced by low concentrations of nicotine. Nature, 383: 713–716.

Gu, Q. and Singer, W. (1993) Effects of intracortical infusion of anticholinergic drugs on neuronal plasticity in kitten striate cortex. Eur. J. Neurosci., 5: 475–485.

Haj-Dahmane, S. and Andrade, R. (1996) Muscarinic activation of a voltage-dependent cation nonselective current in rat association cortex. J. Neurosci., 16: 3848–3861.

Hajos, N., Papp, E.C., Acsady, L., Levey, A.I. and Freund, T.F (1998) Distinct interneuron types express m2 muscarinic receptor immunoreactivity on their dendrites or axon terminals in the hippocampus. Neuroscience, 82: 355–376.

Haley, J.E., Delmas, P., Offermanns, S., Abogadie, F.C., Simon, M.I., Buckley, N.J. and Brown, D.A.. (2000) Muscarinic inhibition of calcium current and M current in Galpha q-deficient mice. J. Neurosci., 20: 3973–3979.

Halliwell, J.V. and Adams, P.R. (1982) Voltage-clamp analysis of muscarinic excitation in hippocampal neurons. Brain Res., 250: 71–92.

Hammer, R., Berrie, C.P., Birdsall, N.J., Burgen, A.S. and Hulme, E.C. (1980) Pirenzepine distinguishes between different subclasses of muscarinic receptors. Nature, 283: 90–92.

Hasselmo, M.E. and Barkai, E. (1995) Cholinergic modulation of activity-dependent synaptic plasticity in the piriform cortex and associative memory function in a network biophysical simulation. J. Neurosci., 15: 6592–6604.

Hasselmo, M.E. and Schnell, E (1994) Laminar selectivity of the cholinergic suppression of synaptic transmission in rat hippocampal region CA1: computational modeling and brain slice physiology. J. Neurosci., 14: 3898–3914.

Hounsgaard, J. (1978) Presynaptic inhibitory action of acetylcholine in area CA1 of the hippocampus. Exp. Neurol., 62: 787–797.

Hunt, R. and Taveau, R. de M. (1906) On the physiological action of cetain cholin derivatives and new methods for detecting cholin. Br. Med. J., 2: 1788–1791.

Ji, D. and Dani, J.A. (2000) Inhibition and disinhibition of pyramidal neurons by activation of nicotinic receptors on hippocampal interneurons. J. Neurophysiol., 83: 2682–2690.

Jones, S. and Yakel, J.L. (1997) Functional nicotinic ACh receptors on interneurones in the rat hippocampus. J. Physiol. (Lond.), 504: 603–610.

Klink, R. and Alonso, A. (1997) Ionic mechanisms of muscarinic depolarization in entorhinal cortex layer II neurons. J. Neurophysiol., 77: 1829–1843.

Köhler, C., Chan-Palay, V. and Wu, J.Y. (1984) Septal neurons containing glutamic acid decarboxylase immunoreactivity project to the hippocampal region in the rat brain. Anat. Embryol. (Berl.), 169: 41–44.

Krnjević, K. (1969) Central cholinergic pathways. Fed. Proc., 28: 113–120.

Krnjević, K. and Phillis, J.W. (1963a) Acetylcholine-sensitive cells in the cerebral cortex. J. Physiol. (Lond.), 166: 296–327.

Krnjević, K. and Phillis, J.W. (1963b) Pharmacological properties of acetylcholine-sensitive cells in the cerebral cortex. J. Physiol. (Lond.), 166: 328–350.

Krnjević, K., Pumain, R. and Renaud, L. (1971) The mechanism of excitation by acetylcholine in the cerebral cortex. J. Physiol. (Lond.), 215: 247–268.

Krnjević, K., Reiffenstein, R.J. and Ropert, N. (1981) Disinhibitory action of acetylcholine in the rat's hippocampus: extracellular observations. Neuroscience, 6: 2465–2474.

Krnjević, K. and Reinhardt, W. (1979) Choline excites cortical neurons. Science, 206: 1321–1323.

Krnjević, K. and Ropert, N. (1981) Septo-hippocampal pathway modulates hippocampal activity by a cholinergic mechanism. Can. J. Physiol. Pharmacol., 59: 911–914.

Krnjević, K. and Silver, A. (1965) A histochemical study of cholinergic fibres in the cerebral cortex. J. Anat., 99: 711–759.

Krnjević, K. and Silver, A. (1966) Acetylcholinesterase in the developing forebrain. J. Anat., 100: 63–89.

Kuzmiski, J.B. and MacVicar, B.A. (2001) Cyclic nucleotide-gated channels contribute to the cholinergic plateau potential in hippocampal CA1 pyramidal neurons. J. Neurosci., 21: 8707–8714.

Lamas, J.A., Selyanko, A.A. and Brown, D.A. (1997) Effects of a cognition-enhancer, linopirdine (DuP 996), on M-type potassium currents (IK(M)) and some other voltage- and ligand-gated membrane currents in rat sympathetic neurons. Eur. J. Neurosci., 9: 605–616.

Levey, A.I., Kitt, C.A., Simonds, W.F., Price, D.L. and Brann, M.R. (1991) Identification and localization of muscarinic acetylcholine receptor proteins in brain with subtype-specific antibodies. J. Neurosci., 11: 3218–3226.

MacIntosh, F.C. and Oborin, P.E. (1953) Release of acetylcholine from intact cerebral cortex. Abstr. XIX Int. Physiol. Congr. pp. 580–581.

MacVicar, B.A. and Tse, F.W. (1989) Local neuronal circuitry underlying cholinergic rhythmical slow activity in CA3 area of rat hippocampal slices. J. Physiol. (Lond.), 417: 197–212.

Madison, D.V., Lancaster, B. and Nicoll, R.A. (1987) Voltage clamp analysis of cholinergic action in the hippocampus. J. Neurosci., 7: 733–741.

Maloney, K.J., Cape, E.G., Gotman, J. and Jones, B.E. (1997) High-frequency gamma electroencephalogram activity in association with sleep–wake states and spontaneous behaviors in the rat. Neuroscience, 76: 541–555.

Meves, H., Schwarz, J.R. and Wulfsen, I. (1999) Separation of M-like current and ERG current in NG108–15 cells. Br. J. Pharmacol., 127: 1213–1223.

Mitchell, J.F. (1963) The spontaneous and evoked release of acetylcholine from the cerebral cortex. J. Physiol. (Lond.), 105: 98–116.

Nakajima, Y., Nakajima, S., Leonard, R.J. and Yamaguchi, K. (1986). Acetylcholine raises excitability by inhibiting the past transient potassium current in cultured hippocampal neurons. Proc. Natl. Acad. Sci. USA, 83: 3022–3026.

Nachmansohn, D. and Machado, A.L. (1943) The formation of acetylcholine. A new enzyme: 'choline acetylase'. J. Neurophysiol., 6: 397–403.

Pan, Z., Selyanko, A.A., Hadley, J.K., Brown, D.A, Dixon, J.E. and McKinnon, D. (2001) Alternative splicing of KCNQ2 potassium channel transcripts contributes to the functional diversity of M-currents. J. Physiol. (Lond.), 531: 347–358.

Petersen, R.C. and Kanow, C. (2001) Mild cognitive impairment—State of the art 2001. Rev. Neurol. (Paris), 157(Suppl. 10): 29.

Peto, R., Lopez, A.D., Boreham, J., Thun, M., Heath, C., Jr. and Doll, R. (1996) Mortality from smoking worldwide. Br. Med. Bull., 52: 12–21.

Prendergast, M.A., Harris, B.R., Mayer, S., Holley, R.C., Hauser, K.F. and Littleton, J.M. (2001) Chronic nicotine exposure reduces N-methyl-D-aspartate receptor-mediated damage in the hippocampus without altering calcium accumulation or extrusion: evidence of calbindin-D28K overexpression. Neuroscience, 102: 75–85.

Quastel, J.H., Tennenbaum, M. and Wheatley, A.H.M. (1936) Choline ester formation in, and choline esterase activities of tissues in vitro. Biochem. J., 30: 1668–1681.

Rasmusson, D.D. (2000) The role of acetylcholine in cortical synaptic plasticity. Behav. Brain Res., 115: 205–218.

Rogan, S. and Lippa, C.F. (2002) Alzheimer's disease and other dementias: a review. Am. J. Alzheimers Dis. Other Demen., 17: 11–17.

Rovira, C., Ben-Ari, Y., Cherubini, E., Krnjević, K. and Ropert, N. (1983) Pharmacology of the dendritic action of acetylcholine and further observations on the somatic disinhibition in the rat hippocampus in situ. Neuroscience, 8: 97–106.

Segal, M. and Auerbach, J.M. (1997) Muscarinic receptors involved in hippocampal plasticity. Life Sci., 60: 1085–1091.

Selyanko, A.A. and Brown, D.A. (1996) Intracellular calcium directly inhibits potassium M channels in excised membrane patches from rat sympathetic neurons. Neuron, 16: 151–162.

Selyanko, A.A., Delmas, P., Hadley, J.K., Tatulian, L., Wood, I.C., Mistry, M., London, B. and Brown, D.A. (2002) Dominant-negative subunits reveal potassium channel families that contribute to M-like potassium currents. J. Neurosci., 22 RC212: 1–5.

Selyanko, A.A., Hadley, J.K. and Brown, D.A. (2001) Properties of single M-type CNQ2/KCNQ3 potassium channels expressed in mammalian cells. J. Physiol. (Lond.), 534: 15–24.

Selyanko, A.A., Hadley, J.K., Wood, I.C., Abogadie, F.C., Delmas, P., Buckley, N.J., London, B. and Brown, D.A. (1999) Two types of K^+ channel subunit, Erg1 and KCNQ2/3, contribute to the M-like current in a mammalian neuronal cell. J. Neurosci., 19: 7742–7756.

Shapiro, M.S., Gomeza, J., Hamilton, S.E., Hille, B., Loose, M.D., Nathanson, N.M., Roche, J.P. and Wess, J. (2001) Identification of subtypes of muscarinic receptors that regulate Ca^{2+} and K^+ channel activity in sympathetic neurons. Life Sci., 68: 2481–2487.

Shapiro, M.S., Roche, J.P., Kaftan, E.J., Cruzblanca, H., Mackie, K. and Hille, B. (2000) Reconstitution of muscarinic modulation of the KCNQ2/KCNQ3 $K^{(+)}$ channels that underlie the neuronal M current. J. Neurosci., 20: 1710–1721.

Shimohama, S., Akaike, A. and Kimura, J. (1996) Nicotine-induced protection against glutamate cytotoxicity. Nicotinic cholinergic receptor-mediated inhibition of nitric oxide formation. Ann. N.Y. Acad. Sci., 777: 356–361.

Shute, C.C. and Lewis, P.R. (1963) Cholinesterase-containing systems of the brain of the rat. Nature, 199: 1160–1164.

Shute, C.C. and Lewis, P.R. (1967) The ascending cholinergic reticular system: neocortical, olfactory and subcortical projections. Brain, 90: 497–520.

Sillito, A.M. and Kemp, J.A. (1983) Cholinergic modulation of the functional organization of the cat visual cortex. Brain Res., 289: 143–155.

Singer, W. (2001) Consciousness and the binding problem. Ann. N.Y. Acad. Sci., 929: 123–146.

Singh, N.A., Charlier, C., Stauffer, D., DuPont, B.R., Leach, R.J., Melis, R., Ronen, G.M., Bjerre, I., Quattlebaum, T., Murphy, J.V., McHarg, M.L., Gagnon, D., Rosales, T.O., Peiffer, A., Anderson, V.E., Leppert, M. (1998) A novel potassium channel gene, KCNQ2 is mutated in an inherited epilepsy of newborns. Nat. Genet., 18: 25–29.

Spehlmann, R. (1963) Acetylcholine and prostigmine electrophoresis at visual cortex neurons. *J. Neurophysiol.*, 26: 127–139.

Stansfeld, C.E., Roper, J., Ludwig, J., Weseloh, R.M., Marsh, S.J., Brown, D.A. and Pongs, O. (1996) Elevation of intracellular calcium by muscarinic receptor activation induces a block of voltage-activated rat ether-a-go-go channels in a stably transfected cell line. Proc. Natl. Acad. Sci. USA, 93: 9910–9914.

Stumpf, C. (1965) Drug action on the electrical activity of the hippocampus. Int. Rev. Neurobiol., 8: 77–138.

Toselli, M., Lang, J., Costa, T. and Lux, H.D. (1989) Direct modulation of voltage-dependent calcium channels by muscarinic activation of a pertussis toxin-sensitive G-protein in hippocampal neurons. Pflügers Arch., 415: 255–261.

Toth, K., Freund, T.F. and Miles, R. (1997) Disinhibition of rat hippocampal pyramidal cells by GABAergic afferents from the septum. J. Physiol. (Lond.), 500: 463–474.

Vanderwolf, C.H. and Stewart, D.J. (1986) Joint cholinergic-serotonergic control of neocortical and hippocampal electrical activity in relation to behavior: effects of scopolamine, ditran, trifluoperazine and amphetamine. Physiol. Behav., 38: 57–65.Spehlmann, R. (1963) Acetylcholine and prostigmine electrophoresis at visual cortex neurons. J.Neurophysiol., 26: 127–139.

Wang, H.S., Pan, Z., Shi, W., Brown, B.S., Wymore, R.S., Cohen, I.S., Dixon, J.E. and McKinnon, D. (1998) KCNQ2 and KCNQ3 potassium channel subunits: molecular correlates of the M-channel. Science, 282: 1890–1893.

Wonnacott, S. (1997) Presynaptic nicotinic ACh receptors. Trends Neurosci., 20: 92–98.

Yang, W.P., Levesque, P.C., Little, W.A., Conder, M.L., Ramakrishnan, P., Neubauer, M.G. and Blanar, M.A. (1998) Functional expression of two KvLQT1-related potassium channels responsible for an inherited idiopathic epilepsy. J. Biol. Chem., 273: 19419–19423.

Zhang, W., Basile, A.S., Gomeza, J., Volpicelli, L.A., Levey, A.I. and Wess, J. (2002) Characterization of central inhibitory muscarinic autoreceptors by the use of muscarinic acetylcholine receptor knock-out mice. J. Neurosci., 22: 1709–1717.

Progress in Brain Research, Vol. 145
ISSN 0079-6123

CHAPTER 6

Functional and molecular characterization of neuronal nicotinic ACh receptors in rat hippocampal interneurons

Jerrel L. Yakel* and Zuoyi Shao

Laboratory of Signal Transduction, National Institute of Environmental Health Sciences, National Institutes of Health, F2-08, 111 T.W. Alexander Drive, P.O. Box 12233, Research Triangle Park, NC 27709, USA

Introduction

Neuronal nicotinic acetylcholine receptors (nAChRs) are ligand-gated ion channels that are widely expressed in the central and peripheral nervous system, where they can mediate fast excitatory synaptic transmission. These receptor channels belong to the superfamily of ligand-gated ion channels, that also includes the serotonin 5-HT$_3$, GABA$_A$, and glycine receptors. The nAChRs are involved in a variety of physiological processes, including cognition and development (Jones et al., 1999). In addition, dysfunctions in nAChRs may be a causative factor in a variety of neurodegenerative diseases, including Alzheimer's disease (AD), Parkinson's disease, epilepsy, schizophrenia, and in aging (Nordberg, 1994; Steinlein et al., 1995; Freedman et al., 1997; Jones et al., 1999; Paterson and Nordberg, 2000; Court et al., 2001; Dani, 2001). In the brain, nAChRs are known to function both at presynaptic (e.g., in the regulation of neurotransmitter release) and postsynaptic sites (Wonnacott, 1997; Jones et al., 1999). Currently, there are at least 11 different nAChR subunits known to be expressed in the rat nervous system; nine of these ($\alpha2$–$\alpha7$ and $\beta2$–$\beta4$) are expressed in the adult rat CNS (McGehee and Role, 1995; Boyd, 1997; Elgoyhen et al., 2001). Like

other ligand-gated ion channels, the subunit composition of nAChRs determines their functional properties. A variety of functional and molecular studies have suggested that the two major subtypes of nAChRs found in the hippocampus are composed of $\alpha7$ (the majority of the α-bungarotoxin binding sites) and $\alpha4\beta2$ (the high affinity nicotine binding sites) subunits (Alkondon and Albuquerque, 1993; Alkondon et al., 1997; Jones and Yakel, 1997; Frazier et al., 1998; McQuiston and Madison, 1999; Ji and Dani, 2000). However as discussed below, the molecular makeup of nAChRs in the brain is most likely to be much more complicated than this relatively simple picture. For example, when comparing the properties of native nAChRs from hippocampal recordings with those of heterologously expressed channels formed from $\alpha7$ and $\alpha4\beta2$ subunits, it is clear that the composition of these native receptors is more complex (McQuiston and Madison, 1999; Sudweeks and Yakel, 2000). To enhance our understanding of the biophysical and pharmacological properties of functional nAChRs, we must understand which of the many different nAChR subunits are expressed in particular regions of interest in the brain, and understand how these subunits assemble to form functional channels.

For decades, it has been known that nicotine enhances cognition (Levin, 1992; Changeux et al., 1998). Substantial evidence suggests that nAChRs are involved in specific cognitive functions in humans and

---

*Corresponding author: Tel.: (919) 541-1407;
Fax: (919) 541-1898; E-mail: yakel@niehs.nih.gov

DOI: 10.1016/S0079-6123(03)45006-1

animal models, such as attention and performance in working and associative memory (Levin and Simon, 1998). Furthermore, dysfunctions in nAChR function may be a factor in a variety of neurodegenerative diseases. For example, in AD, the cholinergic pathways that innervate the hippocampus and cortex undergo substantial degeneration, and the number of nAChRs decreases (Perry et al., 1995; Auld et al., 1998; Francis et al., 1999; Selkoe, 1999; Paterson and Nordberg, 2000; Court et al., 2001). In AD, the extensive accumulation of the β-amyloid peptide ($A\beta_{1-42}$) has been proposed to lead to the progressive loss of cognitive function (Kuo et al., 1996). Interestingly, $A\beta_{1-42}$ has recently been shown to either inhibit nAChR function in rat hippocampal neurons (Liu et al., 2001; Pettit et al., 2001), or to activate (in particular the $\alpha7$ subtype) nAChR function for receptors expressed in *Xenopus* oocytes (Dineley et al., 2002). Lastly, nicotine enhances cognitive function in some AD patients (Nordberg, 1994). Thus, a clearer understanding of the physiological role of nAChRs in the brain regions relevant to cognitive function and neurodegenerative disease is essential, in particular if nAChRs are to become useful therapeutic targets in treating these and possibly other diseases.

In the rat hippocampus, local inhibitory interneurons and principal excitatory neurons receive cholinergic input from the medial septum-diagonal band (MSDB) complex of the basal forebrain (Frotscher and Léránth, 1985; Woolf, 1991). Intrinsic cholinergic interneurons have also been described within the hippocampus (Matthews et al., 1987; Cobb et al., 1999). Hippocampal GABAergic interneurons can coordinate the activity of large numbers of principal cells, and therefore have an important role in the regulation of hippocampal excitability and processing (Jones et al., 1999). Interneurons are thought to be responsible for regulating hippocampal oscillations such as rhythmical slow activity (RSA, or theta rhythm), an important component of the hippocampal EEG. It has been suggested that septal cholinergic and GABAergic projections, fire in rhythmic bursts to entrain hippocampal interneurons, thereby inducing rhythmic inhibition of pyramidal cells (Stewart and Fox, 1990). Intrinsic cholinergic interneurons have also been shown to pattern network activity via nAChRs in rat hippocampal slices (Cobb et al., 1999), and

nAChR activation increases theta activity in freely moving rabbits (Yamamoto, 1998).

Thus, the activation of nAChRs via cholinergic input activates inhibitory interneurons, which in turn contributes to the synchronization of the principal cell firing. These mechanisms are very likely important in understanding the link between the activation of these receptors and cognition. However, as mentioned earlier, the precise role, and even the molecular makeup, of the nAChRs in the hippocampus is not fully known. This is critical information in the understanding of how this class of receptors plays its physiological roles, and also how it may participate in neurodegenerative disease.

Selective expression of functional nAChRs on rat hippocampal interneurons

In the past few years, several groups including our own, have reported that rat hippocampal interneurons selectively express functional nAChRs (Alkondon et al., 1997; Jones and Yakel, 1997; Frazier et al., 1998; McQuiston and Madison, 1999; Ji and Dani, 2000). When recording from neurons in acute rat hippocampal slices (using whole-cell patch-clamp recording techniques), the rapid application of ACh induces a rapid excitatory nAChR-mediated inward current response in interneurons from the CA1 stratum radiatum and dentate gyrus (Fig. 1A and B). For most recordings, the bath solution contained the muscarinic AChR antagonist atropine (10 μM), the glutamate receptor antagonists CNQX and APV (both at 10 μM), the GABA$_A$ receptor antagonists bicuculline (10 μM) or picrotoxin (50 μM), and TTX (1 μM). The application of ACh, however, does not activate responses in the principal excitatory cells (e.g., pyramidal cells in the CA1 or granule cells in the dentate gyrus) (Fig. 1C and D). In these interneurons, the nAChR-mediated responses are partially blocked ($\sim$60–70%) by the $\alpha7$-selective antagonists MLA and α-bungarotoxin (Fig. 2A and B), suggesting that the $\alpha7$ nAChR subunit forms functional nAChRs in rat hippocampal interneurons. The incomplete block by MLA, and the fact that antagonists that block non-$\alpha7$ nAChRs (e.g., DHβE and mecamylamine) also block the nAChR-mediated responses in these interneurons

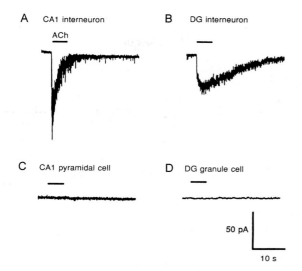

A CA1 interneuron

ACh

B DG interneuron

C CA1 pyramidal cell

D DG granule cell

50 pA

10 s

Fig. 1. Effect of ACh on rat hippocampal neurons. The rapid application of ACh (100 μM for 5 s; indicated by horizontal bars) to neurons held at −70 mV induced inward current responses in a CA1 stratum radiatum interneuron (A) and in a dentate gyrus (DG) interneuron (B), but not in a CA1 pyramidal neuron (C) nor a DG granule cell (D). The bath solution contained atropine (10 μM; muscarinic receptor antagonist), CNQX and APV (both at 10 μM; glutamate receptor antagonists), bicuculline (10 μM; GABA$_A$ receptor antagonist), and TTX (1 μM). Reprinted with permission from Jones and Yakel (1997).

(Fig. 2C), suggests that these interneurons also express non-α7 nAChRs (Alkondon and Albuquerque, 1993; Alkondon et al., 1997; Jones and Yakel, 1997; McQuiston and Madison, 1999; Ji and Dani, 2000; Alkondon and Albuquerque, 2001). The composition of these receptors has been difficult to establish with certainty, due to a lack of pharmacological agents that are specific for most of the nAChR subunits other than α7. However, as discussed earlier, the most likely composition of these non-α7 subtypes has previously been proposed to be α4β2. In addition, heterogeneity in the functional properties of these non-α7 subtypes is apparent when recording from different types of interneurons in different layers of the hippocampal CA1 region (McQuiston and Madison, 1999). To further understand the molecular makeup of nAChRs in these hippocampal interneurons, we have utilized single-cell reverse transcription polymerase chain reaction (RT-PCR) techniques to examine the subunit mRNA expression pattern within individual neurons.

Molecular characterization using single-cell RT-PCR analysis

We combined patch-clamp recording with single-cell RT-PCR analysis to study the molecular and functional properties of nAChRs in rat hippocampal CA1 interneurons. In this way, we could detect subunits that we might not be able to detect in individual living cells from which we recorded, and also to make comparisons between subunit detection and functional properties. It should be kept in mind that the single-cell RT-PCR technique characterizes the subunit mRNA expression pattern, rather than the proteins directly. Although the mRNA expression is not always directly related to functional protein expression, we previously reported a strong correlation between the data obtained by these two techniques, as will be discussed below. Since McQuiston and Madison (1999) had previously shown that the pharmacological properties of the nAChRs in interneurons from the stratum oriens were different from those in the stratum radiatum, we examined interneurons from these two layers. We found that stratum radiatum interneurons expressed (in greater than 20% of cells) the α4, α5, α7 and β2–β4 subunits, and that interneurons from the stratum oriens expressed the α2, α3, α4, α7, β2 and β3 subunits (Fig. 3). As expected, the α4, α7 and β2 subunits were found in both interneuronal types, consistent with the idea of α7 and α4β2 being the major subtypes of functional nAChR channels in these neurons. However, other subunits are likely to participate in forming functional nAChRs, since the properties of the channels in the hippocampus do not precisely match those expected of either homomeric α7 channels or α4β2 channels (McQuiston and Madison, 1999; Sudweeks and Yakel, 2000). Our PCR data suggested the intriguing possibilities that the α5 nAChR subunit might be participating in forming functional channels (most likely with the α4 and β2 subunits) in the stratum radiatum interneurons, and that the α2 nAChR subunit might be forming functional channels in the stratum oriens interneurons. Furthermore, when we compared the functional properties (e.g., kinetics of activation and inactivation) and coexpression patterns of the various subunits, we found that there were correlations between subunit expression patterns and kinetics of

98

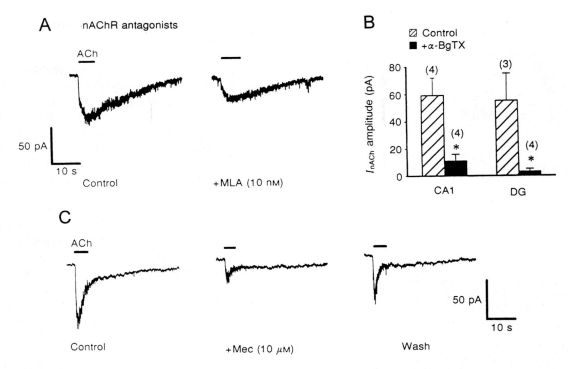

Fig. 2. Block of ACh-induced responses by nAChR antagonists. In CA1 interneurons, the bath application of the α7-selective antagonists methyllycaconitine (MLA, 10 nM) (A) or α-bungarotoxin (α-BgTX, 100 nM, >10 min incubation) (B) blocked the responses, as did the bath application of the broad-spectrum nAChR antagonist mecamylamine (Mec, 10 μM) (C). Reprinted with permission from Jones and Yakel (1997).

the responses. As expected, expression of the α7 subunit was correlated with fast-activating responses, and the α2 and α4 subunits were correlated with medium- and slow-activating responses, respectively. Furthermore, we found evidence to suggest that the β2 subunit might be assembling with the α7 subunit to form a novel type of heteromeric α7-containing nAChR channel, an observation that was recently confirmed (see below) by the expression of recombinant α7β2 nAChRs in heterologous expression systems (Khiroug et al., 2002). These various points will be discussed in more detail below.

The nAChR α2 subunit in the stratum oriens interneurons

Although Wada et al. (1989) and us detected the α2, α3, and α4 subunits in the stratum oriens interneurons, the α2 subunit is likely to be contributing to the slow, non-α7 nAChR-mediated responses in these

neurons. McQuiston and Madison (1999) had previously shown that these responses were insensitive to block by α-CTX MII (a blocker of α3-containing nAChRs), and had a relatively low sensitivity to block by DHβE (which potently blocks α4-containing nAChRs). These pharmacological data, combined with the presence of the α2 subunit and the functional correlation with slower responses in the stratum oriens interneurons (Sudweeks and Yakel, 2000), together, provide evidence that the α2 subunit is likely to be a major functional contributor to the non-α7 nAChR-mediated responses in these stratum oriens interneurons.

The nAChR α5 subunit in the stratum radiatum

In the stratum radiatum, we detected the α5, β3, and β4 subunits (in addition to the α4, α7 and β2 subunits). While the functional implications of the β3 and β4

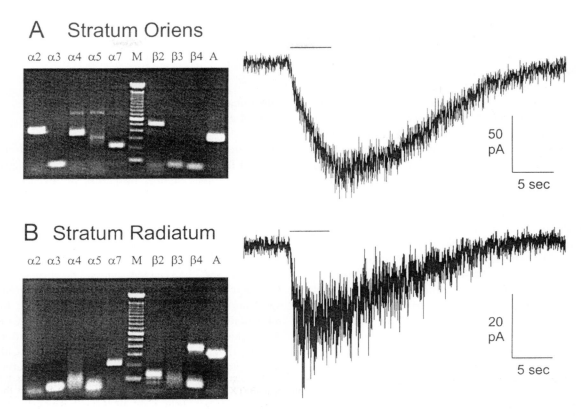

Fig. 3. Detection of nAChR subunits using single-cell RT-PCR and functional properties. In a stratum oriens interneuron (A) and a stratum radiatum interneuron (B), various RT-PCR products for specific nAChR subunits can be seen from the gels (left), with corresponding responses to ACh (100 μM; right). The 'M' denotes the marker lane for use in identifying the sizes of the RT-PCR products, and 'A' signifies the β-actin control. Reprinted with permission from Sudweeks and Yakel (2000).

subunits are presently unclear, the detection of the α5 subunit mRNA is interesting for many reasons. Recent work from the Role laboratory has clearly shown that the chick α5 nAChR subunit forms functional channels, both in native tissue (chick sympathetic neurons) and in heterologous expression systems, along with the chick α4 and β2 subunits (Ramirez-Latorre et al., 1996; Yu and Role, 1998b; Girod et al., 1999). The most notable alteration in functional properties due to the coexpression with the α5 subunit was a significantly larger (nearly double) single channel conductance value, as compared to α4β2 channels (Ramirez-Latorre et al., 1996). Interestingly, as discussed later, we have identified a non-α7 single channel conductance in excised patches from stratum radiatum neurons, with a conductance similar to that expected for α5-containing nAChRs.

Lastly, it was reported that the α5 subunit was coexpressed with the α4 and β2 subunits in a specific subpopulation of interneurons in the neocortex (Porter et al., 1999), and that the α4, α5 and β2 subunits have been coimmunoprecipitated from chick brain (Conroy and Berg, 1998). Therefore, in total, these data suggest that the α5 subunit is likely to be a major functional contributor to the non-α7 nAChR-mediated responses in the stratum radiatum interneurons.

The α7-containing nAChRs in the hippocampus; possible coassembly with the β2 subunit

It had been previously thought that mammalian α7 subunits mostly form homomeric receptors, as no direct evidence for coassembly of any other subunit with the $\alpha7$ subunit had been reported (McGehee and Role, 1995; Chen and Patrick, 1997; Drisdel and Green, 2000). However, multiple functional subtypes of $\alpha7$-containing nAChRs have been reported in rat (Cuevas and Berg, 1998; Cuevas et al., 2000), suggesting the possibility for heteromeric $\alpha7$-containing nAChRs. In the chick, the properties of native $\alpha7$ nAChRs often do not match those of heterologously expressed homomeric $\alpha7$ nAChRs, and $\alpha7$-containing heteromeric chick nAChRs can be formed in heterologous expression systems (Anand et al., 1993; Yu and Role, 1998a; Girod et al., 1999; Palma et al., 1999). While a major proportion of functional nAChRs in rat hippocampal interneurons most likely contain the $\alpha7$ subunit, the properties of these $\alpha7$-containing receptors are not identical to those expected of homomeric $\alpha7$ receptors. In particular, the native $\alpha7$-containing receptors in the hippocampus desensitize more slowly and have a smaller single channel conductance (discussed below) in comparison to the properties of recombinant heterologously expressed receptors (Shao and Yakel, 2000; Sudweeks and Yakel, 2000). This suggests that $\alpha7$-containing nAChRs expressed in rat hippocampal interneurons are not homomeric assemblies of $\alpha7$ subunits.

We previously found that the $\alpha7$ and $\beta2$ subunits were coexpressed in individual rat hippocampal interneurons and were both correlated with fast-activating responses (Sudweeks and Yakel, 2000). Thus, we hypothesized that the $\alpha7$ and $\beta2$ subunits might be coassembling to form a novel type of heteromeric nAChR in these interneurons. To test this, the $\alpha7$ and $\beta2$ subunits were coexpressed in *Xenopus* oocytes and human embryonic kidney (TSA120) cells. Coexpression of the $\beta2$ subunit with the $\alpha7$ subunit significantly slowed the rate of nAChR desensitization and altered pharmacological properties (Fig. 4; Khiroug et al., 2002). Whereas ACh, carbachol, and choline were full or near-full agonists for homomeric $\alpha7$ receptor channels, both carbachol and choline were only partial agonists in oocytes expressing both $\alpha7$ and $\beta2$ subunits. In addition, the EC_{50} values for all three agonists significantly increased when the $\beta2$ subunit was coexpressed with the $\alpha7$ subunit. Furthermore, these two subunits were coimmunoprecipitated from transiently transfected TSA120 cells. These data provide direct evidence that the nAChR $\alpha7$ and $\beta2$ subunits coassemble to form a

functional heteromeric nAChR, which may further help to explain nAChR channel diversity in rat hippocampal interneurons.

Functional characterization using single channel analysis

Since the molecular makeup of nAChRs can be inferred from their functional properties, we investigated the single channel properties of nAChRs in outside-out patches from CA1 stratum radiatum interneurons in the slice. ACh induced the opening of two distinct types of nAChR single channel currents, a 38 pS channel and a 62 pS channel (Fig. 5). MLA, the $\alpha7$-selective antagonist, dramatically reduced the opening of the 38 pS channel, but had no effect on the 62 pS channel. In contrast, DHβE, the broad spectrum nAChR antagonist with a preference for $\alpha4$-containing nAChRs, blocked the opening of the 62 pS channel but not the 38 pS channel. These data suggest that at least two different types of nAChR single channel currents were at play in stratum radiatum interneurons; an $\alpha7$-containing receptor with a single channel conductance of 38 pS, and a non-$\alpha7$ receptor with a single channel conductance of 62 pS.

What might the molecular makeup of these channels be? We can obtain clues from the biophysical, pharmacological, and molecular properties of these channels. For example, as mentioned earlier, the properties of the $\alpha7$-containing nAChRs do not match those expected from homomeric $\alpha7$ nAChRs, mostly due to the slow rate of desensitization and the single channel conductance value. Although we have suggested that the $\alpha7$ and $\beta2$ subunits might be coassembling in the hippocampal interneurons, we do not yet know the single channel properties (in particular the single channel conductance) of these $\alpha7\beta2$ channels. For the non-$\alpha7$/62 pS channel, the conductance value of this channel is not consistent with an $\alpha4\beta2$ channel alone, but is consistent with a channel containing the $\alpha5$ nAChR subunit, along with the $\alpha4$ and $\beta2$ subunits (see above). Current work in the laboratory is further attempting to address more precisely the molecular makeup of these channels. In addition, as the properties of the nAChRs in the stratum oriens

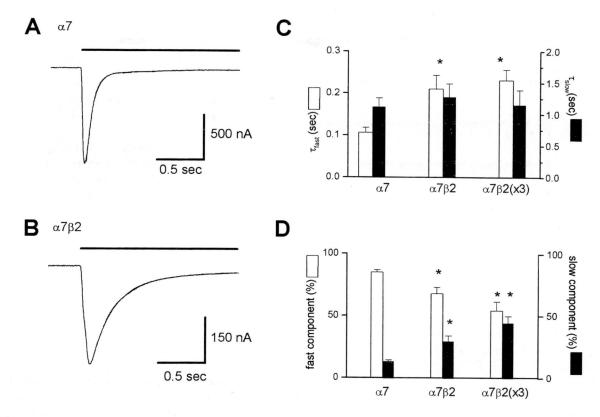

Fig. 4. Coexpression of the $\beta2$ subunit with the $\alpha7$ subunit alters the kinetics of desensitization. The rapid application of carbachol (1 mM; indicated by horizontal bar) induced a rapid and fast desensitizing response in oocytes expressing homomeric $\alpha7$ nAChRs (A). The coexpression of the $\beta2$ with the $\alpha7$ subunit slowed the rate of desensitization (B). The response kinetics were biphasic, and the relative values for the different fits are shown in C and D. The rate of the fast decay phase, and the relative fractions of both phases, were significantly different with the coexpression of the $\beta2$ subunit; the rate of the slow decay phase was not significantly different. Injecting a three times higher amount of $\beta2$ subunit RNA (i.e., $\alpha7\beta2(\times 3)$ channels) with the $\alpha7$ subunit RNA did not significantly alter the kinetics of desensitization versus the oocytes injected with equal amounts of the $\alpha7$ and $\beta2$ subunit RNA. Reprinted with permission from Khiroug et al. (2002).

interneurons are different, it is necessary to investigate the single channel properties of these receptors as well. This information might help to explore the possible presence of the $\alpha2$ subunit in these nAChR channels, and what, if any, effect this subunit may be having on the single channel properties.

Possible role of nAChR function in Alzheimer's disease

Alzheimer's disease is a human neurological disorder characterized by an increasing loss of cognitive function and the presence of extracellular neuritic

plaques composed of the β-amyloid peptide (Aβ_{1-42}). In addition, AD is also accompanied by various deficits in cholinergic transmission in the brain, including the loss of cholinergic neurons in the basal forebrain, decrease in release of ACh, and decrease in choline acetyltransferase activity (Auld et al., 1998; Selkoe, 1999). The link between the molecular correlates of AD and the loss of cognitive function are currently unknown. It has been suggested that impairment of the cholinergic system may occur early in AD and lead to cognitive deficits (Francis et al., 1999; Paterson and Nordberg, 2000). Potential targets in AD pathology are the nAChRs, because they are known to be involved in cognition and AD

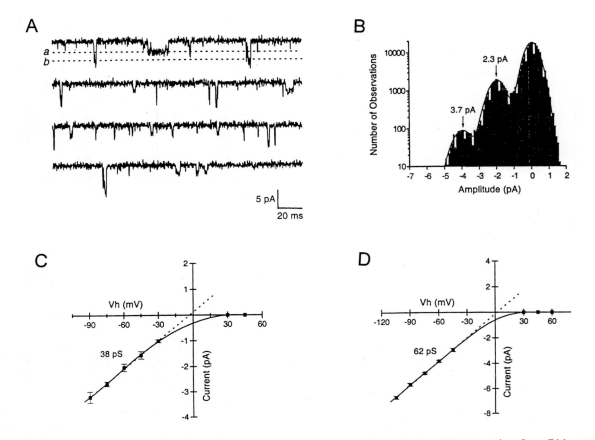

Fig. 5. ACh induces the opening of two distinct single channel conductances in excised outside-out patches. In a CA1 stratum radiatum interneuron, the rapid application of ACh (10 μM) at a holding potential of −60 mV induced the opening of two distinct channels, as seen in the traces (A) or the amplitude histogram (B). Plotting the dependence of the amplitude on the holding potential for these two channels yielded a single-channel conductance for the smaller channel of 38 pS (C), and of 62 pS for the larger channel (D). Reprinted with permission from Shao and Yakel (2000).

patients exhibit decreased numbers of these receptors (Le Novere and Changeux, 1995; Levin and Simon, 1998; Jones et al., 1999; Paterson and Nordberg, 2000). In addition, it was recently reported that $A\beta_{1-42}$ binds to the $\alpha7$ and non-$\alpha7$ subtypes of nAChRs with high affinity (Wang et al., 2000a,b).

To investigate whether there is a link between $A\beta_{1-42}$ and nAChR function, we have examined the effect of $A\beta_{1-42}$ on nAChR currents in the rat hippocampal slice. We found that $A\beta_{1-42}$ inhibits whole-cell and single channel nAChR currents from rat CA1 stratum radiatum interneurons at concentrations as low as 100 nM (Fig. 6; Pettit et al., 2001). This inhibition appears specific for neuronal nAChRs, because $A\beta_{1-42}$ had no effect on glutamate

or serotonin 5-HT$_3$ receptors, and may be a direct effect on the channel rather than an indirect effect via a signal transduction cascade. In addition, the magnitude of $A\beta_{1-42}$ inhibition was dependent on the subtype of nAChR. When investigating the block of the single channel currents by $A\beta_{1-42}$, the non-$\alpha7/62$ pS channel was significantly more inhibited (i.e., 54%) than the $\alpha7$-containing 38 pS channel (i.e., 14%). Thus, chronic inhibition of cholinergic signaling by $A\beta_{1-42}$ and, in particular, the non-$\alpha7$ nAChRs, could contribute to the cognitive deficits associated with AD. The involvement of non-$\alpha7$ nAChRs in cognitive function has long been recognized (Picciotto et al., 1995; Levin and Simon, 1998). In addition, non-$\alpha7$ nAChRs have been shown to pattern the network

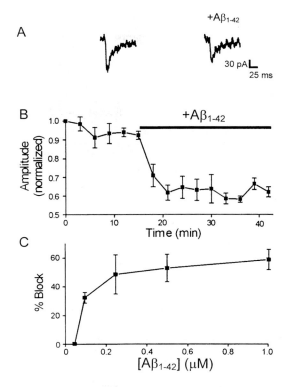

Fig. 6. $A\beta_{1-42}$ blocks current through nAChRs. The bath application of $A\beta_{1-42}$ (2 μM) significantly reduced the amplitude of the nAChR response (A) that was evoked by the local photolysis of caged carbachol. Averaged response amplitude (5 cells) before and after the application of $A\beta_{1-42}$ is shown in B. The dose-dependency of this inhibition is shown in C. Reprinted with permission from Pettit et al. (2001). Copyright 2001 by the Society for Neuroscience.

activity in rodent hippocampus via the regulation of theta activity (Cobb et al., 1999). Lastly, there is a substantial decrease in the number of nAChRs (in particular $\alpha 4$-containing) in AD patients (Paterson and Nordberg, 2000; Court et al., 2001).

One major question about the block of nAChRs by $A\beta_{1-42}$ is what the molecular mechanism of inhibition may be. In addition to the full-length $A\beta_{1-42}$ peptide, we previously showed that the fragment of $A\beta_{1-42}$ including amino acid residues 12–28, $A\beta_{12-28}$, also inhibited these channels (Pettit et al., 2001). Recently, we have begun investigating whether other regions of $A\beta_{1-42}$ might also inhibit nAChR function. Propionyl-amyloid β-protein (31–34) amide (Pr-$A\beta_{31-34}$) is a small fragment of $A\beta_{1-42}$ which

previously was reported to block the neurotoxic effects associated with $A\beta_{1-42}$ on cholinergic neurons of the rat magnocellular nucleus basalis in vivo (Harkany et al., 1999). We tested whether Pr-$A\beta_{31-34}$ might be able to block the inhibitory effect of $A\beta_{1-42}$ on nAChRs. We found instead that Pr-$A\beta_{31-34}$ (> 50 nM) mimicked the affect of $A\beta_{1-42}$ by blocking the function of the nAChRs, in a rapid, dose-dependent, and reversible manner (Fig. 7A, left). At a concentration of 1 μM, the percent block was $41 \pm 7\%$ (mean $\pm$ SE; 4 cells; Fig. 7A, right). The $A\beta_{1-42}$ peptide fragment including residues 31–35, $A\beta_{31-35}$, also blocked nAChRs, in a rapid and dose-dependent manner with a similar potency to Pr-$A\beta_{31-34}$ (Fig. 7B, left). At a concentration of 1 μM, $A\beta_{31-35}$ reduced the amplitude of responses by $38 \pm 6\%$ (5 cells; Fig. 7B, right). These data further support the notion that β-amyloid peptides directly block the function of nAChRs in rat hippocampal interneurons, and indicate that multiple regions of $A\beta_{1-42}$ (i.e., residues 31–34/35 and 12–28; Pettit et al., 2001) inhibit functional nAChR channels.

Since Pr-$A\beta_{31-34}$ was previously reported to block the toxic effects of $A\beta_{1-42}$, we examined whether Pr-$A\beta_{31-34}$ might influence the interaction between $A\beta_{1-42}$ and nAChRs. We used various concentrations of Pr-$A\beta_{31-34}$ (i.e., 50–1000 nM), which were coapplied with a single dose of $A\beta_{1-42}$ (250 nM); this concentration of $A\beta_{1-42}$ was previously reported to block nAChRs by $51 \pm 14\%$ (3 cells; Pettit et al., 2001). Pr-$A\beta_{31-34}$ did not significantly reduce the inhibition of nAChRs by $A\beta_{1-42}$, since the coapplication of both still significantly reduced the amplitude of nAChR-mediated responses (Fig. 7C). The extent of block by high doses of Pr-$A\beta_{31-34}$ (i.e., 1 μM) and $A\beta_{1-42}$ (250 nM) was significantly larger than that of $A\beta_{1-42}$ alone; however, the block was not significantly larger than at maximal (i.e., 1 μM) doses of $A\beta_{1-42}$. This is further evidence that Pr-$A\beta_{31-34}$ is mimicking the effect of $A\beta_{1-42}$, and not likely inhibiting nAChRs via a mechanism different than $A\beta_{1-42}$. Understanding the mechanism whereby the small synthetic β-amyloid peptide fragments, Pr-$A\beta_{31-34}$ and $A\beta_{31-35}$, rapidly inhibit nAChR channels in rat hippocampal slice interneurons might lead to the elucidation of mechanisms involved in the cognitive deficits associated with Alzheimer's disease, and the development of therapeutic agents to treat patients with this disease.

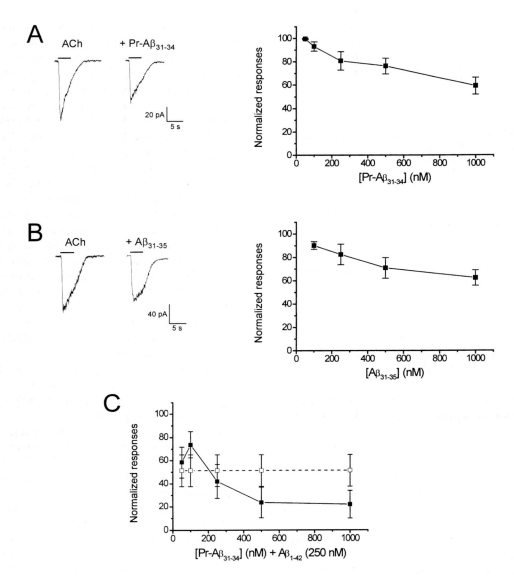

Fig. 7. Block of nAChR response by small β-amyloid peptide fragments. (A) ACh (1 mM) application (horizontal bar) induced an inward current response that was blocked (by 28%) by coapplication with Pr-Aβ$_{31-34}$ (500 nM; left, for traces). The dose-inhibition curve (averaged from 3 to 6 cells) is shown on the right. (B) ACh response was also blocked (by 22%) by coapplication with Aβ$_{31-35}$ (500 nM; left, for traces), with the dose-inhibition curve (4–6 cells) on the right. (C) Inhibition of ACh responses by coapplication of varying concentrations of Pr-Aβ$_{31-34}$ (50–1000 nM) and Aβ$_{1-42}$ (250 nM) (filled squares); each point represents the average from 3 to 7 cells. The block by Aβ$_{1-42}$ (250 nM) alone is also shown (open squares and dotted line).

Determining the subunit dependence of block of nAChRs by Aβ$_{1-42}$ is also likely to be very important in further understanding of the cognitive decline in AD. As mentioned above, we have observed that the non-α7 nAChRs are particularly sensitive to block by Aβ$_{1-42}$, although other labs have reported that the α7 nAChRs might be important targets as well (Wang et al., 2000a,b; Dineley et al., 2002; Liu et al., 2002). Nevertheless, knowing which subunits are the targets of Aβ$_{1-42}$ and its molecular mechanism of interaction (e.g., the site of action, whether through direct binding or via an open channel blocking mechanism), will certainly help in understanding any putative role of nAChRs in AD and in the development of drugs to treat AD.

Current and future directions

A major focus of the current work in the nAChR field is to continue to elucidate the molecular makeup of functional nAChRs in various regions of the brain and nervous system, and for our group, in particular, those in rat hippocampal interneurons. This task is made difficult for several reasons, not the least of which is the lack of specific pharmacological tools for each nAChR subunit. The known molecular diversity of the nAChRs in the brain, and the known diversity of interneuronal types within the hippocampus are also complicating factors. Nevertheless, there has been enormous progress in the past few years, so the prospects for further advances remain excellent. Once we have a more complete picture of the molecular makeup, it will be important to learn more about the physiological role that these receptors play in the regulation of hippocampal excitability, from the short-term effects on synaptic transmission, to the longer-term effects on synaptic plasticity, and perhaps the cognitive mechanisms as well. For example, nAChRs are permeable to calcium to an extent which is dependent upon the particular subunit present. A variety of intracellular signal transduction cascades are regulated by calcium, directly and/or indirectly. These mechanisms can in turn feedback upon the nAChR itself and regulate its function. Such mechanisms are likely to be very important in understanding how nAChRs participate in regulating hippocampal excitability and cognition, and/or perhaps the role that these receptors are playing in neurodegenerative disease.

References

Alkondon, M. and Albuquerque, E.X. (1993) Diversity of nicotinic acetylcholine receptors in rat hippocampal neurons. I. Pharmacological and functional evidence for distinct structural subtypes. J. Pharmacol. Exp. Ther., 265: 1455–1473.

Alkondon, M. and Albuquerque, E.X. (2001) Nicotinic acetylcholine receptor $\alpha7$ and $\alpha4\beta2$ subtypes differentially control GABAergic input to CA1 neurons in rat hippocampus. J. Neurophysiol., 86: 3043–3055.

Alkondon, M., Pereira, E.F., Barbosa, C.T. and Albuquerque, E.X. (1997) Neuronal nicotinic acetylcholine receptor activation modulates gamma-aminobutyric acid release from CA1 neurons of rat hippocampal slices. J. Pharmacol. Exp. Ther., 283: 1396–1411.

Anand, R., Peng, X. and Lindstrom, J. (1993) Homomeric and native $\alpha7$ acetylcholine receptors exhibit remarkably similar but non-identical pharmacological properties, suggesting that the native receptor is a heteromeric protein complex. FEBS Lett., 327: 241–246.

Auld, D.S., Kar, S. and Quirion, R. (1998) β-amyloid peptides as direct cholinergic neuromodulators: a missing link? Trends Neurosci., 21: 43–49.

Boyd, R.T. (1997) The molecular biology of neuronal nicotinic acetylcholine receptors. Crit. Rev. Toxicol., 27: 299–318.

Changeux, J.P., Bertrand, D., Corringer, P.J., Dehaene, S., Edelstein, S., Lena, C., Le Novere, N., Marubio, L., Picciotto, M., Zoli, M. (1998) Brain nicotinic receptors: structure and regulation, role in learning and reinforcement. Brain Res. Rev., 26: 198–216.

Chen, D. and Patrick, J.W. (1997) The α-bungarotoxin-binding nicotinic acetylcholine receptor from rat brain contains only the $\alpha7$ subunit. J. Biol. Chem., 272: 24024–24029.

Cobb, S.R., Bulters, D.O., Suchak, S., Riedel, G., Morris, R.G. and Davies, C.H. (1999) Activation of nicotinic acetylcholine receptors patterns network activity in the rodent hippocampus. J. Physiol., 518: 131–140.

Conroy, W.G. and Berg, D.K. (1998) Nicotinic receptor subtypes in the developing chick brain: appearance of a species containing the $\alpha4$, $\beta2$, and $\alpha5$ gene products. Mol. Pharmacol., 53: 392–401.

Court, J., Martin-Ruiz, C., Piggott, M., Spurden, D., Griffiths, M. and Perry, E. (2001) Nicotinic receptor abnormalities in Alzheimer's disease. Biol. Psychiatry, 49: 175–184.

Cuevas, J. and Berg, D.K. (1998) Mammalian nicotinic receptors with $\alpha7$ subunits that slowly desensitize and rapidly recover from α-bungarotoxin blockade. J. Neurosci., 18: 10335–10344.

Cuevas, J., Roth, A.L. and Berg, D.K. (2000) Two distinct classes of functional $\alpha7$-containing nicotinic receptor on rat superior cervical ganglion neurons. J. Physiol., 525: 735–746.

Dani, J.A. (2001) Overview of nicotinic receptors and their roles in the central nervous system. Biol. Psychiatry, 49: 166–174.

Dineley, K.T., Bell, K.A., Bui, D. and Sweatt, J.D. (2002) β-amyloid peptide activates $\alpha7$ nicotinic acetylcholine receptors expressed in *Xenopus* oocytes. J. Biol. Chem., 277: 25056–25061.

Drisdel, R.C. and Green, W.N. (2000) Neuronal α-bungarotoxin receptors are $\alpha7$ subunit homomers. J. Neurosci., 20: 133–139.

Elgoyhen, A.B., Vetter, D.E., Katz, E., Rothlin, C.V., Heinemann, S.F. and Boulter, J. (2001) $\alpha10$: A determinant of nicotinic cholinergic receptor function in mammalian vestibular and cochlear mechanosensory hair cells. P.N.A.S., 98: 3501–3506.

Francis, P.T., Palmer, A.M., Snape, M. and Wilcock, G.K. (1999) The cholinergic hypothesis of Alzheimer's disease: a review of progress. J. Neurol. Neurosurg. Psychiatry, 66: 137–147.

Frazier, C.J., Rollins, Y.D., Breese, C.R., Leonard, S., Freedman, R. and Dunwiddie, T.V. (1998) Acetylcholine activates an α-bungarotoxin-sensitive nicotinic current in rat hippocampal interneurons, but not pyramidal cells. J. Neurosci., 18: 1187–1195.

Freedman, R., Coon, H., Myles-Worsley, M., Orr-Urtreger, A., Olincy, A., Davis, A., Polymeropoulos, M., Holik, J., Hopkins, J., Hoff, M., Rosenthal, J., Waldo, M.C., Reimherr, F., Wender, P., Yaw, J., Young, D.A., Breese, C.R., Adams, C., Patterson, D., Adler, L.E., Kruglyak, L., Leonard, S. and Byerley, W. (1997) Linkage of a neurophysiological deficit in schizophrenia to a chromosome 15 locus. P.N.A.S., 94: 587–592.

Frotscher, M. and Léránth, C. (1985) Cholinergic innervation of the rat hippocampus as revealed by choline acetyltransferase immunocytochemistry: a combined light and electron microscopic study. J. Comp. Neurol., 239: 237–246.

Girod, R., Crabtree, G., Ernstrom, G., Ramirez-Latorre, J., McGehee, D., Turner, J. and Role, L. (1999) Heteromeric complexes of α5 and/or α7 subunits. Effects of calcium and potential role in nicotine-induced presynaptic facilitation. Ann. N.Y. Acad. Sci., 868: 578–590.

Harkany, T., Abraham, I., Laskay, G., Timmerman, W., Jost, K., Zarandi, M., Penke, B., Nyakas, C. and Luiten, P.G. (1999) Propionyl-IIGL tetrapeptide antagonizes β-amyloid excitotoxicity in rat nucleus basalis. NeuroReport, 10: 1693–1698.

Ji, D. and Dani, J.A. (2000) Inhibition and disinhibition of pyramidal neurons by activation of nicotinic receptors on hippocampal interneurons. J. Neurophysiol., 83: 2682–2690.

Jones, S. and Yakel, J.L. (1997) Functional nicotinic ACh receptors on interneurones in the rat hippocampus. J. Physiol., 504: 603–610.

Jones, S., Sudweeks, S. and Yakel, J.L. (1999) Nicotinic receptors in the brain: correlating physiology with function. Trends Neurosci., 22: 555–561.

Khiroug, S.S., Harkness, P.C., Lamb, P.W., Sudweeks, S.N., Khiroug, L., Millar, N.S. and Yakel, J.L. (2002) Rat nicotinic ACh receptor α7 and β2 subunits co-assemble to form functional heteromeric nicotinic receptor channels. J. Physiol., 540: 425–434.

Kuo, Y.M., Emmerling, M.R., Vigo-Pelfrey, C., Kasunic, T.C., Kirkpatrick, J.B., Murdoch, G.H., Ball, M.J. and Roher, A.E. (1996) Water-soluble Aβ (N-40, N-42) oligomers in normal and Alzheimer disease brains. J. Biol. Chem., 271: 4077–4081.

Le Novere, N. and Changeux, J.P. (1995) Molecular evolution of the nicotinic acetylcholine receptor: an example of multigene family in excitable cells. J. Mol. Evol., 40: 155–172.

Levin, E.D. (1992) Nicotinic systems and cognitive function. Psychopharmacology, 108: 417–431.

Levin, E.D. and Simon, B.B. (1998) Nicotinic acetylcholine involvement in cognitive function in animals. Psychopharmacology, 138: 217–230.

Liu, Q., Kawai, H. and Berg, D.K. (2001) β-Amyloid peptide blocks the response of α7-containing nicotinic receptors on hippocampal neurons. P.N.A.S., 98: 4734–4739.

Matthews, D.A., Salvaterra, P.M., Crawford, G.D., Houser, C.R. and Vaughn, J.E. (1987) An immunocyto-chemical study of choline acetyltransferase-containing neurons and axon terminals in normal and partially deafferented hippocampal formation. Brain Res., 402: 30–43.

McGehee, D.S. and Role, L.W. (1995) Physiological diversity of nicotinic acetylcholine receptors expressed by vertebrate neurons. Ann. Rev. Physiol., 57: 521–546.

McQuiston, A.R. and Madison, D.V. (1999) Nicotinic receptor activation excites distinct subtypes of interneurons in the rat hippocampus. J. Neurosci., 19: 2887–2896.

Nordberg, A. (1994) Human nicotinic receptors–their role in aging and dementia. Neurochem. Int., 25: 93–97.

Palma, E., Maggi, L., Barabino, B., Eusebi, F. and Ballivet, M. (1999) Nicotinic acetylcholine receptors assembled from the α7 and β3 subunits. J. Biol. Chem., 274: 18335–18340.

Paterson, D and Nordberg, A. (2000) Neuronal nicotinic receptors in the human brain. Prog. Neurobiol., 61: 75–111.

Perry, E.K., Morris, C.M., Court, J.A., Cheng, A., Fairbairn, A.F., McKeith, I.G., Irving, D., Brown, A. and Perry, R.H. (1995) Alteration in nicotine binding sites in Parkinson's disease, Lewy body dementia and Alzheimer's disease: possible index of early neuropathology. Neuroscience, 64: 385–395.

Pettit, D.L., Shao, Z. and Yakel, J.L. (2001) β-Amyloid$_{1-42}$ peptide directly modulates nicotinic receptors in the rat hippocampal slice. J. Neurosci., 21: RC120.

Picciotto, M.R., Zoli, M., Lena, C., Bessis, A., Lallemand, Y., LeNovere, N., Vincent, P., Pich, E.M., Brulet, P., Changeux, J.P. (1995) Abnormal avoidance learning in mice lacking functional high-affinity nicotine receptor in the brain. Nature, 374: 65–67.

Porter, J.T., Cauli, B., Tsuzuki, K., Lambolez, B., Rossier, J. and Audinat, E. (1999) Selective excitation of subtypes of neocortical interneurons by nicotinic receptors. J. Neurosci., 19: 5228–5235.

Ramirez-Latorre, J., Yu, C.R., Qu, X., Perin, F., Karlin, A. and Role, L. (1996) Functional contributions of alpha5 subunit to neuronal acetylcholine receptor channels. Nature, 380: 347–351.

Selkoe, D.J. (1999) Translating cell biology into therapeutic advances in Alzheimer's disease. Nature, 399: A23–31.

Shao, Z. and Yakel, J.L. (2000) Single channel properties of neuronal nicotinic ACh receptors in stratum radiatum interneurons of rat hippocampal slices. J. Physiol., 527: 507–513.

Steinlein, O.K., Mulley, J.C., Propping, P., Wallace, R.H., Phillips, H.A., Sutherland, G.R., Scheffer, I.E. and Berkovic, S.F. (1995) A missense mutation in the neuronal nicotinic acetylcholine receptor α-4 subunit is associated with autosomal dominant nocturnal frontal lobe epilepsy. Nature Genet., 11: 201–203.

Stewart, M. and Fox, S.E. (1990) Do septal neurons pace the hippocampal theta rhythm?. Trends Neurosci., 13: 166–168.

Sudweeks, S.N. and Yakel, J.L. (2000) Functional and molecular characterization of neuronal nicotinic ACh receptors in rat CA1 hippocampal neurons. J. Physiol., 527: 515–528.

Wada, E., Wada, K., Boulter, J., Deneris, E., Heinemann, S., Patrick, J. and Swanson, L.W. (1989) Distribution of alpha 2, alpha 3, alpha 4, and beta 2 neuronal nicotinic receptor subunit mRNAs in the central nervous system: a hybridization histochemical study in the rat. J. Comp. Neurol., 284: 314–335.

Wang, H.-Y., Lee, D.H.S., D'Andrea, M.R., Peterson, P.A., Shank, R.P. and Reitz, A.B. (2000a) Amyloid$_{1-42}$ binds to α7 nicotinic acetylcholine receptor with high affinity. J. Biol. Chem., 275: 5626–5632.

Wang, H.Y, Lee, D.H, Davis, C.B and Shank, R.P. (2000b) Amyloid peptide Aβ(1–42) binds selectively and with picomolar affinity to alpha7 nicotinic acetylcholine receptors. J. Neurochem., 75: 1155–1161.

Wonnacott, S. (1997) Presynaptic nicotinic ACh receptors. Trends Neurosci., 20: 92–98.

Woolf, N.J. (1991) Cholinergic systems in mammalian brain and spinal cord. Prog. Neurobiol., 37: 475–524.

Yamamoto, J. (1998) Effects of nicotine, pilocarpine, and tetrahydroaminoacridine on hippocampal theta waves in freely moving rabbits. Eur. J. Pharmacol., 359: 133–137.

Yu, C.R. and Role, L.W. (1998a) Functional contribution of the α7 subunit to multiple subtypes of nicotinic receptors in embryonic chick sympathetic neurones. J. Physiol., 509: 651–665.

Yu, C.R. and Role, L.W. (1998b) Functional contribution of the α5 subunit to neuronal nicotinic channels expressed by chick sympathetic ganglion neurones. J. Physiol., 509: 667–681.

Progress in Brain Research, Vol. 145
ISSN 0079-6123

CHAPTER 7

The nicotinic acetylcholine receptor subtypes and their function in the hippocampus and cerebral cortex

Manickavasagom Alkondon[1,*] and Edson X. Albuquerque[1,2]

[1]*Department of Pharmacology and Experimental Therapeutics, University of Maryland School of Medicine, Baltimore, MD 21201, USA*
[2]*Departamento de Farmacologia Básica e Clínica, Instituto de Ciências Biomédicas, Centro de Ciências da Saúde, Universidade Federal do Rio de Janeiro, Rio de Janeiro, RJ 21944, Brazil*

Abstract: Nicotinic acetylcholine receptors (nAChRs) are widely distributed in the central nervous system and have been implicated in multiple behavioral paradigms and pathological conditions. Nicotinic therapeutic interventions require an extensive characterization of native nAChRs including mapping of their distribution and function in different brain regions. Here, we describe the roles played by different nAChRs in affecting neuronal activity in the hippocampus and cerebral cortex. At least three distinct functional nAChR subtypes ($\alpha7$, $\alpha4\beta2$, $\alpha3\beta4$) can be detected in the hippocampal region, and in many instances a single neuron type is found to be influenced by all three nAChRs. Further, it became clear that GABAergic and glutamatergic inputs to the hippocampal interneurons are modulated via different subtypes of nAChRs. In the cerebral cortex, GABAergic inhibition to the layer V pyramidal neurons is enhanced predominantly via activation of $\alpha4\beta2$ nAChR and to a minor extent via activation of $\alpha7$ nAChR. Such diversity offers pathways by which nicotinic drugs affect brain function.

Keywords: acetylcholine; choline; nicotine; methyllycaconitine; dihydro-β-erythroidine; mecamylamine; GABA; glutamate

Introduction

Nicotinic acetylcholine receptors (nAChRs) are diverse groups of membrane proteins present in muscle, ganglia and neurons of the central nervous system (CNS) (Lukas and Bencherif, 1992; Albuquerque et al., 1997). Earlier binding studies have detected only two subtypes in the brain: (i) a low-affinity receptor labeled by $[^{125}I]\alpha$-bungarotoxin, and (ii) a high-affinity receptor labeled by $[^3H]$nicotine or $[^3H]$acetylcholine (Clark et al., 1985; Marks et al., 1986; Flores et al., 1992). However, molecular biological studies have identified at least nine α subunits

($\alpha2$–$\alpha10$) and three β subunits ($\beta2$–$\beta4$) (Lindstrom, 1996; Elgoyhen et al., 2001) in the CNS. Although, these high number of nAChR subunits raise the possible existence of a variety of structurally divergent native nAChR subtypes, prior to 1990s, there were only a few studies on neuronal nAChRs (Aracava et al., 1987; Lipton et al., 1987). Soon thereafter, several electrophysiological studies surfaced in the literature, identifying and characterizing native nAChR subtypes in various brain regions such as rat hippocampus (Alkondon and Albuquerque, 1990; Alkondon et al., 1992; Zorumski et al., 1992; Alkondon and Albuquerque, 1993; see reviews by Albuquerque et al., 1995; Gray et al., 1996; Albuquerque et al., 1997), rat medial habenula and interpeduncular nucleus (Mulle and Changeux, 1990;

*Corresponding author: Tel.: 410-706-3563;
Fax: 410-706-3991 E-mail: malko001@umaryland.edu

DOI: 10.1016/S0079-6123(03)45007-3

Léna et al., 1993), porcine hypophyseal intermediate lobe cells (Zhang and Feltz, 1990), chick lateral spiriform nucleus (McMahon et al., 1994), rat midbrain neurons (Pidoplichko et al., 1997) and mouse amygdala neurons (Barazangi and Role, 2001). Pharmacological analysis of acetylcholine (ACh)-gated currents in cultured hippocampal neurons indicated the presence of at least three different nAChR subtypes giving rise to kinetically-distinct nicotinic whole-cell currents (Alkondon and Albuquerque, 1993; Castro and Albuquerque, 1995). A rapidly decaying nicotinic current (named as type IA current), sensitive to blockade by methyllycaconitine (MLA) or α-bungarotoxin (α-BGT), a slowly decaying current (named as type II current), sensitive to blockade by dihydro-β-erythroidine (DHβE), and a very slowly decaying current, sensitive to blockade by low concentration of mecamylamine (MEC), have been described (see review by Albuquerque et al., 1995, 1997). In situ hybridization analysis and immunocytochemical studies substantiated the initial conclusion that type IA and type II currents are mediated, respectively, by α7 and α4β2 nAChR subunits (Alkondon et al., 1994; Barrantes et al., 1995; Zarei et al., 1999). It was inferred based on the sensitivity to MEC that the type III currents are mediated by α3β4 nAChRs, and recent studies by other investigators (Xiao et al., 1998; Papke et al., 2001) corroborate that MEC is more selective for α3β4 nAChR than for other known subunit combinations. Further, the studies by Zoli et al. (1998) extended the classification of nicotinic current types even further to include type IV currents in the mouse brain, and provided evidence for the involvement of other nAChR subunits.

Recently, choline, the product of hydrolysis of ACh, has received much attention as a nicotinic pharmacophore after the initial discovery that this endogenous agent can activate and desensitize some native and reconstituted nAChRs (Mandelzys et al., 1995; Papke et al., 1996; Alkondon et al., 1997b). These findings have rekindled the interest of the early attempts by Krnjević and others (see Krnjević and Reinhardt, 1979) to recognize the excitatory role of microiontophoretically injected choline in the cortical regions of the brain. Choline is an agonist like ACh at the α7 nAChR, and is a partial agonist at the α3β4 nAChR (Papke et al., 1996; Albuquerque, et al.,

1997; Alkondon et al., 1999). At low micromolar concentrations, choline desensitizes α7 nAChRs present in hippocampal neurons (Alkondon et al., 1997b, 1999). Choline-activated single channel α7 nAChR-mediated events have kinetics that are indistinguishable from those of the events induced by ACh (Mike et al., 2000). Yet, choline differs from ACh in being less potent (one-tenth potency), and allowing the receptors to recover from desensitization at a rapid rate (Mike et al., 2000).

There is ample evidence from various groups of investigators that α7 and α4β2 nAChRs are affected in various illnesses such as schizophrenia (Breese et al., 2000) and Alzheimer's disease (Whitehouse et al., 1988; Nordberg, 1999; Perry et al., 2000; Albuquerque et al., 2001), and such reports justify the need to characterize the properties and function of various nAChRs in the brain. The discovery of the allosteric potentiating ligand (APL) site in the nAChRs (Pereira et al., 1993, 2002) offers a unique advantage in that an impaired nicotinic function, as it occurs in Alzheimer's disease, can be corrected by the use of APLs. Galantamine, the first APL recommended for the treatment of Alzheimer's disease, has been shown recently (Santos et al., 2002) to strengthen glutamatergic and GABAergic synaptic transmission based on nAChR-dependent mechanisms.

In this brief study, we discuss recent findings from our laboratory on the effects of ACh, choline and nicotine on the nAChRs present in the neurons of brain slices from the cortex and the hippocampus, and additionally report some corroborative new findings. Several common principles emerge from the analysis of our results on the role of nAChRs in the brain neuronal activity.

A single neuron type can be modulated by as many as three nAChR subtypes

Recent electrophysiological findings from several laboratories have indicated the presence of α7 and α4β2 nAChRs on interneurons in hippocampal slices (Alkondon et al., 1997a; Jones and Yakel, 1997; Frazier et al., 1998b; Alkondon et al., 1997a, 1999; McQuiston and Madison, 1999; Sudweeks and Yakel, 2000). In general, interneurons located in the

CA1 layers stratum radiatum (SR), stratum oriens (SO), and stratum lacunosum moleculare (SLM) express $\alpha7$ nAChRs, whereas the interneurons in CA1 layers SO and SLM also express $\alpha4\beta2$ nAChRs. We further observed that 28% of SR interneurons contain $\alpha4\beta2$ nAChRs.

Here, we present evidence to support the concept that several nAChR subtypes could influence a single neuron type (see Fig. 1). Application of choline to a CA1 SR interneuron induced a burst of action potentials (recorded as action currents) in the cell-attached patch configuration (Fig. 1Ba), or a nicotinic inward current in the whole-cell configuration (Fig. 1Bb). This current was inhibited by 10 nM MLA (Fig. 1Bc), indicating that choline-induced currents and action potentials are the result of activation of somatodendritic $\alpha7$ nAChRs in the SR interneuron. In the continuous presence of MLA and $GABA_A$ receptor antagonist bicuculline (10 μM), application of ACh to the same interneuron induced an inward nicotinic current that decayed much more slowly than the current induced by choline in the absence of MLA (Fig. 1Bd). In addition to the slow nicotinic inward current, ACh also increased the frequency of spontaneous postsynaptic currents (PSCs) that remained superimposed on the slow nicotinic inward current (Fig. 1Bd). These spontaneous PSCs are mediated via AMPA-type glutamate receptors because they had an average decay-time constant of about 2 ms (see averaged trace in Fig. 1Be) and were inhibited by CNQX (see Alkondon and Albuquerque, 2002). To evaluate the pharmacological properties of the nAChR that affected the frequency of glutamate PSCs, recordings

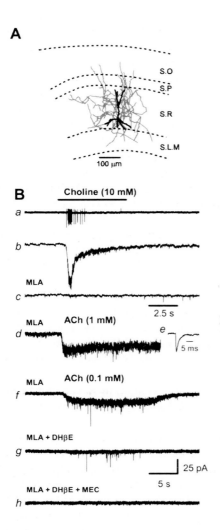

A

B

Choline (10 mM)

a

b

MLA

c

MLA   ACh (1 mM)   2.5 s

d   e   5 ms

MLA   ACh (0.1 mM)

f

MLA + DHβE

g   25 pA

5 s

MLA + DHβE + MEC

h

Fig. 1. Modulation of single neuron type by distinct nAChR subtypes. (A) Sample neurolucida drawing of a CA1 stratum radiatum (SR) interneuron from a rat hippocampal slice. (B) Pharmacological analysis indicates that three nAChR subtypes influence SR interneurons. A sequence of electrophysiological recordings made from two SR interneurons in the presence of the muscarinic receptor antagonist atropine (1 μM) and the $GABA_A$ receptor antagonist bicuculline (10 μM). In the first interneuron, U-tube application of choline induced a burst of action currents in the cell-attached recording mode (a). In the same interneuron, choline induced a macroscopic inward current in the whole-cell patch configuration (b). A bath exposure to 10 nM MLA inhibited completely choline-evoked current (c). In the same interneuron, in the continuous presence of MLA, ACh (1 mM) induces a slowly decaying inward nicotinic current and enhances the frequency of spontaneous glutamate PSCs (d). The trace average obtained from several PSCs reveal a rapid decay with a time constant of about 2 ms (e). In the second interneuron, in the presence of MLA, a lower concentration of ACh (0.1 mM) induces an inward nicotinic current and increases the frequency of glutamate PSCs (f). DHβE (10 μM), when included along with MLA in the bath solution, suppressed ACh-induced slow nicotinic current, but did not affect the ACh-enhanced frequency of the PSCs (g). Mecamylamine (MEC, 1 μM) inhibits ACh-triggered glutamatergic PSCs (h). Recordings were done at -60 mV pipette potential in cell-attached conditions and at -68 mV in whole-cell configurations. Rats at postnatal days 17–19 were used.

were made from another CA1 SR interneuron in the continuous presence of MLA and bicuculline. Application of ACh induced slowly decaying nicotinic inward current associated with increased frequency of glutamate PSCs (Fig. 1Bf). A bath exposure of the slice to DHβE (10 μM) inhibited the nicotinic inward current, but had only a marginal blocking effect on ACh-induced increase in the frequency of glutamate PSCs (Fig. 1Bg). These findings suggest that slow nicotinic currents are the result of the activation of $\alpha4\beta2$ nAChRs. The inclusion of MEC in addition to MLA and DHβE in the bath solution inhibited the ability of ACh to increase the frequency of glutamate PSCs (Fig. 1Bh), suggesting that a third nAChR subtype, presumably an $\alpha3\beta4$ subtype, contributed to modulate glutamate release onto the interneuron studied (see also Alkondon et al., 2003). Altogether, the results presented in Fig. 1 and those reported earlier (Alkondon et al., 1999, 2003; Alkondon and Albuquerque, 2002) demonstrate that the activity of a single type of neuron (for example, CA1 SR interneuron) can be modulated by at least three distinct nAChR subtypes.

Activation of nAChRs enhances GABAergic inhibition to hippocampal pyramidal neurons and to interneurons: layer-dependent and nAChR subtype-dependent effects

The expression of nAChR subtypes in the somatodendritic regions of hippocampal GABAergic interneurons implies that nAChR activation would excite these neurons and thereby enhance GABAergic inhibition in the hippocampal neurocircuitry. However, the overall influence of nAChR-dependent GABAergic inhibition on the hippocampal neuronal output can be appreciated only when the neurons that receive the nAChR-sensitive GABAergic inputs are identified. One way to know which neuron type received the largest number of nAChR-sensitive GABAergic synapses is to determine the maximum strength of nAChR-activated GABAergic inhibition in various neuron types in the hippocampus. To this end, we have undertaken a systematic study to measure the strength, a combination of frequency and amplitude, of GABAergic PSCs imposed on various identified neuron types in the hippocampal slice during a short exposure to nicotinic agonists (see Alkondon and Albuquerque, 2001). We used the U-tube application device to deliver agonists to a large area surrounding and including the neuron under study. In this method, the applied agonist could activate both somatodendritic nAChRs at the GABAergic neurons synapsing onto the neuron under study and most preterminal/terminal nAChRs at the GABAergic synapses made onto that neuron under study. Thus, when near saturating concentrations of the agonists (for example, 10 mM choline to activate $\alpha7$ nAChR, 1 mM ACh to activate both $\alpha7$ and $\alpha4\beta2$ nAChRs) were applied in this manner in the presence of the muscarinic receptor antagonist atropine (1 μM), it was possible to study a large percentage (40–70%) of the maximal nAChR-dependent GABAergic inhibition in the neurons studied (Alkondon and Albuquerque, 2001).

Figure 2 shows sample recordings of GABAergic PSCs induced by the $\alpha7$ nAChR agonist choline in CA1 and CA3 pyramidal neurons and in the CA1 SLM interneurons. The inclusion of biocytin in the pipette solution allowed us to reconstruct the image of the neuron by fixing the slices at the end of each experiment. This procedure validated the proper identification of the neuron type being evaluated. As illustrated in Fig. 2, choline enhances the frequency and amplitude of GABAergic PSCs in all three neuron types. The varying magnitude of choline-elicited GABAergic PSCs observed among the three neurons suggests that the $\alpha7$ nAChR activation would implement different degrees of GABAergic inhibition to each neuron type. Using the GABA PSCs net charge analysis, it was possible to compare the impact of $\alpha7$ nAChR activation among various neuron types in the CA1 region (Alkondon and Albuquerque, 2001). As shown in Fig. 3, the maximum net charge was observed in SLM interneuron followed by CA1 pyramidal neuron and SR interneuron. Interneurons in the pyramidal layer and in the SO were the least affected by nAChR-dependent GABAergic inhibition.

To evaluate the contribution of $\alpha4\beta2$ nAChR-dependent GABAergic inhibition, we subtracted the net charge of choline-induced PSCs from that of ACh-induced PSCs, where the difference was a measure of the $\alpha4\beta2$ nAChR-dependent activity.

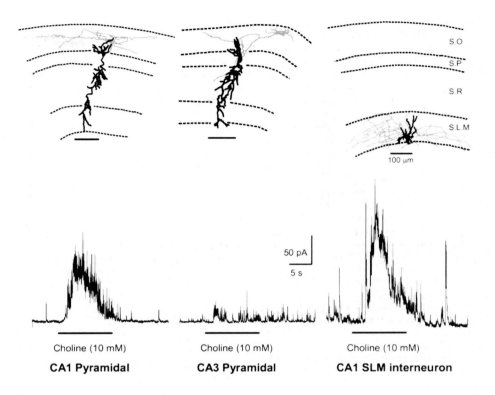

Fig. 2. The level of $\alpha7$ nAChR-dependent GABAergic inhibition varies among neuron types. Top row: Sample neurolucida drawings of biocytin-filled neurons from hippocampal slices. Bottom row: Samples of choline-evoked GABAergic PSCs recorded from the neurons corresponding to the drawings shown in the top row. Recordings were done at 0 mV using cesium methanesulfonate-containing pipette solution. Rats at postnatal days 9 (pyramidal neurons) and 15 (SLM interneurons) were used. Though not shown here, these responses were sensitive to blockade by 10 μM bicuculline.

This analysis was undertaken assuming that at the test concentrations used in this study, choline and ACh are equally effective in activating $\alpha7$ nAChRs and that choline does not activate $\alpha4\beta2$ nAChRs (Alkondon et al., 1999; Mike et al., 2000). The subtraction analysis indicated that the contribution of $\alpha4\beta2$ nAChR-induced PSCs is highest in pyramidal neurons with the rank order being SPI < SO < SR < SLM < SPP (Fig. 3). Thus, the results depicted in Fig. 3 show that there is a layer-dependent GABAergic inhibition exerted by both $\alpha7$ and $\alpha4\beta2$ nAChRs in the CA1 region of the hippocampus.

The layer-specific effects of nAChRs described above suggest that nAChR-dependent GABAergic inhibition can modify the impact of afferent signals at the level of dendrites but not the cell soma of the CA1 pyramidal neurons. For instance, GABAergic inter-neurons in the pyramidal layer send axonal projections specifically to the soma and axon initial segments of pyramidal neurons and are, therefore, in a position to affect the timing and initiation of action potentials in the pyramidal neurons and thereby affect their overall output (Buhl et al., 1996; Miles et al., 1996). However, the presence of low numbers of somatodendritic nAChRs in the SPI interneurons (McQuiston and Madison, 1999), and the negligible GABAergic inhibition exerted by nAChR activation in the SPI interneurons (Fig. 3; Alkondon and Albuquerque, 2001) suggest that the nAChRs would have the least influence in controlling the pyramidal neuron output. On the other hand, interneurons in both SR and SLM send axonal projections to pyramidal neuron dendrites, and are

114

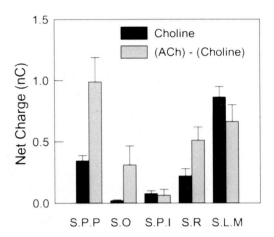

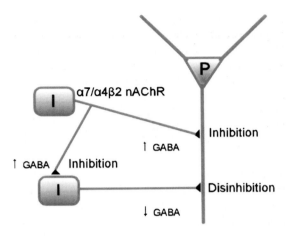

Fig. 3. nAChR-sensitive GABAergic inhibition varies among various neurons depending on their location in different CA1 layers. Net charge of choline- or ACh-triggered GABAergic PSCs in different neurons is plotted. Choline (10 mM) or ACh (1 mM) was applied over the cell somata and dendrites via a U-tube. S.P.P, stratum pyramidale pyramidal neuron; S.O, stratum oriens interneuron; S.P.I, stratum pyramidale interneuron; S.R, stratum radiatum interneuron; S.L.M, stratum lacunosum moleculare interneuron. Data adapted from Alkondon and Albuquerque (2001).

Fig. 4. nAChR activation can result in either inhibition or disinhibition. Proposed scheme illustrates a neurocircuitry involving a pyramidal neuron (P) and two interneurons (I).

in a position to modulate the impact of excitatory afferent inputs from both intrinsic and extrinsic sources. Therefore, nAChR-triggered GABAergic PSCs can mediate a direct inhibition of pyramidal cell dendrites (Fig. 4), a mechanism that can filter out extraneous signals arriving at the dendrites. However, the nAChR-triggered GABAergic PSCs at the SR interneurons can exert a disinhibition at the pyramidal neurons (Fig. 4), a mechanism that can help amplify the effect of excitatory signals. Thus, both the location and the timing of nicotinic cholinergic activity would be critical determinants of the final outcome of the events that takes place at pyramidal neurons. In fact, a recent study (Ji et al., 2001) provided evidence to show that synaptic potentiation or depression depends on the timing of nAChR activation in the hippocampus.

The large magnitude of GABAergic PSCs induced by α7 nAChR activation at the SLM interneuron deserves special mention, because α7 nAChR activation can cause neuronal hyperpolarization via GABAergic mechanism, a signal that is able to

trigger rebound burst firing in SLM interneurons. Indeed, a recent study (Perkins, 2002) indicated that focal applications of GABA to CA1-SLM can cause giant GABA-mediated PSCs in CA3 pyramidal neurons. This study suggested that GABA may directly or indirectly excite SLM interneuron, which in turn recruit a network response by its extensive interconnectivity to other interneurons. Burst firing in SLM interneurons suppresses spikes in pyramidal neurons evoked by stimulation of Schaffer collaterals (Dvorak-Carbone and Schuman, 1999), and this allows selective activation of the pyramidal cells via the perforant pathway. Such selective inhibition of intrinsic (Schaffer collateral) but not extrinsic (perforant path) afferent activity helps in selecting between encoding and retrieval modes of associative memory systems (Hasselmo and Schnell, 1994; Wallenstein and Hasselmo, 1997; Paulsen and Moser, 1998).

Activation of nAChRs enhances GABAergic inhibition to cortical neurons

The nAChR-dependent enhancement of GABAergic inhibition is a widespread phenomena, as this mechanism has been observed not only in the hippocampus but also in many other brain regions in rat and other species. For example, nAChR-triggered GABA release has been reported to occur in the rat interpeduncular nucleus, rat dorsal motor nucleus of the vagus, CA1 field of the rat hippocampus, rat spinal cord interneurons, mouse thalamus, chick lateral spiriform nucleus and mouse brain synaptosomes (McMahon et al., 1994; Alkondon et al., 1997a; Bertolino et al., 1997; Léna and Changeux, 1997; Lu et al., 1998; Fucile et al., 2002). More importantly, the nAChR-triggered GABAergic PSCs have been recorded from the interneurons of human cerebral cortical slices (Alkondon et al., 2000b). Whole-cell patch-clamp recordings from the interneurons of human cerebral cortical slices exhibit $\alpha7$ or $\alpha4\beta2$ nAChR-mediated inward currents and increased frequency of spontaneous GABAergic PSCs when the neurons are exposed to the nicotinic agonists (Alkondon et al., 2000b). The $\alpha4\beta2$ nAChR appears to be the predominant type involved in the modulation of GABAergic inhibition to the human cerebral cortical interneurons, and the outcome of this action can be a disinhibition of pyramidal neurons in the cortical neurocircuitry (Alkondon et al., 2000b; see also Fig. 4).

We expanded the study on cortical neurons using rat brain slices with an intent to (i) verify that nicotinic mechanisms present in the human cortex also exist in the rat brain, and (ii) ascertain that cortical pyramidal neurons, like the hippocampal pyramidal neurons, encounter nAChR-dependent GABAergic inhibition. Experiments were carried out on layer V pyramidal neurons in rat cortical brain slices taken from the temporal region. As illustrated in Fig. 5, application of either choline or ACh enhances the frequency and amplitude of GABAergic PSCs. The magnitudes of ACh-induced responses were 10- to 20-fold larger than those of choline-induced responses ($n = 5$ neurons). This observation indicates that $\alpha4\beta2$ nAChR plays a predominant role compared to $\alpha7$ nAChR in modulating GABAergic inhibition onto cortical pyramidal neurons. Taken together with the previous reports that cortical regions contain an abundance of high-affinity nicotine sites (representing the $\alpha4\beta2$ nAChR subtype) (Marks et al., 1986; Flores et al., 1992), our published results on human cortical interneurons (Alkondon et al., 2000b) and present results on cortical pyramidal neurons (see Fig. 5) support the concept that $\alpha4\beta2$ nAChR-dependent

GABAergic inhibition plays an important role in cortical function.

GABAergic inhibition can be enhanced by low levels of ACh and nicotine-possible implications

In the cerebral cortex and in the hippocampus, cholinergic afferents project in a diffuse manner (Mesulam et al., 1983; Frotscher and Léránth, 1985; Woolf, 1991; Schäfer et al., 1998). In both regions, unlike the glutamatergic and GABAergic afferents, cholinergic fibers form direct synaptic contacts with the postjunctional region of only a minor fraction of the total number of terminals present (Mrzljak et al., 1995; Umbriaco et al., 1995; but see Turrini et al., 2001). These anatomical data are consistent with the physiological observations that a direct nicotinic synaptic transmission has only been demonstrated in a few brain regions (Alkondon et al., 1998; Frazier et al., 1998a; Hefft et al., 1999). Predominantly, the cholinergic varicosities do not make synaptic contacts with nearby neuronal elements, and in many instances are found close to other excitatory or inhibitory axon terminals without making synaptic contacts (Umbriaco et al., 1994, 1995; see also chapter 2). The prevalence of such anatomical features in the CNS provided the clue that ACh could diffuse and act on receptors present some distance away from the original site of release (Descarries et al., 1997). It has been estimated that such diffused ACh could exist at a concentration in the nM to μM range (Vizi and Lábos, 1991). To verify that nAChR-dependent GABA release could be activated by low agonist concentrations, which are typically available during volume transmission (Fuxe and Agnati, 1991; Descarries et al., 1997), the following experiments were performed. In the first experiment, 10 μM ACh was applied to see whether it could increase the frequency of GABAergic PSCs in the recorded neurons. Indeed, a short application of 10 μM ACh enhanced the frequency of GABAergic PSCs in the SR interneurons (Alkondon et al., 1999; Alkondon and Albuquerque, 2001; Santos et al., 2002). Considering that ACh was applied to the slices in the absence of an acetylcholinesterase inhibitor, the effective concentration that led to the increase in

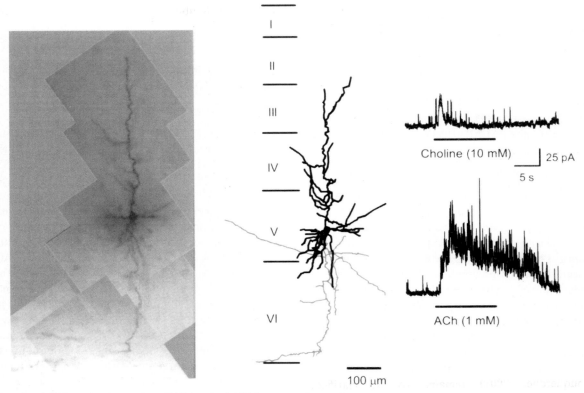

Fig. 5. nAChR activation exerts GABAergic inhibition onto layer V temporal cortical pyramidal neurons. Left panel: Photomicrograph of a biocytin-filled layer V pyramidal neuron in a cortical slice from a 15-day-old rat. Middle panel: Neurolucida tracing of the same neuron shows the location of the cell somata, dendrites and axon. Right panel: In the same neuron, U-tube application of choline or ACh induces GABAergic PSCs of varying magnitude. Similar type of responses were obtained in other layer V cortical pyramidal neurons. Recordings were done at 0 mV using cesium methanesulfonate-containing pipette solution.

the frequency of GABAergic PSCs could be estimated to be in the low micromolar range similar to those present during a nonsynaptic transmission. The observation that a low concentration of ACh (10 μM) increased the frequency of GABAergic PSCs in the absence of initiation of action potentials in the SR interneuron suggested that the nAChRs activated were present close to the axon terminals (Alkondon and Albuquerque, 2001). The finding that focally applied ACh (1 mM, 25 ms) at the cell somata enhanced the frequency of GABAergic PSCs in the recorded SR interneurons provided further support that this action is mediated via activation of nAChRs present at the GABAergic synapses but not at the cell body and dendrites of other interneurons (Alkondon and Albuquerque, 2001). Due to the fact that ACh

failed to induce GABAergic PSCs in the presence of the sodium channel blocker tetrodotoxin, it was concluded that the nAChRs must be located at the preterminal regions rather than at the terminal presynaptic sites (Alkondon et al., 1999; Alkondon and Albuquerque, 2001).

Considering that the preterminal nAChRs present at the GABAergic axon are sensitive to low ACh concentrations, in the second set of trials, we tested whether other exogenous nicotinic agents could also activate these nAChRs. The alkaloid nicotine is a commonly abused drug that reaches arterial blood concentrations up to 500 nM immediately after a cigarette smoke inhalation (Henningfield et al., 1993). Bath-applied nicotine at concentrations ranging between 250 nM and 1000 nM caused a concentration-dependent increase in the frequency of GABAergic PSCs in SR interneurons (Alkondon et al., 2000a). This effect appears to arise from the activation of $\alpha4\beta2$ nAChR but not $\alpha7$ nAChR, because DHβE instead of MLA inhibited nicotine-induced increases in the frequency of GABAergic PSCs (Alkondon et al., 2000a). However, a continuous exposure to low levels of nicotine for several minutes results in desensitization of both $\alpha7$ and $\alpha4\beta2$ nAChRs that consequently leads to a reduction in the nAChR-dependent GABAergic inhibition (Alkondon et al., 2000a; Almeida et al., 2000).

Concluding remarks

Recent work on brain slices from rat and human confirmed that $\alpha7$, $\alpha4\beta2$ and $\alpha3\beta4$ nAChRs can play distinct roles in modulating GABAergic and glutamatergic transmission, depending on the target neuron as well as on the nAChR subtype involved. The functional significance of the finding that a high degree of $\alpha7$ nAChR-dependent GABAergic inhibition is noted at the SLM interneuron (Alkondon and Albuquerque, 2001) remains to be clarified. Meanwhile, the $\alpha4\beta2$ nAChR appears to modulate GABAergic function at very low agonist concentrations, allowing the possibility that this mechanism can be activated tonically by the levels of ACh present during diffuse transmission, or by levels of nicotine normally encountered in cigarette smokers.

Acknowledgments

The authors would like to thank Ms. Mabel Zelle and Mrs. Barbara Marrow for their technical assistance. We thank Dr. E.F.R. Pereira for her helpful suggestions on the manuscript and Mrs. Bhagavathy Alkondon for preparation of brain slices, biocytin processing and neuronal drawings. The work was supported by United States Public Health Service Grant NS-25296.

Abbreviations

nAChR	nicotinic acetylcholine receptor
CNS	central nervous system
ACh	acetylcholine
MLA	methyllycaconitine
α-BGT	α-bungarotoxin
DHβE	dihydro-β-erythroidine
MEC	mecamylamine
SR	stratum radiatum
SO	stratum oriens
SLM	stratum lacunosum moleculare
SPI	stratum pyramidale interneuron
SPP	stratum pyramidale pyramidal neuron
PSCs	postsynaptic currents
APL	allosteric potentiating ligand

References

Albuquerque, E.X., Pereira, E.F.R., Castro, N.G. and Alkondon, M. (1995) Neuronal nicotinic receptors: function, modulation and structure. Semin. Neurosci., 7: 91–101.

Albuquerque, E.X., Alkondon, M., Pereira, E.F.R., Castro, N.G., Schrattenholz, A., Barbosa, C.T.F., Bonfante-Cabarcas, R., Aracava, Y., Eisenberg, H.M. and Maelicke, A. (1997) Properties of neuronal nicotinic acetylcholine receptors: pharmacological characterization and modulation of synaptic function. J. Pharmacol. Exp. Ther., 280: 1117–1136.

Albuquerque, E.X., Santos, M.D., Alkondon, M., Pereira, E.F.R. and Maelicke, A. (2001) Modulation of nicotinic receptor activity in the central nervous system: a novel approach to the treatment of Alzheimer disease. Alzheimer Disease and Associated Disorders, 15(Suppl. 1): S19–S25.

Alkondon, M. and Albuquerque, E.X. (1990) α-Cobratoxin blocks the nicotinic acetylcholine receptor in rat hippocampal neurons. Eur. J. Pharmacol., 191: 505–506.

Alkondon, M. and Albuquerque, E.X. (1993) Diversity of nicotinic acetylcholine receptors in rat hippocampal neurons. I. Pharmacological and functional evidence for distinct structural subtypes. J. Pharmacol. Exp. Ther., 265: 1455–1473.

Alkondon, M. and Albuquerque, E.X. (2001) Nicotinic acetylcholine receptor $\alpha7$ and $\alpha4\beta2$ subtypes differentially control GABAergic input to CA1 neurons in rat hippocampus. J. Neurophysiol., 86: 3043–3055.

Alkondon, M. and Albuquerque, E.X. (2002) A non-$\alpha7$ nicotinic acetylcholine receptor modulates excitatory input to hippocampal CA1 interneurons. J. Neurophysiol., 87: 1651–1654.

Alkondon, M., Pereira, E.F.R., Wonnacott, S. and Albuquerque, E.X. (1992) Blockade of nicotinic currents in hippocampal neurons defines methyllycaconitine as a potent and specific receptor antagonist. Mol. Pharmacol., 41: 802–808.

Alkondon, M., Reinhardt, S., Lobron, C., Hermsen, B., Maelicke, A. and Albuquerque, E.X. (1994) Diversity of nicotinic acetylcholine receptors in rat hippocampal neurons. II. The rundown and inward rectification of agonist-elicited whole-cell currents and identification of receptor subunits by *in situ* hybridization. J. Pharmacol. Exp. Ther., 271: 494–506.

Alkondon, M., Pereira, E.F.R., Barbosa, C.T.F. and Albuquerque, E.X. (1997) Neuronal nicotinic acetylcholine receptor activation modulates γ-aminobutyric acid release from CA1 neurons of rat hippocampal slices. J. Pharmacol. Exp. Ther., 283: 1396–1411.

Alkondon, M., Pereira, E.F.R., Cortes, W.S., Maelicke, A. and Albuquerque, E.X. (1997) Choline is a selective agonist of α7 nicotinic acetylcholine receptors in the rat brain neurons. Eur. J. Neurosci., 9: 2734–2742.

Alkondon, M., Pereira, E.F.R. and Albuquerque, E.X. (1998) α-Bungarotoxin- and methyllycaconitine-sensitive nicotinic receptors mediate fast synaptic transmission in interneurons of rat hippocampal slices. Brain Res., 810: 257–263.

Alkondon, M., Pereira, E.F.R., Eisenberg, H.M. and Albuquerque, E.X. (1999) Choline and selective antagonists identify two subtypes of nicotinic acetylcholine receptors that modulate GABA release from CA1 interneurons in rat hippocampal slices. J. Neurosci., 19: 2693–2705.

Alkondon, M., Pereira, E.F.R., Almeida, L.E.F., Randall, W.R. and Albuquerque, E.X. (2000) Nicotine at concentrations found in cigarette smokers activates and desensitizes nicotinic acetylcholine receptors in CA1 interneurons of rat hippocampus. Neuropharmacology, 39: 2726–2739.

Alkondon, M., Pereira, E.F.R., Eisenberg, H.M. and Albuquerque, E.X. (2000) Nicotinic receptor activation in human cerebral cortical interneurons: a mechanism for inhibition and disinhibition of neuronal networks. J. Neurosci., 20: 66–75.

Alkondon, M., Pereira, E.F.R. and Albuquerque, E.X. (2003) NMDA and AMPA receptors contribute to the nicotinic chlorinergic excitation of CA1 interneurons in the rat hippocampus. *J. Neurophysiol.*, (In press).

Almeida, L.E.F., Pereira, E.F., Alkondon, M., Fawcett, W.P., Randall, W.R. and Albuquerque, E.X. (2000) The opioid antagonist naltrexone inhibits activity and alters expression of α7 and α4β2 nicotinic receptors in hippocampal neurons: implications for smoking cessation programs. Neuropharmacology, 39: 2740–2755.

Aracava, Y., Deshpande, S.S., Swanson, K.L., Rapoport, H., Wonnacott, S., Lunt, G. and Albuquerque, E.X. (1987) Nicotinic acetylcholine receptors in cultured neurons from the hippocampus and brain stem of the rat characterized by single channel recording. FEBS Lett., 222: 63–70.

Barazangi, N. and Role, L.W. (2001) Nicotine-induced enhancement of glutamatergic and GABAergic synaptic transmission in the mouse amygdala. J. Neurophysiol., 86: 463–474.

Barrantes, G.E., Rogers, A.T., Lindstrom, J. and Wonnacott, S. (1995) Alpha-bungarotoxin binding sites in rat hippocampal and cortical cultures: initial characterization, colocalization with alpha 7 subunits and up-regulation by chronic nicotine treatment. Brain Res., 672: 228–236.

Bertolino, M., Kellar, K.J., Vicini, S. and Gillis, R.A. (1997) Nicotinic receptor mediates spontaneous GABA release in the rat dorsal motor nucleus of the vagus. Neuroscience, 79: 671–681.

Breese, C.R., Lee, M.J., Adams, C.E., Sullivan, B., Logel, J., Gillen, K.M., Marks, M.J., Collins, A.C. and Leonard, S. (2000) Abnormal regulation of high affinity nicotinic receptors in subjects with schizophrenia. Neuropsychopharmacol., 23: 351–364.

Buhl, E.H., Szilágyi, T., Halasy, K. and Somogyi, P. (1996) Physiological properties of anatomically identified basket and bistratified cells in the CA1 area of the rat hippocampus *in vitro*. Hippocampus, 6: 294–305.

Castro, N.G. and Albuquerque, E.X. (1995) The α-bungarotoxin-sensitive hippocampal nicotinic acetylcholine receptor channel has a high calcium permeability. Biophy. J., 68: 516–524.

Clark, P., Schwartz, R., Paul, S., Pert, C. and Pert, A. (1985) Nicotinic binding in rat brain: autoradiographic comparison of [^{3}H]acetylcholine, [^{3}H]nicotine, and [^{125}I] α-bungarotoxin. J. Neurosci., 5: 1307–1315.

Descarries, L., Gisiger, V. and Steriade, M. (1997) Diffuse transmission by acetylcholine in the CNS. Prog. Neurobiol., 53: 603–625.

Dvorak-Carbone, H. and Schuman, E.M. (1999) Patterned activity in stratum lacunosum moleculare inhibits CA1 pyramidal neuron firing. J. Neurophysiol., 82: 3213–3222.

Elgoyhen, A.B., Vetter, D.E., Katz, E., Rothlin, C.V., Heinemann, S.F. and Boulter, J. (2001) α10: a determinant of nicotinic cholinergic receptor function in mammalian vestibular and cochlear mechanosensory hair cells. Proc. Natl. Acad. Sci. USA, 98: 3501–3506.

Flores, C., Rogers, S., Pabreza, L., Wolfe, B. and Kellar, K. (1992) A subtype of nicotinic cholinergic receptor in rat brain is composed of α4 and β2 subunits and is upregulated by chronic nicotine treatment. Mol. Pharmacol., 41: 31–37.

Frazier, C.J., Buhler, A.V., Weiner, J.L. and Dunwiddie, T.V. (1998) Synaptic potentials mediated via alpha-bungarotoxin-sensitive nicotinic acetylcholine receptors in rat hippocampal interneurons. J. Neurosci., 18: 8228–8235.

Frazier, C.J., Rollins, Y.D., Breese, C.R., Leonard, S., Freedman, R. and Dunwiddie, T.V. (1998) Acetylcholine activates an α-bungarotoxin-sensitive nicotinic current in rat hippocampal interneurons, but not pyramidal cells. J. Neurosci., 18: 1187–1195.

Frotscher, M. and Léránth, C. (1985) Cholinergic innervation of the rat hippocampus as revealed by choline acetyltransferase immunocytochemistry: Combined light and electron microscopic study. J. Comp. Neurol., 239: 237–246.

Fucile, S., Lax, P. and Eusebi, F. (2002) Nicotine modulates the spontaneous synaptic activity in cultured embryonic rat spinal cord interneurons. J. Neurosci. Res., 67: 329–336.

Fuxe, K. and Agnati, L.F. (1996). Two principal modes of electrochemical communication in the brain: volume versus wiring transmission. In: K. Fuxe and L.F. Agnati (Eds.),

Volume Transmission in the Brain: Novel Mechanisms in Neural Transmission. Raven Press, New York, pp. 1–9.

Gray, R., Rajan, A.S., Radcliffe, K.A., Yakehiro, M. and Dani, J.A. (1996) Low concentrations of nicotine enhance glutamatergic synaptic transmission in the hippocampus. Nature, 383: 713–716.

Hasselmo, M.E. and Schnell, E. (1994) Laminar selectivity of the cholinergic suppression of synaptic transmission in rat hippocampal region CA1: computational modeling and brain slice physiology. J. Neurosci., 14: 3898–3914.

Hefft, S., Hulo, S., Bertrand, D. and Muller, D. (1999) Synaptic transmission at nicotinic acetylcholine receptors in rat hippocampal organotypic cultures and slices. J. Physiol. (Lond.), 515: 769–776.

Henningfield, J.E., Stapleton, J.M., Benowitz, N.L., Grayson, R.F. and London, E.D. (1993) Higher levels of nicotine in arterial blood than in venous blood after cigarette smoking. Drugs and Alcohol Dependence, 33: 23–29.

Ji, D., Lape, R. and Dani, J.A. (2001) Timing and location of nicotinic activity enhances or depresses hippocampal synaptic plasticity. Neuron, 31: 131–141.

Jones, S. and Yakel, J.L. (1997) Functional nicotinic ACh receptors on interneurons in the rat hippocampus. J. Physiol. (Lond.), 504: 603–610.

Krnjević, K. and Reinhardt, W. (1979) Choline excites cortical neurons. Science, 206: 1321–1323.

Léna, C. and Changeux, J.-P. (1997) Role of Ca^{2+} ions in nicotinic facilitation of GABA release in mouse thalamus. J. Neurosci., 17: 576–585.

Léna, C., Changeux, J.P. and Mulle, C. (1993) Evidence for "preterminal" nicotinic receptors on GABAergic axons in the rat interpeduncular nucleus. J. Neurosci., 13: 2680–2688.

Lindstrom, J. (1996). Neuronal nicotinic acetylcholine receptors. In: T. Narahashi (Ed.), Ion Channels. Plenum Press, New York, pp. 377–450.

Lipton, S.A., Aizenman, E. and Loring, R.H. (1987) Neuronal nicotinic acetylcholine responses in solitary mammalian retinal ganglion cells. Pflügers Arch., 410: 37–43.

Lu, Y., Grady, S., Marks, M.J., Picciotto, M., Changeux, J.-P. and Collins, A. (1998) Pharmacological characterization of nicotinic receptor-stimulated GABA release from mouse brain synaptosomes. J. Pharmacol. Exp. Ther., 287: 648–657.

Lukas, R.J. and Bencherif, M. (1992) Heterogeneity and regulation of nicotinic acetylcholine receptors. Int. Rev. Neurobiol., 34: 25–131.

Mandelzys, A., De Koninck, P. and Cooper, E. (1995) Agonist and toxin sensitivities of ACh-evoked currents on neurons expressing multiple nicotinic ACh receptor subunits. J. Neurophysiol., 74: 1212–1221.

Marks, M.J., Stitzel, J.A., Romm, E., Weher, J.M. and Collins, A.C. (1986) Nicotinic binding sites in the rat and mouse brain: Comparison of acetylcholine, nicotine and α-bungarotoxin. Mol. Pharmacol., 30: 427–436.

McMahon, L.L., Yoon, K.W. and Chiappinelli, V.A. (1994) Nicotinic receptor activation facilitates GABAergic neurotransmission in the avian lateral spiriform nucleus. Neuroscience, 59: 689–698.

McQuiston, A.R. and Madison, D.V. (1999) Nicotinic receptor activation excites distinct subtypes of interneurons in the rat hippocampus. J. Neurosci., 19: 2887–2896.

Mesulam, M.M., Mufson, E.J., Wainer, B.H. and Levey, A.I. (1983) Central cholinergic pathways in the rat: and overview based on an alternative nomenclature (Ch1-Ch6). Neuroscience, 10: 1185–1201.

Mike, A., Castro, N.G. and Albuquerque, E.X. (2000) Choline and acetylcholine have similar kinetic properties of activation and desensitization on the α7 nicotinic receptors in rat hippocampal neurons. Brain Res., 882: 155–168.

Miles, R., Tóth, K., Gulyás, A.I., Hájos, N. and Freund, T.F. (1996) Differences between somatic and dendritic inhibition in the hippocampus. Neuron, 16: 815–823.

Mrzljak, L., Pappy, M., Leranth, C. and Goldman-Rakic, P.S. (1995) Cholinergic synaptic circuitry in the Macaque prefrontal cortex. J. Comp. Neurol., 357: 603–617.

Mulle, C. and Changeux, J.-P. (1990) A novel type of nicotinic receptor in the rat central nervous system characterized by patch-clamp techniques. J. Neurosci., 10: 169–175.

Nordberg, A. (1999) PET studies and cholinergic therapy in Alzheimer's disease. Rev. Neurol. (Paris), 155: 853–863.

Papke, R.L., Bencherif, M. and Lippiello, P. (1996) An evaluation of neuronal nicotinic acetylcholine receptor activation by quaternary nitrogen compounds indicates that choline is selective for the α7 subtype. Neurosci. Lett., 213: 201–204.

Papke, R.L., Sanberg, P.R. and Shytle, R.D. (2001) Analysis of mecamylamine stereoisomers on human nicotinic receptor subtypes. J. Pharmacol. Exp. Ther., 297: 646–656.

Paulsen, O. and Moser, E.I. (1998) A model of hippocampal memory encoding and retrieval: GABAergic control of synaptic plasticity. Trends Neurosci., 21: 273–278.

Pereira, E.F.R., Hilmas, C., Santos, M.D., Alkondon, M., Maelicke, A. and Albuquerque, E.X. (2002) Unconventional ligands and modulators of nicotine receptors. J. Neurobiol., 53: 479–500.

Pereira, E.F.R., Reinhardt-Maelicke, S., Schrattenholz, Z., Maelicke, A. and Albuquerque, E.X. (1993) Identification and characterization of a new agonist site on nicotinic acetylcholine receptors of cultured hippocampal neurons. J. Pharmacol. Exp. Ther., 265: 1474–1491.

Perkins, K.L. (2002) GABA application to hippocampal CA3 or CA1 stratum lacunosum-moleculare excites an interneuron network. J. Neurophysiol., 87: 1404–1414.

Perry, E., Martin-Ruiz, C., Lee, M., Griffiths, M., Johnson, M., Piggott, M., Haroutunian, V., Buxbaum, J.D., Nasland, J., Davis, K. (2000) Nicotinic receptor subtypes in human brain aging, Alzheimer and Lewy body disease. Eur. J. Pharmacol., 393: 215–222.

Pidoplichko, V., DeBiasi, M., Williams, J.T. and Dani, J.A. (1997) Nicotine activates and desensitizes midbrain dopamine neurons. Nature, 390: 401–404.

Santos, M.D., Alkondon, M., Pereira, E.F.R., Aracava, Y., Eisenberg, H.M., Maelicke, A. and Albuquerque, E.X. (2002) The nicotinic allosteric potentiating ligand galantamine

120

facilitates synaptic transmission in the mammalian central nervous system. Mol. Pharmacol., 61: 1222–1234.

Schäfer, M.K.H., Eiden, L.E. and Weihe, E. (1998) Cholinergic neurons and terminal fibers revealed by immunocytochemistry for the vesicular acetylcholine transporters. I. Central nervous system. Neuroscience, 84: 331–359.

Sudweeks, S.N. and Yakel, J.L. (2000) Functional and molecular characterization of neuronal nicotinic ACh receptors in rat CA1 hippocampal neurons. J. Physiol. (Lond.), 527: 515–528.

Turrini, P., Casu, M.A., Wong, T.P., De Koninck, Y., Ribeiro-da-Silva, A. and Cuello, A.C. (2001) Cholinergic nerve terminals establish classical synapses in the rat cerebral cortex: synaptic pattern and age-related atrophy. Neuroscience, 105: 277–285.

Umbriaco, D., Watkins, K.C., Descarries, L., Cozzari, C. and Hartman, B.K. (1994) Ultrastructural and morphometric features of the acetylcholine innervation in adult rat parietal cortex. An electron microscopic study in serial sections. J. Comp. Neurol., 348: 351–373.

Umbriaco, D., Garcia, S., Beaulieu, C. and Descarries, L. (1995) Relational features of acetylcholine, noradrenaline, serotonin and GABA axon terminals in the stratum radiatum of adult rat hippocampus (CA1). Hippocampus, 5: 605–620.

Vizi, E.S. and Lábos, E. (1991) Non-synaptic interactions at presynaptic level. Prog. Neurobiol., 37: 145–163.

Wallenstein, G.V. and Hasselmo, M.E. (1997) GABAergic modulation of hippocampal population activity: sequence learning, place field development, and the phase precession effect. J. Neurophysiol., 78: 393–408.

Whitehouse, P., Martino, A., Marcus, K., Zweig, R., Singer, H., Price, D. and Kellar, K. (1988) Reduction in acetylcholine and nicotine binding in several degenerative diseases. Arch. Neurol., 45: 722–724.

Woolf, N.J. (1991) Cholinergic systems in mammalian brain and spinal cord. Prog. Neurobiol., 37: 475–524.

Xiao, Y., Mayer, E.L., Thompson, J.M., Surin, A., Wroblewski, J. and Kellar, K.J. (1998) Rat $\alpha3/\beta4$ subtype of neuronal nicotinic acetylcholine receptor stably expressed in a transfected cell line: pharmacology of ligand binding and function. Mol. Pharmacol., 54: 322–333.

Zarei, M.M., Radcliffe, K.A., Chen, D., Patrick, J.W. and Dani, J.A. (1999) Distribution of nicotinic acetylcholine receptor $\alpha7$ and $\beta2$ subunits on cultured hippocampal neurons. Neuroscience, 88: 755–764.

Zhang, Z.W. and Feltz, P. (1990) Nicotinic acetylcholine receptors in porcine hypophyseal intermediate lobe cells. J. Physiol. (Lond.), 422: 83–101.

Zoli, M., Léna, C., Picciotto, M.R. and Changeux, J.-P. (1998) Identification of four classes of brain nicotinic receptors using beta2 mutant mice. J. Neurosci., 18: 4461–4472.

Zorumski, C.F., Thio, L.L., Isenberg, K.E. and Clifford, D.B. (1992) Nicotinic acetylcholine currents in cultured postnatal rat hippocampal neurons. Mol. Pharmacol., 41: 931–936.

Progress in Brain Research, Vol. 145
ISSN 0079-6123

CHAPTER 8

Studies of muscarinic neurotransmission with antimuscarinic toxins

Lincoln T. Potter*, Donna D. Flynn, Jing-Sheng Liang and Mark H. McCollum

*Department of Molecular and Cellular Pharmacology, University of Miami School of Medicine,
P.O. Box 016189, Miami, FL 33101, USA*

Abstract: M_1 and M_4 muscarinic receptors are the most prevalent receptors for acetylcholine in the brain, and m1-toxin1 and m4-toxin are the most specific ligands yet found for their extracellular faces. Both toxins are antagonists. These toxins and their derivatives with biotin, radioiodine and fluorophores are useful for studying M_1- and M_4-linked neurotransmission. We have used the rat striatum for many studies because this tissue expresses exceptionally high concentrations of both receptors, the striatum regulates movement, and movement is altered by antimuscarinic agents, M_1-knockout and M_4-knockout. These toxins and their derivatives may also be used for studies of M_1 and M_4 receptors in the hippocampus and cortex.

Keywords: muscarinic receptors; toxin; striatum; acetylcholine; dendritic spines

Introduction

A recent review of snake toxins that bind specifically to individual subtypes of muscarinic receptors should be consulted for the history of this field, and for the amino acid sequences of the multiple toxins that bind to muscarinic receptors (Potter, 2001). MT7 (Nasman et al., 2000) has recently been shown to have the same sequence as m1-toxin1 (Potter et al., 1997; Krajewski et al., 2001), and MT3 is the same toxin as m4-toxin.

Muscarinic mechanisms in the brain

Our approach to the study of muscarinic mechanisms in the brain is based on the following ideas. (1) Acetylcholine is an important neurotransmitter in the brain. (2) Acetylcholine activates five subtypes of muscarinic receptors (M_1–M_5) and several varieties of nicotinic receptors, each of which is distributed heterogeneously in the brain (Levey, 1993; McGehee and Role, 1995; Dani, 2001). (3) Nicotine acts in the CNS, so we have information about the therapeutic potential of nicotine and its analogs. Nicotine acts on nerve terminals to increase the release of several neurotransmitters, it increases sensory input to the cortex, and its action to release dopamine slows the onset and diminishes the severity of Parkinson's disease (McGehee and Role, 1995; Kelton et al., 2000; Levin and Rezvani, 2000; Dani, 2001). (4) We know much less about what the activation of M_1–M_5 receptors does in the brain (or the therapeutic potential of specific agonists or antagonists for each receptor), because there have been no specific drugs for these receptors that are useful in vivo, either in animals or humans. (5) Esterase inhibitors are given widely to humans with Alzheimer's disease to rejuvenate cognition, but their mechanism of action is unknown (Barnes et al., 2000). The effect of increased levels of acetylcholine could be on M_1, M_2, M_3, M_4, M_5 and/or nicotinic receptors. (6) The most prevalent acetylcholine receptors in the brain are M_1 and M_4 receptors (Levey, 1993). There is growing

*Corresponding author: Tel.: 305-243-6912; Fax: 305-243-4555; E-mail: lpotter@miami.edu

DOI: 10.1016/S0079-6123(03)45008-5

evidence that M_1-activation increases (Marino et al., 1998; Rouse et al., 1999), and M_4-activation decreases the excitatory effects of glutamate in multiple parts of the brain, although these are not the only receptor coupling mechanisms. (7) Because of the prevalence of M_1 receptors, the excitatory effects of M_1-activation, and the fact that nonspecific muscarinic antagonists diminish cognition, there is a widespread belief that an M_1-agonist should improve memory in dementia. In support of this idea, M_1-Knockout mice show severe memory losses charcteristic of cortical/hippocampal dysfunction (Anagnostaras et al., 2003). (8) Both M_1- and M_4-knockout mice are hyperactive (Gomeza et al., 1999; Gerber et al., 2001; Miyakawa et al., 2001), and nonspecific muscarinic antagonists improve movement in Parkinson's disease, especially tremors (Fahn et al., 1990). Yet neither nonspecific muscarinic antagonists nor esterase inhibitors are notable for their effects on normal movement.

We conclude from the foregoing that we need better agonists and antagonists to examine the individual roles of M_1–M_5 receptors in the CNS.

Development of antimuscarinic toxins and their derivatives to study M_1 and M_4 receptors

We have discovered or synthesized the following ligands for studies of muscarinic receptors. (1) m1-Toxin1 is an antagonist (Cuevas et al., 1997; Marino et al., 1998; Rouse et al., 1999) that binds specifically and irreversibly to M_1 muscarinic receptors in vitro and in vivo (Liang et al., 2001; Potter, 2001). It is the most selective ligand known for characterizing M_1-neurotransmission. Antibodies against M_1 receptors are also highly selective, but they bind to the intracellular side of M_1 receptors (Levey, 1993) and cannot serve as antagonists. The fact that m1-toxin1 binds irreversibly means that only a few micrograms of the toxin are necessary for most experiments, the toxin can be used to occlude M_1 receptors so that other receptors can be studied easily (Potter and Purkerson, 1995; Purkerson and Potter, 1998), and the toxin can be used to assess the behavioral effects of acute M_1-blockade in localized regions of the brain (Liang et al., 2001). (2) m4-Toxin is an antagonist (Cuevas and Adams, 1997) that binds reversibly to M_4 receptors with an affinity ($K_d = 0.96$ nM) that is 102-

fold higher than that for M_1 receptors (Potter, 2001). It is the most selective ligand known for characterizing M_4-neurotransmission, and is completely specific for M_4 receptors after M_1 receptors are occluded with m1-toxin1 (Potter and Purkerson, 1995; Purkerson and Potter, 1998). Since m4-toxin binds reversibly, it is necessary to use substantial amounts in vitro (e.g., 10 nM to block 91% of M_4 receptors), and the duration of M_4-blockade is limited in vivo. (3) Derivatives of both toxins with biotin have the same binding properties as the native toxins. Biotinylated m1-toxin1 is useful for purifying M_1 receptors by affinity chromatography on avidin resins (McCollum et al., 2001). Biotinylated m4-toxin has been used with avidin-fluorophores for localizing M_4 receptors (Santiago and Potter, 2001), but will probably be superceded by fluorescent m4-toxin for anatomical studies. (4) ^{125}I-m1-Toxin1 and ^{125}I-m4-toxin bind like the native toxins (Liang and Potter, unpublished). They are useful for assays of M_1 receptors in membranes and cryostat sections (McCollum et al., 2001), and for the localization of M_1 receptors on the extracellular surfaces of neurons after injection of the toxin in vivo. (5) Conjugates of m1-toxin1 and m4-toxin with fluorophores are under study, and appear to be promising ligands. A fluorescent derivative of m1-toxin1 (MT-7) has been shown to yield good confocal images of the localization of M_1 receptors even at neuromuscular junctions in muscles, where the receptor level is very low (Minic et al., 2002). (6) m1-Toxin1, m4-toxin and their derivatives bind in vitro at 4–37°C in a wide variety of media, including digitonin solutions, and all bind specifically in vivo. The bound toxins are readily immobilized by perfusion fixation. (7) A site-directed mutant of m1-toxin1 (F38I) binds specifically but reversibly to M_1 receptors (Krajewski et al., 2001). Another mutant toxin (K65E) binds specifically and irreversibly, but does not slow the dissociation of ^{3}H-N-methylscopolamine (Max et al., 1993).

Development of new agonists to study M_1 receptors

Spalding et al. (2002) have discovered a highly unusual agonist that acts on an ectopic activation site of M_1 receptors. Because of its M_1-specificity, it is likely to provide a great deal of new information about M_1-linked neurotransmission.

Cholinergic mechanisms in the striatum

We have focused much of our attention on the striatum for the following reasons. (1) The rat caudate-putamen has more M_4 receptors than any other tissue (64 pmol/g; 51% of the total muscarinic receptors), and nearly as many M_1 receptors (53 pmol/g; 42%) as the hippocampus (Potter and Purkerson, 1995; Purkerson and Potter, 1998). It also has exceptional levels of choline acetyltransferase, acetylcholine and acetylcholinesterase, and a few nicotinic receptors (Graybiel and Ragsdale, 1993). (2) Most of the M_1 and M_4 receptors in the striatum are on the dendritic spines of striatal projection neurons (Hersch et al., 1994). Figure 1 shows the wiring of two sets of these projection neurons into 'direct and indirect' brain circuits that modulate movement (Smith et al., 1998; Wichmann and DeLong, 1998). The spines of both sets of projection neurons are heavily innervated by glutaminergic neurons from the cortex, by cholinergic interneurons, and by dopaminergic neurons from the substantia nigra compacta (Fig. 2) (Smith and Bolam, 1990; Hersch et al., 1994, 1995; Smith et al., 1998). Other neurotransmitters are present but not as important. M_1 receptors are coexpressed with glutamate receptors at the tips of the spines (Hersch et al., 1994), and there is some evidence that there are more M_1 receptors on indirect than direct neurons (Harrison et al., 1996). In contrast, M_4 and D_1 receptors are preferentially distributed on the spines of projection neurons in the direct pathway, and D_2 receptors are preferentially distributed on projection neurons in the indirect pathway (Fig. 2) (Hersch et al., 1994, 1995; Harrison et al., 1996; Santiago and Potter, 2001). (3) The striatum is unusual in that many of the autoreceptors for acetylcholine on cholinergic nerve terminals are M_4 receptors instead of M_2 receptors (Zhang et al., 2002). So anything that causes M_4-blockade must increase the release of acetylcholine. (4) The number of M_4 receptors on the surface of striatal neurons is regulated by endogenous and exogenous cholinergic activity (Bernard et al., 1999). (5) Many of the nicotinic receptors in the striatum are on dopaminergic

nerve terminals. Nicotine increases the release of dopamine (Wonnacott et al., 2000; Dani, 2001), and so must acetylcholine, by acting on nicotinic receptors. (6) The striatum regulates movement, which is easier to study than memory in animals (Chase et al., 1998). (7) Direct and indirect striatal projection pathways have opposite feedback effects in the cortex (Fig. 1). Anything that produces a direct > indirect imbalance (including loss of indirect projection neurons early in Huntington's disease, excessive use of levodopa or dopamine agonists for Parkinson's disease, upregulation of dopamine receptors in tardive dyskinesia, and loss of one subthalamic nucleus in hemiballismus) can yield excessive movement. Anything that produces an indirect > direct imbalance (including loss of nigrostriatal dopaminergic neurons in Parkinson's disease, blockade of dopamine receptors by anti-schizophrenic drugs, and depletion of dopamine by reserpine) can yield inadequate movement. (8) Unilateral lesions of the nigrostriatal tract in rodents produce an excellent and widely-studied model of Parkinson's disease ('hemi-PD') (Ungerstedt, 1976). (9) Simultaneous retrograde tracing of striatonigral projection neurons in the direct pathway, and of striatopallidal projection neurons in the indirect pathway, with fluorescent microbeads, shows that these neurons lie side by side with overlapping dendritic fields, yet there is only 4% overlap of their projections (Santiago and Potter, 2001). (10) Antimuscarinic agents are used to treat or to delay the onset of several hypokinetic disorders, including Parkinson's disease and Parkinsonism induced by dopamine antagonists (Fahn et al., 1990). The receptor sites and subtypes that account for the effectiveness of antimuscarinic agents in these disease states are not known. (11) M_1-knockout increases dopamine in the striatum of normal animals, and it causes hyperactivity (Gerber et al., 2001; Miyakawa et al., 2001). Both of these effects appear due to diminished activity of GABAergic striatonigral neurons that inhibit nigrostriatal dopaminergic neurons in the substantia nigra compacta. (12) M_4-knockout produces some increase in movement, and an increased responsiveness to dopamine (Gomeza et al., 1999). These effects presumably depend upon increased acetylcholine release from cholinergic nerve terminals in the striatum, and dis-inhibition of direct projection

124

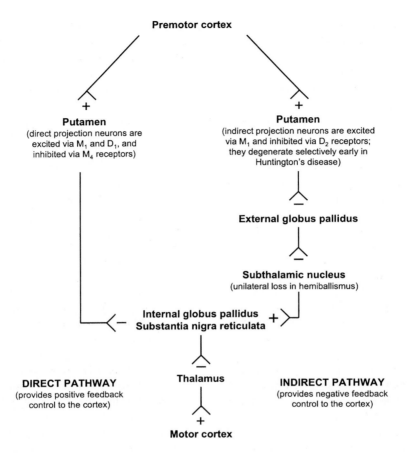

Premotor cortex

Putamen
(direct projection neurons are
excited via M_1 and D_1, and
inhibited via M_4 receptors)

Putamen
(indirect projection neurons are excited
via M_1 and inhibited via D_2 receptors;
they degenerate selectively early in
Huntington's disease)

External globus pallidus

Subthalamic nucleus
(unilateral loss in hemiballismus)

Internal globus pallidus
Substantia nigra reticulata

Thalamus

DIRECT PATHWAY
(provides positive feedback
control to the cortex)

INDIRECT PATHWAY
(provides negative feedback
control to the cortex)

Motor cortex

Fig. 1. Striatal projection neurons in the human brain. This figure demonstrates how two sets of striatal projection neurons can have opposite functions because of their wiring into direct and indirect neuronal circuits that exert opposite feedback effects in the motor cortex (Smith et al., 1998; Wichmann and DeLong, 1998). Plus signs indicate excitatory neurotransmission, whereas minus signs indicate inhibitory neurotransmission. The activity in these pathways is believed to be balanced, normally. This balance is very sensitive to dopamine levels, since dopamine excites the direct pathway via D_1 receptors, increasing movement, and it inhibits the indirect pathway via D_2 receptors, also increasing movement. Anything that decreases dopamine (e.g., Parkinson's disease) leads to an indirect > direct pathway imbalance and less movement, and anything that increases dopamine (e.g., excess levodopa) leads to a direct > indirect pathway imbalance and more movement. Cholinergic effects on these two pathways are even more complex, as noted in Fig. 2.

neurons (see Fig. 2). Both mechanisms should be beneficial in Parkinson states, even when dopamine levels are very low. (13) Since antimuscarinic agents are useful for hypokinetic disorders, one might expect that muscarinic agents or esterase inhibitors could be used to treat hyperkinetic disorders resulting from a direct > indirect pathway imbalance. But the non-specific effects of esterase inhibitors have not proved very therapeutic. (14) Nicotine is useful to treat or to delay the onset of Parkinson's disease (Louis and

Clarke, 1998; Kelton et al., 2000), presumably because it acts on dopaminergic nerve terminals to increase the release of dopamine. (15) Hemi-PD decreases the numbers of dendritic spines on striatal projection neurons (Ingham et al., 1998) and increases striatal M_1 receptors (Dawson et al., 1991). Since dopamine produces opposite effects on direct and indirect projection neurons (Figs. 1 and 2), hemi-PD may produce opposite effects on their spines. Other clinically relevant situations that are

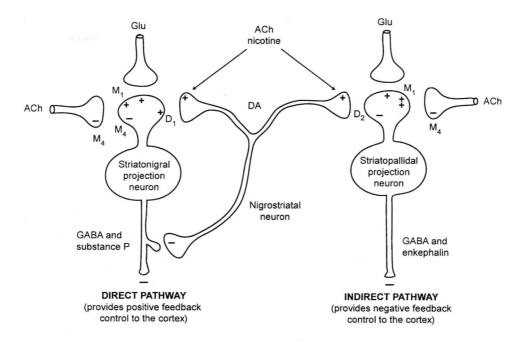

Fig. 2. A working model of the cholinergic control of direct and indirect striatal projection neurons in the rat. This cartoon of the innervation of dendritic spines on striatal projection neurons is based on data from many sources. Glu = glutamate, DA = dopamine, ACh = acetylcholine. Acetylcholine has at least five major effects, all of which are important. (1) Acetylcholine activates M_1 receptors which are coexpressed with glutamate receptors on dendritic spines (Hersch et al., 1994), to facilitate excitatory glutaminergic input. There appear to be more M_1 receptors on indirect than direct projection neurons (Harrison et al., 1996), so it is possible that M_1-activation tips the balance of activity of projection neurons towards indirect > direct. (2) Activation of striosome-based striatonigral neurons that go to the substantia nigra compacta reduces dopamine release from nigrostriatal neurons in the striatum (Graybiel et al., 2000; Gerber et al., 2001), thereby promoting indirect > direct activity. (3) Acetylcholine activates postsynaptic M_4 receptors that are preferentially expressed on direct striatonigral neurons (Hersch et al., 1994; Harrison et al., 1996; Santiago and Potter, 2001), causing inhibition and promoting indirect > direct activity. (4) Acetylcholine activates alpha4beta2 nicotinic receptors on dopaminergic nerve terminals and increases dopamine release (Sharples et al., 2000), thereby promoting direct > indirect activity. (5) Acetylcholine activates presynaptic M_4 (and some M_2) autoreceptors on cholinergic nerve terminals to decrease acetylcholine release (Zhang et al., 2002), with effects mediated by each of the above mechanisms. We believe that these complex interactions can be partially clarified in normal and hemi-PD rats by determining the effects of M_1- and M_4-blockade on movement, and by determining the effects of increasing or decreasing dopamine or acetylcholine on the spines themselves.

likely to alter dendritic spines and their M_1 receptors include the chronic administration of nicotine, esterase inhibitors and antimuscarinic agents.

It is evident that the effects of acetylcholine within the striatum and on movement are very complex. We believe that acute M_1-blockade, M_4-blockade, and studies of M_1 and M_4 receptors on dendritic spines can help to resolve a number of questions about the roles of each receptor in the striatum, and possibly lead to new therapies for motor disorders of basal ganglia origin. Examples of these approaches are noted in the following sections.

Effect of unilateral striatal M_1-knockdown on spontaneous movement

Unilateral intrastriatal injections of m1-toxin1 can produce nearly complete blockade of M_1 receptors in one whole rat striatum for several days ('M_1-knockdown') (Liang et al., 2001). We examined the effects of right M_1-knockdown on spontaneous forearm movements in normal rats and detected no change (Liang et al., 2001). But the expected result was increased left movement (based on the ability of M_1-knockout to increase dopamine), and the method used (Schallert and Tillerson, 1999) was not sufficient to show hyperactivity. So three further sets of experiments are planned. First, we will document hyperactivity after bilateral M_1-knockdown in normal animals. Animals with various degrees of right hemi-PD will then be tested to see whether improvements in spontaneous left forearm movement are dependent upon the presence of dopamine, as expected. Finally, rats with haloperidol-induced Parkinsonism will be tested with right M_1-knockdown to see whether left movement is improved. These studies will demonstrate whether improvement of movement due to M_1-blockade is dependent upon dopamine. These studies will be followed by parallel work with animals with M_4-knockdown. Based on Fig. 2, we anticipate that intrastriatal M_4-blockade will improve movement even when dopamine is deficient.

Effect of intrastriatal M_1-knockdown on pilocarpine-induced seizures

Large doses of the nonspecific muscarinic agonist, pilocarpine, produce tonic-clonic seizures in rodents. These seizures have been considered as a useful model for human epilepsy, and they resemble the convulsive effects of potent esterase inhibitors (Turski et al., 1989). Pilocarpine-induced seizures are absent in M_1-knockout mice (Hamilton et al., 1997), which implies a critical role for M_1 receptors in their initiation. The seizures are diminished by intrastriatal scopolamine or D_2 agonists (Turski et al., 1989) and by intrastriatal D_1 antagonists (Bourne, 2001), implying an important role for the striatum. We have investigated the effects of pilocarpine after bilateral striatal M_1-knockdown (Gutierrez-Ford and Potter, 2002). The rats were given 380 mg/kg of pilocarpine (after 1 mg/kg of N-methylscopolamine to protect peripheral muscarinic receptors), and were observed for 90 min. Extensive intrastriatal M_1-blockade did not alter the initiation or progression of pilocarpine-induced seizures. Thus striatal M_1 receptors are not required for the production of these seizures, and the seizures presumably depend upon the net excitatory effects of pilocarpine on M_1–M_5 receptors in the motor cortex. Blockade of intrastriatal M_1 receptors should decrease the activity of both direct and indirect projection neurons, and unrestrained M_4-inhibition of the direct neurons by pilocarpine should independently decrease the activity of direct projection neurons. Thus diminishing the activity of both sets of striatal projection neurons does not control seizures induced by the action of pilocarpine in the cortex.

Confocal microscopy of muscarinic receptors on dendritic spines

It is clear that most of the M_1 receptors in the brain are co-localized with glutamate receptors at the tips of dendritic spines in the cortex, hippocampus and striatum, that M_1-activation potentiates excitatory NMDA currents, and that M_1-knockout causes severe cortical-hippocampal memory deficits and hyperactivity. Thus M_1-rich spines are crucial for normal cognition and movement. We have begun to examine the specific characteristics of these spines in the rat hippocampus and striatum with fluorescent m1-toxin1 and confocal microscopy (McCollum et al., 2003). So far it is clear that the fluorescent toxin saturates M_1 receptors in vivo, that toxin-receptor complexes are stable in perfusion-fixed tissue, and that M_1-rich dendritic spines are beautifully imaged by confocal microscopy. We intend to use this approach to address the following questions. How does the administration of donepezil (an esterase inhibitor used to treat dementia) (Barnes et al., 2000), or atropine (a common muscarinic antagonist) (Wall et al., 1992) affect the numbers and luminosities of M_1-rich spines in the hippocampus? How does hemi-PD, the administrator of levodopa (Simon et al., 2000), a dopaminergic agonist (Watts, 1997), nicotine (Kelton et al., 2000) or haloperidol affect the numbers and luminosities of M_1-rich spines on striatonigral and striatopallidal neurons?

References

Anagnostaras, P.B., Murphy, G.C., Hamilton, S.E., Mitchell, S.L., Ranama, N.P. and Nathanson, W.M. (2003) Selective cognitive dysfunction in acetylcholine M_1 muscarinic mutant mice. Nature Neurosci., 6: 51–58.

Barnes, C.A., Meltzer, J., Houston, F., Orr, G., McGann, K. and Wenk, G.L. (2000) Chronic treatment of old rats with donepezil or galantamine: effects on memory, hippocampal plasticity and nicotinic receptors. Neuroscience, 99: 17–23.

Bernard, V., Levey, A.I. and Bloch, B. (1999) Regulation of the subcellular distribution of m4 muscarinic acetylcholine receptors in striatal neurons *in vivo* by the cholinergic environment: evidence for regulation of cell surface receptors by endogenous and exogenous stimulation. J. Neurosci., 19: 10237–10249.

Bourne, J.A.. (2001) SCH 23390: the first selective D(1)-like receptor antagonist. CNS Drug Rev. Winter, 7: 399–414.

Chase, T.N., Oh, J.D. and Blanchet, P.J. (1998) Neostriatal mechanisms in Parkinson's disease. Neurology, 51(Suppl. 1): S30–S35.

Cuevas, J. and Adams, D.J. (1997) M4 muscarinic receptor activation modulates calcium channel currents in rat intracardiac neurons. J. Neurophysiol., 78: 1903–1912.

Cuevas, J., Harper, A.A., Trequattrini, C. and Adams, D.J. (1997) Passive and active membrane properties of isolated rat intracardiac neurons: regulation by H- and M-currents. J. Neurophysiol., 78: 1890–1902.

Dani, J.A. (2001) Overview of nicotinic receptors and their roles in the central nervous system. Biol. Psychiatry, 49: 166–174.

Dawson, T.M., Dawson, V.L., Gage, F.H., Fisher, L.J., Hunt, M.A. and Wamsley, J.K. (1991) Downregulation of muscarinic receptors in the rat caudate-putamen after lesioning of the ipsilateral nigrostriatal dopamine pathway with 6-hydroxydopamine: normalization by fetal mesencephalic transplants. Brain Res., 540: 145–152.

Fahn, S., Burke, R. and Stern, Y. (1990) Antimuscarinic drugs in the treatment of movement disorders. Prog. Brain Res., 84: 389–397.

Gerber, D.J., Sotnikova, T.D., Gainetdinov, R.R., Huang, S.Y., Caron, M.G. and Tonegawa, S. (2001) Hyperactivity, elevated dopaminergic transmission, and response to amphetamine in M1 muscarinic acetylcholine receptor-deficient mice. Proc. Nat. Acad. Sci. USA, 98: 15312–15317.

Gomeza, J., Zhang, L., Kostenis, E., Felder, C., Bymaster, F., Brodkin, J., Shannon, H., Xia, B., Deng, C., Wess, J. (1999) Enhancement of D1 dopamine receptor-mediated locomotor stimulation in M4 muscarinic acetylcholine receptor knock-out mice. Proc. Nat. Acad. Sci. USA, 96: 10483–10488.

Graybiel, A.M. and Ragsdale, C.W., Jr. (1993) Biochemical anatomy of the striatum. In: P.C. Emson (Ed.), *Chemical Neuroanatomy*. Raven Press, NY, pp. 427–504.

Graybiel, A.M., Canales, J.J. and Capper-Loup, C. (2000) Levodopa-induced dyskinesias and dopamine-dependent stereotypies: a new hypothesis. Trends Neurosci., 23(Suppl. 10): S71–S77.

Gutierrez-Ford, C. and Potter, L.T. (2002) Role of striatal muscarinic receptors in pilocarpine-induced seizures. Soc. Neurosci. Abstr., 27: 912.7.

Hamilton, S.E., Loose, M.D., Qi, M., Levey, A.I., Hille, B., McKnight, G.S., Idzerda, R.L. and Nathanson, N.M. (1997) Disruption of the m1 receptor gene ablates muscarinic receptor-dependent M current regulation and seizure activity in mice. Proc. Nat. Acad. Sci. USA, 94: 13311–13316.

Harrison, M.B., Tissot, M. and Wiley, R.G. (1996) Expression of m1 and m4 muscarinic receptor mRNA in the striatum following a selective lesion of striatonigral neurons. Brain Res., 734: 323–326.

Hersch, S.M., Gutekunst, C.A., Rees, H.D., Heilman, C.J. and Levey, A.I. (1994) Distribution of m1-m4 muscarinic receptor proteins in the rat striatum: light and electron microscopic immunocytochemistry using subtype-specific antibodies. J. Neurosci., 14: 3351–3363.

Hersch, S.M., Ciliax, B.J., Gutekunst, C.A., Rees, H.D., Heilman, C.J., Yung, K.K., Bolam, J.P., Ince, E., Yi, H., Levey, A.I. (1995) Electron microscopic analysis of D1 and D2 dopamine receptor proteins in the dorsal striatum and their synaptic relationships with motor corticostriatal afferents. J. Neurosci., 15: 5222–5237.

Ingham, C.A., Hood, S.H., Taggart, P. and Arbuthnott, G.W. (1998) Plasticity of synapses in the rat neostriatum after unilateral lesion of the nigrostriatal dopaminergic pathway. J. Neurosci., 18: 4732–4743.

Kelton, M.C., Kahn, H.J., Conrath, C.L. and Newhouse, P.A. (2000) The effects of nicotine on Parkinson's disease. Brain Cogn., 43: 274–282.

Krajewski, J.L., Dickerson, I.M. and Potter, L.T. (2001) Site-directed mutagenesis of m1-toxin1: two residues responsible for high affinity binding to M1 muscarinic receptors. Mol. Pharmacol., 60: 725–731.

Levey, A.I. (1993) Immunological localization of m1-m5 muscarinic acetylcholine receptors in peripheral tissues and brain. Life Sci., 52: 441–448.

Levin, E.D. and Rezvani, A.H. (2000) Development of nicotinic drug therapy for cognitive disorders. Eur. J. Pharmacol., 393: 141–146.

Liang, J.S., Santiago, M.P. and Potter, L.T. (2001) Sustained unilateral blockade of rat striatal M1 muscarinic receptors with m1-toxin, in vivo. Brain Res., 921: 211–218.

Louis, M. and Clarke, P.B. (1998) Effect of ventral tegmental 6-hydroxydopamine lesions on the locomotor stimulant action of nicotine in rats. Neuropharmacology, 37: 1503–1513.

Marino, M., Rouse, S.T., Levey, A.I., Potter, L.T. and Conn, P.J. (1998) Activation of the genetically defined m1 muscarinic receptor potentiates N-methyl-D-aspartate (NMDA) receptor currents in hippocampal pyramidal cells. Proc. Nat. Acad. Sci. USA, 95: 11465–11470.

Max, S.I., Liang, J.S. and Potter, L.T. (1993) Stable allosteric binding of m1-toxin to m1 muscarinic receptors. Mol. Pharmacol., 44: 1171–1175.

McCollum, M.H., Flynn, D.D. and Potter, L.T. (2001) Biotinylated m1-toxin. Soc. Neurosci. Abstr., 27: 810.16.

McCollum, M.H., Flynn, D.D., Liang, J.-L. and Potter, L.T. (2003) Fluorescent m1-toxin1: a probe for studying compensatory changes in dendritic spines rich in M_1 muscarinic receptors. Soc. Neurosci. Abstr. 29: in press.

McGehee, D.S. and Role, L.W. (1995) Physiological diversity of nicotinic acetylcholine receptors expressed by vertebrate neurons. Ann. Rev. Physiol., 57: 521–546.

128

Minic, J., Molgo, J., Karlsson, E. and Krejci, E. (2002) Regulation of acetylcholine release by muscarinic receptors at the mouse neuromuscular junction depends on the activity of acetylcholinesterase. Eur. J. Neurosci., 15: 439–448.

Miyakawa, T., Yamada, M., Duttaroy, A. and Wess, J. (2001) Hyperactivity and intact hippocampus-dependent learning in mice lacking the M1 muscarinic acetylcholine receptor. J. Neurosci., 15: 5239–5250.

Nasman, J., Jolkkonen, M., Ammoun, S., Karlsson, E. and Akerman, K.E.O. (2000) Recombinant expression of a selective blocker of M_1 muscarinic receptors. Biochem. Biophys. Res. Comm., 271: 435–439.

Potter, L.T. (2001) Snake toxins that bind specifically to individual subtypes of muscarinic receptors. Life Sci., 68: 2541–2547.

Potter, L.T. and Purkerson, S.L. (1995) Pharmacology of striatal muscarinic receptors. In: M. Ariano and J. Surmeier (Eds.), Molecular and Cellular Mechanisms of Neostriatal Function. R.G. Landes Co., New York, pp. 241–254.

Potter, L.T., Krajewski, J.L. and Dickerson, I.M. (1997) Cloning and expression of the cDNA for an m1-toxin. Life Sci., 60: 1205.

Purkerson, S.L. and Potter, L.T. (1998) Use of antimuscarinic toxins to facilitate studies of striatal m4 muscarinic receptors. J. Pharmacol., 284: 707–713.

Rouse, S.T., Marino, M.J., Potter, L.T., Conn, P.J. and Levey, A.I. (1999) Muscarinic receptor subtypes involved in hippocampal circuits. Life Sci., 64: 501–509.

Santiago, M.P. and Potter, L.T. (2001) Biotinylated m4-toxin demonstrates more M4 muscarinic receptor protein on direct than indirect striatal projection neurons. Brain Res., 894: 12–20.

Schallert, T. and Tillerson, J.L. (1999) Development of behavioral outcome measures for preclinical Parkinson's research. In: D.F. Emerlich, R.L. Dean and P.R. Sanberg (Eds.), Central Nervous System Diseases. Innovative Animal Models from Lab to Clinic. Humana Press, Inc, Clifton, NJ, pp. 131–169.

Sharples, C.G., Kaiser, S., Soliakov, L., Marks, M.J., Collins, A.C., Washburn, M., Wright, E., Spencer, J.A., Gallagher, T., Whiteaker, P. and Wonnacott, S. (2000) UB-165: a novel nicotinic agonist with subtype selectivity implicates the alpha4beta2 subtype in the modulation of dopamine release from rat striatal synaptosomes. J. Neurosci., 20: 2783–2791.

Simon, N., Mouchet, J. and Bruguerolle, B. (2000) Effects of seven-day continuous infusion of L-DOPA on daily rhythms in the rat. Eur. J. Pharmacol., 401: 79–83.

Smith, Y., Bevan, M.D., Shink, B. and Bolam, J.P. (1998) Microcircuitry of the direct and indirect pathways of the basal ganglia. Neuroscience, 86: 353–387.

Smith, D.A. and Bolam, J.P. (1990) The neural network of the basal ganglia as revealed by the study of synaptic connections of identified neurones. Trends Neurol. Sci., 13: 259–2265.

Spalding, T.A., Trotter, C., Skjaerbaek, N., Messier, T.L., Currier, E.A., Burstein, E.S., Li, D., Hacksell, U. and Brann, M.R. (2002) Discovery of an ectopic activation site on the M1 muscarinic receptor. Mol. Pharmacol., 61: 1297–1302.

Turski, L., Ikonomidou, C., Turski, W.A., Bortolotto, C.A. and Cavalheiro, E.A. (1989) Review: cholinergic mechanisms and epileptogenesis. The seizures produced by pilocarpine: a novel experimental model of intractable epilepsy. Synapse, 2: 154–171.

Ungerstedt, U. (1976) 6-Hydroxydopamine-induced degeneration of the nigrostriatal dopamine pathway: the turning syndrome. Pharmacol. Ther., 2: 37–40.

Wall, S.J., Yasuda, R.P., Li, M., Ciesla, W. and Wolfe, B.B. (1992) Differential regulation of subtypes m1-m5 muscarinic receptors in forebrain by chronic atropine administration. J. Pharmacol., 262: 584–588.

Watts, R.L. (1997) The role of dopamine agonists in early Parkinson's disease. Neurology, 49(Suppl 1): S34–S48.

Wichmann, T. and DeLong, M.R. (1998) Models of basal ganglia function and pathophysiology of movement disorders. Neurosurg. Clin. N. Amer., 9: 223–236.

Wonnacott, S., Kaiser, S., Mogg, Soliakov, A.L. and Jones, I.W. (2000) Presynaptic nicotinic receptors modulating dopamine release in the rat striatum. Eur. J. Pharmacol., 393: 51–58.

Zhang, W., Basile, A.S., Gomeza, J., Volpicelli, L.A., Levey, A.I. and Wess, J. (2002) Characterization of central inhibitory muscarinic autoreceptors by the use of muscarinic acetylcholine receptor knockout mice. J. Neurosci., 22: 1709–1717.

Cortical properties and functions modulated by acetylcholine

Progress in Brain Research, Vol. 145
ISSN 0079-6123

CHAPTER 9

Hyperactivation of developing cortical circuits by acetylcholine and the ontogeny of abnormal cognition and emotion: findings and hypothesis

Alejandro Peinado* and D. Paola Calderon

*Department of Neuroscience, Albert Einstein College of Medicine,
1300 Morris Park Ave, Bronx, NY 10461, USA*

Perhaps the most exciting possibility for the future is the extension of this type of work [on the role of experience during a critical period in the development of the visual system] to other systems besides sensory. Experimental psychologists and psychiatrists both emphasize the importance of early experience on subsequent behavior patterns—could it be that deprivation of social contacts or the existence of other abnormal emotional situations early in life may lead to a deterioration of connections in some as yet unexplored part of the brain?

David Hubel (1967).

Background and hypothesis

Activity and the development of cortical circuits

It is now a well-established fact that neuronal activity plays an important role in the development of sensory pathways. Beginning with the work of Hubel

*Corresponding author: Tel.: 718 430 3681;
Fax: 718 430 8821; E-mail: peinado@aecom.yu.edu

DOI: 10.1016/S0079-6123(03)45009-7

and Wiesel almost five decades ago, a wealth of information has accumulated in support of this notion and progress has been made toward elucidating many of the mechanisms involved (Wiesel, 1982; Katz and Shatz, 1996; Crair, 1999; Zhang and Poo, 2001). Although it may not be perfectly clear whether activity is involved in the initial specification of connections, in the maintainance of previously formed connections, or in both (see Katz and Crowley, 2002), there is no doubt that reducing activity at critical times results in connectivity that is highly abnormal. The same is likely true of excessive activation during development (see Lynch et al., 2000) although, by comparison to activity-deprivation, this aspect of activity-dependence has barely been studied.

That the developing nervous system is capable of generating organized activity by itself, be it instead of or in addition to sensory-driven activity (O'Donovan, 1999; Ben-Ari, 2001), further adds to the notion that activity is an integral part of the complex program of development by which connectivity patterns emerge. The evidence suggests that the overall level of activity in developing networks needs to be carefully maintained within an appropriate window, particularly during the time when the bulk of circuitry is being laid down.

From what we know so far about activity-dependent mechanisms and their role in circuit formation, it is reasonable to infer that other systems besides sensory also depend on activity for the proper formation of their circuitry. With the exception of some studies on hippocampus, however, activity-dependent development in nonsensory brain regions remains a poorly explored area of neural development. As a result, we know very little about the vulnerabilities of nonsensory areas to factors that can alter their levels of activation during development. This is unfortunate considering the potential relevance of this knowledge to the etiology of disorders that affect higher cognitive functions. Prominent among these disorders is schizophrenia, a disorder widely regarded to involve neural development (Lewis and Levitt, 2002).

When considering the possibility of abnormal levels of activation, cortical areas such as the frontal regions, the premotor or supplementary motor areas, and the cingulate areas are of particular interest for two reasons: First, functional abnormalities in these areas in mature brain are generally associated with a range of neuropsychiatric disorders (Cohen et al., 1987; Weinberger et al., 1992; Carter et al., 1997; Curtis et al., 1998; Bradshaw and Sheppard, 2000; Holcomb et al., 2000), and therefore the occurrence of developmental events that may derail the process of circuit formation in these areas is of great importance clinically. Second, these areas have the ability to exhibit sustained neuronal firing in the absence of sustained sensory input, a property thought to underlie the ability of neural circuits in these areas to hold information in 'working' (short-term) memory (Durstewitz et al., 2000). It follows that if this property is present in some form during early circuit formation and can be driven to excess through a combination of abnormal genetic and/or environmental factors, these areas could be selectively vulnerable to the deleterious effects of excessive activation. According to this view, lower level sensory and motor areas, those closer to the peripheral end of the perception–action cycle and not as directly involved in holding information in working memory, would be less vulnerable insofar as they are less likely to generate sustained firing. Our analysis of developmental processes in cortex has as an important aim defining the extent and nature of potential differences between cortical areas.

Cholinergic modulation and excessive activity in developing networks

Could abnormal cholinergic function during circuit formation be one possible scenario under which excessive activation occurs in immature higher cortical circuits? We consider this likely since activation of muscarinic receptors typically enhances neuronal excitability. More importantly, muscarinic receptors seem to be directly involved in enabling sustained firing. Thus in frontal cortex in vitro, neurons switch to a sustained firing mode in response to muscarinic activation (Haj-Dahmane and Andrade, 1996, 1998). This type of firing is similar to that seen in single units in vivo during delayed matching tasks requiring working memory. Moreover, muscarinic blockade in the medial prefrontal cortex of rats interferes with the performance of a delayed matching task (Broersen et al., 1994, 1995), suggesting again an important role for muscarinic receptors in sustained firing.

If present, what effect could excessive cholinergic activity have during development? One possibility at the local circuit level is that, particularly in areas of cortex that subserve various types of working memory, increased cholinergic activity will cause abnormally high levels of neuronal firing. This in turn could disrupt the process of competition necessary for synapse selection and lead to formation of spurious connections. Excessive spurious connections might lead later in life to excessive pruning of synapses, resulting in an impoverished neuropil, a common anatomical finding in the cerebral cortex of schizophrenics (Feinberg, 1982; Selemon and Goldman-Rakic, 1999; Thompson et al., 2001).

It is interesting that muscarinic cholinergic hyperactivity may be one feature of some or all patients with schizophrenia (Tandon et al., 1991). Whether this implies increased signaling through *cortical* muscarinic receptors in schizophrenics is not known. Many other regions could be involved. Garcia-Rill et al. (1995), for example, found a very significant increase in the number of cholinergic mesopontine neurons in schizophrenics. However, since mesopontine neurons innervate and depolarize thalamocortical neurons but not cortical neurons, it is unclear what this means for a possible cortical cholinergic excess. Further studies of network activation in developing cortex will clarify the extent to which excessive signaling via other modulatory transmitter receptors (e.g., metabotropic glutamate receptors) can also translate into higher overall activity levels. And further postmortem analysis of schizophrenic brains may help clarify the extent of cortical cholinergic abnormalities in this disorder.

Another key question that needs to be examined, but one that is unlikely to be addressed through postmortem analysis of brains of psychiatric patients, is whether cholinergic hyperactivity was already present during the period of circuit formation in those patients whose schizophrenic symptoms correlate with cholinergic excess. It is already clear that anatomically the cholinergic innervation of the cortex is present from very early on (Mechawar and Descarries, 2001; Mechawar et al., 2002). This implies that it could play an important role developmentally. In principle therefore, an early abnormality in this system could disrupt the processes responsible for setting up the correct wiring in key cortical areas, particularly those containing a high proportion of neurons capable of sustained firing under muscarinic control. Such a disruption could persist into adulthood in the form of a functional deficit arising from altered connectivity. Here we present some findings that bear on how this disruption might come about. Some of these findings have been described previously (Peinado, 2000).

Materials and methods

The experiments described here were done on 300–400 μm-thick slices obtained from Long–Evans rats pups aged 0–15 days. The plane of section was coronal, sagital, or horizontal and, since they were taken from whole brains, included as much cortex as could be obtained in each plane. Slices were incubated in artificial CSF (aCSF) containing fura-2 AM (5 μg/ml) for 1–2 h at 30°C. The composition of the aCSF, modified according to MacGregor et al. (2001), to prevent neuronal swelling was (in mM): NaCl 109, KCl 2.5, KH_2PO_4 1.25, $MgCl–6H_2O$ 1, $NaHCO_3$ 35, Glucose 10, HEPES 20, and $CaCl_2$ 2. The pH was 7.4 after bubbling CO_2. Slices were then placed in a temperature-controlled (29 ± 1°C) perfusion chamber (rate: 2–3 ml/min; volume: 200–400 μl)

on a microscope (Zeiss Axioskop Fixed Stage) stage and viewed with 10X, 20X, 40X or 63X water-immersion objectives. A digital camera (Princeton Instruments Princeton NJ, 512 × 512 Frame Transfer, 1 MHz) was used for optical recordings. A 0.5X adapter was sometimes placed in the microscope's trinocular port in order to view a larger field (up to 1.5 mm × 1.5 mm) with the CCD camera. IPLab software (Scanalytics, VA) running on a Macintosh computer was used for image acquisition and analysis. Camera was operated on a 5 × 5 binning mode. All fura-2 activity was recorded at a single excitation wavelength using a 380 ± 5 nm band-pass filter. Emission fluorescence was filtered with a 400-nm longpass filter.

For imaging of voltage signals, slices were stained for 5 min with the voltage-sensitive dye di-4-ANEPPS (Fluhler et al., 1985) (0.66 mg/ml in aCSF) while in the perfusion chamber, and only after a wave-expressing region had been located using fura-2 imaging. Wave activity was imaged immediately, after a 5 min wash period, to minimize artifactual signals attributable to gradual internalization of the dye and photodynamic damage (Schaffer et al., 1994). For dual fura-2/di-4-ANEPPS recordings excitation filters (380 ± 5 and 545 ± 5 nm) were alternated every 0.5 sec, and a single emission filter (590 nm longpass) was used. Neutral density filters were used to equalize fura-2 and di-4-ANEPPS signal intensities and to minimize photobleaching.

For whole-cell current-clamp recordings, patch pipettes were pulled from borosilicate glass (3–5 MΩ) and filled with an internal solution containing (in mM): K gluconate 110, KCl 20, HEPES 10, EGTA 10, $CaCl_2$ 1, $MgCl_2$ 1, pH 7.2; osmolarity was adjusted to 307 mOsm with sucrose. Signals were amplified with an Axopatch 200A patch-clamp amplifier (Axon Instruments Union City, CA) and acquired using Synapse (Synergy Research Silver Spring, MD) and an ITC-16 computer interface (Instrutech Corporation Port Washington, NY).

Drugs were applied in the bath through a gravity perfusion system. For brief applications drug perfusion was synchronized to imaging using Synapse and began 10 s after the start of imaging to allow acquisition of baseline fluorescence values. To determine the minimum perfusion time required to reach the desired agonist concentration in the

134

bath, separate experiments were carried out in parallel using a water-soluble fluorescent dye (NBD-methylglucamine, M.W. 356; Molecular Probes Eugene, OR) dissolved in the aCSF at typical agonist concentrations. The change in fluorescence as a function of time was measured at the chamber with the CCD camera (Fig. 1). A 20-s perfusion time was found to be necessary for the dye fluorescence in the bath to reach a steady state level. Most agonist applications were therefore limited to 20 s. Actual concentrations inside the slice were not measured but are presumed to be lower than in the bath. The interval between applications was 10 min or more to ensure near complete wash of drug from the slice.

Findings and discussion

Cholinergic stimulation of early postnatal cortex: waves of activation

Under conditions of normal aCSF, electrical recordings in cortical brain slices at any age show very little, if any, spontaneous electrical activity. The same is true of neonatal brain slices examined using fluorescent calcium dyes and imaging techniques, in

which a few individual neurons may exhibit spontaneous changes in intracellular $[Ca^{2+}]$, but there is no large-scale patterned activation (see Garaschuk et al., 2000 for a notable exception). Although this virtual absence of activity is clearly not a state which is likely to exist in vivo, it does provide an opportunity to examine, free of ongoing activation, the effect of modulatory transmitters, which are likely to be present in vivo, but whose afferents have been removed along with other excitatory afferents during the preparation of a slice. One such group of afferents that is inevitably missing from a slice is the cholinergic innervation originating in the basal forebrain.

To understand the possible effects of acetylcholine (ACh) on immature cortical circuits, we have examined the response of cortical neurons to cholinergic stimulation. We have done this by imaging large-scale activity at low magnification and recording changes in membrane potential with whole cell patch clamp in response to different cholinergic agonists during the first postnatal week. Although the experiments employ the lowest agonist concentrations capable of eliciting the responses under investigation as observed in a slice preparation, it is unclear whether such responses can occur in vivo under normal conditions.

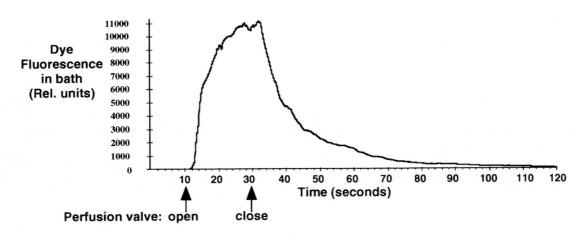

Fig. 1. Time course of the cholinergic agonist superfusion in bath. To optically measure the rate of rise and decay in the concentration of cholinergic agonists during superfusion of a slice, experiments were simulated using a small molecular weight fluorescent dye (NBD-methylglucamine, Molecular Probes) in the superfusate instead of the cholinergic agonist. A dual channel perfusion setup was used with one reservoir containing dye-aCSF (20 μM dye; comparable to drug concentration used during actual agonist experiments). At the time indicated by the first arrowhead, the normal aCSF was turned off and the perfusion was switched to the dye-containing aCSF. The record is an average of ten traces showing the time course of changes in dye concentration during a 20 s application.

It is possible that they reflect a response more likely to occur as a result of slight abnormalities in any of a number of steps along the signal transduction cascade involved in cholinergic signaling.

What our studies have shown is that carbachol can induce a large transient increase in intracellular calcium, particularly in upper layer cortical neurons (Fig. 2). The calcium transient has two components. One is a low amplitude signal that occurs in direct response to the increased levels of agonist in the bath and at more or less the same time in all neurons throughout the slice. The second component rides on the first, is much larger in amplitude, has a fast rise and exhibits a lag at one cortical site relative to another site located some distance away. This lag is due to the fact that this second component reflects an event that occurs as a wave of activation that spreads

along the horizontal dimension of cortex at about 125 μm/s. The low amplitude component outlasts the relatively brief (typically 20 s) application of the agonist. Likewise, the wave component will often initiate long after (sometimes more than a minute after) the agonist washout from the bath was begun, suggesting that the trigger for this event may not be entirely under the control of processes set in motion by the agonist, but may also depend on a second independent and possibly stochastic event occurring in the network of neurons.

A likely candidate for this trigger event is the depolarization of a group of neurons above some threshold, possibly simply the action potential threshold. To test whether depolarization of a local group of neurons is sufficient to initiate the wave event induced when a cholinergic agonist is applied to a

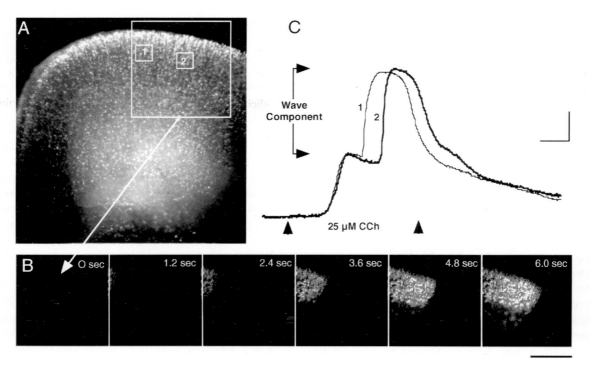

Fig. 2. The response to cholinergic stimulation of immature cortex is a wave-like activation of cortical networks. (A) Typical low magnification image of a coronal P5 cortical slice stained with the calcium indicator dye Fura-2 AM, showing the region from which a sequence of images was obtained for analysis of neuronal responses to bath-applied carbachol (CCh) as in B. (B) Select images (at 1.2 s intervals) from an experimental sequence processed to highlight the lateral advance of a wave of activation. White areas reflect high calcium relative to baseline. Scale bar, 300 μm. (C) Fluorescence changes obtained at two separate sites (shown in A) reflect the changes in intracellular calcium in response to agonist (applied during period between arrowheads). The wave component of the response rides on a lower amplitude calcium transient that, in contrast to the wave component, occurs simultaneously throughout the slice. Scale bar: 10 s, 5% ΔF/F.

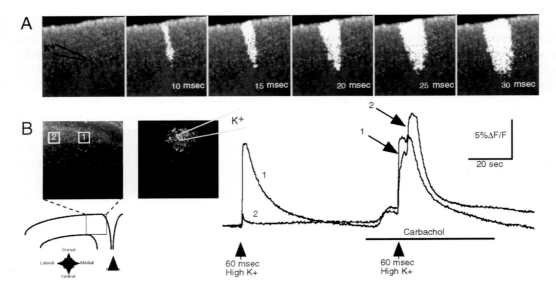

Fig. 3. Local injection of high potassium saline is not sufficient to induce wave-like responses in the absence of a cholinergic agonist. (A) Spatial extent of the effect of injecting 126 mM K^+ aCSF into the slice through a small tip pipette in a P3 cortex slice bathed in normal aCSF. From left to right the processed images show the effects of progressively longer pressure pulses. White region maps cells exhibiting a greater than 3% change in fluorescence. Pial surface is at top. First image shows injection site. (B) Experiment showing the effect of bath-applied carbachol (25 μM) on the response to high potassium injection. Leftmost image shows regions of interest (ROI) from which traces at right were obtained. Second image shows area responding to high potassium in the absence of carbachol. Traces show response to potassium before and during carbachol at both ROIs. Notice large difference in ROI #2 as wave propagates from injection site during carbachol stimulation. P5 slice. Region imaged shown on diagram at left.

slice, we made small pressure injections of high potassium (high K^+) aCSF into the slice (Fig. 3A). Our results show that, in the absence of the cholinergic agonist, depolarizing even relatively large groups of neurons with high K^+ aCSF, a procedure that causes large changes in neuronal calcium near the injection site, was unable to induce a wave event, suggesting that local depolarization of neurons is not sufficient to initiate these events. Previously we have shown that large-scale depolarization of the slice by bath-applied glutamate is also unable to induce a wave event (Peinado, 2000). In contrast, local injection of high K^+ aCSF in combination with bath application of a cholinergic agonist results in a wave of activation that spreads beyond the region directly affected by the high K^+ and mimics the spontaneously initiated wave in every respect (Fig. 3B). The fact that in such experiments the initiation of the wave event is timed precisely to the injection suggests that depolarization may be the limiting factor in wave initiation when cholinergic stimulation alone is used.

Initial observations indicated that the wave component could be recorded at some, but not all, cortical sites, in coronal slices prepared from the middle third (in the antero-posterior dimension) of the brain. Thus based on experiments where agonist was applied to the slice repeatedly while recording from the same site, two types of sites were identified: those in which waves of activation could be elicited repeatedly and those in which no waves could be induced despite repeated agonist application. To better define the identity of responsive and non-responsive areas, we recently did a more detailed mapping of these areas. The results are summarized below. First, however, we describe some aspects of the physiology of waves.

Waves of network activation: Physiology and pharmacology

Two approaches were used to determine if neuronal electrical activity was present and necessary for the calcium transients to occur. Membrane potential activity was recorded in parallel with the calcium imaging by using either a voltage sensitive dye (di-4-ANEPPS) to detect large-scale changes, or whole cell recordings to measure changes in single neurons. These experiments revealed that depolarization and firing occur during the wave component of the cholinergic response (Fig. 4). Moreover, a requirement for action potential firing in generating the wave component, although not the lower amplitude component, was demonstrated by the fact that the sodium channel blocker tetrodotoxin abolishes that part of the response (Peinado, 2000).

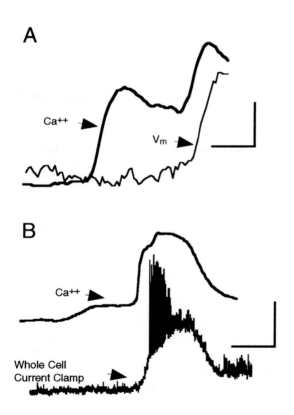

Fig. 4. Membrane depolarization and firing accompany the wave component of the cholinergic response. (A) Simultaneous calcium and voltage recording using Fura-2 and the voltage sensitive dye di-4-ANEPPS, respectively, shows that depolarization occurs during the wave component of the response. Scale bar: 20 s, 5%ΔF/F (Ca^{++}), 0.5% ΔF/F (V$_m$). (B) Simultaneous calcium and whole cell current clamp recording shows that firing occurs during the rising phase of the wave component. P6 slice. Scale bar: 10 s, 10%ΔF/F, 25 mV.

Although carbachol activates both muscarinic and nicotinic receptors, wave responses were blocked by the broad-spectrum muscarinic antagonist, atropine, and identical responses were obtained with the selective muscarinic agonist, muscarine. Pharmacological blockade of specific receptor subtypes was then used to demonstrate that the response appears to be mediated by M1 and M3 type receptors (Peinado, 2000), both of which are known to signal via a cascade involving phosophoinositide metabolism (Caulfield, 1993; Loffelholz, 1996). It is unclear, however, what conductance(s) is (are) affected by muscarinic activation that leads to the observed response. Two likely possibilities are the closure of a potassium conductance or the opening of a cation nonselective conductance. The closure of a potassium conductance would explain the increase in input resistance first described by Krnjević et al. (1971). However both of these mechanisms have been invoked to account for the depolarization and the ensuing firing observed in cortical neurons in response to muscarinic activation (McCormick and Prince, 1986; Haj-Dahmane and Andrade, 1996).

It is unclear also by what mechanism activity propagates from one neuron to another during a wave event. Experiments in which, ionotropic glutamate receptor antagonists were perfused on the slice prior to and during application of a muscarinic agonist, indicate that propagation is unaffected by blockade of these receptors. Thus AMPA and NMDA receptors, which are the most likely candidates for the spread of excitation in cortex, appear not to be necessary for this phenomenon to occur.

Another intercellular signaling mechanism that could be involved in propagation of muscarinic-evoked activity is the neurotransmitter GABA. Although there is evidence that GABA may act to excite neurons in early postnatal brain (Cherubini et al., 1991), our preliminary experiments, blocking GABAA receptors with bicuculline or its chloride channel with picrotoxin, have failed to produce a consistent effect on muscarine-induced waves of activity that would suggest a role for GABA in wave propagation (unpublished observations). Other mechanisms for transmission of excitation, such as gap junctions, glycine receptors, and metabotropic glutamate receptors, still need to be tested for their role in propagation of waves following muscarinic activation.

A final important feature of the muscarine-induced activity waves in cortex is the fact that their expression appears to be restricted to a fairly narrow temporal window, mostly from about P3 to P7 in rat. After this time, concentrations of muscarine as high as 100 μM were not able to induce the wave component of activation, even during prolonged (1 min) applications of the agonist.

Muscarinic activation of developing cortical networks: area specificity

In our initial studies of muscarinic activation in neonatal cortex, we imaged random locations throughout the entire extent of coronal slices. The slices were taken from the brain's midsection along the anterior-posterior axis, and in most cases included the entire dorso-ventral and medio-lateral extent of the hemisphere. As mentioned, not all regions of cortex responded to muscarinic stimulation with a wave of activation. Many regions only exhibited the low amplitude part of the response described above, but not the wave. An early indication that cortex may be divided into regions that support wave events and regions that do not came from calcium imaging sequences, in which wave propagation was observed to terminate at the same location when imaged repeatedly over a long period (Fig. 5).

To reveal a consistent pattern in the expression of wave responses, we performed a more systematic mapping of the areas exhibiting activity, combining low magnification imaging with successive imaging of adjacent fields. Our results show that at P3 waves are able to invade most areas of a slice. Starting at P4, however, wave responses become more restricted spatially. In coronal sections of brain taken from midsection, there are two areas that consistently exhibit wave responses. One, which we have selected for most subsequent studies, is the most dorsal strip of cortex, an area that most likely corresponds to FR2, also known as the medial agranular cortex (AGm). This area may perform functions similar to what the supplementary motor cortex performs in primates and other higher order mammals (Uylings and van Eden, 1990). Studies have shown involvement

of this area in organizing behavior and attention (Burcham et al., 1997). AGm is highly interconnected with many cortical areas, including the cingulate cortex medially and the motor cortex laterally; it thus occupies a key position at the interface between motivation and action (Reep et al., 1987). Typically, in a coronal slice, the wave response in this area initiates medially and travels laterally for about 500–1000 μm, at which point the propagation ends abruptly. Beyond this point, in the motor and somatosensory areas, the response to muscarine consists of the low amplitude nonwave component alone (Fig. 5).

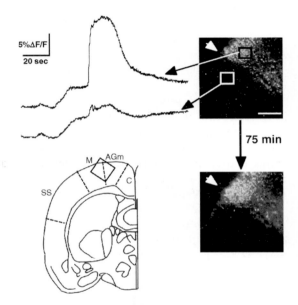

Fig. 5. Sharp stable boundaries separate areas that generate a wave response to cholinergic stimulation from areas that do not. Processed images taken from sequences recorded 75 min apart show the spatial extent of cholinergically-induced waves originating in the dorsal (AGm) region of cortex in a P5 slice. Light area is the region where cells exhibited large calcium transients. Traces at left show fluorescence changes sampled on both sides of the boundary. Notice that the low amplitude calcium transient, but not the wave component, is present in the dark region, where the wave fails to invade. This failure of waves to invade lateral cortex is typical of slices taken at P4 and older ages. Diagram at lower left shows the place from which images were taken (square) and the approximate location of cortical areas. However, the precise location of boundaries between areas cannot be determined with certainty in immature slices. C, cingulated cortex; M, motor cortex; SS, somatosensory cortex. Arrowheads mark the same location on the slice in both images. Scale bar, 100 μm.

Experiments in which a vertical cut was made in the slice lateral to the point of initiation, but within the wave-expressing region, suggest that the medial initiation site is necessary for expression of this activity in the rest of the region; as if a pacemaker of sorts is located medially within this region. More needs to be done, however, to define the spatial extent of this putative pacemaker, as well as to determine what cells or cellular properties are found in this region that endow it with the ability to initiate this activation. One possibility consistent with some experiments is that cells in this area are simply depolarized more easily to threshold by the muscarinic stimulation. This view is supported by the observation that highly localized injections of high potassium aCSF can, in the presence of muscarine (although not in its absence), elicit a wave of activation from anywhere in this dorsal region (Fig. 3).

Another region where wave activation takes place in coronal slices is a strip of lateral cortex. At the level from which slices are obtained, this area is ventral to the somatosensory cortex and may correspond to the agranular insular cortex, believed to contain higher order (associational) cortical regions involved in working memory for food reward (Ragozzino and Kesner, 1999). With slices that have been obtained from regions closer to the anterior pole of the brain, or those that have been prepared in the horizontal or sagital planes, it has been possible to observe very extensive wave-like activation in the prefrontal cortex. In contrast, imaging in sagital slices that include the occipital pole (visual cortex) reveals that cholinergic stimulation does not induce a wave of activation in this sensory region.

Thus, the pattern that has emerged so far from these mapping studies is one that suggests some correspondence between areas that exhibit this form of hyperactivation in response to muscarinic stimulation and areas that in adult cortex have a preponderance of neurons with the ability to fire in a sustained mode.

Because the wave-like events exhibited by these areas in response to muscarinic agonists involve both action potential firing and large changes in intracellular calcium occurring simultaneously in large numbers of neurons during a period of high synaptic plasticity, the potential exists for them to have profound effects on the emerging pattern of connectivity in vivo.

Future directions: Can muscarinic hyperactivation occur during cortical development in vivo?

We are currently exploring, biochemically and electrophysiologically, whether the type of muscarinic activation described here produces lasting changes in cortical networks in vitro. Ultimately, however, we are interested in two key questions concerning this type of activation: Does it occur in vivo, and if so, does it occur as part of normal development or only as a result of some combination of abnormal genetic and environmental factors? Our bias, based on the premise that the observed response constitutes hyperactivation, and that too much activation is likely to be damaging to emerging neural circuits, is that this kind of massive synchronous activation is not likely to be part of normal cortical development. What factors might give rise to this kind of activity? One intriguing possibility among environmental factors that may result in muscarinic hyperactivity in cortex is perinatal stress. The possibility that stress leads to abnormalities in developing cortical circuits has received some attention recently. We believe it is worth investigating in greater detail, particularly in light of growing evidence for a potential link between stress early in development and mental illness later in life.

One experimental paradigm that has been used extensively to study neonatal stress is that of maternal separation, a manipulation whereby pups are removed and deprived of contact with their mother for varying lengths of time during postnatal development (Lehmann and Feldon, 2000). Maternal separation has been shown to lead to altered emotional and cognitive capacity in adult animals and is viewed as an animal model for some aspects of psychiatric disorders in humans. It is known that this type of manipulation of the animal's environment has immediate as well as long-term effects on neuroendocrine systems, as well as on the ability of the hypothalamic–pituitary–adrenal axis to respond to stress. It is not unreasonable therefore to suspect that the changes in cognitive and emotional behaviors, detected in animals following these manipulations are, at least in part, related to permanent alterations in the biochemistry of their stress response (Ladd et al., 2000; Bremne and Vermetten, 2001). At the same time, our understanding of such abnormal emotional and cognitive behaviors would benefit from greater knowledge of the functional changes occurring in specific brain circuits. In particular, knowledge about how circuits in cortical areas such as the frontal cortex are affected would be useful. Frontal cortex is not only directly involved in generating such behaviors, but may also be involved in regulating hypothalamic–pituitary–adrenal axis activity, as suggested recently by Sullivan and Gratton (2002).

A number of changes in brain chemistry following different maternal separation protocols have been described, including changes in levels of NGF in several brain areas (Cirulli et al., 2000), decreased turnover of dopamine in medial prefrontal cortex (Matthews et al., 2001), and increases in NMDA receptor expression in several limbic areas (Ziabreva et al., 2000). Specific changes in cholinergic function have not been reported to our knowledge, but could be present, for example as a consequence of altered NGF levels. Should altered cholinergic function occur during the period in which we observe hyperactivation, it could lead to an excess of spurious synchronous calcium transients through-out these networks. This, in turn, could lead to lasting circuit abnormalities, and these could be responsible for altered emotional and cognitive behavior later in life.

It is noteworthy that there are specific behavioral effects that arise when maternal separation is performed within the time window during which we see muscarinic hyperactivation, but do not arise when the separation is done later. An example of this is the learning of active avoidance by adult rats that were briefly separated early in life. Lehmann et al. (1999), using the active avoidance paradigm, have described a pronounced difference in the ability of adult rats to learn active avoidance depending on whether they were separated on postnatal day 4 (MS4) or a few days later on day 9 (MS9): MS4 rats showed almost no avoidance responses even after 8 blocks of 10 acquisition trials, whereas MS9 rats learned at a rate that was as good as, and possibly faster than, control (not separated) rats. Clearly, the precise timing of a

stressful event determines how the developing nervous system will be impacted.

In light of the known role of medial cortical areas in active avoidance behaviors (Gabriel et al., 1991), many interesting questions arise regarding a possible link between maternal separation and the ability of these circuits to respond to muscarinic stimulation with an exaggerated response early in life. For example, are slices from animals that were separated at P4 more or less sensitive to muscarinic agonists when examined immediately after the separation period? Can the effect of separation on active avoidance learning be mitigated by pharmacological interventions that reduce or eliminate this hyper-activation, or is the effect made more pronounced? Can the altered behavior be induced by local infusion of muscarinic agonists during P4 in lieu of maternal separation? Answering these questions could be an important first step toward establishing or disproving a causal link between the vulnerability of select cortical regions to hyperactivation early in life and the fact that function in those regions is altered in many psychiatric disorders.

Concluding remarks

We have attempted to outline a hypothesis that can begin to explain some of the ontogenetic causes of emotional and cognitive disturbances characteristic of psychiatric disorders in terms of deviations in activity-dependent processes that may occur during the initial wiring of specific cortical circuits. We postulate that a combination of proximal factors generating excessive activation of circuits related to cognition and emotion, if present during a specific developmental period that also coincides with a period when these circuits are exceedingly plastic, could irreversibly alter the developmental program required for proper connectivity in these circuits, such that their functional capacity is rendered suboptimal thereafter. We further postulate that a key element of the vulnerability of specific cognitive and emotional circuits is the ability of their neurons to exhibit the property of sustained firing, a property believed to underlie the ability of these circuits to implement short-term or 'working', memory. We have not explicitly discussed the primary causes, be they genetic or environmental, that may be at the root of various psychiatric disorders, and which include such possibilities as viral infection, perinatal trauma, immunologic disorders, hereditary encephalopathy, toxin exposure and primary metabolic disease (Marenco and Weinberger, 2000). We believe that identifying the mechanisms that are potentially most directly linked to the specific pathology of psychiatric disorders will ultimately prove useful as we attempt to sort through the significance of the various potential primary causes. We suggest that one of these mechanisms, though not necessarily the only mechanism, is hyperactivity of specific cortical networks by abnormalities in cholinergic or other modulatory transmitter systems during a critically important period when features of the basic circuitry are being established and refined in an activity-dependent manner.

References

Ben-Ari, Y. (2001) Developing networks play a similar melody. Trends Neurosci., 24: 353–360.

Bradshaw, J.L., Sheppard, D.M. (2000) The neurodevelopmental frontostriatal disorders: evolutionary adaptiveness and anomalous lateralization. Brain Lang., 73: 297–320.

Bremne, J.D., Vermetten, E. (2001) Stress and development: behavioral and biological consequences. Dev. Psychopathol., 13: 473–489.

Broersen, L.M., Heinsbroek, R.P., de Bruin, J.P., Joosten, R.N., van Hest, A. and Olivier, B. (1994) Effects of local application of dopaminergic drugs into the dorsal part of the medial prefrontal cortex of rats in a delayed matching to position task: comparison with local cholinergic blockade. Brain Res., 645: 113–122.

Broersen, L.M., Heinsbroek, R.P., de Bruin, J.P., Uylings, H.B. and Olivier, B. (1995) The role of the medial prefrontal cortex of rats in short-term memory functioning: further support for involvement of cholinergic, rather than dopaminergic mechanisms. Brain Res., 674: 221–229.

Burcham, K.J., Corwin, J.V., Stoll, M.L. and Reep, R.L. (1997) Disconnection of medial agranular and posterior parietal cortex produces multimodal neglect in rats. Behav. Brain Res., 86: 41–47.

Carter, C.S., Mintun, M., Nichols, T. and Cohen, J.D. (1997) Anterior cingulate gyrus dysfunction and selective attention deficits in schizophrenia: [^{15}O]H$_2$O PET study during single-trial Stroop task performance. Am. J. Psychiatry, 154: 1670–1675.

Caulfield, M.P. (1993) Muscarinic receptors—characterization, coupling and function. Pharmacol. Ther., 58: 319–379.

Cherubini, E., Gaiarsa, J.L. and Ben-Ari, Y. (1991) GABA: an excitatory transmitter in early postnatal life. Trends Neurosci., 14: 515–519.

Cirulli, F., Alleva, E., Antonelli, A. and Aloe, L. (2000) NGF expression in the developing rat brain: effects of maternal separation. Brain Res. Dev. Brain Res., 123: 129–134.

Cohen, R.M., Semple, W.E., Gross, M., Nordahl, T.E., DeLisi, L.E., Holcomb, H.H., King, A.C., Morihisa, J.M. and Pickar, D. (1987) Dysfunction in a prefrontal substrate of sustained attention in schizophrenia. Life Sci., 40: 2031–2039.

Crair, M.C. (1999) Neuronal activity during development: permissive or instructive? Curr. Opin. Neurobiol., 9: 88–93.

Curtis, V.A., Bullmore, E.T., Brammer, M.J., Wright, I.C., Williams, S.C., Morris, R.G., Sharma, T.S., Murray, R.M. and McGuire, P.K. (1998) Attenuated frontal activation during a verbal fluency task in patients with schizophrenia. Am. J. Psychiatry, 155: 1056–1063.

Durstewitz, D., Seamans, J.K. and Sejnowski, T.J. (2000) Neurocomputational models of working memory. Nature Neurosci. Suppl. 3: 1184–1191.

Feinberg, I. (1982) Schizophrenia: caused by a fault in programmed synaptic elimination during adolescence. J. Psychiat. Res., 17: 319–334.

Fluhler, E., Burnham, V.G. and Loew, L.M. (1985) Spectra, membrane binding, and potentiometric responses of new charge shift probes. Biochemistry, 24: 5749–5755.

Gabriel, M., Kubota, Y., Sparenborg, S., Straube, K. and Vogt, B.A. (1991) Effects of cingulate cortical lesions on avoidance learning and training-induced unit activity in rabbits. Exp. Brain Res., 86: 585–600.

Garaschuk, O., Linn, J., Eilers, J. and Konnerth, A. (2000) Large-scale oscillatory calcium waves in the immature cortex. Nat. Neurosci., 3: 452–459.

Garcia-Rill, E., Biedermann, J.A., Chambers, T., Skinner, R.D., Mrak, R.E., Husain, M. and Karson, C.N. (1995) Mesopontine neurons in schizophrenia. Neuroscience, 66: 321–335.

Haj-Dahmane, S., Andrade, R. (1996) Muscarinic activation of a voltage-dependent cation nonselective current in rat association cortex. J. Neurosci., 16(12): 3848–3861.

Haj-Dahmane, S., Andrade, R. (1998) Ionic mechanism of the slow after depolarization induced by muscarinic receptor activation in rat prefrontal cortex. J. Neurophysiol., 80(3): 1197–1210.

Holcomb, H.H., Lahti, A.C., Medoff, D.R., Weiler, M., Dannals, R.F. and Tamminga, C.A. (2000) Brain activation patterns in schizophrenic and comparison volunteers during a matched-performance auditory recognition task. Am. J. Psychiatry, 157: 1634–1645.

Hubel, D.H. (1967) Eleventh Bowditch lecture. Effects of distortion of sensory input on the visual system of kittens. Physiologist, 10: 17–45.

Katz, L.C., Crowley, J.C. (2002) Development of cortical circuits: lessons from ocular dominance columns. Nat. Rev. Neurosci., 3: 34–42.

Katz, L.C., Shatz, C.J. (1996) Synaptic activity and the construction of cortical circuits. Science, 274: 1133–1138.

Krnjevic, K., Pumain, R. and Renaud, L. (1971) The mechanism of excitation by acetylcholine in the cerebral cortex. J. Physiol., 215(1): 247–268.

Ladd, C.O., Huot, R.L., Thrivikraman, K.V., Nemeroff, C.B., Meaney, M.J. and Plotsky, P.M. (2000) Long-term behavioral and neuroendocrine adaptations to adverse early experience. Prog. Brain Res., 122: 81–103.

Lehmann, J., Feldon, J. (2000) Long-term biobehavioral effects of maternal separation in the rat: consistent or confusing? Rev. Neurosci., 11(4): 383–408.

Lehmann, J., Pryce, C.R., Bettschen, D. and Feldon, J. (1999) The maternal separation paradigm and adult emotionality and cognition in male and female Wistar rats. Pharmacol. Biochem. Behav., 64: 705–715.

Lewis, D.A., Levitt, P. (2002) Schizophrenia as a disorder of neurodevelopment. Annu. Rev. Neurosci., 25: 409–432.

Loffelholz, K. (1996) Muscarinic receptors and cell signaling. Prog. Brain Res., 109: 191–194.

Lynch, M., Sayin, U. et al. (2000) Long-term consequences of early postnatal seizures on hippocampal learning and plasticity. Eur. J. Neurosci., 12(7): 2252–2264.

MacGregor, D.G., Chesler, M. and Rice, M.E. (2001) HEPES prevents brain edema in rat brain slices. Neurosci. Lett., 303: 141–144.

Marenco, S., Weinberger, D.R. (2000) The neurodevelopmental hypothesis of schizophrenia: following a trail of evidence from cradle to grave. Dev. Psychopathol., 12: 501–527.

Matthews, K., Dalley, J.W., Matthews, C., Tsai, T.H. and Robbins, T.W. (2001) Periodic maternal separation of neonatal rats produces region- and gender-specific effects on biogenic amine content in postmortem adult brain. Synapse, 40: 1–10.

McCormick, D.A., Prince, D.A. (1986) Mechanisms of action of acetylcholine in the guinea-pig cerebral cortex in vitro. J. Physiol., 375: 169–194.

Mechawar, N, Descarries, L. (2001) The cholinergic innervation develops early and rapidly in the rat cerebral cortex: a quantitative immunocytochemical study. Neuroscience, 108: 555–567.

Mechawar, N., Watkins, K.C. and Descarries, L. (2002) Ultrastructural features of the acetylcholine innervation in the developing parietal cortex of rat. J. Comp. Neurol., 443: 250–258.

O'Donovan, M.J. (1999) The origin of spontaneous activity in developing networks of the vertebrate nervous system. Curr. Opin. Neurobiol., 9: 94–104.

Peinado, A. (2000) Traveling slow waves of neural activity: a novel form of network activity in developing neocortex. J. Neurosci., 20(2): RC54.

Ragozzino, M.E., Kesner, R.P. (1999) The role of the agranular insular cortex in working memory for food reward value and allocentric space in rats. Behav. Brain Res., 98: 103–112.

Reep, R.L., Corwin, J.V., Hashimoto, A. and Watson, R.T. (1987) Efferent connections of the rostral portion of medial agranular cortex in rats. Brain Res. Bull., 19: 203–221.

Schaffer, P., Ahammer, H., Muller, W., Koidl, B. and Windisch, H. (1994) Di-4-ANEPPS causes photodynamic damage to isolated cardiomyocytes. Pflugers Arch., 426: 548–551.

Selemon, L.D., Goldman-Rakic, P.S. (1999) The reduced neuropil hypothesis: a circuit based model of schizophrenia. Biol. Psychiatry, 45: 17–25.

Sullivan, R.M., Gratton, A. (2002) Prefrontal cortical regulation of hypothalamic-pituitary-adrenal function in the rat and implications for psychopathology: side matters. Psychoneuroendocrinology, 27: 99–114.

Tandon, R., Shipley, J.E., Greden, J.F., Mann, N.A., Eisner, W.H. and Goodson, J.A. (1991) Muscarinic cholinergic hyperactivity in schizophrenia. Relationship to positive and negative symptoms. Schizophr. Res., 4: 23–30.

Thompson, P.M., Vidal, C., Giedd, J.N., Gochman, P., Blumenthal, J., Nicolson, R., Toga, A.W. and Rapoport, J.L. (2001) Mapping adolescent brain change reveals dynamic wave of accelerated gray matter loss in very early-onset schizophrenia. Proc. Natl. Acad. Sci. USA, 98: 11650–11655.

Uylings, H.B., van Eden, C.G. (1990) Qualitative and quantitative comparison of the prefrontal cortex in rat and in primates, including humans. Prog. Brain Res., 85: 31–62.

Weinberger, D.R., Berman, K.F., Suddath, R. and Torrey, E.F. (1992) Evidence of dysfunction of a prefrontal-limbic network in schizophrenia: a magnetic resonance imaging and regional cerebral blood flow study of discordant monozygotic twins. Am. J. Psychiatry, 149: 870–879.

Wiesel, T.N. (1982) Postnatal development of the visual cortex and the influence of environment. Nature, 299: 583–591.

Zhang, L.I, Poo, M.M. (2001) Electrical activity and development of neural circuits. Nat. Neurosci., 4: Suppl. 1207–1214.

Ziabreva, I., Schnabel, R. and Braun, K. (2000) Parental deprivation induces N-methyl-D-aspartate-receptor upregulation in limbic brain areas of Octodon degus: protective role of the maternal call. Neural Plast., 7: 233–244.

Progress in Brain Research, Vol. 145
ISSN 0079-6123

CHAPTER 10

Synaptic mechanisms and cholinergic regulation in auditory cortex

Raju Metherate* and Candace Y. Hsieh

*Department of Neurobiology and Behavior, University of California,
2205 McGaugh Hall, Irvine, CA 92697-4450, USA*

Introduction

A single neuron in primary auditory cortex (ACx) may receive hundreds of synaptic inputs. However, not all synapses are equal—behavioral state, prior experience and acoustic context all help determine which inputs most effectively excite the neuron. Developmental manipulations or behavioral training, for example, can produce dramatic changes in frequency receptive fields that may be transient or long-lasting; at least some of the changes, presumably, are due to synaptic modifications at the cortical level. To understand the capabilities of ACx and how it functions in different contexts, it is important to understand how changes in synaptic gain may alter the range of effective inputs to a cell. Acetylcholine (ACh) plays a variety of roles—e.g., in arousal, attention, and learning—and understanding its specific synaptic actions is critical to understanding each role. In this article, we will review recent work on cholinergic actions of nicotinic and muscarinic receptors (nAChRs and mAChRs) related to our efforts to understand the development and function of ACx.

Cortical processing of auditory inputs

Direct and indirect evidence from a number of studies indicate that neurons in ACx receive synaptic inputs from a much wider range of frequencies than the range that overtly excites the neuron. Extracellular recordings of local field potentials (which reflect synaptic activity) and intracellular recordings of synaptic activity (Fig. 1) indicate that synaptic receptive fields (EPSPs and IPSPs) are remarkably broad spectrally, considerably broader than spike-based receptive fields (Ribaupierre et al., 1972; Volkov and Galazjuk, 1991; Eggermont, 1996; Kaur et al., 2002). Optical imaging of ACx activity (intrinsic signal imaging or voltage-sensitive dyes) shows that while peaks of tone-evoked activity are organized tonotopically, weaker responses to single tones can activate the entire ACx (Bakin et al., 1996; Hess and Scheich, 1996; Horikawa et al., 1996). Thus, a single tone can, in principle, evoke EPSPs in neurons throughout ACx. Extracellular unit studies have shown that a cortical neuron's discharge to stimuli within its receptive field can be modified by concurrent stimuli outside the receptive field (Oonishi and Katsuki, 1965; Nelken et al., 1994a,b), and in some cases neurons can be excited solely by complex stimuli whose spectrum lies well outside the receptive field (Whitfield and Evans, 1965; Schulze and Langner,

*Corresponding author: Tel.: 949-824-6141;
Fax: 949-824-2447; E-mail: rmethera@uci.edu

DOI: 10.1016/S0079-6123(03)45010-3

144

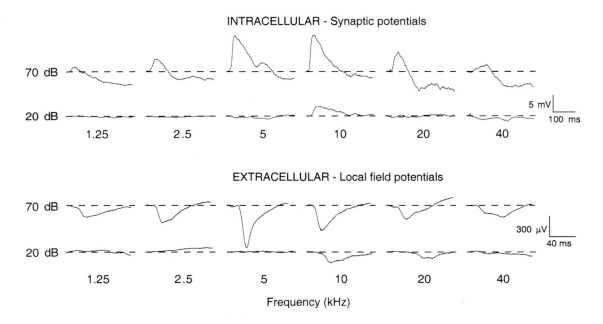

Fig. 1. Intracellular synaptic potentials (top) and extracellular field potentials (bottom) evoked in response to a broad, five-octave range of frequencies at 70 dB sound pressure level. Tone duration 100 ms beginning at trace onset (10 ms rise and fall ramps). Intracellular potentials were obtained using whole-cell recording from a layer 3 neuron in primary ACx. Extracellular potentials were obtained subsequently using a micropipette lowered to the same site as the intracellular recording. Data from urethane-anesthetized rat.

1999) ('receptive field' here refers to the classical receptive field as defined by spike responses to pure tones). At least in some cases, single ACx neurons may integrate information over the entire audible spectrum (Schulze and Langner, 1999).

The neural architecture underlying broad spectral convergence in ACx is only partly understood. Thalamic relay neurons in the primary auditory thalamus—the ventral division of the medial geniculate (MGv)—project to ACx neurons with similar 'best', or characteristic frequencies (Imig and Morel, 1984; Winer et al., 1999; Miller et al., 2001), and are responsible for characteristic frequency and near-characteristic frequency-induced spiking. Thalamo-cortical synapses are excitatory and glutamatergic (Kharazia and Weinberg, 1994), and physiological studies on a brain slice containing the lemniscal auditory thalamocortical pathway (Fig. 2) have shown that the monosynaptic thalamocortical projection to layer 4 activates both α-amino-3-hydroxy-5-methyl-isoxazole-4-propionic acid receptors (AMPARs) and

N-methyl-D-aspartate receptors (NMDARs) (Cruik-shank et al., 2002). Intracortical synapses can mediate long-distance excitation and local inhibition, and may contribute to broad subthreshold convergence in ACx (Matsubara and Phillips, 1988; Ojima et al., 1991; Wallace et al., 1991; Kaur et al., 2002). Thus, ACx neurons may receive strong, narrowband thalamocortical inputs and broader intracortical inputs that can contribute to spectrotemporal integration.

The synaptic receptive field of a neuron in ACx—produced by integrated thalamocortical and intracortical activity—may be an important target of cholinergic regulation. Cortical plasticity, including reorganization of topographic maps and single cell receptive fields, can involve cholinergic inputs (Bakin and Weinberger, 1996; Kilgard and Merzenich, 1998; Shulz et al., 2000) and differential strengthening and weakening of existing synapses (Weinberger, 1995; Gilbert, 1998). Rapid, transient changes in receptive fields that result from shifting attention (Desimone

145

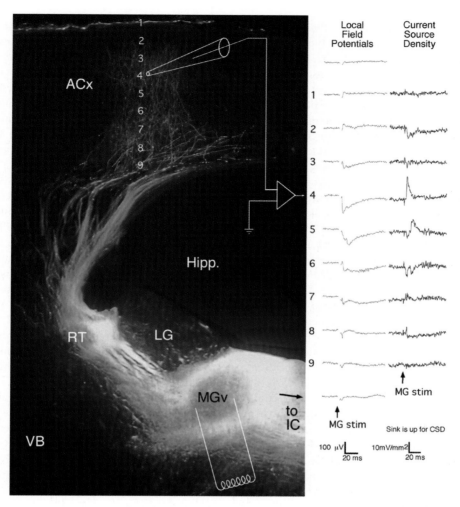

Fig. 2. Anatomy and physiology of an auditory thalamocortical brain slice. MGv stimulation elicits local field potentials recorded in 150 μm steps through the cortical layers (recording sites and corresponding responses labeled 1–9). Current source density analysis reveals major current sink in layer 4 (position 4). Placement of Di-I at the stimulation site subsequently traces the thalamocortical pathway, including dense arborization in layer 4 at the site of the major current sink. Also visible is an increase in fluorescence as thalamocortical fibers cross RT, reflecting axon collaterals and possibly terminal arbors in the auditory portion of RT (Rouiller and Welker, 1991). Abbreviations include Hipp., hippocampus; IC, inferior colliculus; LG, lateral geniculate; RT, reticular nucleus of the thalamus; VB, ventrobasal nucleus. Location of VB is approximate. Figure modified from Cruikshank et al. (2002); brightness adjusted differently for cortex and thalamocortical pathway (fluorescence in pathway is stronger than in cortex).

and Duncan, 1995) may also depend on cholinergic modulation of synaptic strength (Shulz et al., 2000), given likely cholinergic involvement in mechanisms of attention (Hasselmo, 1995). The specific synaptic mechanisms that mediate changes such as these are not known; this review will describe potential mechanisms of cholinergic regulation in ACx function and development.

Muscarinic regulation of ACx neurons and synapses

Many in vivo studies of cholinergic modulation in sensory cortex have demonstrated a role for mAChRs in enhancing responsiveness to afferent inputs. Microapplication of ACh excites cortical neurons, an effect blocked by muscarinic antagonists such as atropine (Krnjević and Phillis, 1963), whereas in early studies nicotinic agonists typically had inconsistent effects (but, see below). Muscarinic actions also enhance responses to sensory stimuli (Sillito and Kemp, 1983; Metherate et al., 1988; McKenna et al., 1989), even after responses were depressed by lesioning cortically-projecting cholinergic neurons in the basal forebrain (Sato et al., 1987). The increased responsiveness resulted, at least in part, from increased postsynaptic membrane resistance due to decreased K^+ conductance (Krnjević et al., 1971). Many studies using in vitro preparations have confirmed and extended these findings, showing that mAChRs activation can decrease several K^+ currents, including the Ca^{++}-activated after hyperpolarization (AHP) current, the M-current, and the leak current (Halliwell and Adams, 1982; McCormick and Prince, 1986; Madison et al., 1987). More recent studies suggest that mAChRs can also activate a nonselective cation current in cortical neurons (Haj-Dahmane and Andrade, 1996).

Synaptic release of ACh in ACx produces effects consistent with the results of agonist application. Frequency receptive fields are modified similarly by anticholinesterase or muscarinic agonist (Ashe et al., 1989), and activation of cholinergic synapses in vitro increases excitability via slow depolarization, increased membrane input resistance, and decreased AHPs (Metherate et al., 1992; Cox et al., 1994). Electrical or chemical stimulation in vivo of cortically-projecting basal forebrain neurons also produces EEG activation (Fig. 3), an effect blocked by intracortical microapplication of atropine (Metherate et al., 1992). Intracellular (whole-cell) recordings in vivo show that EEG activation is associated with the abolition of rhythmic long-duration hyperpolarizations (Fig. 3A), and the K^+ channel blocker Cs^+ introduced via the patch pipette similarly reduces rhythmic hyperpolarizations (Fig. 3B) (Metherate and Ashe, 1993b). These findings provide direct support for the suggestion that EEG activation results from muscarinic blockade of rhythmic, K^+-mediated AHPs (Buzsaki et al., 1988). One consequence of EEG activation for sensory processing is implied by the finding that basal forebrain stimulation also enhances, via cortical mAChRs, cortical EPSPs evoked by MGv stimulation (Fig. 3C, D) (Metherate and Ashe, 1993a), as

well as responses to acoustic stimuli (Edeline et al., 1994). These in vivo studies extend agonist studies and demonstrate functional implications of muscarinic cellular actions.

More recently, we have examined cholinergic modulation of synaptic transmission using the auditory thalamocortical slice preparation (Fig. 2). With this preparation, we were able to selectively activate thalamocortical vs. long-distance intracortical pathways, and determine the effects of the cholinergic agonist carbachol. These studies showed that carbachol, acting at mAChRs, suppresses intracortical glutamatergic EPSPs, while having less or no effect on thalamic-evoked EPSPs (Fig. 4) (Hsieh et al., 2000). These effects are seen with the neuron at resting membrane potential, so that voltage-dependent postsynaptic actions of ACh are minimal. Reduced EPSPs likely reflect muscarinic reduction of neurotransmitter release (Hounsgaard, 1978; Valentino and Dingledine, 1981; Segal, 1982, 1989) and the differential effect on intracortical vs. thalamocortical EPSPs may reflect differential distribution of presynaptic mAChRs at intracortical vs. thalamocortical synapses (Sahin et al., 1992). These effects of mAChR activation in ACx are similar to those reported for intrinsic vs. extrinsic afferents to olfactory cortex (Hasselmo and Bower, 1992) (also see chapter in this volume).

The results described above clearly support the notion that ACh regulates synaptic activity in ACx, but a comparison of Figs. 3 and 4 reveals an unexpected inconsistency. Nucleus basalis stimulation facilitates MG-evoked EPSPs (Fig. 3), just as the numerous studies cited above describe muscarinic agonist-mediated enhancement of sensory responses. However, carbachol has little effect on MG-evoked EPSPs in the thalamocortical slice (Fig. 4). There are obvious technical differences between the two studies, e.g., in vivo vs. in vitro, and synaptic ACh vs. exogenous agonist; however, a possibly more important difference lies in the level of membrane depolarization. Neurons recorded in vivo have more depolarized membrane potentials than neurons in vitro (Metherate and Ashe, 1993b), especially given spontaneous membrane potential fluctuations (Fig. 3); as a result, they will exhibit more strongly activated voltage-dependent K^+ currents. Muscarinic blockade of these currents will increase

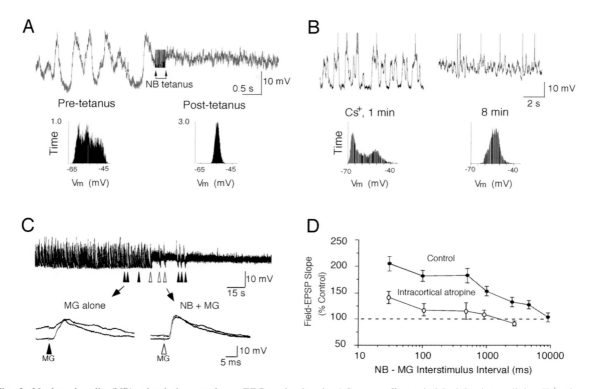

Fig. 3. Nucleus basalis (NB) stimulation produces EEG activation in ACx—an effect mimicked by intracellular K^+ channel blockade—and enhances EPSPs elicited by MG stimulation. All data from urethane-anesthetized rats. (A) Intracellular (whole-cell) recording shows spontaneous large-amplitude low-frequency membrane potential fluctuations before NB stimulation and small-amplitude high-frequency fluctuations after. Resting potential −65 mV; whole-cell recordings in this and other figures are not corrected for junction potential of approximately −10 mV. Insets show amplitude histograms of membrane potential fluctuations. Loss of long-lasting hyperpolarizations with NB stimulation is evident in raw data and amplitude histograms. (B) Whole-cell recording with Cs^+ in the patch pipette results in intracellular dialysis with Cs^+, partial blockade of K^+ channels, and loss of long-lasting hyperpolarizations, mimicking effect of NB stimulation. Recordings are 1 min and 8 min after rupture of the membrane to establish whole-cell recording configuration; spikes are truncated. Corresponding amplitude histograms shown in insets, initial resting potential −70 mV. (C) Another intracellular recording at a slower time scale shows large-amplitude low-frequency membrane potential fluctuations before NB stimulation and small-amplitude high-frequency fluctuations after (sharp microelectrode recording, K^+-acetate in pipette, resting potential −80 mV). The MG is stimulated before NB stimulation (dark triangles), with NB stimulation (light triangles; NB tetanus precedes MG stimulus by 50 ms), and after NB stimulation (dark triangles). Higher resolution traces of MG-evoked EPSPs shown below demonstrate that NB stimulation enhances EPSPs (responses to two MG stimuli are overlaid in each condition). Responses to MG stimulation after NB pairing (second set of dark triangles) remained enhanced above control values (not shown, see Metherate and Ashe, 1993a). (D) NB stimulation enhances the local field potential (field-EPSP) in ACx evoked by MG stimulation, an effect that is reduced by iontophoresis of atropine near the recording electrode. Control data (mean ± SEM) from 17 experiments, atropine data from 4 of those experiments. Data in (A) and (B) modified from Metherate and Ashe (1993b), copyright 1993 by the Society for Neuroscience, and data in (C) and (D) modified from Metherate and Ashe (1993a), copyright 1993 by Synapse.

postsynaptic excitability, whereas such actions will be less pronounced at more hyperpolarized resting potentials in vitro. Thus, postsynaptic actions may dominate in vivo, while presynaptic actions dominate in vitro unless neurons are explicitly depolarized (as is routinely done to demonstrate postsynaptic effects McCormick and Prince, 1986; Metherate et al., 1992). These data emphasize the importance of using different approaches to understand the functional implications of synaptic mechanisms.

148

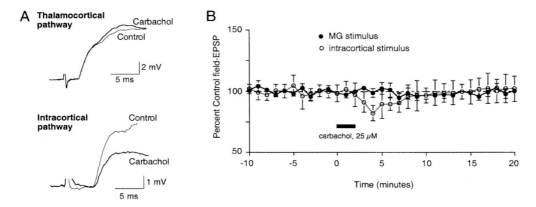

Fig. 4. Differential effects of carbachol on thalamocortical vs. intracortical synaptic transmission in the auditory thalamocortical slice. Thalamocortical pathway activated by stimulation within the pathway (A) or the MG itself (B). Thalamocortical stimuli were alternated with stimulation of intracortical pathways (stimulation within the cortex up to 1 mm from the recording electrode), and then the effects of carbachol determined by bath application. (A) Carbachol (10 μM) has little effect on thalamocortical EPSP but suppresses intracortically-elicited EPSP in the same layer 4 cell. (B) Average data (± SEM) of carbachol's effect on local field potential (field-EPSP) in layer 4 shows suppression of intracortical pathway ($n = 7$) with little effect on thalamocortical pathway ($n = 12$). Effect of carbachol is blocked by atropine (not shown). Data in (A) and (B) modified from Hsieh et al. (2000); copyright 2000, reproduced with permission from Elsevier Science.

It is tempting to speculate on the functional implications of differential regulation of thalamocortical vs. intracortical synapses in ACx. A straightforward interpretation is that muscarinic modulation serves to favor responses to external stimuli over ongoing cortical activity. Selective processing of sensory inputs during arousal or attention could underlie the widely hypothesized function of ACh to increase the 'signal-to-noise' ratio of sensory responses over ongoing cortical activity (Sillito, 1993; Hasselmo, 1995). This simple scenario becomes more complicated with the proposal outlined in the previous section, that frequency receptive fields in ACx neurons result from an integration of thalamocortical and intracortical inputs. Selective muscarinic suppression of noncharacteristic frequency responses (mediated by intracortical synapses), and increased postsynaptic responsiveness to characteristic frequency and near-characteristic frequency inputs (mediated by thalamocortical synapses), could sharpen receptive fields, again supporting an 'enhance signal-to-noise' role for ACh. A similar argument applies to the temporal sequence of sensory-evoked responses: initial (thalamocortical) components of the response will be favored over later (intracortical) components, as observed recently in somatosensory cortex (Dancause et al., 2001). As these examples demonstrate, knowledge of the synaptic consequences of mAChR activation will facilitate an understanding of systems-level sensory processing.

Role of nAChRs in ACx development

The lack of clear effects of nicotine on cortical neurons in early studies (Krnjević and Phillis, 1963) may have been due, in part, to unrecognized rapid receptor desensitization (a feature of neuronal nAChRs, Zhang et al., 1994). More recently, the role of nAChRs in cortex is being revisited and studies have identified important presynaptic actions that increase neurotransmitter release as well as postsynaptic actions that mediate rapid depolarization (Gray et al., 1996; Roerig et al., 1997; Frasier et al., 1998; Alkondon et al., 2000). A prominent role of nAChRs in sensory cortex appears during early postnatal development, and we will review this work next.

During postnatal development of primary auditory, somatosensory, and visual neocortices in the rat, there is a dramatic increase in the expression of the cholinergic enzyme acetylcholinesterase (AChE) (Kristt, 1979; Prusky et al., 1988; Robertson et al.,

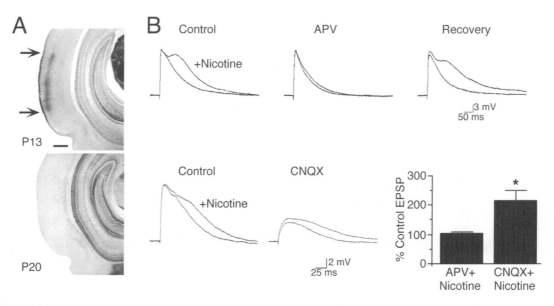

Fig. 5. Transient expression of AChE (A) and nicotine-induced selective enhancement of NMDAR-mediated EPSP (B) during postnatal development of ACx. A. Dense AChE staining of primary ACx (between arrows) occurs in layers 1 and 3/4 of a P13 rat, but is gone at P20. Calibration bar, 500 μm. B. Upper traces: Nicotine (25 μM) applied via micropressure ejection to the apical dendrite of P13 neuron enhances the late-EPSP magnitude, an effect reversibly blocked by the NMDAR antagonist APV (50 μM). Lower traces: In another cell, the nicotine effect is not blocked by the AMPA/kainate receptor antagonist CNQX (20 μM). Average data ($\pm$ SEM) show that APV blocks the effect of nicotine ($n = 5$) whereas CNQX does not ($n = 4$). Data in (A) and (B) modified from Aramakis and Metherate (1998), copyright 1998 by the Society for Neuroscience.

1991). In auditory cortex, the increased expression occurs in the thalamorecipient layers 3–4 beginning on postnatal day (P) 3, reaches peak intensity at P8–10 , and declines to low (adult) levels by P23 (Fig. 5A) (Robertson et al., 1991). Other studies have revealed a parallel increase in nicotine binding sites and the expression of $\alpha 7$ nAChRs in developing sensory cortex (Prusky et al., 1988; Fuchs, 1989; Broide et al., 1995, 1996). Some studies suggest a close relationship between nAChRs and developing thalamocortical (glutamatergic) afferents (Prusky et al., 1988; Broide et al., 1996); however, ultrastructural analysis indicates that $\alpha 7$ nAChRs can be associated with glutamate synapses in all cortical layers (Levy and Aoki, 2002). Together, these findings imply a transient cholinergic function during the first few postnatal weeks (but do not preclude important nAChR functions in the adult; Gioanni et al., 1999; Levy and Aoki, 2002).

We tested this in ACx by recording from layer 3/4 neurons while pressure-ejecting nicotine onto their apical dendrites (visualization of neurons using IR-DIC optics allows delivery of agonist directly to presumed synaptic sites, thus avoiding receptor desensitization) (Aramakis and Metherate, 1998). While nicotine generally does not produce direct postsynaptic responses, it often modifies the neuron's response to afferent stimulation of glutamatergic synapses (Fig. 5B). Importantly, nicotine selectively enhances the late, NMDAR component of EPSPs, but does not affect the earlier AMPAR component. The nicotine enhancement of glutamate EPSPs appears to be due to presynaptic nAChRs whose activation increases evoked glutamate release (e.g., responses to nicotine alone are infrequent) and is blocked by methyllycaconitine citrate (MLA), indicating the specific involvement of $\alpha 7$ nAChRs. The nicotine effect is observed most frequently during postnatal Week 2 and is rare during Week 4 (Fig. 6A). (Yet, nAChRs are prevalent and functional in adult cortex; Gioanni et al., 1999; Levy and Aoki, 2002; it is not clear what factors underlie this apparent discrepancy.) It is interesting to note that while nicotinic modulation of EPSPs is most frequently

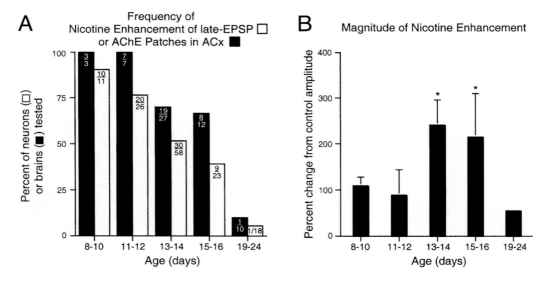

Fig. 6. Developmental time course of AChE staining and effectiveness of nicotine to enhance NMDAR-mediated EPSPs (A) and magnitude of nicotine effect (B). (A) The percentage of brains with AChE-positive staining in ACx declines with increasing age, as does the percentage of neurons whose EPSPs are modified by nicotine. Numbers in each histogram column indicate 'n'. (B) Magnitude of nicotine effect (mean ± SEM percent increase) is greater at P13–16 than at P8–12 (pairwise *t*-tests, $p < 0.05$; same neurons as in (A) but only cells affected by nicotine are included in analysis). Data in (A) modified from Aramakis and Metherate (1998), copyright 1998 by the Society for Neuroscience.

observed early in Week 2 (Fig. 6A), the effects observed in fewer neurons during late Week 2 and early Week 3 are stronger (Fig. 6B).

The transient expression and function of α7 nAChRs during development of ACx suggest the existence of a 'critical period' of heightened sensitivity to sensory experience, especially given the proximity to the onset of hearing. In rat ACx, the first three postnatal weeks is a time of rapid development of neural circuitry. Thalamic afferents innervate the cortex and its six layer structure is established early in Week 1 (Robertson et al., 1991; Ignacio et al., 1995). Rats begin to hear near the middle or end of Week 2, and the cortical evoked potential develops very rapidly during Week 3 (but does not mature fully for several weeks) (Crowley and Hepp-Reymond, 1966; Iwasa and Potsic, 1982; Blatchley et al., 1987). Glutamatergic EPSPs in ACx are of small amplitude and long duration early in Week 2, but rapidly become larger and faster (shorter latencies to peak and shorter durations) (Aramakis et al., 2000). Neuronal intrinsic membrane properties also develop rapidly during this period (Metherate and Aramakis, 1999). Thus, Weeks

2–3—when nAChRs regulate NMDAR-mediated EPSPs—is a period of rapid change during which developing synapses may be particularly sensitive to external influences. It is noteworthy that the largest magnitude nicotine effects occur at or immediately after the onset of hearing (Fig. 6B).

The functional importance of transient nAChR-mediated regulation of NMDAR EPSPs remains unclear, but may relate to experience-dependant maturation of glutamate synapses. The data summarized above suggest the synaptic arrangement in Fig. 7. We propose that the glutamate synapses regulated by presynaptic nAChRs have only NMDARs postsynaptically, whereas other glutamate synapses have AMPARs and/or NMDARs post-synaptically but no presynaptic nAChRs. This arrangement accounts for the apparent presynaptic action of nicotine at α7 nAChRs to selectively regulate EPSPs mediated by NMDARs, but not AMPARs (Fig. 5) (Aramakis and Metherate, 1998). Pure-NMDAR synapses, often called 'silent synapses', occur in developing sensory cortex and are thought to be converted to mature synapses with

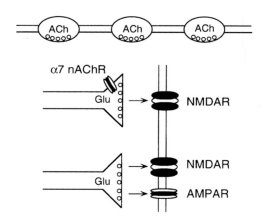

Fig. 7. Hypothetical regulation of pure-NMDA synapses by presynaptic α7 nAChRs. Cholinergic receptors may be activated by 'diffuse transmission' from ACh synapses (Mechawar et al., 2002). See text for details.

activity- (depolarization-) dependent insertion of AMPARs (Isaac et al., 1997; Rumpel et al., 1998). (Note that 'silent' synapses is a misnomer, at least for ACx: at the relatively depolarized resting potentials that are normal at young ages, pure-NMDAR-mediated responses are small but clear; (Aramakis and Metherate, 1998; Metherate and Aramakis, 1999). We propose that release of ACh (e.g., during attention to acoustic stimuli) may increase glutamate release at active synapses and facilitate their conversion to mature synapses with AMPARs. Such maturation would be expected to result in the loss of presynaptic nAChRs as well.

A clear implication of the hypothesis illustrated in Fig. 7 is that manipulating nAChR function should also affect development of glutamate synapses. We have confirmed this by determining the effects of chronic nicotine exposure (CNE; systemic injections of 1–2 mg/kg nicotine hydrogen tartrate twice daily for 5–9 days). We first found that CNE during Week 2—but not during Weeks 1 or 4—alters glutamate EPSP development dramatically (Fig. 8A) (Aramakis et al., 2000). EPSPs after CNE have longer durations and unusual small fluctuations that may indicate disrupted release mechanisms or hyperexcitable synaptic circuitry. Consistent with the hypothesized location of nAChRs at pure-NMDAR synapses, CNE affects only the NMDAR component of EPSPs—roughly doubling the magnitude of this component—and has no affect on the AMPAR component (Fig. 8B).

The effects of CNE during Week 2 are long-lasting, being evident well into Week 4, although CNE during Week 4 itself has no effect (Fig. 8B). Thus, CNE during the putative critical period in Week 2 has striking effects on EPSP development.

Another implication of the hypothesis is that CNE may affect NMDAR composition. During normal development of sensory cortex, NMDARs, which consist of an NR1 subunit in combination with one or more NR2 subunits, progress from containing mostly NR2B subunits to also containing NR2A subunits (Monyer et al., 1994; Flint et al., 1997; Cao et al., 2000). This NMDAR maturation may be related to synaptic plasticity during critical periods (Nase et al., 1999; Roberts and Ramoa, 1999; Philpot et al., 2001, but see Barth and Malenka, 2001). If nAChRs regulate activity at NMDARs, then CNE might be expected to alter NMDAR subunit development. In ACx, levels of NR2B mRNA are high at birth and remain relatively high, whereas NR2A expression is quite low at birth but increases over several weeks; both levels tend to peak around P10–21 and then decline slightly through adulthood (Hsieh et al., 2002a). CNE during Week 2 significantly increases levels of NR2A mRNA for several days, but has little effect on NR2B levels (Hsieh et al., 2002b). These data raise the interesting possibility that CNE may accelerate NMDAR development. However, whether CNE accelerates, or merely disrupts, development remains to be seen.

It will be important to relate nAChR studies in the rat to human development, especially in light of the possible consequences on brain development of nicotine exposure from cigarette smoking. Human third trimester development resembles postnatal Week 2 in the rat in several ways, including thalamocortical innervation, the appearance of transient AChE expression in ACx, and the onset of hearing (Krmpotic-Nemanic et al., 1980, 1983). Further, CNE studies in our laboratory have used nicotine doses designed to mimic blood levels in smokers (Isaac and Rand, 1972; Murrin et al., 1987; Henningfield et al., 1993). These studies raise the possibility that maternal smoking may alter fetal ACx development, with potentially long-lasting consequences. Notably, infants born to mothers who smoke during pregnancy show decreased auditory habituation and ability to orient to a sound (Saxton, 1978; Picone et al., 1982). These

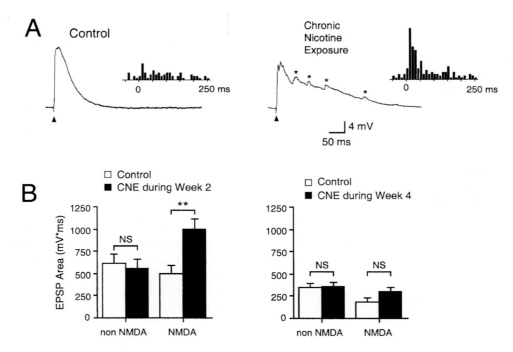

Fig. 8. Chronic nicotine exposure (CNE) during postnatal Week 2 alters EPSP development by altering NMDAR component. EPSPs were recorded in brain slices prepared ~15 h after 5–8 days of CNE. A. Left: Representative EPSP from a control P13 neuron; the inset shows the time of occurrence of miniature fluctuations riding on the EPSP (inset shows cumulative data from 22 control neurons; 0 ms is time of stimulus; y-axis is number of events). Right: CNE during Week 2 increases duration of EPSP and greatly enhances occurrence of miniature fluctuations (inset shows cumulative data from 22 CNE neurons, same y-axis as control histogram). B. Left: CNE during Week 2 doubled the magnitude of the NMDAR component of EPSPs (determined by subtracting the response in APV from control response), but did not affect the magnitude of the nonNMDAR EPSP (response remaining in APV). Right: CNE during Week 4 did not affect either component. Data in (A) and (B) modified from Aramakis et al. (2000), copyright 2000 by the Society for Neuroscience.

children subsequently show auditory-related cognitive deficits as they mature, despite apparent normal hearing range and thresholds (Sexton et al., 1990; McCartney et al., 1994; Fried et al., 1997). Thus, it is possible that nicotine-induced changes in perinatal development may lead to auditory deficits in older children and adults. Future animal studies should examine the consequences of early CNE on subsequent auditory function.

Conclusion

The studies reviewed here clearly demonstrate important roles for the cholinergic system in acoustic information processing. Both mAChRs and nAChRs are implicated, and thus far the role of nAChRs is more prominent during early postnatal development.

A challenge for future studies will be to integrate the wide variety of cellular and synaptic actions to achieve an understanding of auditory processing under different conditions.

Acknowledgments

Research in our laboratory has been supported by NIDCD (DC02967), NIDA (DA12929), NSF (IBN 9510904), and the California Tobacco-Related Disease Research Program (8RT-0059).

Abbreviations

Ach	acetylcholine
AChE	acetylcholinesterase

ACx   auditory cortex

AMPAR α-amino-3-hydroxy-5-methyl-isoxazole-4-propionic acid receptors

CNE   chronic nicotine exposure

EPSP   excitatory postsynaptic potential

IPSP   inhibitory postsynaptic potential

mAChR   muscarinic acetylcholine receptor

MGv   ventral division of the medial geniculate body

nAChR   nicotinic acetylcholine receptor

NB   nucleus basalis

NMDAR N-methyl-D-aspartate receptors

References

Alkondon, M., Pereira, E.F., Almeida, L.E., Randall, W.R. and Albuquerque, E.X. (2000) Nicotine at concentrations found in cigarette smokers activates and desensitizes nicotinic acetylcholine receptors in CA1 interneurons of rat hippocampus. Neuropharmacology, 39: 2726–2739.

Aramakis, V.B. and Metherate, R. (1998) Nicotine selectively enhances NMDA receptor-mediated synaptic transmission during postnatal development in sensory neocortex. J. Neurosci., 18: 8485–8495.

Aramakis, V.B., Hsieh, C.Y., Leslie, F.M. and Metherate, R. (2000) A critical period for nicotine-induced disruption of synaptic development in rat auditory cortex. J. Neurosci., 20: 6106–6116.

Ashe, J.H., McKenna, T.M. and Weinberger, N.M. (1989) Cholinergic modulation of frequency receptive fields in auditory cortex: II. Frequency-specific effects of anticholinesterases provide evidence for a modulatory action of endogenous ACh. Synapse, 4: 44–54.

Bakin, J.S., Kwon, M.C., Masino, S.A., Weinberger, N.M. and Frostig, R.D. (1996) Suprathreshold auditory cortex activation visualized by intrinsic signal optical imaging. Cereb. Cortex, 6: 120–130.

Bakin, J.S. and Weinberger, N.M. (1996) Induction of a physiological memory in the cerebral cortex by stimulation of the nucleus basalis. Proc. Natl. Acad. Sci. U.S.A., 93: 11219–11224.

Barth, A.L. and Malenka, R.C. (2001) NMDAR EPSC kinetics do not regulate the critical period for LTP at thalamocortical synapses. Nat. Neurosci., 4: 235–236.

Blatchley, B.J., Cooper, W.A. and Coleman, J.R. (1987) Development of auditory brainstem response to tone pip stimuli in the rat. Brain Res., 429: 75–84.

Broide, R.S., O'Connor, L.T., Smith, M.A., Smith, J.A.M. and Leslie, F.M. (1995) Developmental expression of α7 neuronal nicotinic receptor messenger RNA in rat sensory cortex and thalamus. Neuroscience, 67: 83–94.

Broide, R.S., Robertson, R.T. and Leslie, F.M. (1996) Regulation of alpha-7 nicotinic acetylcholine receptors in the developing rat somatosensory cortex by thalamocortical afferents. J. Neurosci., 16: 2956–2971.

Buzsaki, G., Bickford, R.G., Ponomareff, G., Thal, L.J., Mandel, R. and Gage, F.H. (1988) Nucleus basalis and thalamic control of neocortical activity in the freely moving rat. J. Neurosci., 8: 4007–4026.

Cao, Z., Lickey, M.E., Liu, L., Kirk, E. and Gordon, B. (2000) Postnatal development of NR1, NR2A and NR2B immunoreactivity in the visual cortex of the rat. Brain Res., 859: 26–37.

Cox, C.L., Metherate, R. and Ashe, J.H. (1994) Modulation of cellular excitability in neocortex: muscarinic receptor and second messenger-mediated actions of acetylcholine. Synapse, 16: 123–136.

Crowley, D.E. and Hepp-Reymond, M.-C. (1966) Development of cochlear function in the ear of the infant rat. J. Comp. Physiol. Psych., 62: 427–432.

Cruikshank, S.J., Rose, H.J. and Metherate, R. (2002) Auditory thalamocortical synaptic transmission, in vitro. J. Neurophysiol., 87: 361–384.

Dancause, N., Dykes, R.W., Miasnikov, A.A. and Agueev, V. (2001) Atropine-sensitive and -insensitive components of the somatosensory evoked potential. Brain Res, 910: 67–73.

Desimone, R. and Duncan, J. (1995) Neural mechanisms of selective visual attention. Ann. Rev. Neurosci., 18: 193–222.

Edeline, J.-M., Hars, B., Maho, C. and Hennevin, E. (1994) Transient and prolonged facilitation of tone-evoked responses induced by basal forebrain stimulations in the rat auditory cortex. Exp. Brain Res., 97: 373–386.

Eggermont, J.J. (1996) How homogeneous is cat primary auditory cortex? Evidence from simultaneous single-unit recordings. Auditory Neurosci., 2: 79–96.

Flint, A.C., Maisch, U.S., Weishaupt, J.H., Kriegstein, A.R. and Monyer, H. (1997) NR2A subunit expression shortens NMDA receptor synaptic currents in developing neocortex. J. Neurosci., 17: 2469–2476.

Frasier, C.J., Buhler, A.V., Weiner, J.L. and Dunwiddie, T.V. (1998) Synaptic potentials mediated via alpha-bungarotoxin-sensitive nicotinic acetylcholine receptors in rat hippocampal interneurons. J. Neurosci., 18: 8228–8235.

Fried, P.A., Watkinson, B. and Siegel, L.S. (1997) Reading and language in 9- to 12-year olds prenatally exposed to cigarettes and marijuana. Neurotoxicol. Teratol., 19: 171–183.

Fuchs, J.L. (1989) [^{125}I]α-Bungarotoxin binding marks primary sensory areas of developing rat neocortex. Brain Res., 501: 223–234.

Gilbert, C.D. (1998) Adult cortical dynamics. Physiol. Rev., 78: 467–485.

Gioanni, Y., Rougeot, C., Clarke, P.B., Lepouse, C., Thierry, A.M. and Vidal, C. (1999) Nicotinic receptors in the rat prefrontal cortex: increase in glutamate release and facilitation of mediodorsal thalamo-cortical transmission. Eur. J. Neurosci., 11: 18–30.

Gray, R., Rajan, A.S., Radcliffe, K.A., Yakehiro, M. and Dani, J.A. (1996) Hippocampal synaptic transmission enhanced by low concentrations of nicotine. Nature, 383: 713–716.

Haj-Dahmane, S. and Andrade, R. (1996) Muscarinic activation of a voltage-dependent cation nonselective current in rat association cortex. J. Neurosci., 16: 3848–3861.

Halliwell, J.V. and Adams, P.R. (1982) Voltage-clamp analysis of muscarinic excitation in hippocampal neurons. Brain Res., 250: 71–92.

Hasselmo, M.E. and Bower, J.M. (1992) Cholinergic suppression specific to intrinsic not afferent fiber synapses in rat piriform (olfactory) cortex. J. Neurophysiol., 67: 1222–1229.

Hasselmo, M.E. (1995) Neuromodulation and cortical function: modeling the physiological basis of behavior. Behav Brain Res, 67: 1–27.

Henningfield, J.E., Stapleton, J.M., Benowitz, N.L., Grayson, R.F. and London, E.D. (1993) Higher levels of nicotine in arterial than in venous blood after cigarette smoking. Drug Alcohol Depend., 33: 23–29.

Hess, A. and Scheich, H. (1996) Optical and FDG mapping of frequency-specific activity in auditory cortex. NeuroReport, 7: 2643–2647.

Horikawa, J., Hosokawa, Y., Kubota, M., Nasu, M. and Taniguchi, I. (1996) Optical imaging of spatiotemporal patterns of glutamatergic excitation and GABAergic inhibition in the guinea-pig auditory cortex in vivo. J. Physiol. (Lond.), 497: 629–638.

Hounsgaard, J. (1978) Presynaptic inhibitory action of acetylcholine in area CA1 of the hippocampus. Exp. Neurol., 62: 787–797.

Hsieh, C.Y., Cruikshank, S.J. and Metherate, R. (2000) Differential modulation of auditory thalamocortical and intracortical synaptic transmission by cholinergic agonist. Brain Res., 880: 51–64.

Hsieh, C.Y., Chen, Y., Leslie, F.M. and Metherate, R. (2002a) Postnatal development of NR2A and NR2B mRNA expression in rat auditory cortex and thalamus. J. Assoc. Res. Otolarnygol. (in press).

Hsieh, C.Y., Leslie, F.M. and Metherate, R. (2002b) Nicotine exposure during a postnatal critical period alters NR2A and NR2B mRNA expression in rat auditory forebrain. Dev. Brain Res., 133: 19–25.

Ignacio, M.P., Kimm, E.J., Kageyama, G.H., Yu, J. and Robertson, R.T. (1995) Postnatal migration of neurons and formation of laminae in rat cerebral cortex. Anat. Embryol. (Berl.), 191: 89–100.

Imig, T.J. and Morel, A. (1984) Topographic and cytoarchitectonic organization of thalamic neurons related to their targets in low-, middle-, and high-frequency representations in cat auditory cortex. J. Comp. Neurol., 227: 511–539.

Isaac, P.F. and Rand, M.J. (1972) Cigarette smoking and plasma levels of nicotine. Nature, 236: 308–310.

Isaac, J.T.R., Crair, M.C., Nicoll, R.A. and Malenka, R.C. (1997) Silent synapses during development of thalamocortical inputs. Neuron, 18: 269–280.

Iwasa, H. and Potsic, W.P. (1982) Maturational change of early, middle, and late components of the auditory evoked responses in rats. Otolaryngol. Head Neck Surg., 90: 95–102.

Kaur, S., Lazar, R., Liang, K. and Metherate, R. (2002) Organization of functional connectivity in rat primary auditory cortex: a potential mechanism for broad spectral integration. Soc. Neurosci. Abstr., 28: A458.

Kharazia, V.N. and Weinberg, R.J. (1994) Glutamate in thalamic fibers terminating in layer IV of primary sensory cortex. J. Neurosci., 14: 6021–6032.

Kilgard, M.P. and Merzenich, M.M. (1998) Cortical map reorganization enabled by nucleus basalis activity. Science, 279: 1714–1718.

Kristt, D.A. (1979) Development of neocortical circuitry: histochemical localization of acetylcholinesterase in relation to the cell layers of rat somatosensory cortex. J. Comp. Neurol., 186: 1–15.

Krmpotic-Nemanic, J., Kostovic, I., Kelovic, Z. and Nemanic, D. (1980) Development of acetylcholinesterase (AChE) staining in human fetal auditory cortex. Acta Otolaryngol., 89: 388–392.

Krmpotic-Nemanic, J., Kostovic, I., Kelovic, Z., Nemanic, D. and Mrzljak, L. (1983) Development of the human fetal auditory cortex: growth of afferent fibers. Acta Anat., 116: 69–73.

Krnjević, K. and Phillis, J.W. (1963) Acetylcholine-sensitive cells in the cerebral cortex. J. Physiol. (Lond.), 166: 296–327.

Krnjević, K., Pumain, R. and Renaud, L. (1971) The mechanism of excitation by acetylcholine in the cerebral cortex. J. Physiol. (Lond.), 215: 247–268.

Levy, R.B. and Aoki, C. (2002) Alpha7 nicotinic acetylcholine receptors occur at postsynaptic densities of AMPA receptor-positive and -negative excitatory synapses in rat sensory cortex. J. Neurosci., 22: 5001–5015.

Madison, D.V., Lancaster, B. and Nicoll, R.A. (1987) Voltage clamp analysis of cholinergic action in the hippocampus. J. Neurosci., 7: 733–741.

Matsubara, J.A. and Phillips, D.P. (1988) Intracortical connections and their physiological correlates in the primary auditory cortex (AI) of the cat. J. Comp. Neurol., 268: 38–48.

McCartney, J.S., Fried, P.A. and Watkinson, B. (1994) Central auditory processing in school-age children prenatally exposed to cigarette smoke. Neurotoxicol. Teratol., 16: 269–276.

McCormick, D.A. and Prince, D.A. (1986) Mechanism of action of acetylcholine in the guinea-pig cerebral cortex in vitro. J. Physiol. (Lond.), 375: 169–194.

McKenna, T.M., Ashe, J.H. and Weinberger, N.M. (1989) Cholinergic modulation of frequency receptive fields in auditory cortex: I. Frequency-specific effects of muscarinic agonists. Synapse, 4: 30–43.

Mechawar, N., Watkins, K.C. and Descarries, L. (2002) Ultrastructural features of the acetylcholine innervation in the developing parietal cortex of rat. J. Comp. Neurol., 443: 250–258.

Metherate, R., Tremblay, N. and Dykes, R.W. (1988) The effects of acetylcholine on response properties of cat somatosensory cortical neurons. J. Neurophysiol., 59: 1231–1251.

Metherate, R., Cox, C.L. and Ashe, J.H. (1992) Cellular bases of neocortical activation: modulation of neural oscillations by the nucleus basalis and endogenous acetylcholine. J. Neurosci., 12: 4701–4711.

Metherate, R. and Ashe, J.H. (1993a) Nucleus basalis stimulation facilitates thalamocortical synaptic transmission in rat auditory cortex. Synapse, 14: 132–143.

Metherate, R. and Ashe, J.H. (1993b) Ionic flux contributions to neocortical slow waves and nucleus basalis-mediated activation: whole-cell recordings in vivo. J. Neurosci., 13: 5312–5323.

Metherate, R. and Aramakis, V.B. (1999) Intrinsic electrophysiology of neurons in thalamorecipient layers of developing rat auditory cortex. Dev. Brain Res., 115: 131–144.

Miller, L.M., Escabi, M.A., Read, H.L. and Schreiner, C.E. (2001) Functional convergence of response properties in the auditory thalamocortical system. Neuron, 32: 151–160.

Monyer, H., Burnashev, N., Laurie, D.J., Sakmann, B. and Seeburg, P.H. (1994) Developmental and regional expression in the rat brain and functional properties of four NMDA receptors. Neuron, 12: 529–540.

Murrin, L.C., Ferrer, J.R., Zeng, W.Y. and Haley, N.J. (1987) Nicotine administration to rats: methodological considerations. Life Sci., 40: 1699–1708.

Nase, G., Weishaupt, J., Stern, P., Singer, W. and Monyer, H. (1999) Genetic and epigenetic regulation of NMDA receptor expression in the rat visual cortex. Eur. J. Neurosci., 11: 4320–4326.

Nelken, I., Prut, Y., Vaadia, E. and Abeles, M. (1994a) Population responses to multifrequency sounds in the cat auditory cortex: one- and two-parameter families of sounds. Hear Res., 72: 206–222.

Nelken, I., Prut, Y., Vaddia, E. and Abeles, M. (1994b) Population responses to multifrequency sounds in the cat auditory cortex: four-tone complexes. Hear Res., 72: 223–236.

Ojima, H., Honda, C.N. and Jones, E.G. (1991) Patterns of axonal collateralization of identified supragranular pyramidal neurons in the cat auditory cortex. Cerebral Cortex, 1: 80–94.

Oonishi, S. and Katsuki, Y. (1965) Functional organization and integrative mechanism on the auditory cortex of the cat. Jap. J. Physiol., 15: 342–365.

Philpot, B.D., Sekhar, A.K., Shouval, H.Z. and Bear, M.F. (2001) Visual experience and deprivation bidirectionally modify the composition and function of NMDA receptors in visual cortex. Neuron, 29: 157–169.

Picone, T.A., Allen, L.H., Olsen, P.N. and Ferris, M.E. (1982) Pregnancy outcome in North American women. II. Effects of diet, cigarette smoking, stress, and weight gain on placentas, and on neonatal physical and behavioral characteristics. Am. J. Clin. Nutr., 36: 1214–1224.

Prusky, G.T., Arbuckle, J.M. and Cynader, M.S. (1988) Transient concordant distributions of nicotinic receptors and acetylcholinesterase activity in infant rat visual cortex. Brain Res., 467: 154–159.

Ribaupierre, F.d., Goldstein, M.H. and Yeni-Komshian, G. (1972) Intracellular study of the cat's primary auditory cortex. Brain Res., 48: 185–204.

Roberts, E.B. and Ramoa, A.S. (1999) Enhanced NR2A subunit expression and decreased NMDA receptor decay time at the onset of ocular dominance plasticity in the ferret. J. Neurophysiol., 81: 2587–2591.

Robertson, R.T., Mostamand, F., Kageyama, G.H., Gallardo, K.A. and Yu, J. (1991) Primary auditory cortex in the rat: transient expression of acetylcholinesterase in developing geniculocortical projections. Dev. Brain Res., 58: 81–95.

Roerig, B., Nelson, D.A. and Katz, L.C. (1997) Fast synaptic signaling by nicotinic acetylcholine and serotonin 5-HT3 receptors in developing visual cortex. J. Neurosci., 17: 8353–8362.

Rouiller, E.M. and Welker, E. (1991) Morphology of corticothalamic terminals arising from the auditory cortex of the rat: a Phaseolus vulgaris-leucoagglutinin (PHA-L) tracing study. Hear Res., 56: 179–190.

Rumpel, S., Hatt, H. and Gottmann, K. (1998) Silent synapses in the developing rat visual cortex: Evidence for postsynaptic expression of synaptic plasticity. J. Neurosci., 18: 8863–8874.

Sahin, M., Bowen, W.D. and Donoghue, J.P. (1992) Location of nicotinic and muscarinic cholinergic and mu-opiate receptors in rat cerebral neocortex: evidence from thalamic and cortical lesions. Brain Res., 579: 135–147.

Sato, H., Hata, Y., Hagihara, K. and Tsumoto, T. (1987) Effects of cholinergic depletion on neuron activities in the cat visual cortex. J. Neurophysiol., 58: 781–794.

Saxton, D. (1978) The behavior of infants whose mothers smoke in pregnancy. Early Human Dev., 2: 363–369.

Schulze, H. and Langner, G. (1999) Auditory cortical responses to amplitude modulations with spectra above frequency receptive fields: evidence for wide spectral integration. J. Comp. Physiol. [A], 185: 493–508.

Segal, M. (1982) Multiple action of acetylcholine at a muscarinic receptor studied in the rat hippocampal slice. Brain Res., 246: 77–87.

Segal, M. (1989) Presynaptic cholinergic inhibition in hippocampal cultures. Synapse, 4: 305–312.

Sexton, M., Fox, N.L. and Hebel, J.R. (1990) Prenatal exposure to tobacco: II. Effects on cognitive functioning at age three. Int. J. Epidemiol., 19: 72–77.

Shulz, D.E., Sosnik, R., Ego, V., Haidarliu, S. and Ahissar, E. (2000) A neuronal analogue of state-dependent learning. Nature, 403: 549–553.

Sillito, A.M. and Kemp, J.A. (1983) Cholinergic modulation of the functional organization of the cat visual cortex. Brain Res., 289: 143–155.

Sillito, A.M. (1993) The cholinergic neuromodulatory system: an evaluation of its functional roles. Prog. Brain Res., 98: 371–378.

Valentino, R.J. and Dingledine, R. (1981) Presynaptic inhibitory effect of acetylcholine in the hippocampus. J. Neurosci., 7: 784–792.

Volkov, I.O. and Galazjuk, A.V. (1991) Formation of spike response to sound tones in cat auditory cortex neurons:

interaction of excitatory and inhibitory effects. Neuroscience, 43: 307–321.

Wallace, M.N., Kitzes, L.M. and Jones, E.G. (1991) Intrinsic inter- and intralaminar connections and their relationship to the tonotopic map in cat primary auditory cortex. Exp. Brain Res., 86: 527–544.

Weinberger, N.M. (1995) Dynamic regulation of receptive fields and maps in the adult sensory cortex. Ann. Rev. Neurosci., 18: 129–158.

Whitfield, I.C. and Evans, E.F. (1965) Responses of auditory cortical neurons to stimuli of changing frequency. J. Neurophysiol., 28: 655–672.

Winer, J.A., Kelly, J.B. and Larue, D.T. (1999) Neural architecture of the rat medial geniculate body. Hear Res., 130: 19–41.

Zhang, Z., Vijayaraghavan, S. and Berg, D.K. (1994) Neuronal acetylcholine receptors that bind alpha-bungarotoxin with high affinity function as ligand-gated ion channels. Neuron, 12: 167–177.

Progress in Brain Research, Vol. 145
ISSN 0079-6123

CHAPTER 11

Activity, modulation and role of basal forebrain cholinergic neurons innervating the cerebral cortex

Barbara E. Jones*

Department of Neurology and Neurosurgery, McGill University, Montreal Neurological Institute,
3801 University Street, Montreal, QC H3A 2B4, Canada

Abstract: The basal forebrain constitutes the ventral extra-thalamic relay from the brainstem activating system to the cerebral cortex. Cholinergic neurons form an important contingent of this relay, yet represent only a portion of the cortically projecting and other basal forebrain neurons, which include GABAergic neurons. By recording, labeling and identifying neurons by their neurotransmitter first in vitro and then in vivo, we have determined that the cholinergic neurons have different physiological and pharmacological properties than other codistributed neurons. Cholinergic neurons discharge at higher rates during cortical activation than during cortical slow wave activity, and are excited by transmitters released from brainstem afferent neurons, including glutamate from reticular formation, noradrenaline (NA) from locus coeruleus, and orexin and histamine from posterior hypothalamus. In contrast, particular GABAergic neurons discharge at higher rates during cortical slow wave activity and are inhibited by NA. When NA is administered into the basal forebrain in naturally sleeping-waking rats, it elicits an increase in fast gamma EEG activity and diminution of slow delta EEG activity while promoting waking and eliminating slow wave sleep (SWS). Cholinergic neurons also have the capacity to discharge in rhythmic bursts when activated by particular agonists, notably neurotensin (NT). When NT is administered into the basal forebrain, it stimulates theta activity in addition to gamma while promoting waking and paradoxical sleep (PS). By increasing discharge and firing in rhythmic bursts in response to transmitters of the activating systems, cholinergic neurons can thus stimulate cortical activation with gamma and theta activity along with the states of waking and PS. Colocalized GABAergic basal forebrain neurons which are inhibited by transmitters of the arousal systems can oppose these actions and promote delta activity and SWS.

Introduction

The cholinergic neurons that innervate the cerebral cortex are distributed through the basal forebrain within the nuclei of the diagonal band of Broca (DBB), magnocellular preoptic (MCPO), substantia innominata (SI) and globus pallidus (GP) in the rat (Fig. 1) (Rye et al., 1984; Gritti et al., 1993). They lie in the path of fibers ascending to the forebrain within the ventral pathway from the reticular formation (Fig. 2) (Jones and Yang, 1985; Jones, 1995). They

*Tel.: 514-398-1913; Fax: 514-398-5871;
E-mail: barbara.jones@mcgill.ca

DOI: 10.1016/S0079-6123(03)45011-5

thus form the extrathalamic relay from the brainstem activating system to the cerebral cortex (Moruzzi and Magoun, 1949; Starzl et al., 1951). The cholinergic neurons act upon the cortex by release from their terminals of acetylcholine (ACh) that potently excites cortical neurons and stimulates cortical activation (Krnjević, 1967; McCormick and Prince, 1986; McCormick, 1992). ACh is released in the cortex maximally in association with cortical activation during the states of waking and paradoxical sleep (PS) (Celesia and Jasper, 1966; Jasper and Tessier, 1971; Marrosu et al., 1995). Antagonism of ACh effects in the cortex by administration of the muscarinic receptor antagonist, scopolamine, results in a

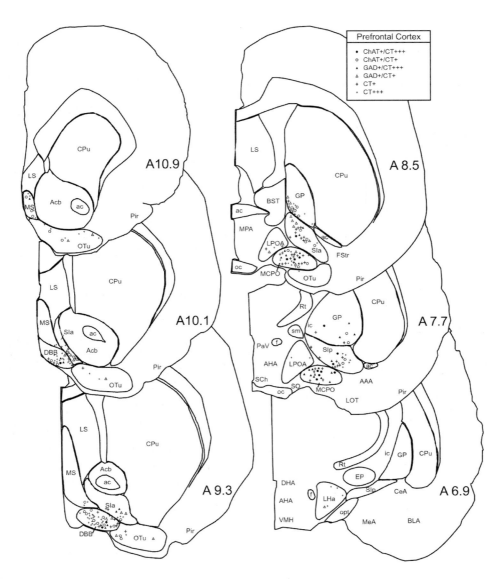

Fig. 1. The distribution of cholinergic, GABAergic and other noncholinergic neurons that project to the cortex. The projection neurons were retrogradely labeled following injections of cholera toxin (CT) into the prefrontal cortex and marked as single-immunolabeled for CT (CT+, crosses from the ChAT-CT-series) or double-immunolabeled for CT and choline-acetyl transferase (ChAT/CT, circles) or glutamic acid decarboxylase (GAD/CT, triangles). Darkly retrogradely labeled cells (thick crosses and solid symbols) are distinguished from lightly retrogradely labeled cells (thin crosses and open symbols). See list for abbreviations. The figure is adapted from Gritti et al. (1997).

loss of cortical activation and dramatic changes in these states (Longo, 1966; Jones, 1993). Yet, lesions of the basal forebrain have been reported to produce changes in all sleep–wake states, including slow wave sleep (SWS), and their associated electroencephalographic (EEG) activities (Stewart et al., 1984; Szymusiak and McGinty, 1986b; Buzsaki et al., 1988). Moreover, recording studies aimed at characterizing the discharge profile of cholinergic neurons have encountered many different cell types having different activity profiles in relation to EEG activity and sleep–wake states in the basal forebrain

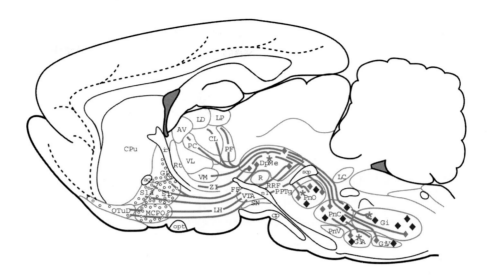

Fig. 2. Ascending projections from neurons of the brainstem reticular formation (diamonds or stars) into the forebrain course along a dorsal pathway into the thalamus and a ventral pathway through the hypothalamus up to the basal forebrain cholinergic cell area. The cholinergic cells (open circles) project to the cerebral cortex in a relatively widespread manner thus serving as an important relay to the cortex of the ascending reticular activating system. The majority of neurons in the reticular formation are thought to utilize glutamate as a neurotransmitter (Jones, 1995). They include those with ascending projections to the forebrain (gray diamonds), descending projections to the spinal cord (black diamonds) and bifurcating projections to the forebrain and spinal cord (stars). See list for abbreviations. The figure is adapted from Jones (1995).

cholinergic cell area (Detari and Vanderwolf, 1987; Buzsaki et al., 1988; Szymusiak and McGinty, 1989). This heterogeneity is not surprising given that the cholinergic cells are distributed among many other noncholinergic cells that include numerous GABAergic neurons, some of which also project to the cerebral cortex and may play very different roles than the cholinergic cells (Fig. 1) (Gritti et al., 1997). To understand the role of the cholinergic and noncholinergic, including GABAergic, basal forebrain neurons, it is thus necessary to record from individual neurons that can be identified immunohistochemically.

In vitro study of the intrinsic properties of cholinergic neurons

In collaboration with Michel Muhlethaler and his students in Geneva, and Angel Alonso in Montreal, we examined the intrinsic properties of identified cholinergic neurons first in brain slices. The cholinergic neurons were distinctive among all cells recorded in the basal forebrain due to the presence of prominent calcium currents, including a

low threshold spike that endows them with the capacity to discharge rhythmically in bursts (Khateb et al., 1992). Although noncholinergic neurons also showed rhythmic properties, they did so in a different manner and through entirely different mechanisms (Alonso et al., 1996). Together with certain noncholinergic neurons, it thus appeared from in vitro studies that the cholinergic cells would have the capacity to modulate the cerebral cortex in a rhythmic manner. However, whether such rhythmic bursting and cortical modulation would occur with cortical activation, as in the septo-hippocampal system (Petsche et al., 1965; Alonso et al., 1987; Dragoi et al., 1999), or during slow wave activity, as in the thalamo-cortical system (Steriade et al., 1994), could only be determined by in vivo studies.

In vivo study of the discharge profile of cholinergic and GABAergic neurons

Across the sleep–waking cycle, multiple cortical rhythms are observed as a function of sleep–wake state and behavior, as evident in the rat (Fig. 3)

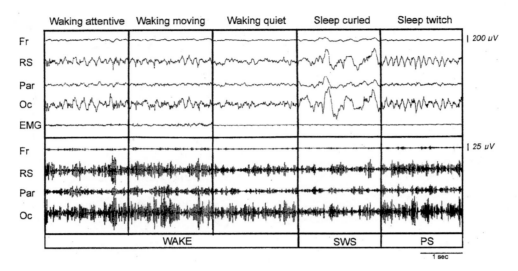

Fig. 3. Unfiltered (top rows) and high frequency filtered (for gamma: 30.5–58.0 Hz, bottom rows) EEG associated with different behaviors and sleep–wake states in the rat. The electromyogram (EMG) from the neck muscles is also shown. The EEG during attentive and active waking and during PS is characterized by a rhythmic slow activity or theta, which is most prominent on the retrosplenial cortex (RS) and occipital cortex (Oc), and high amplitude gamma activity. During SWS, the EEG is characterized by irregular delta wave activity with low amplitude gamma. See list for abbreviations. The figure is adapted from Maloney et al. (1997).

(Maloney et al., 1997). During attentive or active waking and PS, high frequency gamma activity (30–60 Hz) is maximal and rides upon rhythmic slow activity or theta (4–10 Hz), which is most prominent on limbic cortex (see retrosplenial, RS in Fig. 3). During quiet waking, gamma is low and theta is absent. During slow wave sleep (SWS), gamma is minimal and irregular slow waves of a delta frequency (1–3 Hz) are prominent. In order to determine how cholinergic neurons might modulate these different cortical activities, we applied the technique of juxtacellular labeling (Pinault, 1996) to units recorded in association with EEG activity in urethane-anesthetized rats (Manns et al., 2000a). Under these conditions, the EEG activity is characterized by irregular slow waves that are similar to the delta waves of natural SWS in the undisturbed animal; but it is also somewhat reactive since with somatosensory stimulation (pressure applied to the tail), it changes from irregular delta-like activity to rhythmic theta-like activity with increased gamma activity (despite the absence of any behavioral response). Recorded neurons labeled with Neurobiotin that were choline-acetyltransferase (ChAT)-immunoreactive (Fig. 4) demonstrated distinctive discharge

properties in association with these stimulation-induced EEG changes (Fig. 5). As also found by others (Duque et al., 2000), all cholinergic neurons increased their rate of discharge with cortical activation. In addition, we found that the majority (∼ 75%) also changed their pattern of discharge from an irregular single spike mode to a rhythmic bursting mode. The rhythmic discharge was correlated with the rhythmic theta-like activity on limbic cortex. These results suggest that all cholinergic basal forebrain neurons discharge maximally and many discharge in bursts with cortical activation, such that they may modulate the cortex in a rhythmic manner to stimulate theta along with gamma activity during states of active waking and PS.

Among noncholinergic cells, GABAergic cells are in the majority more active during irregular slow wave cortical activity than during stimulation-induced cortical activation in urethane-anesthetized rats (Manns et al., 2000b). As locally projecting neurons, such cells could exert an inhibitory influence on nearby cholinergic cells to decrease cortical activation, or as projection neurons to cortex (evidenced by antidromic activation), others could exert an inhibitory influence directly onto cortical

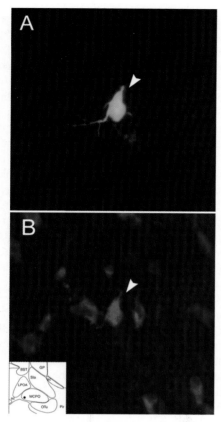

Fig. 4. A cholinergic neuron that was recorded in association with EEG activity in a urethane-anesthetized rat (see Fig. 5). The cell was juxtacellularly labeled with Neurobiotin (Nb, revealed with green Cy2-conjugated streptavidin in A) and immunostained for ChAT (revealed with red Cy3-conjugated secondary antibodies in B). As shown in inset, the cell was located in the MCPO. See list for abbreviations. The figure is adapted from Manns et al. (2000a).

cells to dampen their activity in association with SWS. Accordingly, certain GABAergic neurons are potentially sleep-promoting neurons that would correspond to SWS-active cells recorded in unanesthetized animals (Szymusiak and McGinty, 1986a; Szymusiak et al., 2000).

In vitro and in vivo study of the modulation of basal forebrain neurons

Our in vitro studies indicated that neuromodulators differentially affect cholinergic and noncholinergic

neurons in the basal forebrain and some promote rhythmic discharge in the cholinergic cells. *Noradrenaline* (NA) is contained in fibers originating in the locus coeruleus that arborize in the basal forebrain (Semba et al., 1988; Jones and Cuello, 1989), as well as projecting directly to the cerebral cortex (Jones and Moore, 1977; Jones and Yang, 1985). Contacted by the noradrenergic fibers, the cholinergic cells therein are depolarized and excited by NA through α_1 adrenergic receptors (Fort et al., 1995; Zaborszky and Cullinan, 1996; Smiley et al., 1999). In vitro, the depolarization was most often associated with tonic discharge. This action would suggest that the cholinergic neurons may be tonically active during waking, when the locus coeruleus neurons are active (Aston-Jones and Bloom, 1981). Noncholinergic cells show different responses to NA including one subgroup that is hyperpolarized and inhibited by NA (Fort et al., 1998), similar to putative sleep-promoting cells in the preoptic area (Gallopin et al., 2000). In contrast to the cholinergic neurons, these noncholinergic cells could be inhibited during waking and become active during sleep, when the locus coeruleus neurons become inactive. They could correspond to the GABAergic cells recorded in the anesthetized animals that are active during irregular slow wave activity and inhibited during stimulation-evoked cortical activation (above, (Manns et al., 2000b)). According to immunohisto-chemical evidence, these GABAergic possibly sleep-active cells bear α_2 adrenergic receptors (Hou et al., 2001). During waking, cholinergic and particular GABAergic basal forebrain neurons would thus be reciprocally modulated by NA, the cholinergic neurons excited and the sleep-active GABAergic neurons inhibited. NA would thus have the capacity to stimulate cortical activation while preventing the onset of SWS through actions upon the different basal forebrain neurons. To assess the actual effects of modulation of the cholinergic and particular GABAergic neurons by NA upon EEG activity and state, we administered NA by microinjection into the basal forebrain (Fig. 6) (Cape and Jones, 1998). Injected during the day, when the animals are normally asleep, NA eliminated delta slow wave activity and SWS. It also suppressed PS, and resulted in a significant increase in gamma EEG activity (Fig. 7). These results would indicate that cholinergic

162

Nb+/ChAT+ MCPO cell

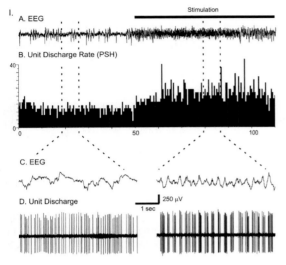

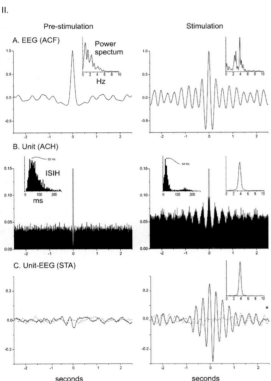

neurons may act in tandem with noradrenergic neurons during waking to promote cortical activation while preventing sleep.

Contained in afferents arising mainly from the dorsal raphe nucleus, *serotonin* (5-hydroxytrypta-mine, 5-HT) hyperpolarizes and inhibits cholinergic basal forebrain neurons (Semba et al., 1988; Jones and Cuello, 1989; Khateb et al., 1993). When injected into the basal forebrain during the day, when the animals are normally asleep, serotonin produces a decrease in gamma activity and theta while not preventing delta activity or SWS (Cape and Jones, 1998). PS is suppressed. The results indicate that serotonin, which was once posited to play a role in generating SWS (Jouvet, 1972), may do so by attenuating activity in cholinergic basal forebrain neurons, indirectly reducing cortical gamma and theta activity and thus possibly facilitating the onset of SWS from a quiet waking state, when the serotonergic neurons are active (McGinty and Harper, 1976). This inhibition would be lifted during SWS and the entry into PS, when the serotonergic neurons become silent and the cholinergic basal forebrain neurons would again become active.

Similar to NA, other transmitters of the ascending activating systems, including *histamine* and the more recently discovered, *orexin* (*hypocretin*), which are contained in cells of the posterior hypothalamus, depolarize and excite the cholinergic neurons (Khateb et al., 1995a; Eggermann et al., 2001).

slow rate (B and D). During rhythmic slow (theta-like) activity accompanied by higher gamma activity (I, right panels) that was evoked by somatosensory stimulation (A, expanded in C), the unit discharged in a rhythmic bursting fashion at a higher rate (B and D). For EEG (A) and unit discharge (B) in I., time is in seconds indicated under unit discharge rate and frequency in Hz on y axis. By examining the auto-correlation function (ACF) of the EEG (II, A) and the autocorrelation histogram (ACH) of the unit, it was apparent that the EEG and unit discharge were rhythmic during stimulation. Two peaks in the interspike interval histogram (ISIH) reveal the rhythmic bursting spike pattern of the unit. By the spike triggered average (STA), it was evident that the unit rhythmic burst discharge was significantly cross-correlated with the rhythmic slow (theta-like) activity of the EEG, suggesting that the unit could be modulating the EEG in this rhythmic activity. The unit-to-EEG correlation was significantly different (*) from the randomized spike train to EEG function (in gray). The figure was taken from Manns et al. (2000a).

Fig. 5. Discharge properties of an identified cholinergic neuron (shown in Fig. 4) in association with EEG activity in a urethane-anesthetized rat. During the irregular slow wave activity (I, left panels) of the undisturbed condition (A, expanded in C), the unit discharged in a tonic manner at a

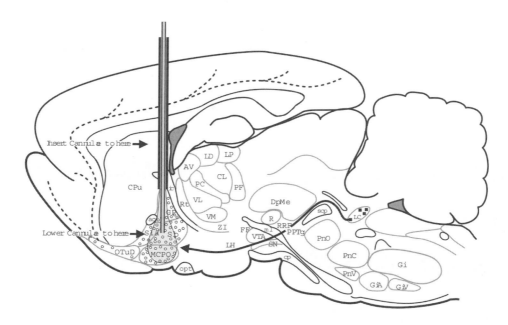

Fig. 6. Microinjection of transmitters or their agonists into the basal forebrain cholinergic cell area was performed in the rat using a remotely driven system that allowed insertion of the drug-filled cannulae (one each side) into permanently fixed outer cannulae ('Insert Cannulae to here') prior to the experiment and then lowering of the cannulae into the basal forebrain cholinergic cell area ('Lower Cannulae to here') immediately preceding the injection (see Fig. 7). This procedure allowed full relaxation of the animal following placement of the cannulae while preventing desensitization of receptors by drug leakage into the site prior to the injection. In addition to other substances described (see text), NA was injected into the basal forebrain to mimic the effect of that transmitter that is contained in fibers innervating the cholinergic neurons from the locus coeruleus (LC, see Fig. 7). The figure was adapted from Cape and Jones (1998).

Interestingly, these modulators do not affect the putative sleep active cells in the basal forebrain or adjacent preoptic area (Fort et al., 1998; Gallopin et al., 2000; Eggermann et al., 2001). Microinjections of orexin into the basal forebrain cholinergic cell area stimulate cortical activation and waking (Espana et al., 2001).

From the effects of receptor agonists, it is clear that *glutamate*, which is contained in the neurons of the ascending reticular activating system (Fig. 2) (Jones, 1995), excites cholinergic neurons and through NMDA receptors, can also drive them into rhythmic bursting discharge (Khateb et al., 1995b). Microinjections of NMDA into the basal forebrain result in increased gamma and theta activity on the cortex in association with an active waking state (Cape and Jones, 2000). SWS and PS are eliminated. However, these effects cannot be attributed uniquely to excitation of cholinergic neurons since glutamate and its agonists act upon multiple cell types.

Peptides have the capacity to act very selectively upon target neurons. The cholinergic basal forebrain neurons selectively bear receptors to *neurotensin* (NT) (Szigethy and Beaudet, 1987), a peptide that is contained in afferent fibers originating in part from neurons in the posterior hypothalamus and midbrain as well as others in the forebrain (Morin and Beaudet, 1998). By in vitro studies, it was found that NT stimulated rhythmic bursting discharge in the cholinergic neurons without affecting other noncholinergic neurons in the region (Alonso et al., 1994). Upon in vivo administration by microinjection into the basal forebrain (as in Fig. 5), NT was also found to be selectively internalized in cholinergic neurons and to induce a bursting discharge in Neurobiotin-filled and identified cholinergic neurons in the urethane-anesthetized rat (Cape et al., 2000). NT microinjection in the freely moving rat produced an elimination of cortical slow wave activity or delta and the appearance of theta activity in association

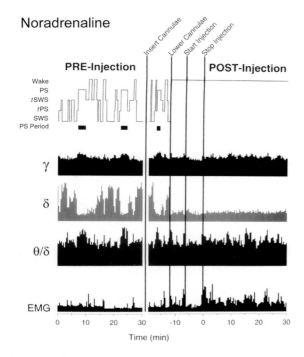

Fig. 7. The effect of microinjection of (150 mM) noradrenaline (NA) into the basal forebrain cholinergic cell area is shown upon sleep–wake state (hypnogram, top row), EEG activities (spectral band amplitudes from retrosplenial cortex, middle rows) and EMG amplitude (bottom row). As compared to Ringer (in the postinjection period, not shown) or to no substance (as seen in the preinjection period on the left), NA stimulated cortical activation that was marked by a loss of delta activity and a significant increase in gamma activity. SWS was prevented and waking stimulated during the day when the rat normally sleeps the majority of the time. The figure was adapted from Cape and Jones (1998).

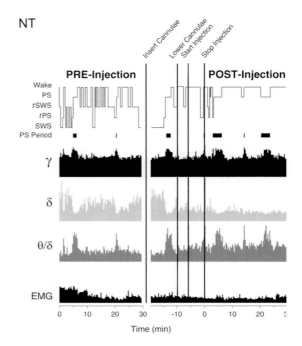

Fig. 8. The effect of microinjection of (1 mM) Neuroten (NT) into the basal forebrain cholinergic cell area is sho upon sleep–wake state (hypnogram, top row), EEG activit (spectral band amplitudes from retrosplenial cortex, mid rows) and EMG amplitude (bottom row). As compared Ringer (in the postinjection period, not shown) or to substance (as seen in the preinjection period on the left), I stimulated cortical activation that was particularly marked enhanced theta activity. Delta was suppressed along with SW While the animal remained relatively quiet, it alternated direc between states of waking and PS. The figure was adapted fro Cape et al. (2000).

with moderately high gamma activity (Fig. 8). This EEG pattern was associated with an alternation of waking and PS states in absence of any intervening SWS during the time of day when rats sleep the majority of the time. There was a dose-dependent decrease in delta activity and a reciprocal increase in theta and gamma activity along with waking and PS (Fig. 9). The results suggest that selective activation and promotion of rhythmic burst discharge of cholinergic basal forebrain neurons prevents SWS and stimulates theta and gamma cortical activity along with the states of waking and PS.

Most recently, we have been able to record basal forebrain neurons using micropipettes for labeling

with Neurobiotin in naturally waking-sleeping ra under head-restraint. Preliminary results ha revealed possible cholinergic neurons that dischar in bursts during active waking and PS in associatic with theta activity (Lee et al., 2002). These cells we highly active during active waking, virtually sile during SWS and maximally active during PS. Th profile of activity is coherent with the profile acetylcholine (ACh) release from the cortex acrc these states (Marrosu et al., 1995). We await defi tive identification of these cells by immunohist chemistry to confirm that cholinergic basal forebra neurons discharge in rhythmic bursts to modula cortical activity during active waking and PS. V also found cells that were maximally active duri

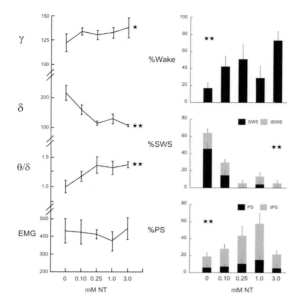

Fig. 9. The dose-dependent effect of Neurotensin (NT) upon
EEG activities (spectral band amplitudes) and EMG amplitude
(left panel) and sleep–wake states (right panel). NT produced a
dose-dependent increase in gamma and theta activity with a
reciprocal decrease in delta activity. It simultaneously produced
a dose-dependent increase in waking and PS with a reciprocal
decrease in SWS. One to three stars indicate statistically
significant effects (p 0.05, 0.01 or 0.001). The figure was
adapted from Cape et al. (2000).

SWS and that are suspected to be GABAergic
neurons, which may inhibit the cholinergic cells or
directly dampen cortical activity during SWS.

Regulation of EEG activity and sleep–wake states by basal forebrain neurons

What is the significance of the particular discharge of
cholinergic neurons for modulating EEG activity and
sleep–wake states? As it likely occurs during active
waking and PS, the bursting discharge by cholinergic
neurons should produce a maximal release of ACh in
the cerebral cortex (Lisman, 1997). Through a
prolonged depolarization mediated by muscarinic
receptors, ACh excites pyramidal cells to discharge in
a fast tonic firing mode and to cease discharging in a
slow burst firing mode usually associated with SWS
(Krnjević, 1967; McCormick, 1992; Metherate et al.,
1992). ACh also produces a fast depolarization of

interneurons through both muscarinic and nicotinic
receptors and inhibits a certain class of interneurons
(McCormick and Prince, 1986; Xiang et al., 1998;
Porter et al., 1999). Accordingly, the phasic release of
ACh that would occur with theta-like bursting
discharge could modulate theta-like activity in a
temporally precise manner by fast muscarinic and
nicotinic actions upon interneurons while promoting
high frequency activity by additional prolonged
muscarinic actions upon pyramidal neurons. Such
rhythmic modulation would parallel that in the
hippocampus, where inhibitory interneurons pace
the pyramidal neurons by inhibitory postsynaptic
potentials (IPSPs), to produce a theta oscillation with
faster gamma waves riding upon the rhythmic slow
waves (Soltesz and Deschenes, 1993). Just as the
hippocampus depends upon the phasic input from the
septum to generate theta (Petsche et al., 1962), so
other cortical regions may depend upon the
cholinergic basal forebrain neurons for phasic input
to generate rhythmic slow activity. Such rhythmic
driving would provide a template for coherent
activity in functionally linked circuits across widely
distributed visceral/limbic, somatic and associative
cortical fields. It could simultaneously enhance
coherence in the gamma frequency range that has
been shown to be important for binding among
neurons across such fields (Gray et al., 1989; Munk
et al., 1996). Modulation at a theta frequency can
also maximize synaptic plasticity (Larson et al., 1986;
Huerta and Lisman, 1993). Stimulation of basal
forebrain has been shown to facilitate plasticity and
associative learning in the cortex (Kilgard and
Merzenich, 1998; McLin et al., 2002). Thus, during
cortical activation that occurs during the states of
waking and PS, rhythmic bursting by cholinergic
neurons could enhance excitability while pacing
cortical interneurons and pyramidal cells, facilitate
coherence among distributed networks of cortical
projection neurons and promote synaptic plasticity
and learning in those networks.

As reviewed above, the cholinergic neurons are
modulated by the transmitters of the brainstem and
posterior hypothalamic activating systems, and thus
serve as an important extrathalamic relay for these
systems to the cerebral cortex. They are thus enlisted
in activation to sensory stimuli, including somatosensory, auditory and visual, that excite the brainstem

reticular formation. In addition, however, they are also positioned to receive sensory input that arrives in the forebrain from the olfactory system (Paolini and McKenzie, 1997) and to evoke cortical activation in the forebrain in response to this input, independent or in advance of that stimulated by the brainstem systems. The cortical activation that can be evoked in the *cerveau isolé* preparation by olfactory stimulation (Villablanca, 1965) may therefore be mediated by the basalo-cortical system. The return of spontaneous cortical activation in the chronic *cerveau isolé* preparation (Villablanca, 1965) may also be mediated by the basalo-cortical system that would comprise the core of autochthonous forebrain mechanisms for cyclical cortical activation independent of the brainstem. Evidence here and from other pharmacological studies suggests that the cholinergic basal forebrain neurons play a particularly important role in that system. Selective activation of the cholinergic neurons by NT has revealed their capacity both to stimulate cortical activation, including theta and gamma activity, and to generate the states of waking and PS while preventing SWS (Cape et al., 2000). By excitation of other selective local and distant neuronal circuits, the cholinergic neurons thus have a potent influence in both modulating cortical activity and determining the state of the animal. Particular GABAergic neurons (Manns et al., 2000b) display the capacity to oppose the cholinergic neurons locally or within the cerebral cortex to dampen cortical activity and to promote SWS. Accordingly, the basal forebrain cholinergic and noncholinergic neurons have the capacity to regulate cortical activation and sleep–wake states and to do so in response to but also independent of the brainstem activating systems.

Acknowledgments

The author's laboratory was supported by grants from the Canadian Institute of Health Research (CIHR, 13458) and National Institute of Mental Health (NIMH, RO1 MH60119-01A1). I thank Lynda Mainville, Ivana Gritti, Ian Manns, Edmund Cape and Angel Alonso in Montreal, and Michel Muhlethaler (along with Asaid Khateb, Patrice Fort, Mauro Serafin, Daniele Machard and Emmanuel Eggermann) in Geneva, who all contributed to the research reviewed in this chapter. I also thank Elida Arriza and Naomi Takeda for graphic and secretarial assistance, respectively.

Abbreviations

AAA	anterior amygdaloid area
ac	anterior commissure
Acb	nucleus accumbens
ACF	autocorrelation function
ACh	acetylcholine
ACH	autocorrelation histogram
AHA	anterior hypothalamic area
AV	anteroventral thalamic nucleus
BLA	basolateral amygdaloid nucleus
BST	bed nucleus of the stria terminalis
CeA	central amygdaloid nucleus
CL	centrolateral thalamic nucleus
cp	cerebral peduncle
Cpu	caudate putamen
DBB	diagonal band of Broca nucleus
DHA	dorsal hypothalamic area
DpMe	deep mesencephalic nucleus
EEG	electroencephalogram
EMG	electromyogram
EP	entopeduncular nucleus
f	fornix
FF	fields of Forel
Fr	frontal cortex
FStr	fundus striati
Gi	gigantocellular reticular nucleus
GiA	gigantocellular reticular nucleus, alpha part
GiV	gigantocellular reticular nucleus, ventral part
GP	globus pallidus
ic	internal capsule
ISIH	interspike interval histogram
LC	locus coeruleus nucleus
LD	laterodorsal thalamic nucleus
LH	lateral hypothalamus
LHa	lateral hypothalamic area
LOT	lateral olfactory tract nucleus
LP	lateral posterior thalamic nucleus
LPOA	lateral preoptic area
LS	lateral septum
MCPO	magnocellular preoptic nucleus
MeA	medial amygdaloid nucleus
ml	medial lemniscus

MPA	medial preoptic area
MS	medial septum
NA	noradrenaline
Nb	Neurobiotin
NT	neurotensin
oc	optic chiasm
Oc	occipital cortex
opt	optic tract
OTu	olfactory tubercle
OTuD	olfactory tubercle deep layers
Par	parietal cortex
PaV	paraventricular hypothalamic nucleus
PC	paracentral thalamic nucleus
PF	parafascicular thalamic nucleus
Pir	piriform cortex
PnC	pontine reticular nucleus, caudal part
PnO	pontine reticular nucleus, oral part
PnV	pontine reticular nucleus, ventral part
PPTg	pedunculopontine tegmental nucleus
PS	paradoxical sleep
R	red nucleus
Rt	reticular thalamic nucleus
RRF	retrorubral field
RS	retrosplenial cortex
scp	superior cerebellar peduncle
SCh	suprachiasmatic nucleus
SIa	substantia innominata pars anterior
SIp	substantia innominata pars posterior
sm	stria medullaris
SN	substantia nigra
SO	nucleus supraopticus
STA	spike triggered average
SWS	slow wave sleep
TPS	transition to paradoxical sleep
TSWS	transition to slow wave sleep
VL	ventrolateral thalamic nucleus
VM	ventromedial thalamic nucleus
VMH	ventromedial hypothalamic nucleus
VTA	ventral tegmental area
ZI	zona incerta

References

Alonso, A., Gaztelu, J.M., Buno, W., Jr. and Garcia-Austt, E. (1987) Cross-correlation analysis of septohippocampal neurons during theta rhythm. Brain Res., 413: 135–146.

Alonso, A., Faure, M.-P. and Beaudet, A. (1994) Neurotensin promotes oscillatory bursting behavior and is internalized in basal forebrain cholinergic neurons. J. Neurosci., 14: 5778–5792.

Alonso, A., Khateb, A., Fort, P., Jones, B.E. and Muhlethaler, M. (1996) Differential oscillatory properties of cholinergic and non-cholinergic nucleus Basalis neurons in guinea pig brain slice. Eur. J. Neurosci., 8: 169–182.

Aston-Jones, G. and Bloom, F.E. (1981) Activity of norepinephrine-containing locus coeruleus neurons in behaving rats anticipates fluctuations in the sleep–waking cycle. J. Neurosci., 1: 876–886.

Buzsaki, G., Bickford, R.G., Ponomareff, G., Thal, L.J., Mandel, R. and Gage, F.H. (1988) Nucleus basalis and thalamic control of neocortical activity in the freely moving rat. J. Neurosci., 8: 4007–4026.

Cape, E.G. and Jones, B.E. (1998) Differential modulation of high frequency gamma electroencephalogram activity and sleep–wake state by noradrenaline and serotonin microinjections into the region of cholinergic basalis neurons. J. Neurosci., 18: 2653–2666.

Cape, E.G. and Jones, B.E. (2000) Effects of glutamate agonist versus procaine microinjections into the basal forebrain cholinergic cell area upon gamma and theta EEG activity and sleep–wake state. Eur. J. Neurosci., 12: 2166–2184.

Cape, E.G., Manns, I.D., Alonso, A., Beaudet, A. and Jones, B.E. (2000) Neurotensin-induced bursting of cholinergic basal forebrain neurons promotes gamma and theta cortical activity together with waking and paradoxical sleep. J. Neurosci., 20: 8452–8461.

Celesia, G.G. and Jasper, H.H. (1966) Acetylcholine released from cerebral cortex in relation to state of activation. Neurology, 16: 1053–1064.

Detari, L. and Vanderwolf, C.H. (1987) Activity of identified cortically projecting and other basal forebrain neurones during large slow waves and cortical activation. Brain Res., 437: 1–8.

Dragoi, G., Carpi, D., Recce, M., Csicsvari, J. and Buzsaki, G. (1999) Interactions between hippocampus and medial septum during sharp waves and theta oscillation in the behaving rat. J. Neurosci., 19: 6191–6199.

Duque, A., Balatoni, B., Detari, L. and Zaborszky, L. (2000) EEG correlation of the discharge properties of identified neurons in the basal forebrain. J. Neurophysiol., 84: 1627–1635.

Eggermann, E., Serafin, M., Bayer, L., Machard, D., Sanit-Mleux, B., Jones, B.E. and Muhlethaler, M. (2001) Orexins/hypocretins excite basal forebrain cholinergic neurones. Neuroscience, 108: 177–181.

Espana, R.A., Baldo, B.A., Kelley, A.E. and Berridge, C.W. (2001) Wake-promoting and sleep-suppressing actions of hypocretin (orexin): basal forebrain sites of action. Neuroscience, 106: 699–715.

Fort, P., Khateb, A., Pegna, A., Muhlethaler, M. and Jones, B.E. (1995) Noradrenergic modulation of cholinergic nucleus basalis neurons demonstrated by *in vitro* pharmacological and immunohistochemical evidence in the guinea pig brain. Eur. J. Neurosci., 7: 1502–1511.

Fort, P., Khateb, A., Serafin, M., Muhlethaler, M. and Jones, B.E. (1998) Pharmacological characterization and differentiation of non-cholinergic nucleus basalis neurons *in vitro*. NeuroReport, 9: 1–5.

Gallopin, T., Fort, P., Eggermann, E., Cauli, B., Luppi, P.H., Rossier, J., Audinat, E., Muhlethaler, M. and Serafin, M. (2000) Identification of sleep-promoting neurons *in vitro*. Nature, 404: 992–995.

Gray, C.M., Konig, P., Engel, A.K. and Singer, W. (1989) Oscillatory responses in cat visual cortex exhibit inter-columnar synchronization which reflects global stimulus properties. Nature, 338: 334–337.

Gritti, I., Mainville, L. and Jones, B.E. (1993) Codistribution of GABA- with acetylcholine-synthesizing neurons in the basal forebrain of the rat. J. Comp. Neurol., 329: 438–457.

Gritti, I., Mainville, L., Mancia, M. and Jones, B.E. (1997) GABAergic and other non-cholinergic basal forebrain neurons project together with cholinergic neurons to meso- and iso-cortex in the rat. J. Comp. Neurol., 383: 163–177.

Hou, Y.P., Manns, I.D. and Jones, B.E. (2001) Alpha-2 adrenergic receptors on GABAergic basal forebrain neurons that discharge maximally during cortical slow wave activity. Actas de Fisiologia, 7: 183.

Huerta, P.T. and Lisman, J.E. (1993) Heightened synaptic plasticity of hippocampal CA1 neurons during a cholinergically induced rhythmic state. Nature, 364: 723–725.

Jasper, H.H. and Tessier, J. (1971) Acetylcholine liberation from cerebral cortex during paradoxical (REM) sleep. Science, 172: 601–602.

Jones, B.E. and Moore, R.Y. (1977) Ascending projections of the locus coeruleus in the rat. II. Autoradiographic study. Brain Res., 127: 23–53.

Jones, B.E. and Yang, T.-Z. (1985) The efferent projections from the reticular formation and the locus coeruleus studied by anterograde and retrograde axonal transport in the rat. J. Comp. Neurol., 242: 56–92.

Jones, B.E. and Cuello, A.C. (1989) Afferents to the basal forebrain cholinergic cell area from pontomesencephalic—catecholamine, serotonin, and acetylcholine—neurons. Neuroscience, 31: 37–61.

Jones, B.E. (1993). The organization of central cholinergic systems and their functional importance in sleep–waking states. In: A.C. Cuello (Ed.), Cholinergic Function and Dysfunction Prog. Brain Res, Vol. 98, Elsevier, Amsterdam, pp. 61–71.

Jones, B.E. (1995). Reticular formation. Cytoarchitecture, transmitters and projections. In: G. Paxinos (Ed.), The Rat Nervous System. Academic Press, Australia, New South Wales, pp. 155–171.

Jouvet, M. (1972) The role of monoamines and acetylcholine-containing neurons in the regulation of the sleep–waking cycle. Ergeb. Physiol., 64: 165–307.

Khateb, A., Muhlethaler, M., Alonso, A., Serafin, M., Mainville, L. and Jones, B.E. (1992) Cholinergic nucleus basalis neurons display the capacity for rhythmic bursting

activity mediated by low threshold calcium spikes. Neuroscience, 51: 489–494.

Khateb, A., Fort, P., Alonso, A., Jones, B.E. and Muhlethaler, M. (1993) Pharmacological and immunohisto-chemical evidence for a serotonergic input to cholinergic nucleus basalis neurons. Eur. J. Neurosci., 5: 541–547.

Khateb, A., Fort, P., Pegna, A., Jones, B.E. and Muhlethaler, M. (1995a) Cholinergic nucleus Basalis neurons are excited by histamine *in vitro*. Neuroscience, 69: 495–506.

Khateb, A., Fort, P., Serafin, M., Jones, B.E. and Muhlethaler, M. (1995b) Rhythmical bursts induced by NMDA in cholinergic nucleus basalis neurones *in vitro*. J. Physiol. (Lond.), 487.3: 623–638.

Kilgard, M.P. and Merzenich, M.M. (1998) Cortical map reorganization enabled by nucleus basalis activity. Science, 279: 1714–1718.

Krnjević, K. (1967) Chemical transmission and cortical arousal. Anesthesiology, 28: 100–104.

Larson, J., Wong, D. and Lynch, G. (1986) Patterned stimulation at the theta frequency is optimal for the induction of hippocampal long-term potentiation. Brain Res., 368: 347–350.

Lee, M.G., Manns, I.D., Alonso, A. and Jones, B.E. (2002) Sleep–wake discharge profile of basal forebrain neurons recorded and labeled by the juxtacellular method in head-restrained rats. Soc. Neurosci. Abst. Online, 672.4.

Lisman, J.E. (1997) Bursts as a unit of neural information: making unreliable synapses reliable. Trends Neurosci, 20: 38–43.

Longo, V.G. (1966) Behavioral and electroencephalographic effects of atropine and related compounds. Pharmacol. Rev., 18: 965–996.

Maloney, K.J., Cape, E.G., Gotman, J. and Jones, B.E. (1997) High frequency gamma electroencephalogram activity in association with sleep–wake states and spontaneous behaviors in the rat. Neuroscience, 76: 541–555.

Manns, I.D., Alonso, A. and Jones, B.E. (2000a) Discharge properties of juxtacellularly labeled and immunohistochemically identified cholinergic basal forebrain neurons recorded in association with the electroencephalogram in anesthetized rats. J. Neurosci., 20: 1505–1518.

Manns, I.D., Alonso, A. and Jones, B.E. (2000b) Discharge profiles of juxtacellularly labeled and immunohistochemically identified GABAergic basal forebrain neurons recorded in association with the electroencephalogram in anesthetized rats. J. Neurosci., 20: 9252–9263.

Marrosu, F., Portas, C., Mascia, S., Casu, M.A., Fa, M., Giagheddu, M., Imperato, A. and Gessa, G.L. (1995) Microdialysis measurement of cortical and hippocampal acetylcholine release during sleep–wake cycle in freely moving cats. Brain Res., 671: 329–332.

McCormick, D.A. and Prince, D.A. (1986) Mechanisms of action of acetylcholine in the guinea-pig cerebral cortex *in vitro*. J. Physiol., 375: 169–194.

McCormick, D.A. (1992) Neurotransmitter actions in the thalamus and cerebral cortex and their role in neuromodulation of thalamocortical activity. Progr. Neurobiol., 39: 337–388.

McGinty, D. and Harper, R.M. (1976) Dorsal raphe neurons: depression of firing during sleep in cats. Brain Res., 101: 569–575.

McLin, D.E., 3rd, Miasnikov, A.A. and Weinberger, N.M. (2002) Induction of behavioral associative memory by stimulation of the nucleus basalis. Proc. Natl. Acad. Sci. USA, 99: 4002–4007.

Metherate, R., Cox, C.L. and Ashe, J.H. (1992) Cellular bases of neocortical activation: modulation of neural oscillations by the nucleus basalis and endogenous acetylcholine. J. Neurosci., 12: 4701–4711.

Morin, A.J. and Beaudet, A. (1998) Origin of the neurotensinergic innervation of the rat basal forebrain studied by retrograde transport of cholera toxin. J. Comp. Neurol., 391: 30–41.

Moruzzi, G. and Magoun, H.W. (1949) Brain stem reticular formation and activation of the EEG. Electroenceph. Clin. Neurophysiol, 1: 455–473.

Munk, M.H.J., Roelfsema, P.R., Konig, P., Engel, A.K. and Singer, W. (1996) Role of reticular activation in the modulation of intracortical synchronization. Science, 272: 271–274.

Paolini, A.G. and McKenzie, J.S. (1997) Intracellular recording of magnocellular preoptic neuron responses to olfactory brain. Neuroscience, 78: 229–242.

Petsche, H., Stumpf, C. and Gogolak, G. (1962) The significance of the rabbit's septum as a relay station between the midbrain and the hippocampus. I. The control of hippocampus arousal activity by the septum cells. Electroenceph. Clin. Neurophysiol., 14: 202–211.

Petsche, H., Gogolak, G. and van Zwieten, P.A. (1965) Rhythmicity of septal cell discharges at various levels of reticular excitation. Electroenceph. Clin. Neurophysiol., 19: 25–33.

Pinault, D. (1996) A novel single-cell staining procedure performed in vivo under electrophysiological control: morpho-functional features of juxtacellularly labeled thalamic cells and other central neurons with biocytin or Neurobiotin. J. Neurosci. Methods, 65: 113–136.

Porter, J.T., Cauli, B., Tsuzuki, K., Lambolez, B., Rossier, J. and Audinat, E. (1999) Selective excitation of subtypes of neocortical interneurons by nicotinic receptors. J. Neurosci., 19: 5228–5235.

Rye, D.B., Wainer, B.H., Mesulam, M.-M., Mufson, E.J. and Saper, C.B. (1984) Cortical projections arising from the basal forebrain: a study of cholinergic and noncholinergic components employing combined retrograde tracing and immunohistochemical localization of choline acetyltransferase. Neuroscience, 13: 627–643.

Semba, K., Reiner, P.B., McGeer, E.G. and Fibiger, H.C. (1988) Brainstem afferents to the magnocellular basal forebrain studied by axonal transport, immunohistochemistry and electrophysiology in the rat. J. Comp. Neurol., 267: 433–453.

Smiley, J.F., Subramanian, M. and Mesulam, M.M. (1999) Monoaminergic-cholinergic interactions in the primate basal forebrain. Neuroscience, 93: 817–829.

Soltesz, I. and Deschenes, M. (1993) Low- and high-frequency membrane potential oscillations during theta activity in CA1 and CA3 pyramidal neurons of the rat hippocampus under ketamine-xylazine anesthesia. J. Neurophysiol., 70: 97–116.

Starzl, T.E., Taylor, C.W. and Magoun, H.W. (1951) Ascending conduction in reticular activating system, with special reference to the diencephalon. J. Neurophysiol., 14: 461–477.

Steriade, M., Contreras, D. and Amzica, F. (1994) Synchronized sleep oscillations and their paroxysmal developments. Trends Neurosci., 17: 199–208.

Stewart, D.J., MacFabe, D.F. and Vanderwolf, C.H. (1984) Cholinergic activation of the electrocorticogram: Role of the substantia innominata and effects of atropine and quinuclidinyl benzilate. Brain Res., 322: 219–232.

Szigethy, E. and Beaudet, A. (1987) Selective association of neurotensin receptors with cholinergic neurons in the rat basal forebrain. Neurosci. Lett., 83: 47–52.

Szymusiak, R. and McGinty, D. (1986a) Sleep-related neuronal discharge in the basal forebrain of cats. Brain Res., 370: 82–92.

Szymusiak, R. and McGinty, D. (1986b) Sleep suppression following kainic acid-induced lesions of the basal forebrain. Exp. Neurol., 94: 598–614.

Szymusiak, R. and McGinty, D. (1989) Sleep–waking discharge of basal forebrain projection neurons in cats. Brain Res. Bull, 22: 423–430.

Szymusiak, R., Alam, N. and McGinty, D. (2000) Discharge patterns of neurons in cholinergic regions of the basal forebrain during waking and sleep. Behav. Brain Res., 115: 171–182.

Villablanca, J. (1965) The electrocorticogram in the chronic cerveau isole cat. Electroenceph. Clin. Neurophysiol., 19: 576–586.

Xiang, Z., Huguenard, J.R. and Prince, D.A. (1998) Cholinergic switching within neocortical inhibitory networks. Science, 5379: 985–988.

Zaborszky, L. and Cullinan, W.E. (1996) Direct catecholaminergic-cholinergic interactions in the basal forebrain. I. Dopamine-beta-hydroxylase- and tyrosine hydroxylase input to cholinergic neurons. J. Comp. Neurol., 374: 535–554.

control of the microvascular tone as regard)

Progress in Brain Research, Vol. 145
ISSN 0079-6123

CHAPTER 12

Cholinergic modulation of the cortical microvascular bed

Edith Hamel*

Laboratory of Cerebrovascular Research, Department of Neurology & Neurosurgery, Montreal Neurological Institute, McGill University, 3801 University Street, Montreal, QC H3A 2B4, Canada

Introduction

Acetylcholine (ACh) has been assigned important roles in cortical function under both physiological and pathophysiological conditions. This includes a control of the microvascular tone as originally suggested from the diffuse increases in cortical cerebral blood flow (CBF) observed following stimulation of basal forebrain neurons. Some constituents of this neurovascular system have recently been elucidated. Indeed, the cholinergic parasympathetic cerebrovascular innervation does not extend beyond the Virchow-Robin space of penetrating arteries and cannot explain the cholinergic-mediated changes in local perfusion observed following activation of cholinergic pathways/neurons located within the brain parenchyma. Similarly the significant increases in CBF observed in the ipsilateral fronto-parietal cortex upon electrical or chemical stimulation of basal forebrain neurons (Arneric, 1989; Biesold et al., 1989; Lacombe et al., 1989) cannot be attributed to extra-cerebral blood vessels, as the diameter of pial vessels did not change following stimulation of basal forebrain neurons (Adachi et al., 1992). In contrast, it was shown by in vivo confocal microscopy that a majority of intracortical microvessels dilated in response to this

stimulation (Thomas et al., 1997), thus suggesting that basal forebrain neurons can modulate the diameter of intracortical microvessels. As will be described in this paper, there is strong evidence that basal forebrain ACh neurons are primarily but not exclusively involved in this regulation.

Cholinergic component of the blood flow response to basal forebrain stimulation

Physiological studies measuring cortical CBF have shown that the increases in cortical perfusion noted after stimulation of basal forebrain neurons are largely but not exclusively imputable to cholinergic mechanisms. For instance, the perfusion changes are accompanied by increases in ACh release and ACh levels that are approximately twofold in the frontoparietal cortex as compared to prestimulus levels (Kurosawa et al., 1989). These increases are proportional to stimulus intensity and frequency and also occur after microinjection of glutamate in the basal forebrain, suggesting activation of ACh neurons. The increases in cortical perfusion are potentiated by administration of the ACh esterase inhibitors physostigmine and heptylphysostigmine (Linville et al., 1992; Peruzzi et al., 2000), and are partially blocked by the muscarinic receptor (mAChR) antagonists atropine and scopolamine

*Tel.: 514-398-8928; Fax: 514-398-8106;
E-mail: edith.hamel@mcgill.ca

DOI: 10.1016/S0079-6123(03)45012-7

(Biesold et al., 1989; Dauphin et al., 1991). When administered together with the nicotinic receptor antagonist mecamylamine, atropine blocks a greater part of the CBF increase following basal forebrain stimulation, suggesting that both receptor populations are implicated.

Nitrergic component of the blood flow response to basal forebrain stimulation

The CBF increases in cortical perfusion induced by stimulation of the basal forebrain are attenuated in a dose-dependent manner by nonselective inhibitors of nitric oxide (NO) production, such as L-N^G-nitroarginine (L-NNA) (Raszkiewicz et al., 1992). In contrast, 7-nitroindazole (7-NI), a more selective inhibitor of neuronal nitric oxide synthase (NOS), has been reported to be ineffective (Zhang et al., 1995; Iadecola and Zhang, 1996). As coapplication of atropine and L-NNA on the cerebral cortex does not further decrease the blood flow response attenuated by L-NNA alone, the two components are most likely interdependent, possibly through muscarinic activation of endothelial NO production (Zhang et al., 1995). As will be presented below, we found that intracortical arterioles dilate following ACh application, a response mediated by muscarinic receptor activation and NO production (Elhusseiny and Hamel, 2000).

Basal forebrain ACh neurons project to cortical microvessels and NO neurons

Using the anterograde transport of *Phaseolus vulgaris leucoagglutinin* (PHA-L) microiontophoretically injected in the basal forebrain, we showed that its neurons project broadly not only to the cortical neuropile but also to the arterioles and capillaries within the cerebral cortex (Fig. 1) (Vaucher and Hamel, 1995). Various neurotransmitters, sometimes colocalized, are found within basal forebrain neurons, including ACh, NO, γ-aminobutyric acid (GABA) and galanin (Gritti et al., 1993; Kubota et al., 1994; Sugaya and McKinney, 1994; Sobreviela et al., 1998). We have demonstrated, using immunocytochemistry and basal forebrain lesions, that ACh and NO-synthesizing basal forebrain neurons, which

correspond in part to the same population of neurons (Sugaya and McKinney, 1994; Sobreviela et al., 1998), provide an important portion of perivascular fibers to cortical microvessels (Vaucher and Hamel, 1995; Tong and Hamel, 1999). In contrast, perivascular GABA nerve fibers in the cerebral cortex originate almost exclusively from local GABA interneurons (Vaucher et al., 2000). These encompass different sub-populations of neurons that costore either neuropeptide Y (NPY) and NOS, or vasoactive intestinal polypeptide (VIP) and, occasionally, ACh, in addition to peptides such as somatostatin (SS), cholecystokinin (CCK) and different calcium binding proteins (for a review, Estrada and DeFelipe, 1998).

We also found that basalocortical ACh and NO-synthesizing fibers project to cortical NO interneurons in both rat and man (Vaucher et al., 1997; Tong and Hamel, 1999, 2000). These NO interneurons, which belong to a distinct sub-population of GABA interneurons that colocalize somatostatin and NPY (Kubota et al., 1994), associate extensively with the cortical microvascular bed (see Estrada and DeFelipe, 1998). They are thus strategically positioned to serve as local integrator of basalocortical ACh pathways (Kawaguchi, 1997). However, the CBF increases observed following basal forebrain stimulation are reportedly independent of intracortical neurons (Linville et al., 1993), suggesting that most of the cortical flow response is due to direct vascular effects from basal forebrain afferents, at least under the stimulation paradigms used in these studies.

Intracortical neurons as a local relay for integrating local perfusion to neuronal activity

Several populations of interneurons within the cerebral cortex send projections to neighboring microvessels and could modify microvascular tone. In this respect, a population of neurons with activity related to spontaneous waves of CBF has recently been identified in the cerebral cortex and is suspected of transducing neuronal signals into vasomotor responses (Golanov et al., 1994, 2000). These neurons could be involved in both the propagation and restriction of the neurally-mediated changes in local

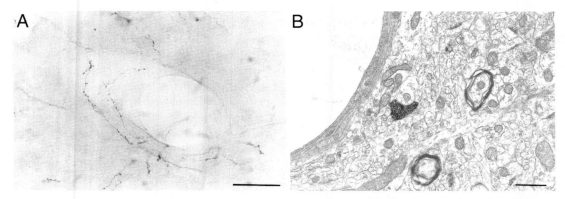

Fig. 1. (A) PHA-L-immunoreactive fibers in the cerebral cortex can be seen to surround an intracortical arteriole in the frontal cortex (Bar: 50 μm). (B) Perivascular ACh nerve terminal (labeled with an antibody against choline acetyltransferase (ChAT)) in the rat fronto-parietal cortex. The terminal is located within 1 μm from the vessel basal lamina (Bar: 1 μm). Reproduced with permission from Vaucher and Hamel (1995).

CBF and could correspond to specific sub-populations of GABA interneurons. These could act as local integrators of cortical perfusion as a function of neuronal activity, in keeping with their role of local integrator of cortical function (Kawaguchi and Kubota, 1997). For instance, cortical NO (vasodilator) neurons as well as those containing the vasoactive peptides VIP (vasodilator) or NPY (vasoconstrictor) are intimately associated with cortical microvessels. Moreover, the average size of perivascular NOS, VIP and NPY nerve terminals in the cerebral cortex (X ± s.e.m., 0.53 ± 0.4, 0.56 ± 0.4 and 0.58 ± 0.2 μm², respectively) (Chédotal et al., 1994; Abounader and Hamel, 1997; Tong and Hamel, 2000), corresponds to that of GABAergic nerve terminals (0.50 ± 0.3 μm²) (Vaucher et al., 2000), in keeping with the possibility that they belong to the same or distinct sub-populations of GABA neurons. In this respect, our results also showed that cortical GABA nerve terminals are particularly numerous around intracortical blood vessels (Fig. 2) (Vaucher et al., 2000), possibly highlighting their role in the regulation of cerebrovascular functions. In contrast, most perivascular ACh cortical terminals were much smaller (0.32 ± μm²) (Chédotal et al., 1994; Vaucher and Hamel, 1995), and they largely (>60%) originated from the basal forebrain (Vaucher et al., 1997; Tong and Hamel, 2000). This observation suggests that cortical ACh interneurons are not the main regulator of local CBF or, alternatively, that only those that colocalize with VIP

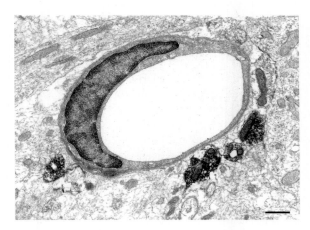

Fig. 2. Numerous GABA (immunostained for GAD 67) nerve terminals and dendrites (weakly labeled on the top right of the vessel) surrounding a capillary in the cerebral cortex of rat. Bar: 1 μm. Reproduced with permission from Vaucher et al. (2000).

might fulfill this role. GABA interneurons thus appear strategically positioned to translate incoming neuronal signals from various sub-cortical afferents (Price et al., 1993; Meng et al., 1995; Morales and Bloom, 1997) into adapted local neuronal and vascular responses, including those originating from basal forebrain ACh neurons (Kawaguchi, 1997) (Fig. 3). In this respect, the presence of M2 muscarinic ACh receptors (mAChRs) on intracortical NO neurons (Moro et al., 1995; Smiley et al., 1998), and of nicotinic receptors on a specific population of cortical GABA interneurons containing VIP (Porter

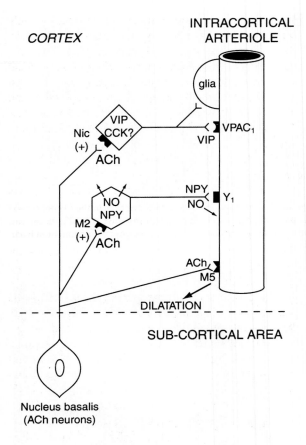

CORTEX

INTRACORTICAL
ARTERIOLE

Nucleus basalis
(ACh neurons)

Fig. 3. Hypothetical representation of basal forebrain ACh neurons projections to cortical microvessels and to different sub-populations of GABA interneurons containing either VIP or NO and acting as local neurovascular relay neurons. ACh: acetylcholine; NO: nitric oxide; VIP: vasoactive intestinal polypeptide; NPY: neuropeptide Y; CCK: cholecystokinin; Y_1: NPY-Y_1 receptor; $VPAC_1$: VIP type 1 receptor; M_5: M_5 mAChR.

et al., 1999) could explain the modulation of the cortical blood flow response elicited by basal forebrain stimulation with muscarinic and nicotinic receptor antagonists, and by nitric oxide synthase inhibitors. Whether the effective vasoactive substance is GABA, as previously documented in the hippocampus (Fergus and Lee, 1997), NO, a vasoactive neuropeptide costored within GABA neurons, or a substance released from surrounding activated perivascular astrocytes (Angulo et al., 2001) remains to be established.

We recently observed, in isolated superfused living cortical brain slices, that patch clamp stimulation of single, cortical bipolar GABA interneurons with irregular spiking activity typical of a subset of cortical VIP interneurons (IS neurons, Porter et al., 1998, 1999) could result in dilation of neighboring microarterioles. In contrast, GABA interneurons with regular fast spiking activity (FS cells, Cauli et al., 2000) could elicit transient contraction of adjacent microvessels. Stimulation of most interneurons did not exert any effect on the local microvessels (Cauli et al., 2003). These preliminary results are exciting as they suggest that activation of specific populations of cortical interneurons may be accompanied by localized changes in microarteriole tone, which would be compatible with a spatial and temporal regulation of local perfusion as a function of local neuronal activity.

Neurotransmitters in perivascular terminals: vasomotor effects on intracortical arterioles

It has been known for several years that intracortical microvessels and capillaries are endowed with mAChRs (for review Dauphin and MacKenzie, 1995). However, only recently were they identified, and the vasomotor responses they mediate in the cerebromicrovascular bed unraveled by the improvement in pharmacological and molecular tools. Our pharmacological studies of mAChR binding sites in isolated human cortical capillary and microvessel fractions have indicated the presence of heterogenous receptors corresponding predominantly to M1, M3 and M5 mAChR subtypes (Linville and Hamel, 1995). Due to the limitation in the pharmacological tools, we performed molecular investigations and showed that endothelial cells express mRNA for M2 and M5 mAChRs, while the smooth muscle cells had high level of expression for the M1 and M3 in addition to M2 and M5 mAChRs (Elhusseiny et al., 1999). However, we suggested that the latter two receptors were most likely due to the contaminating (~ 10–15%) endothelial cells that are inherent to the primary smooth muscle cell cultures used in our investigation. Indeed, the M2 mAChR was unable to

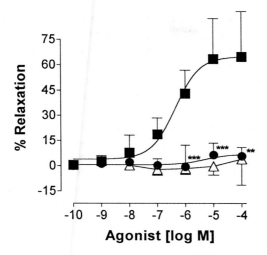

Fig. 4. Concentration-dependent dilation to ACh (squares) and nicotine (triangle) in bovine intracortical arterioles ($n = 4$–10, diameter ~ 47 μm) at spontaneous tone. The ACh-induced dilation was inhibited by L-NNA (10^{-4} M) (circles) ($p < 0.001$). Cortical arterioles were dissected from the parenchyma, mounted in closed sac between two glass micropipettes (diameter 30–40 μm) in an arteriograph chamber, pressurized (60 mmHg) and intraluminal diameter measured on line by videomicroscopy. After stabilization (45–60 min), compounds were added extraluminally to the superfusion solution at the desired concentrations. Changes in diameter were measured and dilation expressed relative to maximal dilation induced by papaverine (10^{-4} M). Modified from Elhusseiny and Hamel (2000).

couple to its signaling pathway in smooth muscle cells, while it effectively inhibited the forskolin-induce cAMP production in endothelial cells (Elhusseiny et al., 1999). The M2 mAChR subtype has also been shown by immunocytochemistry to be present in brain capillary endothelial cells (Smiley et al., 1998). When investigating the vasomotor responses elicited by application of ACh on isolated, superfused and pressurized intracortical arterioles from bovine and human, we found that only a dilation could be observed, and this whether the vessels were tested at basal tone or after preconstriction (Elhusseiny and Hamel, 2000) (Fig. 4). When mAChR antagonists were tested for their ability to block the ACh-induced dilation, the order of potency (pIC_{50} values) was 4-DAMP (9.2 ± 0.3) $\gg$ pirenzepine (6.7 ± 0.4) > AF-DX 384 (5.9 ± 0.2). Linear regression analysis

indicated a significant and positive correlation only with the M5 mAChR type ($r = 0.997$, $p < 0.05$), suggesting that this receptor subtype was the mediator of the ACh-induced dilation in cortical microarterioles. Interestingly, a recent study showed that ACh-mediated dilation of large cerebral blood vessels is abolished in M5 mAChR knockout mice (Yamada et al., 2001). Together with our findings in intracortical microarteries, these results suggest that the M5 mAChR subtype plays an important role in regulating the tonus of cerebral blood vessels and microvessels and, possibly, in the perfusion response to cholinergic stimulation. In contrast, despite the fact that nicotinic binding sites have previously been reported by one group of investigators (Kalaria et al., 1994), nicotine was shown to exert no vasomotor response in isolated microvessels (Elhusseiny and Hamel, 2000) (Fig. 4).

The possibility that other neuroeffectors contained within cortical interneurons, such as GABA, VIP, NPY or NO, might also be involved in the integrated vascular responses following neuronal activation of intracortical neurons has to be considered. In this respect, NPY-Y1 receptors which mediate contraction in large cerebral arteries (Abounader et al., 1995) have been shown to be expressed in cortical microarterioles (Abounader et al., 1999), and more specifically in those that are surrounded by NPY-immunoreactive fibers (Bao et al., 1997). Recently obtained results suggest that VIP receptors of the type 1 (VPAC1 receptors) are expressed in smooth muscle cells of intracortical arterioles, similar to their localization in cerebral arteries where they mediate dilation (Fahrenkrug et al., 2000). Although the vasomotor effects of these peptide receptors have not been tested in intracortical arterioles, VIP and NPY have been shown to respectively mediate dilation and constriction of penetrating arterioles isolated from the rat first segment of the middle cerebral artery and most likely corresponding to the lenticulostriate arteries (Dacey et al., 1988). Similar effects in intracortical arterioles, if observed, would provide strong arguments for a role of GABA interneurons in the adaptation of local CBF to neuronal activity. Moreover, GABA itself could be involved in the regulation of microvascular tone as reported on hippocampus microvessels (Fergus and Lee, 1997).

176

Cerebrovascular dysfunctions in Alzheimer's disease

Perfusion deficits have been characterized in specific cortical subdivisions of the cerebral cortex in Alzheimer's disease, most particularly in the temporal and parietal cortex (Johnson et al., 1998). These have been partially mimicked in the cortex of rats with a selective lesion of basal forebrain ACh neurons (Waite et al., 1999). Moreover, CBF (Niwa et al., 2002) or ACh-mediated cerebral dilations (Iadecola et al., 1999) are greatly reduced in transgenic mice that over-express the amyloid-β precursor protein. An intrinsic microvascular pathology (amyloid deposition, endothelial cell thinning, smooth muscle cell and pericyte degeneration, and thickening of the basal lamina) is apparent in the vast majority of Alzheimer patients (Vinters et al., 1994) and could partly account for these deficits. However, we recently found that cortical microvessels and NO-synthesizing interneurons in neuropathologically confirmed cases of Alzheimer's disease exhibit a severe cholinergic denervation (Tong and Hamel, 1999), indicating that two important aspects of microvascular tone regulation are severely impaired in this neurodegenerative pathology.

Summary

Cortical microvessels receive a cholinergic input that originates primarily from basal forebrain neurons which, upon stimulation, induce significant increases in cortical perfusion together with a dilation of intracortical microvessels. Heterogeneous mAChRs have been detected in cortical microvessels with expression of M2 and M5 subtypes in endothelial cells, while M1 and M3, and possibly M5 mAChR subtypes, were expressed in smooth muscle cells. Application of ACh to isolated and pressurized microarterioles, whether at basal tone or pharmacologically preconstricted, elicited only a dilation. This response was dependent on NO production, and was mediated by a mAChR, the pharmacology of which correlated best with the M5 receptor subtype. ACh afferents also project to intracortical neurons that synthesize NO and VIP. These correspond to distinct sub-populations of GABA interneurons which were found to send numerous projections to local microvessels. Preliminary results suggest expression of the VPAC1 receptor in the smooth muscle cells of intracortical arterioles, where it could mediate dilation as it does in cerebral arteries. Together these results indicate that basal forebrain ACh fibers can directly affect the cortical microvascular bed, but further suggest that specific populations of GABA interneurons could serve as a functional relay to adapt perfusion to locally increased neuronal activity. In confirmed cases of Alzheimer's disease, we found a severe ACh denervation of both cortical microvessels and NO neurons, suggesting that two important regulators of cortical perfusion are dysfunctional in this pathology.

Acknowledgments

The author would particularly like to thank R. Abounader, A. Chédotal, A. Elhusseiny, D. Linville, M. St-Georges, X.-K. Tong and E. Vaucher for their excellent work and effort in the realization of the studies cited in this review. The collaboration of Ms. I. Ferezou, and Drs. B. Cauli, B. Lambolez and J. Rossier (Laboratoire de Neurobiologie, ESPCI, Paris) is also acknowledged for the work on the superfused living brain slices. Funding was provided by the Medical Research Council of Canada (now CIHR, grant MOP-53334), and a Blaise Pascal International Research Chair from la Région de l'Ile de France.

References

Abounader, R. and Hamel, E. (1997) Associations between Neuropeptide Y nerve terminals and intraparenchymal microvessels in rat and human cerebral cortex. J. Comp. Neurol., 388: 444–453.

Abounader, R., Villemure, J.-G. and Hamel, E. (1995) Characterization of neuropeptide Y (NPY) receptors in human cerebral arteries with selective agonists and the new Y1 antagonist BIBP 3226. Brit. J. Pharmacol., 116: 2245–2250.

Abounader, R., Elhusseiny, A., Cohen, Z., Olivier, A., Stanimirovic, D., Quirion, R. and Hamel, E. (1999) Expression of neuropeptide Y receptors mRNA and protein in human brain vessels and cerebromicrovascular cells in culture. J. Cereb. Blood Flow Metab., 19: 155–163.

Adachi, T., Baramidze, D.G. and Sato, A. (1992) Stimulation of the nucleus basalis of Meynert increases cortical cerebral blood flow without influencing diameter of the pial artery in rats. Neurosci. Lett., 143: 173–176.

Angulo, M.C., Zonta, M., Gobbo, S., Pozzan, T. and Carmignoto, G. (2001) Astrocytes regulate the coupling between cortical blood flow and neuronal activity. Soc. Neurosci. Abstr., 27: 505.21.

Arneric, S.P. (1989). Cortical cerebral blood flow is modulated by cholinergic basal forebrain neurons: effects of ibotenic acid lesions and electrical stimulation. In: Seylaz I.J. and MacKenzie E.T. (Eds.), Neurotransmission and Cerebrovascular Function. Elsevier Science Publishers B.V., pp. 381–384.

Bao, L., Kopp, J., Zhang, X., Xu, Z.Q., Zhang, L.F., Wong, H., Walsh, J. and Hokfelt, T. (1997) Localization of neuropeptide Y Y1 receptors in cerebral blood vessels. Proc. Natl. Acad. Sci. USA, 94: 12661–12666.

Biesold, D., Inanami, O., Sato, A. and Sato, Y. (1989) Stimulation of the nucleus basalis of Meynert increases cerebral cortical blood flow in rats. Neurosci. Lett., 98: 39–44.

Cauli, B., Porter, J.T., Tsuzuki, K., Lambolez, B, Rossier, J., Quenet, B. and Audinat, E. (2000) Classification of fusiform neocortical interneurons based on unsupervised clustering. Proc. Natl. Acad. Sci. USA, 97: 6144–6149.

Cauli, B., Tong, X., Lamboez, B., Rossier, J. and Harmel, E. (2003) Stimulation of distinct populations of GABA interneurons elicit changes in local microvessels diameter in the rat cerebral cortex. Soc. Neurosci. Abstr., in press.

Chedotal, A., Umbriaco, D., Descarries, L., Hartman, B.K. and Hamel, E. (1994) Light and electron microscopic immunocytochemical analysis of the neurovascular relationships of choline acetyltransferase (ChAT) and vasoactive intestinal polypeptide (VIP) nerve terminals in the rat cerebral cortex. J. Comp. Neurol., 343: 57–71.

Dacey, R.G., Jr., Bassett, J.E. and Takayasu, M. (1988) Vasomotor responses of rat intracerebral arterioles to vasoactive intestinal peptide, substance P, neuropeptide Y, and bradykinin. J. Cereb. Blood Flow Metab., 8: 254–261.

Dauphin, F. and MacKenzie, E.T. (1995) Cholinergic and vasoactive intestinal polypeptidergic innervation of the cerebral arteries. Pharmac. Ther., 67: 385–417.

Dauphin, F., Lacombe, P., Sercombe, R., Hamel, E. and Seylaz, J. (1991) Hypercapnia and stimulation of the substantia innominata increase rat frontal cortical blood flow by different cholinergic mechanisms. Brain Res., 553: 75–83.

Elhusseiny, A. and Hamel, E. (2000) Muscarinic—but not nicotinic—acetylcholine receptors mediate a nitric oxide-dependent dilation in brain cortical arterioles: A possible role for the M5 receptor subtype. J. Cereb. Blood Flow Metab., 20: 298–305.

Elhusseiny, A., Cohen, Z., Olivier, A., Stanimirovic, D.B. and Hamel, E. (1999) Functional acetylcholine muscarinic receptor subtypes in human brain microcirculation: Identification and cellular localization. J. Cereb. Blood Flow Metab., 19: 794–802.

Estrada, C. and DeFelipe, J. (1998) Nitric oxide-producing neurons in the neocortex: Morphological and functional relationship with intraparenchymal microvasculature. Cerebral Cortex, 8: 193–203.

Fahrenkrug, J., Hannibal, J., Tams, J. and Georg, B. (2000) Immunohistochemical localization of the VIP1 receptor (VPAC1R) in rat cerebral blood vessels: Relation to PACAP and VIP containing nerves. J. Cereb. Blood Flow Metab., 20: 1205–1214.

Fergus, A. and Lee, K.S. (1997) GABAergic regulation of cerebral microvascular tone in the rat. J. Cereb. Blood Flow Metab., 17: 992–1003.

Golanov, E.V., Yamamoto, S. and Reis, D.J. (1994) Spontaneous waves of cerebral blood flow associated with a pattern of electrocortical activity. Am. J. Physiol., 266: R204–R214.

Golanov, E.V., Christensen, J.R.C. and Reis, D.J. (2000) Electrophysiological properties of cortical neurons mediating elevations in cerebral blood flow (RCBF) evoked by excitation of neurons of rostral ventrolateral medulla. Soc. Neurosci. Abstr., 26: 648.2.

Gritti, I., Mainville, L. and Jones, B. (1993) Codistribution of GABA- with acetylcholine-synthesizing neurons in the basal forebrain of the rat. J. Comp. Neurol., 329: 438–457.

Iadecola, C. and Zhang, F. (1996) Permissive and obligatory roles of NO in cerebrovascular responses to hypercapnia and acetylcholine. Am. J. Physiol., 271: R990–R1001.

Iadecola, C., Zhang, F., Niwa, K., Eckman, C., Turner, S.K., Fischer, E., Younkin, S., Borchelt, D.R., Hsiao, K.K., Carlson, G.A. (1999) SOD1 rescues cerebral endothelial dysfunction in mice overexpressing amyloid precursor protein. Nature Neurosci., 2: 157–161.

Johnson, K.A., Jones, K., Holman, B.L., Becker, J.A., Spiers, P.A., Satlin, A. and Albert, M.S. (1998) Preclinical prediction of Alzheimer's disease using SPECT. Neurology, 50: 1563–1571.

Kalaria, R.N., Homayoun, P. and Whitehouse, P.J. (1994) Nicotinic cholinergic receptors associated with mammalian cerebral vessels. J. Autonomic Nervous System, 49: S3–S7.

Kawaguchi, Y. (1997) Selective cholinergic modulation of cortical GABAergic cell subtypes. J. Neurophysiol., 78: 1743–1747.

Kawaguchi, Y. and Kubota, Y. (1997) GABAergic cell subtypes and their synaptic connections in rat frontal cortex. Cerebral Cortex, 7: 476–486.

Kubota, Y., Hattori, R. and Yui, Y. (1994) Three distinct subpopulations of GABAergic neurons in rat frontal agranular cortex. Brain Res., 649: 159–173.

Kurosawa, M., Sato, A. and Sato, Y. (1989) Stimulation of the nucleus basalis of Meynert increases acetylcholine release in the cerebral cortex in rats. Neurosci Lett., 98: 45–50.

Lacombe, P., Sercombe, R., Verrecchia, C., Philipson, V., MacKenzie, E.T. and Seylaz, J. (1989) Cortical blood flow increases induced by stimulation of the substantia innominata in the unanesthetized rat. Brain Res., 491: 1–14.

Linville, D.G. and Hamel, E. (1995) Pharmacological characterization of muscarinic acetylcholine binding sites in human and bovine cerebral microvessels. Naunyn-Schmiedeberg's Arch. Pharmacol., 352: 179–186.

Linville, D.G., Giacobini, E. and Arneric, S.P. (1992) Heptylphysostigmine enhances basal forebrain control of cortical cerebral blood flow. J. Neurosci. Res., 31: 573–577.

Linville, D.G., Williams, S. and Arneric, S.P. (1993) Basal forebrain control of cortical cerebral blood flow is independent of local cortical neurons. Brain Res., 622: 26–34.

Meng, W., Tobin, J.R. and Busija, D.W. (1995) Glutamate-induced cerebral vasodilation is mediated by nitric oxide through N-methyl-D-aspartate receptors. Stroke, 26: 857–863.

Morales, M. and Bloom, F.E. (1997) The 5-HT$_3$ receptor is present in different subpopulations of GABAergic neurons in the rat telencephalon. J. Neurosci., 17: 3157–3167.

Moro, V., Badaut, J., Springhetti, V., Edvinsson, L., Seylaz, J. and Lasbennes, F. (1995) Regional study of the co-localization of neuronal nitric oxide synthase with muscarinic receptors in the rat cerebral cortex. Neuroscience, 69: 797–805.

Niwa, K., Kazama, K., Younkin, S.G., Carlson, G.A. and Iadecola, C. (2002) Alterations in cerebral blood flow and glucose utilization in mice overexpressing the amyloid precursor protein. Neurobiol. Dis., 9: 61–68.

Peruzzi, P., von Euw, D. and Lacombe, P. (2000) Differentiated cerebrovascular effects of physostigmine and tacrine in cortical areas deafferented from the nucleus basalis magnocellularis suggest involvement of basalocortical projections to microvessels. Ann. N.Y. Acad. Sci., 903: 394–406.

Porter, J.T., Cauli, B., Staiger, J.F., Lambolez, B., Rossier, J. and Audinat, E. (1998) Properties of bipolar VIPergic interneurons and their excitation by pyramidal neurons in the rat neocortex. Eur. J. Neurosci., 10: 3617–3628.

Porter, J.T., Cauli, B., Tsuzuki, K., Lambolez, B., Rossier, J. and Audinat, E. (1999) Selective excitation of subtypes of neocortical interneurons by nicotinic receptors. J. Neurosci., 19: 5228–5235.

Price, R.H., Jr., Mayer, B. and Beitz, A.J. (1993) Nitric oxide synthase neurons in rat brain express more NMDA receptor mRNA than non-NOS neurons. NeuroReport, 4: 807–810.

Raszkiewicz, J.L., Linville, D.G., Kerwin, J.F., Jr., Wagenaar, F. and Arneric, S.P. (1992) Nitric oxide synthase is critical in mediating basal forebrain regulation of cortical cerebral circulation. J. Neurosci. Res., 33: 129–135.

Smiley, J.F., Levey, A.I. and Mesulam, M.-M. (1998) Infracortical interstitial cells concurrently expressing M2-muscarinic receptors, acetylcholinesterase and nicotinamide adenine dinucleotide phosphate-diaphorase in the human and monkey cerebral cortex. Neuroscience, 84: 755–769.

Sobreviela, T., Jaffar, S. and Mufson, E.J. (1998) Tyrosine kinase A, galanin and nitric oxide synthase within basal forebrain neurons in the rat. Neuroscience, 87: 447–461.

Sugaya, K. and McKinney, M. (1994) Nitric oxide synthase gene expression in cholinergic neurons in the rat brain examined by combined immunocytochemistry and in situ hybridization histochemistry. Mol. Brain Res., 23: 111–125.

Thomas, L., Charbonne, R., Gal, N., Borredon, J., Von Euw, D., Seylaz, J. and Lacombe, P. (1997) Microvascular contribution to the dilatory response to stimulation of the rat basal forebrain as studied by in vivo confocal microscopy. J. Cereb. Blood Flow Metab., 17: S754.

Tong, X.K. and Hamel, E. (1999) Regional cholinergic denervation of cortical microvessels and nitric oxide synthase-containing neurons in Alzheimer's disease. Neuroscience, 92: 163–175.

Tong, X.K. and Hamel, E. (2000) Basal forebrain nitric oxide synthase (NOS)-containing neurons project to microvessels and NOS neurons in the rat neocortex: cellular basis for cortical blood flow regulation. Eur. J. Neurosci., 12: 2769–2780.

Vaucher, E. and Hamel, E. (1995) Cholinergic basal forebrain neurons project to cortical microvessels in the rat: Electron microscopic study with anterogradely transported Phaseolus vulgaris leucoagglutinin and choline acetyltransferase immunocytochemistry. J. Neurosci., 15: 7427–7441.

Vaucher, E., Linville, D. and Hamel, E. (1997) Cholinergic basal forebrain projections to nitric oxide synthase-containing neurons in the rat cerebral cortex. Neuroscience, 79: 827–836.

Vaucher, E., Tong, X.K., Cholet, N., Lantin, S. and Hamel, E. (2000) GABA neurons provide a rich input to microvessels but not nitric oxide neurons in the rat cerebral cortex: A means for direct regulation of local cerebral blood flow. J. Comp. Neurol., 421: 161–171.

Vinters, H.V., Secor, D.L., Read, S.L., Frazee, J.G., Tomiyasu, U., Stanley, T.M., Ferreiro, J.A. and Akers, M.-A. (1994) Microvasculature in brain biopsy specimens from patients with Alzheimer's disease: An immunohistochemical and ultrastructural study. Ultrastructural Pathol., 18: 333–348.

Waite, J.J., Holschneider, D.P. and Scremin, O.U. (1999) Selective immunotoxin-induced cholinergic deafferentation alters blood flow distribution in the cerebral cortex. Brain Res., 818: 1–11.

Yamada, M., Lamping, K.G., Duttaroy, A., Zhang, W., Cui, Y., Bymaster, F.P., McKinzie, D.L., Felder, C.C., Deng, C.X., Faraci, F.M., Wess, J. (2001) Cholinergic dilation of cerebral blood vessels is abolished in M(5) muscarinic acetylcholine receptor knockout mice. Proc. Natl. Acad. Sci. USA, 98: 14096–14101.

Zhang, F., Xu, S. and Iadecola, C. (1995) Role of nitric oxide and acetylcholine in neocortical hyperemia elicited by basal forebrain stimulation: Evidence for an involvement of endothelial nitric oxide. Neuroscience, 69: 1195–1204.

Progress in Brain Research, Vol. 145
ISSN 0079-6123

CHAPTER 13

Acetylcholine systems and rhythmic activities during the waking–sleep cycle

Mircea Steriade*

Laboratoire de Neurophysiologie, Faculté de Médecine, Université Laval, Quebec, QC G1K 7P4, Canada

Abstract: The two processes of activation in thalamocortical systems exerted by mesopontine cholinergic neurons are (*a*) a direct depolarization associated with increased input resistance of thalamic relay neurons, which is antagonized by muscarinic blockers, and (*b*) a disinhibition of the same neurons via hyperpolarization of inhibitory thalamic reticular neurons. Low-frequency (< 15 Hz) oscillations during slow-wave sleep, characterized by rhythmic and prolonged hyperpolarizations, are suppressed by brainstem cholinergic neurons and nucleus basalis cholinergic and GABAergic neurons projecting to thalamic and reticular neurons. Fast rhythms (20–60 Hz) appear during the sustained depolarization of thalamic and neocortical neurons during brain-active states that are accompanied by increased release of acetylcholine (ACh) in the thalamus and cerebral cortex. Such fast rhythms also occur during the depolarizing phases of the slow oscillation (0.5–1 Hz) in non-REM sleep. Intracellular recordings of neocortical neurons during natural states of waking and sleep demonstrate stable and increased input resistance of corticocortical and corticothalamic neurons during the sustained depolarization in wakefulness, compared to the depolarizing phase of the slow oscillation in non-REM sleep. Despite the highly increased synaptic inputs along different afferent systems that open many conductances of cortical neurons during wakefulness, the increased input resistance is attributed to the effect of acetylcholine on cortical neurons.

Keywords: mesopontine cholinergic nuclei; nucleus basalis; thalamocortical neurons; cortical neurons; intracellular recordings; input resistance; waking; sleep states; in vivo experiments

Two cholinergic systems: their connections and related issues

Two cholinergic systems have access to, and activate, thalamocortical and neocortical neurons. One of them originates in the pedunculopontine and laterodorsal tegmental (PPT/LDT) cholinergic nuclei and projects to virtually all thalamic nuclei, but has no direct projections to cortex (see Steriade, 2001b). The other originates in nucleus basalis (NB) and projects to cortex and the thalamic reticular nucleus (reviewed in Asanuma, 1997; Semba, 2000). Both these cholinergic systems exert activating effects on

thalamocortical and cortical neurons. Activation is defined as a state of membrane polarization which brings thalamic and cortical neurons closer to their firing threshold, thus ensuring safe synaptic transmission and quick responses to either external stimuli during waking or internal drives during REM sleep (Steriade, 1991).

Brainstem cholinergic neurons activate cortical neurons through a dorsal and a ventral pathway. The dorsal one is a bisynaptic projection, the first relay using acetylcholine (ACh) as neurotransmitter to thalamocortical neurons that, in turn, release glutamate at the cortical level. The ventral projection activates NB neurons. Since ACh inhibits NB cells by hyperpolarization and decreased input resistance (Khateb et al., 1997), PPT/LDT neurons activate

*Tel.: (418) 656 5547; Fax: (418) 656 3236;
E-mail: mircea.steriade@phs.ulaval.ca

DOI: 10.1016/S0079-6123(03)45013-9

NB cells through glutamate (Rasmusson et al., 1994, 1996) that is colocalized with ACh in many PPT/LDT cells (Lavoie and Parent, 1994). The presence of these two parallel activating pathways (from brainstem to cortex, via synaptic relays within the thalamus or NB) is supported by experiments showing that brainstem-induced depolarization of cortical neurons, their enhanced excitability, and replacement of slow oscillations by fast rhythms, can be achieved after extensive ipsilateral lesions of either thalamus or NB (Steriade et al., 1993a) (Fig. 1). Thus, at variance with some assumptions placing exclusive emphasis on one or another cholinergic system, either PPT/LDT or NB cholinergic nuclei are sufficient to activate the cerebral cortex.

Moreover, although the emphasis was placed since the 1980s on cholinergic neurons, in order to specify chemically-coded systems within a structure prior viewed as 'non-specific', non-cholinergic neurons are much more numerous than the cholinergic ones in the brainstem reticular core and most of them, especially the large-sized ones, are glutamatergic (Jones, 2000). In the upper midbrain reticular formation, where there are virtually no cholinergic neurons, cells with antidromically identified projections to the thalamus display precursor signs of increased activity during shifts from slow-wave sleep to either wakefulness or REM sleep (Steriade et al., 1982), and are thus regarded as crucial for changing the state of vigilance toward brain activation. Some glutamate-induced

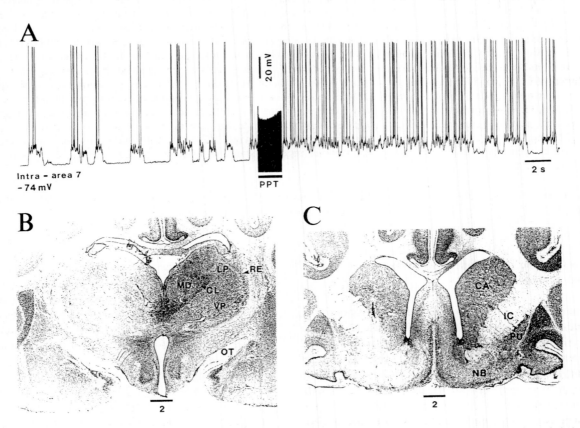

Fig. 1. Cortical activation elicited by stimulation of mesopontine tegmental nuclei may occur after either extensive excitotoxic lesions of the ipsilateral thalamus or excitotoxic lesions of nucleus basalis (NB). Intracellular recordings were performed in cats under urethane anesthesia. A, slow sleep oscillation in area 7 neuron and its disruption, associated with tonic firing, after a pulse-train (1 s, 30 Hz) applied to the mesopontine (PPT) nucleus. B, kainic-acid lesion of the ipsilateral thalamus in the cat whose intracellular recording is depicted in A. C, kainic-acid lesion of the ipsilateral NB; the PPT-elicited activation of cortical neurons was similar to that depicted in A after thalamectomy. Adapted from Steriade et al. (1993a).

excitatory actions consist in depolarization and increased input resistance of thalamocortical neurons, similarly to the effects exerted by ACh (Curró Dossi et al., 1991; McCormick and von Krosigk, 1992), as both these neurotransmitters block a 'leak' K^+ current (reviewed in Steriade et al., 1997). The role of glutamatergic midbrain and medullary neurons is also important in maintaining mesopontine PPT/LDT neurons in a sustained excitatory state, even during waking (Steriade et al., 1990) when monoaminergic neurons are active and exert an inhibitory tone upon them (Lübke et al., 1992; Williams and Reiner, 1993; Leonard and Llinás, 1994). Brainstem and basal forebrain cholinergic neurons also contain nitric oxide (NO) that produces depolarization of thalamocortical neurons and inactivates the hyperpolarization-dependent currents that generate an intrinsic sleep oscillation of these neurons (see below), thus shifting the electrical activity from slow-wave sleep to activated patterns (Pape and Mager, 1992; Pape, 1995). NO is a gaseous messenger implicated in diffuse activation processes. This may be related to the fact that, at least in cerebral cortex, the cholinergic afferents are predominantly non-junctional and rarely form synaptic profiles (Umbriaco et al., 1994; Descarries et al., 1997).

In what follows, I will focus on the actions exerted by the cholinergic systems on rhythmic activities during slow-wave sleep and brain-active states.

From slow-wave sleep oscillatory activities to fast rhythms during brain-active states

Setting brainstem PPT/LDT cholinergic neurons into action results in a blockage of the low-frequency rhythmic activities (< 15 Hz) that characterize slow-wave sleep and the occurrence of fast rhythms (~ 20–60 Hz) that define waking and REM sleep. These two brain-active states should not be qualified as 'EEG-desynchronized', as is often done, because fast rhythms are synchronized over restricted neocortical territories and within corticothalamocortical loops (Steriade et al., 1996a,b) (Fig. 2, A–B). The cellular mechanisms underlying the blockage of slow-wave sleep rhythms and those that generate fast rhythms are discussed below.

Three major oscillations are coalesced during slow-wave sleep and each of them is blocked by brainstem or forebrain cholinergic systems

Three rhythmic electrical activities characterize slow-wave sleep. (a) Spindles (7–15 Hz) are generated by pacemaking activities of GABAergic thalamic reticular neurons that impose rhythmic IPSPs onto thalamocortical neurons, which generate low-threshold spike-bursts transferred to cortex (Steriade et al., 1987a, 1993d). (b) Two components of delta waves (1–4 Hz) are generated in the thalamus, as a result of interplay between two hyperpolarization-dependent currents, I_H and I_T (McCormick and Pape, 1990; Leresche et al., 1991), and in the neocortex also following thalamectomy (Villablanca, 1974; Steriade et al., 1993f). (c) The slow oscillation (~ 0.5–1 Hz) is generated intracortically as it is present in the neocortex of thalamectomized animals (Steriade et al., 1993f) and absent in the thalamus of decorticated preparations (Timofeev and Steriade, 1996).

However, in the intact brain there are no simple network (intracortical or intrathalamic) operations that generate 'pure' rhythms, because the cortical slow oscillation has the virtue of grouping other sleep rhythms generated in the thalamus. The rhythmic depolarizing-hyperpolarizing sequences of the slow oscillation (Fig. 2C; Fig. 3, both *CAT* panels) have initially been described using intracellular recordings from cortical neurons in cats under anesthesia (Steriade et al., 1993e,f). The same pattern was found during natural sleep in animals (Steriade et al., 1996a,b), as well as in EEG recordings during human night sleep (Steriade et al., 1993e; Achermann and Borbély, 1997; Amzica and Steriade, 1997; Simon et al., 2000). During the depolarizing phase of the slow oscillation, synchronous discharges of corticothalamic neurons activate thalamic neurons and thus produce spindles (Fig. 3, *CAT* panels). This combination between the slow oscillation and spindles is termed the K-complex in clinical EEG (Fig. 3, *HUMAN*). As to the slow and delta oscillations, they are quite close in their frequency range but nonetheless distinct, as demonstrated by grouping of delta-wave sequences within the frequency of slow oscillation (Steriade et al., 1993f) and by the typical decrease in delta waves (2–4 Hz) from the first to the second episode during human

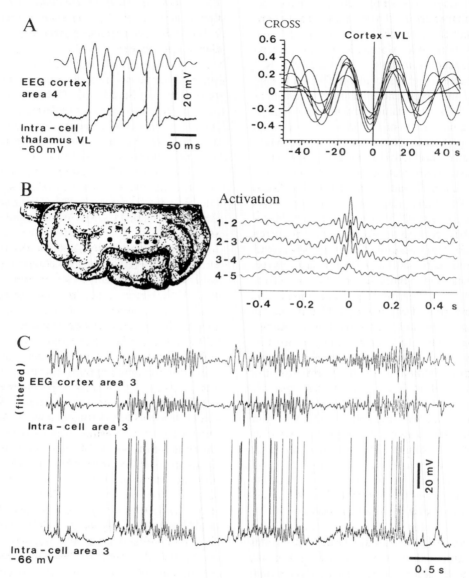

Fig. 2. Activation and fast rhythms of neocortical and thalamic electrical activity. Cats under ketamine-xylazine anesthesia. A, during a brief period of EEG activation, fast rhythms (∼40 Hz) are synchronous in an intracellularly recorded thalamocortical (TC) neuron from the ventrolateral (VL) nucleus (spikes truncated) and negative peaks of depth-EEG are recorded from the depth of area 4. Cross-correlogram between cortex and VL thalamus shows a clear-cut relation, with opposition of phase between depth-negative EEG and intracellular potentials in the TC cell. B, EEG activation induced by a pulse-train (300 Hz) applied to the PPT nucleus. Disruption of the slow sleep-like oscillation was associated with the appearance of fast activity whose amplitude exceeded that of fast waves during sleep-like patterns (not shown). Numbers of recorded cortical foci (1 to 5) correspond to those indicated on the suprasylvian gyrus (areas 5 and 7) of the brain. Cross-correlations between different leads demonstrate synchronization of fast rhythms. C, slow sleep-like oscillation (0.7 Hz) in an intracellularly recorded neuron from cortical area 3, and EEG from the same area. Both intracellular and EEG activities were filtered between 10 and 100 Hz (two top traces). Note fast activity (∼30–40 Hz) during the depolarizing phase of the slow oscillation and coherent activity between cell and field EEG potentials. Adapted from Steriade et al. (1996a,b).

sleep, whereas this decline is not present at the lower frequencies characterizing the slow oscillation (Achermann and Borbély 1997).

To sum up the oscillatory activities that characterize states of vigilance in the living brain are not generated within circumscribed neuronal networks but in interconnected neuronal loops between the cerebral cortex and thalamus, which are under the control of generalized modulatory systems, among them cholinergic ones. This justifies the need for experimental studies conducted in intact-brain preparations.

How do cholinergic systems in the mesopontine reticular core and in the basal forebrain suppress the synchronized slow-wave sleep oscillations generated in thalamocortical systems?

Sleep spindles are blocked by three, non-exclusive mechanisms. Each of them may effectively obliterate this oscillation by acting at the primary site of spindle genesis, the nucleus reticularis thalami (Steriade et al., 1987a), or at the level of target thalamocortical neurons. Stimulation of cholinergic PPT/LDT nuclei in vivo leads to hyperpolarization and increased membrane conductance of thalamic reticular neurons, thus blocking the depolarizing envelope of spindle sequences in these GABAergic cells and preventing the occurrence of spindles (Hu et al., 1989). This synaptic effect, which explains the obliteration of spindles upon arousal, is mimicked by bath application of ACh in slices from reticular nucleus (reviewed in McCormick, 1992). Brainstem cholinergic nuclei may also block spindles by acting directly on thalamocortical neurons, depolarizing them (Fig. 4, A–B), thus preventing the occurrence of low-threshold spike-bursts, which are cardinal signs of spindles and allow them to be transferred from thalamus to cortex. Finally, both cholinergic and GABAergic neurons in NB nucleus project to thalamic reticular neurons (Steriade et al., 1987b; Parent et al., 1988; reviewed in Asanuma, 1997), and either cholinergic or GABAergic actions are inhibitory on these neurons (see above).

The clock-like delta oscillation, which is generated by the interplay between I_H and I_T at hyperpolarized levels of thalamocortical neurons (see above), is also blocked by brainstem cholinergic neurons, which depolarize thalamic relay cells (Fig. 4, A–B) and thus bring them out of the voltage range at which the stereotyped delta rhythm is generated (Steriade et al., 1991a) (Fig. 5).

The cortical slow oscillation is blocked by stimulating the PPT nucleus in acutely prepared animals, through selective erasure of hyperpolarizing phases, often without significant changes in the membrane potential of neocortical cells (Fig. 6, B–C). The PPT effect on cortex is cholinergic in nature (then, it is mediated by NB projections to cortex since thalamocortical projections are glutamatergic) and is abolished by the muscarinic antagonist scopolamine, but not by the nicotinic antagonist mecamylamine (Fig. 6, D–F). In that study (Steriade et al., 1993a), we assumed that the PPT effect on cortically projecting NB cholinergic neurons is glutamatergic, a suggestion that was confirmed experimentally (see above, section 'Two cholinergic systems: their connections and related issues').

The PPT-induced blockage of the cortical slow oscillation by obliteration of hyperpolarizing phases, and its replacement by tonic firing, in acutely prepared animals under different anesthetics was replicated by abolition of slow sleep oscillation during natural brain-activated states of waking and REM sleep. Intracellular recordings in chronically implanted, naturally awake and sleeping cats demonstrate that the first sign during the transition from waking to sleep is the appearance of prolonged and cyclic hyperpolarizations in the frequency range of the slow oscillation (Steriade et al., 2001). On the other hand, the transition from slow-wave sleep to either wakefulness (EEG activation and increased muscular tone; Fig. 7) or REM sleep (EEG activation and muscular atonia; Fig. 8) is invariably associated with abolition of long-lasting hyperpolarizing potentials and appearance of tonic firing. These changes could initially occur without visible changes in the membrane potential that becomes depolarized only a few seconds later (Fig. 7). This time course is different from that displayed by thalamically projecting midbrain reticular (Steriade et al., 1982) and mesopontine cholinergic neurons (Steriade et al., 1990), which increase their firing rates well in advance of the time 0 of brain-activated states. In other words, cortical neurons follow activation processes in brainstem-thalamic systems.

We investigated in naturally awake and sleeping cats the impact of spontaneous synaptic activity on intrinsic neuronal properties, with emphasis on the apparent input resistance (R_{in}), a measure resulting

184

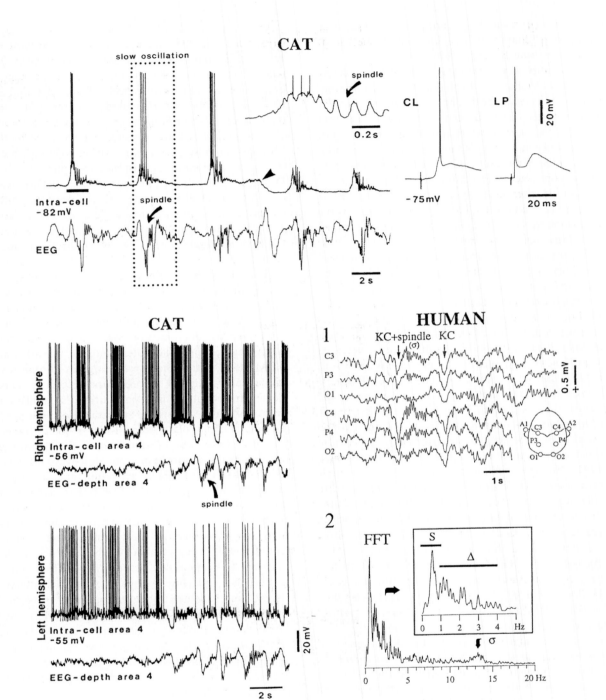

from passive electrical neuronal properties and balanced changes in excitatory and inhibitory inputs from specific and modulatory pathways. This is an important issue because studies in acutely prepared animals predicted a decreased input resistance of neocortical neurons with increases in synaptic inputs during brain-active states, whereas we found an increased input resistance of those neurons during wakefulness, an unexpected result that may be explained by increased release of ACh in cortex during brain activation (Celesia and Jasper, 1966). Let me briefly mention the data.

The R_{in} measured in vivo during relatively quiescent periods is reduced by up to 70% during epochs associated with intense synaptic activity, and increases by up to 70%, approaching the in vitro values after tetrodotoxin (TTX) application in vivo (Paré et al., 1998). In a similar vein, the R_{in} in the isolated cortical slab in vivo, in which the spontaneous activity is drastically reduced, is 49 MΩ, whereas in intact (adjacent) cortical areas of the same animal the value is 22 MΩ (Timofeev et al., 2000). It was then expected that R_{in} would be diminished during the state of wakefulness, when so many conductances are open because of the increased synaptic activity due to inputs from thalamocortical, intracortical, and generalized activating systems. However, this assumption did not take into consideration that ACh released during brain-active states increases the R_{in} of cortical neurons (Krnjević et al., 1971; McCormick 1992; Steriade et al., 1997). In fact, in contrast to the two sleep states, the R_{in} was remarkably stable during the steady state of waking and it reached higher values (31.3 $\pm$ 2.4 MΩ)

than during the depolarizing phase of the slow oscillation in slow-wave sleep (16.8 $\pm$ 2.3 MΩ) (Steriade et al., 2001) (Fig. 9). The increased R_{in} during wakefulness may be related to earlier extracellular recordings showing an enhanced antidromic and synaptic responsiveness of monkey's neocortical neurons during this behavioral state, compared to slow-wave sleep (Steriade et al., 1974).

During slow-wave sleep, R_{in} was almost double during the hyperpolarizing phase of the slow oscillation (30.8 $\pm$ 4.3 MΩ), compared to the depolarizing phase of this oscillation (see above). This indicates that GABAergic processes do *not* mediate the hyperpolarizations of the slow sleep oscillation, as GABA-mediated actions are associated with an increase in membrane conductance. Indeed, recordings with Cl$^-$-filled pipettes showed that the prolonged and cyclic hyperpolarizations during natural slow-wave sleep remained largely unaffected (Steriade et al., 2001; Timofeev et al., 2001). Moreover, none of the intracellularly recorded as well as stained basket (local inhibitory) interneurons fired during the prolonged hyperpolarizing phase of the slow oscillation (Contreras and Steriade, 1995). The increased R_{in} during the hyperpolarizing phase, compared to the depolarizing one, indicates that disfacilitation is the major mechanism underlying the prolonged hyperpolarizing phases during slow-wave sleep (Contreras et al., 1996). The disfacilitation may be explained by a progressive depletion of $[Ca^{2+}]_o$ during the depolarizing phase of the slow oscillation (Massimini and Amzica 2001). This would produce a decrease in synaptic efficacy and an avalanche reaction would

Fig. 3. The cortical slow oscillation groups thalamically generated spindles. *CAT* (top panel), intracellular recording from area 7 (1.5 mm depth) in a cat under urethane anesthesia. Electrophysiological identification (at right) shows orthodromic response to stimulation of thalamic centrolateral (CL) intralaminar nucleus and antidromic response to stimulation of lateroposterior (LP) nucleus. Note the slow oscillation of the neuron and related EEG waves. One cycle of the slow oscillation is framed in dots. Part marked by horizontal bar below the intracellular trace (at left) is expanded above (right) to show spindles following the depolarizing envelope of the slow oscillation. *CAT* (bottom left panel), dual simultaneous intracellular recordings from right and left cortical area 4. Note the spindle during the depolarizing envelope of the slow oscillation and synchronization of EEG when both neurons synchronously displayed prolonged hyperpolarizations. *HUMAN* (bottom right panels), the K-complex (KC) in natural sleep. Scalp monopolar recordings with respect to the contralateral ear are shown (see scheme). Traces display a short episode from stage 3 non-REM sleep. The two arrows point to two K-complexes, consisting of a surface-positive wave, followed (or not) by a sequence of spindle (sigma) waves. Note the synchrony of K-complexes in all recorded sites. Below, frequency decomposition of the electrical activity from C3 lead into three frequency bands: slow oscillation (S, 0 to 1 Hz), delta waves (Δ, 1 to 4 Hz) and spindles (σ, 12–15 Hz). Adapted from Steriade et al. (1993f and 1994, *CAT*) and from Amzica and Steriade (1997, *HUMAN*).

186

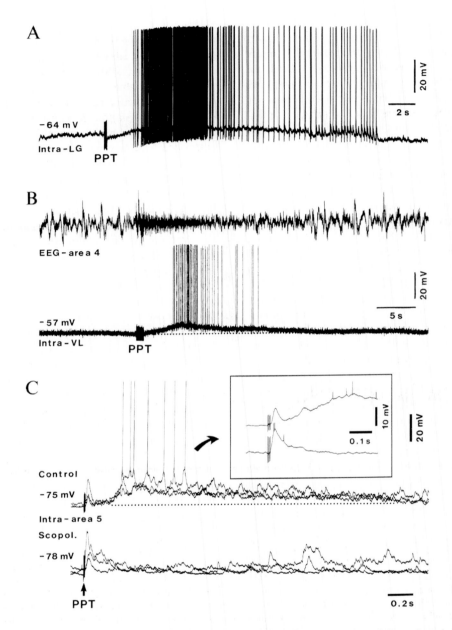

Fig. 4. Prolonged activation of thalamocortical and cortical neurons produced by stimulation of the pedunculopontine tegmental (PPT) cholinergic nucleus. Cats with brainstem transection (A), under urethane anesthesia (B), and under ketamine-xylazine anesthesia (C). A, isolated encephalon (*encéphale isolé*) preparation plus trigeminal deafferentation. Short pulse-train applied to the PPT nucleus produced long-lasting depolarization and tonic firing of thalamic dorsal lateral geniculate neuron. B, PPT pulse-train induced prolonged depolarization and tonic firing in thalamic ventrolateral neuron, as well as cortical EEG activation. C, depolarization of area 5 cortical neuron (depth, 0.8 mm), produced by short pulse-train to PPT, is blocked by scopolamine, a muscarinic antagonist. Dotted line tentatively indicates the baseline. Inset represents average of 15 PPT-evoked responses before and after systemic administration of scopolamine. Membrane potential was −75 mV before, and −78 mV after, systemic administration of scopolamine. A, unpublished data by B. Hu, M. Deschênes and M. Steriade. B–C, adapted from Curró Dossi et al. (1991) and Steriade and Amzica (1996).

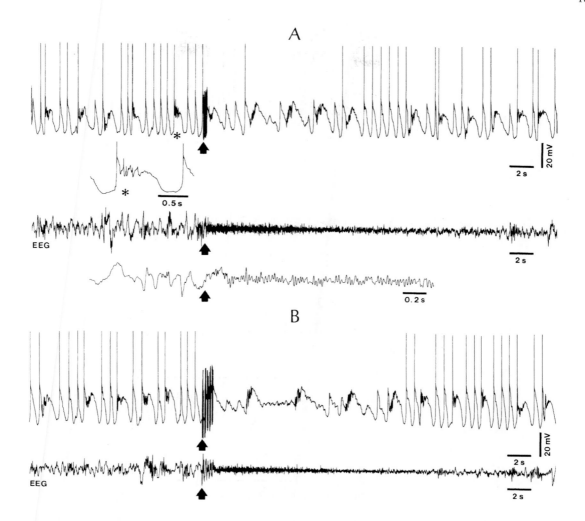

Fig. 5. Suppression of the clock-like delta oscillation in a thalamocortical (TC) neuron by stimulation of the pedunculopontine tegmental (PPT) nucleus, and simultaneous cortical activation with the appearance of fast (∼40 Hz) activity. Cat under urethane anesthesia. Intracellular recording from TC cell in the lateroposterior (LP) nucleus, together with EEG from postcruciate gyrus. A, a pulse-train to the PPT nucleus reduced the rhythmic (∼2 Hz) low-threshold spikes crowned by fast action potentials in TC cell and, simultaneously, induced fast EEG activity. Part marked by asterisk in the intracellular trace is expanded below to show postsynaptic potentials (PSPs) whose origin is the slow cortical oscillation. The EEG before and after the PPT train is expanded below to show ∼40 Hz activity induced by PPT stimulation. B, stronger effects were induced by 5 pulse-trains to PPT nucleus, which completely blocked the clock-delta oscillation of TC cell for ∼15 s, and simultaneously produced a long-lasting EEG activation. Adapted from Steriade et al. (1991a).

eventually lead to the functional disconnection of cortical networks, ultimately producing the disfacilitation that accounts for the hyperpolarizing phase of the slow oscillation.

Other data on the impact of network activities on intrinsic cellular properties are discussed elsewhere (Steriade, 2001a).

Cholinergic actions in the generation of fast rhythms

The first demonstration that the EEG activated response to midbrain reticular stimulation is not only the blockage of spindles and slow waves, but also includes the appearance of fast rhythms

188

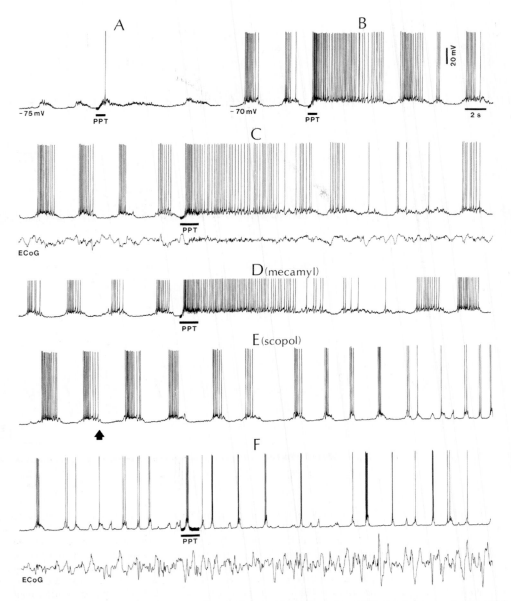

Fig. 6. Muscarinic, but not nicotinic, blockers antagonize cholinergic effects exerted by PPT stimulation on slow cortical oscillation. Cat under urethane anesthesia. Intracellular recording of regular-spiking cortical neuron, antidromically activated from the thalamic centrolateral intralaminar nucleus and synaptically driven from thalamic lateroposterior nucleus. A, rhythmic depolarizing phases at the resting membrane potential (V_m, -75 mV) and the effect of a pulse-train (30 Hz, 0.85 s) to the PPT nucleus. B, suppression of slow cortical oscillation by PPT stimulation at a more depolarized V_m of cortical neuron (-70 mV, under $+0.2$ nA). In C–F, the same depolarizing current was used as in B panel. C, PPT pulse-train (30 Hz), with a double duration (1.7s) led to a suppressing effect on slow cortical oscillation, which was twice as long as in B (note the PPT-induced electrocorticogram (ECoG) response, having a time course similar to that of cellular response). D, mecamylamine (mecamyl) administration (30 μg/kg, i.v.) had no visible effect on PPT-induced disruption of slow oscillation (spikes truncated). E, 20 s after scopolamine (scopol) administration (at arrow, 0.5 mg/kg, i.v.) the number of action potentials of each oscillatory sequence diminished and the frequency of the slow rhythm increased. F (without interruption after E), lack of any effect of PPT stimulation (same parameters as in C–D) on the slow cellular oscillation and on ECoG synchronized rhythms. Adapted from Steriade et al. (1993a).

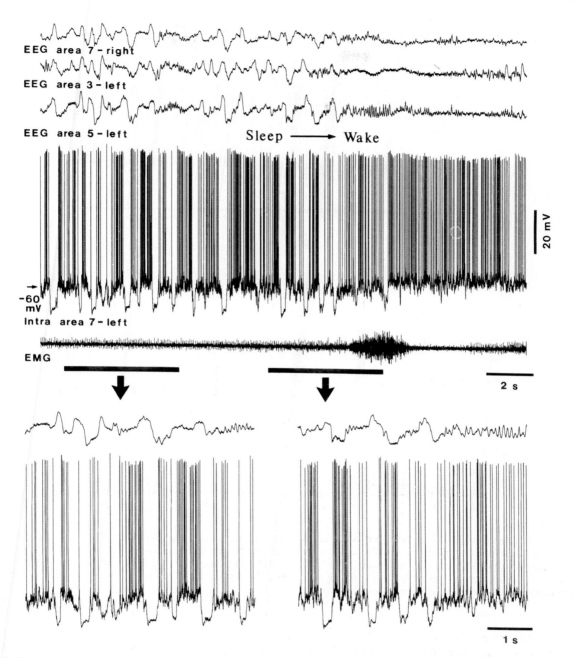

Fig. 7. Transition from natural slow-wave sleep (SWS) to wakefulness is accompanied by obliteration of prolonged hyperpolarizations. Intracellular recording of a regular-spiking (RS) neuron from cortical area 7 in a chronically implanted cat. Five traces depict (from top to bottom): depth-EEG from right area 7 and left areas 3 and 5; intracellular activity of RS neuron from left area 7; and electromyogram (EMG). Two epochs marked by horizontal bars are expanded below (without EMG). Note phasic hyperpolarizations in area 7 neuron, related to depth-positive EEG field potentials, during SWS; tonic firing upon awakening marked by EEG activation and increased muscular tone; and slight depolarization occurring only a few seconds after awakening and blockage of hyperpolarizations. Also note that the firing rate during the depolarizing phases of the slow sleep oscillation is as high as during waking. Adapted from Steriade et al. (2001).

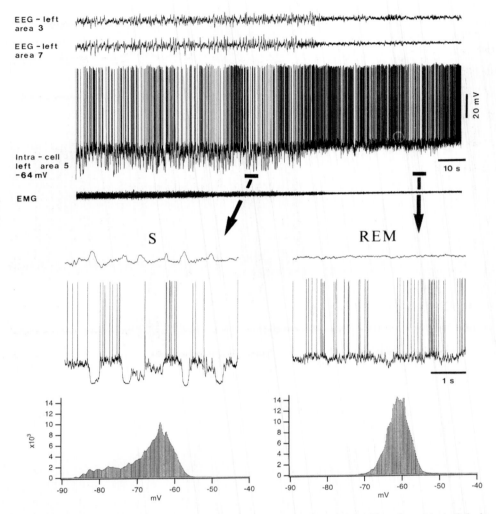

Fig. 8. Cyclic hyperpolarizations characterize neocortical neurons during slow-wave sleep (S) and are blocked during REM sleep. Chronically implanted cat. Intracellular recording of area 5 regular-spiking neuron, together with EEG from areas 3 and 7, and EMG. Periods marked by horizontal bars and arrows are expanded below. The bottom plots show the membrane potential during S (with a tail extending up to −85 mV) and a Gaussian-type histogram during REM sleep, around −60 mV. Also note slight depolarization upon entering REM sleep. Adapted from Steriade et al. (2001).

with a regular acceleration and synchronization ('*accélération synchronisatrice*'), belongs to Bremer et al. (1960; see their Fig. 5), three decades before the excitement about these oscillations that are observed during conditions of increased alertness, focused attention, responses to olfactory and visual stimuli, or occurring spontaneously as a function of thalamic and cortical cells' depolarization (reviewed in Steriade, 2001b).

Besides the two major states of brain alertness (waking and REM sleep), which are conventionally regarded as the only ones associated with spontaneous fast rhythms, the same oscillatory type is superimposed over the depolarizing phase of the slow sleep oscillation, whereas it is selectively abolished during the hyperpolarizing phase of the slow sleep oscillation (see above, Fig. 2C). The fact that fast oscillations appear not only during brain-aroused

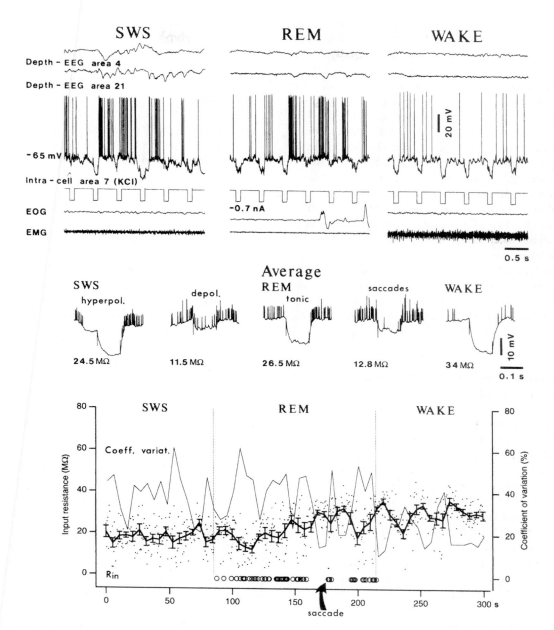

Fig. 9. Apparent input resistance of neocortical neurons during natural states of vigilance. Chronically implanted cat. Upper panel shows three periods of intracellular recording from the same regular-spiking cortical neuron during slow-wave sleep (SWS), REM sleep and waking. Input resistance (R_{in}) was measured by applying 0.1-s hyperpolarizing current pulses, every 0.5 s. Below, averages of responses of this neuron during different epochs in the three states of vigilance (note differences between the hyperpolarizing and depolarizing phases of the slow oscillation in SWS; and between epochs with and without ocular saccades in REM). The plot at bottom shows the dynamic changes of R_{in} during the three states of vigilance, obtained from continuous recording throughout the sleep-waking cycle. Dots represent individual measurements of R_{in}; thick line and standard deviation (SD) bars are the means of R_{in} from every 10 consecutive measurements; thin line is the coefficient of variation from corresponding periods; circles indicate ocular saccades in REM sleep. Note that, during quiet wakefulness, R_{in} increased and that this increase was associated with a decrease in the coefficient of variation. Adapted from Steriade et al. (2001).

192

states, but also over the depolarizing phase of the slow oscillation during anesthesia as well as natural slow-wave sleep (Steriade et al., 1996a,b), indicates that fast oscillations are voltage-dependent, rather than necessarily reflecting high cognitive or conscious processes. Depolarizing current pulses can trigger fast oscillations in cortical (Llinás et al., 1991; Nuñez et al., 1992; Gutfreund et al., 1995) and thalamic (Steriade et al., 1991b; Pinault and Deschênes, 1992; Steriade et al., 1993c) neurons. Cholinergic systems are implicated in the generation of fast oscillations. The depolarization of cortical neurons elicited by stimulation of the mesopontine cholinergic PPT nucleus (see Fig. 4C) and the resulting fast cortical oscillations (Steriade et al., 1991b) are blocked by a muscarinic receptor antagonist. The replacement of slow waves by fast oscillations is an effect relayed by cortically projecting NB cholinergic neurons (Metherate et al., 1992; Metherate and Ashe, 1993) that are modulated differentially by various mono-aminergic systems (Cape and Jones, 1998).

Relations between brainstem cholinergic and locus coeruleus neurons

Although in brain slices ACh and noradrenaline (NA) act on different cell types via pharmacologically distinct receptors, during natural arousal both PPT and locus coeruleus (LC) nuclei are implicated in rather complex interactions.

In vivo, a comparison between the effects induced by brief pulse-trains to LC and those induced by stimulating PPT cholinergic nucleus with the same parameters showed that, although both stimulated structures blocked the cortically generated slow oscillation, the threshold of this effect was lower, and its duration longer, with PPT than with LC stimulation (Fig. 10). The more powerful effect exerted by PPT, compared to LC, stimulation on the cortical slow oscillation may be explained by reciprocal interactions between these two (PPT and LC) structures. LC neurons extend their long dendrites (0.4–0.5 mm) to adjacent areas, up to the PPT nucleus. NA hyperpolarizes cholinergic neurons via α-2 receptors (Lübke et al., 1992; Williams and Reiner, 1992; Leonard and Llinás, 1994), whereas ACh excites LC neurons via m_2 receptors (Egan and

North, 1985). Thus, when PPT is stimulated, LC neurons may be simultaneously activated, whereas LC stimulation may inhibit PPT neurons.

Conclusions

(a) Brainstem cholinergic neurons project to the thalamus, depolarize and increase the apparent input resistance of thalamocortical neurons, and indirectly excite these relay neurons by hyperpolarizing and increasing the membrane conductance of GABAergic thalamic reticular neurons.

(b) Muscarinic receptor blockers antagonize cortical activation processes elicited by mesopontine reticular neurons. The activation effect elicited by brainstem reticular stimulation is relayed by a glutamatergic projection to cortically projecting nucleus basalis cholinergic neurons. None of the two main targets of brainstem cholinergic and glutamatergic neurons, the thalamus and nucleus basalis, is necessary and sufficient for cortical activation, which may occur after extensive lesion of these relay stations.

(c) The three brain rhythms that characterize non-REM sleep (spindles, delta and slow oscillations) are obliterated by brainstem cholinergic, and nucleus basalis cholinergic and GABAergic, actions exerted on thalamocortical, thalamic reticular, and neocortical neurons. The blockage of low-frequency (< 15 Hz) sleep oscillations, which are widely synchronized, is accompanied by the occurrence of fast (20–60 Hz) rhythms, which are synchronized over restricted cortical territories and well defined corticothalamic systems. The fast rhythms appear during the sustained depolarization of thalamic and neocortical neurons in wakefulness and REM sleep, as well as during the depolarizing phases of the slow oscillation in non-REM sleep. Thus, fast rhythms are voltage-dependent and do not necessarily reflect high cognitive and conscious processes.

(d) Intracellular recordings of neocortical neurons during natural states of waking and sleep demonstrate stable and increased input

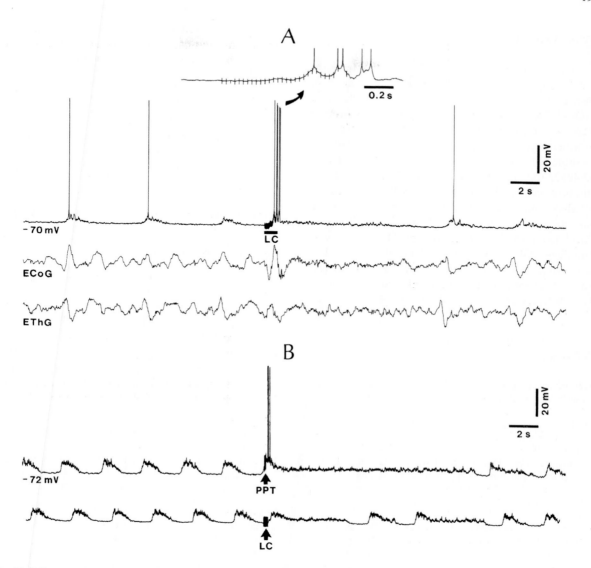

Fig. 10. Effects of locus coeruleus (LC) stimulation on the slow cortical oscillation. Cat under urethane anesthesia. A, regular-spiking, area 5 neuron. LC pulse-train (1 s, 30 Hz; marked by horizontal bar) blocked the slow oscillation and simultaneously produced an activated electrocorticogram (ECoG) and electrothalamogram (EThG) response lasting for ~8 s. B, another area 5 neuron antidromically activated from thalamic centrolateral and lateroposterior nuclei at latencies of 5 and 4 ms, respectively. Stimulation of pedunculopontine tegmental (PPT) cholinergic nucleus (10 stimuli at 100 Hz) blocked the slow oscillation for 15 s. The same parameters of stimulation applied to LC had no effect (not shown); by increasing the duration of LC pulse-train (30 stimuli at 100 Hz), a blocking effect appeared and lasted for 6 s. See comments in text. Adapted from Steriade et al. (1993a).

resistance of these neurons during waking, compared to the depolarizing phase of the slow oscillation in non-REM sleep. As the alert brain is associated with increased synaptic inputs along corticocortical, thalamocortical and generalized modulatory systems, this unexpected effect is attributed to an increased release, during waking, of acetylcholine that is known to increase the input resistance of neocortical neurons.

Acknowledgments

Personal experiments reported in this chapter have been supported by grants from the Canadian Institutes for Health Research (MT-3689 and MOP-36545), Natural Sciences and Engineering Research Council of Canada (170538), Human Frontier Science Program (RG0131), and National Institute of Health of United States (NINDS, 1-R01 NS40522-01). I thank the following Ph.D. students and postdoctoral fellows for their skilful and creative collaboration in experiments performed during the past decade (in chronological order): D. Paré, R. Curró Dossi, A. Nuñez, F. Amzica, D. Contreras, I. Timofeev and F. Grenier. Collaboration with T.J. Sejnowski, M. Bazhenov, A. Destexhe and W. Lytton was instrumental in computational studies on neuronal substrates of waking and sleep states.

References

Achermann, P. and Borbély, A. (1997) Low-frequency (< 1 Hz) oscillations in the human sleep EEG. Neuroscience, 81: 213–222.

Amzica, F. and Steriade, M. (1997) The K-complex: its slow (< 1 Hz) rhythmicity and relation to delta waves. Neurology, 49: 952–959.

Asanuma, C. (1997) Distribution of neuromodulatory inputs in the reticular and dorsal thalamic nuclei. In: M. Steriade, E.G. Jones and D.A. McCormick (Eds.), Thalamus (Vol. 2, Experimental and Clinical Aspects), Elsevier, Oxford, pp. 93–153.

Bremer, F., Stoupel, N. and Van Reeth, P.C. (1960) Nouvelles recherches sur la facilitation et l'inhibition des potentiels évoqués corticaux dans l'éveil réticulaire. Arch. Ital. Biol., 98: 229–247.

Cape, E.G. and Jones, B.E. (1998) Differential modulation of high-frequency γ-electroencephalogram activity and sleep–wake state by noradrenaline and serotonin microinjections into the region of cholinergic basalis neurons. J. Neurosci., 18: 2653–2666.

Celesia, G.G. and Jasper, H.H. (1966) Acetylcholine released from cerebral cortex in relation to state of activation. Neurology, 16: 1053–1064.

Contreras, D. and Steriade, M. (1995) Cellular basis of EEG slow rhythms: a study of dynamic corticothalamic relationships. J. Neurosci., 15: 604–622.

Contreras, D., Timofeev, I. and Steriade, M. (1996) Mechanisms of long-lasting hyperpolarizations underlying slow sleep oscillations in cat corticothalamic networks. J. Physiol. (Lond.), 494: 251–264.

Curró Dossi, R., Paré, D. and Steriade, M. (1991) Short-lasting nicotinic and long-lasting muscarinic depolarizing responses of thalamocortical neurons to stimulation of mesopontine cholinergic nuclei. J. Neurophysiol., 65: 393–406.

Descarries, L., Gisiger, V. and Steriade, M. (1997) Diffuse transmission by acetylcholine in the CNS. Progr. Neurobiol., 53: 603–625.

Egan, T.M. and North, R.A. (1985) Acetylcholine acts on m_2-muscarinic receptors to excite rat locus coeruleus neurones. Br. J. Pharmacol., 85: 733–735.

Gutfreund, Y., Yarom, Y. and Segev, I. (1995) Subthreshold oscillations and resonant frequency in guinea-pig cortical neurons: physiology and modelling. J. Physiol. (Lond.), 483: 621–640.

Hu, B., Steriade, M. and Deschênes, M. (1989) The effects of peribrachial stimulation on reticular thalamic neurons: the blockage of spindle waves. Neuroscience, 31: 1–12.

Jones, B.E. (2000). Basic mechanisms of sleep–wake states. In: M.H. Kryger, T. Toth and W.C. Dement (Eds.), Principles and Practice of Sleep Medicine. Saunders, Philadelphia, pp. 134–154.

Khateb, A., Fort, P., Williams, S., Serafin, M., Jones, B.E. and Mühlethaler, M. (1997) Modulation of cholinergic nucleus basalis neurons by acetylcholine and N-methyl-D-aspartate. Neuroscience, 81: 47–55.

Krnjević, K., Pumain, R. and Renaud, L. (1971) The mechanism of excitation by acetylcholine on the cerebral cortex. J. Physiol. (Lond.), 215: 247–268.

Lavoie, B. and Parent, A. (1994) Pedunculopontine nucleus in the squirrel monkey: distribution of cholinergic and monoaminergic neurons in the mesopontine tegmentum with evidence for the presence of glutamate in cholinergic neurons. J. Comp. Neurol., 344: 190–209.

Leonard, C.S. and Llinás, R.R. (1994) Serotonergic and cholinergic inhibition of mesopontine cholinergic neurons controlling REM sleep: an in vitro electrophysiological study. Neuroscience, 59: 309–330.

Leresche, N., Lightowler, S., Soltesz, I., Jassik-Gerschenfeld, D. and Crunelli, V. (1991) Low-frequency oscillatory activities intrinsic to rat and cat thalamocortical cells. J. Physiol. (Lond.), 441: 155–174.

Llinás, R., Grace, A.A. and Yarom, Y. (1991) In vitro neurons in mammalian cortical layer 4 exhibit intrinsic oscillatory activity in the 10- to 50-Hz frequency range. Proc. Natl. Acad. Sci. USA, 88: 897–901.

Lübke, J.I., Greene, R.W., Semba, K., Kamondi, A., McCarley, R.W. and Reiner, P.B. (1992) Serotonin hyperpolarizes cholinergic low-threshold burst neurons in the rat laterodorsal tegmental nucleus in vitro. Proc. Natl. Acad. Sci. USA, 89: 743–747.

Massimini, M. and Amzica, F. (2001) Extracellular calcium fluctuations and intracellular potentials in the cortex during the slow sleep oscillation. J. Neurophysiol., 85: 1346–1350.

McCormick, D.A. (1992) Neurotransmitter actions in the thalamus and cerebral cortex and their role in neuromodulation of thalamocortical activity. Progr. Neurobiol., 39: 337–388.

McCormick, D.A. and Pape, H.C. (1990) Properties of a hyperpolarization-activated cation current and its role in rhythmic oscillation in thalamic relay neurones. J. Physiol. (Lond.), 431: 291–318.

McCormick, D.A. and von Krosigk, M. (1992) Corticothalamic activation modulates thalamic firing through glutamate metabotropic receptors. Proc. Natl. Acad. Sci. USA, 89: 2774–2778.

Metherate, R. and Ashe, J.H. (1993) Ionic flux contributions to neocortical slow waves and nucleus basalis-mediated activation: whole-cell recordings in vivo. J. Neurosci., 13: 5312–5323.

Metherate, R., Cox, C.L. and Ashe, J.H. (1992) Cellular bases of neocortical activation: modulation of neural oscillations by the nucleus basalis and endogenous acetylcholine. J. Neurosci., 12: 4701–4711.

Nuñez, A., Amzica, F. and Steriade, M. (1992) Voltage-dependent fast (20–40 Hz) oscillations in long-axoned neocortical neurons. Neuroscience, 51: 7–10.

Pape, H.C. (1995) Nitric oxide: an adequate modulatory link between biological oscillators and control systems in the mammalian brain. Seminars Neurosci., 7: 329–340.

Pape, H.C. and Mager, R. (1992) Nitric oxide controls oscillatory activity in thalamocortical neurons. Neuron, 9: 441–448.

Paré, D., Shink, E., Gaudreau, H., Destexhe, A. and Lang, E.J. (1998) Impact of spontaneous synaptic activity on the resting properties of cat neocortical pyramidal neurons in vivo. J. Neurophysiol., 79: 1450–1460.

Parent, A., Paré, D., Smith, Y. and Steriade, M. (1988) Basal forebrain cholinergic and non-cholinergic projections to the thalamus and brainstem in cats and monkeys. J. Comp. Neurol., 277: 281–301.

Pinault, D. and Deschênes, M. (1992) Voltage-dependent 40-Hz oscillations in rat reticular thalamic neurons in vivo. Neuroscience, 51: 245–258.

Rasmusson, D.D., Clow, K. and Szerb, J.C. (1994) Modification of neocortical acetylcholine release and electro-encephalogram desynchronization due to brainstem stimulation by drugs applied to the basal forebrain. Neuroscience, 60: 665–677.

Rasmusson, D.D., Szerb, J.C. and Jordan, J.L. (1996) Differential effects of α-amino-3-hydroxy-5-methyl-4-isoxazole propionic acid and N-methyl-D-aspartate receptor antagonists applied to the basal forebrain on cortical acetylcholine release and EEG desynchronization. Neuroscience, 72: 419–427.

Semba, K. (2000) Multiple output pathways of the basal forebrain: organization, chemical heterogeneity, and roles in vigilance. Behav. Brain Res., 115: 117–141.

Simon, N.R., Mandshanden, I. and Lopes da Silva, F.H. (2000) A MEG study of sleep. Brain Res., 860: 64–76.

Steriade, M. (1991) Alertness, quiet sleep, dreaming. In: A. Peters and E.G. Jones (Eds.), Cerebral Cortex (Vol. 9, Normal and Altered States of Function), Plenum, New York, pp. 279–357.

Steriade, M. (2001a) Impact of network activities on neuronal properties in corticothalamic systems. J. Neurophysiol., 86: 1–39.

Steriade, M. (2001b). The Intact and Sliced Brain. The MIT Press, Cambridge, MA, pp. 1–366.

Steriade, M. and Amzica, F. (1996) Intracortical and corticothalamic coherency of fast spontaneous oscillations. Proc. Natl. Acad. Sci. USA, 93: 2533–2538.

Steriade, M., Deschênes, M. and Oakson, G. (1974) Inhibitory processes and interneuronal apparatus in motor cortex during sleep and waking. I. Background firing and synaptic responsiveness of pyramidal tract neurons and interneurons. J. Neurophysiol., 37: 1065–1092.

Steriade, M., Oakson, G. and Ropert, N. (1982) Firing rates and patterns of midbrain reticular neurons during steady and transitional states of the sleep–waking cycle. Exp. Brain Res., 46: 37–51.

Steriade, M., Domich, L., Oakson, G. and Deschênes, M. (1987a) The deafferented reticularis thalami nucleus generates spindle rhythmicity. J. Neurophysiol., 57: 260–273.

Steriade, M., Parent, A., Paré, D. and Smith, Y. (1987b) Cholinergic and non-cholinergic neurons of cat basal forebrain project to reticular and mediodorsal thalamic nuclei. Brain Res., 408: 372–376.

Steriade, M., Datta, S., Paré, D., Oakson, G. and Curró Dossi, R. (1990) Neuronal activities in brainstem cholinergic nuclei related to tonic activation processes in thalamocortical systems. J. Neurosci., 10: 2541–2559.

Steriade, M., Curró Dossi, R. and Nuñez, A. (1991a) Network modulation of a slow intrinsic oscillation of cat thalamocortical neurons implicated in sleep delta waves: cortical potentiation and brainstem cholinergic suppression. J. Neurosci., 11: 3200–3217.

Steriade, M., Curró Dossi, R., Paré, D. and Oakson, G. (1991b) Fast oscillations (20–40 Hz) in thalamocortical systems and their potentiation by mesopontine cholinergic nuclei in the cat. Proc. Natl. Acad. Sci. USA, 88: 4396–4400.

Steriade, M., Amzica, F. and Nuñez, A. (1993a) Cholinergic and noradrenergic modulation of the slow ($\sim$0.3 Hz) oscillation in neocortical cells. J. Neurophysiol., 70: 1384–1400.

Steriade, M., Contreras, D., Curró Dossi, R. and Nuñez, A. (1993b) The slow ($<$1 Hz) oscillation in reticular thalamic and thalamocortical neurons: scenario of sleep rhythm generation in interacting thalamic and neocortical networks. J. Neurosci., 13: 3284–3299.

Steriade, M., Curró Dossi, R. and Contreras, D. (1993c) Electrophysiological properties of intralaminar thalamocortical cells discharging rhythmic ($\sim$40 Hz) spike-bursts at $\sim$1000 Hz during waking and rapid eye movement sleep. Neuroscience, 56: 1–9.

Steriade, M., McCormick, D.A. and Sejnowski, T.J. (1993d) Thalamocortical oscillation in the sleeping and aroused brain. Science, 262: 679–685.

Steriade, M., Nuñez, A. and Amzica, F. (1993e) A novel slow ($<$1 Hz) oscillation of neocortical neurons in vivo:

depolarizing and hyperpolarizing components. J. Neurosci., 13: 3252–3265.

Steriade, M., Nuñez, A. and Amzica, F. (1993f) Intracellular analysis of relations between the slow (< 1 Hz) neocortical oscillation and other sleep rhythms. J. Neurosci., 13: 3266–3283.

Steriade, M., Contreras, D. and Amzica, F. (1994) Synchronized sleep oscillations and their paroxysmal developments. Trends Neurosci., 17: 199–208.

Steriade, M., Amzica, F. and Contreras, D. (1996a) Synchronization of fast (30–40 Hz) spontaneous cortical rhythms during brain activation. J. Neurosci., 16: 392–417.

Steriade, M., Contreras, D., Amzica, F. and Timofeev, I. (1996b) Synchronization of fast (30–40 Hz) spontaneous oscillations in intrathalamic and thalamocortical networks. J. Neurosci., 16: 2788–2808.

Steriade, M., Jones, E.G. and McCormick, D.A. (1997) Thalamus (Vol. 1, Organisation and Function), Elsevier, Oxford, pp. 1–959.

Steriade, M., Timofeev, I. and Grenier, F. (2001) Natural waking and sleep states: a view from inside neocortical neurons. J. Neurophysiol., 85: 1969–1985.

Timofeev, I. and Steriade, M. (1996) Low-frequency rhythms in the thalamus of intact-cortex and decorticated cats. J. Neurophysiol., 76: 4152–4168.

Timofeev, I., Grenier, F., Bazhenov, M., Sejnowski, T.J. and Steriade, M. (2000) Origin of slow oscillations in deafferented cortical slabs. Cerebr. Cortex, 10: 1185–1199.

Timofeev, I., Grenier, F. and Steriade, M. (2001) Disfacilitation and active inhibition in the neocortex during the natural sleep–wake cycle: an intracellular study. Proc. Natl. Acad. Sci. USA, 98: 1924–1929.

Umbriaco, D., Watkins, K.C., Descarries, L., Cozzari, C. and Hartman, B.K. (1994) Ultrastructural and morphometric features of the acetylcholine innervation in adult rat parietal cortex. An electron microscopic study in serial sections. J. Comp. Neurol., 348: 351–373.

Villablanca, J. (1974). Role of the thalamus in sleep control: sleep–wakefulness studies of chronic cats without the thalamus: the 'athalamic cat'. In: Petre-Quadens O. and Schlag J. (Eds.), Basic Sleep Mechanisms. Academic Press, New York, pp. 51–81.

Williams, J.A. and Reiner, P.B. (1993) Noradrenaline hyperpolarizes identified rat mesopontine cholinergic neurons in vitro. J. Neurosci., 13: 3878–3883.

Progress in Brain Research, Vol. 145
ISSN 0079-6123

CHAPTER 14

Cholinergic mechanisms mediating anesthetic induced altered states of consciousness

Steven B. Backman*, Pierre Fiset and Gilles Plourde

*Department of Anaesthesia, Royal Victoria Hospital, McGill University,
687 Pine Ave. W., Montreal, QC H3A 1A1, Canada*

Background

How does anesthesia produce unconsciousness? A conundrum

General anesthetic drugs produce loss of consciousness and abolish movement in response to noxious stimulation. The hypnotic property of anesthetics has received much less attention than the antinociceptive effect and the literature on the neurophysiological effects of 'subanesthetic' concentrations of general anesthetics is not extensive. (Subanesthetic designates concentration lower than that required for surgical anesthesia—suppression of motor response to pain—yet sufficient to produce profound sedation or unconsciousness). There are several reasons for this. One practical reason is the dose of anesthetic required to prevent movement in response to pain is greater than that required to produce unconsciousness; the loss of consciousness may not be of primary clinical relevance, being viewed only as a step toward surgical anesthesia (Guedel, 1937). The concept of MAC (minimum alveolar concentration of inhaled anesthetic to abolish purposeful movements in response to a surgical stimulus, Eger et al., 1995) increased the emphasis on prevention of response to pain as the most clinically relevant aspect of anesthesia. As such, many studies concerned with mechanisms of

anesthesia employ concentrations well above those required to produce unconsciousness. Another practical reason for the lack of information on the hypnotic property of anesthetics is that consciousness is much more difficult to study and quantify in human (and animal) experiments than are responses to pain (Churchland, 1996). Other factors which make it difficult to understand how anesthetic drugs produce unconsciousness include: (1) sites in the CNS where anesthetics act are not known; (2) central mechanisms that generate the natural states of consciousness remain quite obscure; (3) anesthetics have widely varying chemical structures, so it is difficult to elucidate structure-activity relationships; (4) the numerous effects produced by anesthetics are not antagonised by one specific antagonist, suggesting that anesthetics do not act on a single class of receptors (Lydic and Baghdoyan (1997).

Although general anesthesia may be induced by a wide variety of drugs, it has been appreciated for a long time that their potency correlates extremely well with their oil solubility (Meyer-Overton rule). This suggests that anesthetics produce their effects via a common mechanism of action, for example, by disrupting the structure of the lipid bilayer in neuronal membranes. One implication of the assumption of such a common mechanism is that an understanding of how they produce analgesia will *a priori* allow for an understanding of the mechanism mediating loss of consciousness. However, evidence fails to support a unitary hypothesis of anesthetic action. For example,

---

*Corresponding author: Tel.: (514)-842-1231 Ext. 34883;
E-mail: steven.backman@much.mcgill.ca

DOI: 10.1016/S0079-6123(03)45014-0

with regard to an effect on neuronal lipids, recent studies have demonstrated exceptions to the Meyer-Overton rule. Halogenated volatile drugs have been identified which do not induce anesthesia in rats at partial pressures that well exceed MAC values predicted by their oil solubility (Koblin et al., 1994). Moreover, volatile anesthetics have been shown to inhibit the enzymatic activity of the lipid-free firefly luciferase protein in proportion to their anesthetic potency, suggesting the potential for more complex interactions between anesthetics and neuronal protein-lipid structures (Franks and Lieb, 1994). Recent studies at the cellular and molecular levels have demonstrated that at clinical concentrations, anesthetics (including optical isomers of the same agent) can demonstrate selective effects on specific neuronal ligand-gated channels and second messenger systems (Franks and Lieb, 1993).

Anesthesia and the cholinergic system: implications from sleep research

A host of central neurotransmitter systems subserving GABAergic, glutamatergic, adrenergic, serotonergic and cholinergic transmission are likely affected by anesthetics (Angel, 1993). The effect of anesthetics on cholinergic transmission has been recently reviewed (Lydic and Baghdoyan, 1997; Durieux, 1996; Yamakura et al., 2000; Tassonyi et al., 2002) and it is suggested that alteration of central cholinergic transmission is involved in mediating several effects including sedation and loss of consciousness. Compelling evidence for this suggestion derives from studies on sleep/awake states (Steriade, 1999). Early animal studies demonstrated that acetylcholine (ACh) content of the whole brain varies as a function of arousal (Richter and Crossland, 1949), cortical ACh release changes in phase with the sleep/awake cycle (Celesia and Jasper, 1966), and anesthetics suppress EEG activation and reduce cortical ACh release (Mitchell 1963). In addition, intravenous administration of cholinomimetic drugs in decerebrate animals induces a REM-like state (Matsuzaki et al., 1968). REM sleep is considered to be a 'paradoxical' sleep state characterised by stage I EEG which is activated (as it is during wakefulness), and by rapid eye movements

and muscle atonia. Moreover, administration of physostigmine, a centrally-acting anticholinesterase drug, produces arousal: REM sleep or awakening is evoked in patients in deeper stages of sleep (Sitaram et al., 1976). Anticholinergic drugs, on the other hand, decrease or abolish REM sleep and increase non-REM sleep with enhancement of stage II activity in humans (Sagales et al., 1969).

More recent studies have focused on the later-odorsal and pedunculopontine tegmental nuclei (LDT/PPT) as important regions in the CNS where cholinergic transmission is altered during loss of consciousness. These midbrain nuclei contain cholinergic neurones that project to the medial pontine reticular formation and thalamus, and it has been shown that their energy metabolism is altered in phase with the sleep/awake cycle (Lydic et al., 1991a). Microdialysis studies have demonstrated that REM sleep is accompanied by an increase in ACh release in the medial pontine reticular formation (Lydic et al., 1991b). Injection of carbachol, a muscarinic agonist, into the medial pontine reticular formation produces REM-like sleep (Baghdoyan et al., 1993).

Anesthetic drugs interfere with cholinergic processes associated with altered states of consciousness. For example, administration of opiates (Lydic et al., 1993; Mortazavi et al., 1999), halothane (Keifer et al., 1994, 1996), isoflurane and enflurane (Keifer et al., 1996) are associated with decreased ACh release from neurones in the in the LDT/PPT nuclei projecting to the medial pontine reticular formation. Conversely, drugs that directly effect cholinergic activity also influence on MAC. For example, isoflurane MAC is decreased by hemicholinium-3 (which reduces neuronal content of ACh) and increased by physostigmine (Zucker, 1991). In addition to these effects on MAC, physostigmine increases the end-tidal halothane concentration required to produce a generalized EEG shift from a low to high amplitude pattern corresponding to an anaesthetized state (Roy and Stullken, 1981).

Cholinergic processes, anesthesia and altered states of consciousness

Drugs which specifically affect cholinergic transmission, and which also have an effect on the level of consciousness, have a long history of use by anesthesiologists. Yet, only a few clinical studies assess the role of the cholinergic system in mediating the loss of consciousness produced by anesthetics. Physostigmine decreases the time required for return to consciousness following anesthesia with halothane (Hill et al., 1977) and ketamine (Toro-Matos et al., 1980). Physostigmine also reverses prolonged post-operative somnolence following induction of anesthesia with midazolam (Caldwell and Gross, 1982) and diminishes the time to recover cognitive function following sedation induced by meperidine, propiomazine and scopolamine (Smith et al., 1976). A case report describes a patient who demonstrated delayed arousal following halothane anesthesia and in whom physostigmine produced abrupt awakening (Artru and Hui, 1986). Interestingly, physostigmine has also been reported to antagonize the respiratory depressant effect of morphine (Snir-Mor et al., 1983). Physostigmine has been shown to increase the dose of propofol required to induce loss of consciousness (Fassoulaki et al., 1997). These studies contrast with the lack of effect of physostigmine on the time to recover cognitive function following cessation of anesthesia with enflurane (Kesecioglu et al., 1991), and on the inability of physostigmine to reverse anesthesia or facilitate cognitive recovery in patients anesthetized with sevoflurane (Paraskeva et al., 2002). It is difficult to interpret these findings because of coadministration of several drugs, different endpoints used to assess the level of consciousness, and an unknown, inconsistent and varying level of anesthetic drug at CNS target sites. As these studies were done on patients, it is difficult to account for other confounding variables such as surgical stimulation, temperature and $PaCO_2$.

Scopolamine, a competitive nonselective muscarinic antagonist that crosses the blood brain barrier, is administered preoperatively to induce sedation and amnesia. While this drug is not used alone to produce anesthesia, high doses can produce unconsciousness as part of the central anticholinergic syndrome (Rumack, 1973). It is relevant that this syndrome, also produced by the central antimuscarinic side-effects of antidepressant, antipsychotic, antihistamine, antispasmodic and antiparkinsonian drugs, can be reversed by physostigmine (Granacher and Baldessarini, 1975).

Interaction of anesthetics with cholinergic transmission

It is not known how general anesthetics interfere with cholinergic transmission to produce unconsciousness. One possibility is that they interfere *directly* and there is an extensive literature implicating both nicotinic and muscarinic mechanisms. The nicotinic acetylcholine receptor (nAChR) is one of several ligand-gated ion channels that are extremely sensitive to the inhibitory effects of a variety of anesthetics including etomidate, propofol, ketamine, barbiturates and, particularly, volatile compounds with subtypes of nAChR's demonstrating differential sensitivities (Yamakura et al., 2000; Tassonyi et al., 2002). Anesthetic-induced inhibition of nAChRs may be particularly effective at presynaptic sites, as this receptor type may modulate release of a variety of neuroactive substances including glutamate, GABA, dopamine, norepinephrine, serotonin, and ACh (Sarlord et al., 1997; MacDermott et al., 1999). Interestingly, volatiles predicted to have anesthetic activity based on their oil solubility, but which do not produce anesthesia in animals, also have less effect on nAChRs (Firestone et al., 1994; Raines, 1996; Cardoso et al., 1999). Moreover, isoflurane stereoisomers demonstrate similar potency rank order of effect on MAC (Lysko et al., 1994) and inhibition of nAChRs (Franks and Lieb, 1991). On the other hand, nicotinic antagonists do not appear to produce immobility or hypnosis nor do they decrease the dose of isoflurane required to produce MAC and loss of righting reflex in mice (Flood et al., 2002). It has been suggested that direct anesthetic action on nAChrs may produce effects such as amnesia, inattentiveness and analgesia, which are associated with lower concentration of volatile agents (Flood et al., 2002). With regard to intravenous anesthetics, the S optical isomer of ketamine demonstrates significantly greater inhibitory effects than the R isomer which mirrors the enhanced anesthetic potency of the former compared with the latter (Friederich et al., 2000). Alternatively, a lack of stereoselective inhibitory effect of ketamine on nAChRs has been demonstrated by others (Sasaki et al., 2000), although the contrasting findings may reflect the use of different subtypes of nAChRs (e.g., see Yamakura et al., 2000). Ketamine's inhibitory effect on nAChRs occurs at a clinically relevant dose range, and it has been suggested that this may mediate the analgesia, dysphoria and inattentiveness associated with this drug (Flood and Krasowski, 2000). While thiopental inhibits nAChRs at clinically relevant doses, the S optical isomer has *not* been shown to be more effective than the r isomer, which is the case for their stereoselective potency to induce anesthesia (Downie et al., 2000; Coates et al., 2001; Kamiya et al., 2001). As such, it is suggested that the hypnotic effect produced by barbiturates is not mediated via direct interaction with nAChRs. The inhibitory effect of propofol and etomidate on nAChRs occurs with doses significantly higher than those which are clinically relevant (Flood et al., 1997; Furuya et al., 1999); thus it is argued that direct interaction with nAChRs does not mediate loss of consciousness produced by these drugs.

Muscarinic receptors belong to a superfamily of G-protein coupled receptors, and five subtypes (m1–5) have been cloned. The inhibitory effect of anesthetic drugs on muscarinic processes has been reviewed (Durieux, 1996). Volatile drugs including chloroform, enflurane and isoflurane increase muscarinic antagonist binding by decreasing the ligand-receptor dissociation rate. These drugs also interfere with the muscarinic receptor–G protein interaction (Dennison et al., 1987; Anthony et al., 1988, 1989). Halothane and enflurane inhibit m1 receptor signalling (Lin et al., 1993; Durieux, 1995), and while this receptor subtype does not appear to be inhibited by isoflurane, m3 signalling has been shown to be affected (Nietgen et al., 1998; Do et al., 2001). Muscarinic m2 receptors are inhibited by isoflurane but not by halothane (Magyar and Szabo, 1996). A volatile drug predicted to have anesthetic activity based on its oil solubility, and which interferes with memory and learning but does not produce anesthesia, inhibits m1 receptor signaling (Minami et al., 1997). Ketamine has been shown to inhibit m1 signalling (Durieux and Gregor, 1997), however this effect is not stereoselective, as is the case with nAChRs (see above). Limited data indicate that propofol does not directly effect m1 signalling (Rossi et al., 1996), but may interact with m_2 receptors (Yamamoto et al., 1999).

Of course, anesthetic drugs may interfere with cholinergic transmission indirectly, by altering activity in noncholinergic neuronal systems projecting to cholinergic neurones. There is an extensive literature which demonstrates that anesthetics reduce ACh turnover and release (Ngai et al., 1978; Kikuchi et al., 1998). Cholinergic neurones originating in the brainstem and basal forebrain receive serotonergic (Jones and Cuello, 1989; Luekbe et al., 1992; Honda and Semba, 1994), noradrenergic (Záborsky et al., 1993), dopaminergic and GABAergic (Záborsky et al., 1986) afferent inputs. In this regard, although beyond the scope of this chapter, it is significant that the $GABA_A$ receptor-channel complex is thought to be a likely primary site of action for anesthetics because of its high sensitivity to these agents.

In summary, data from sleep/awake studies suggest that the endogenous changes in level of consciousness are mediated, in part, by altered central cholinergic transmission. There is a substantial literature that demonstrates that anesthetics interfere with cholinergic processes. It therefore seems reasonable to suggest that central cholinergic mechanisms may also mediate the loss of consciousness produced by anesthetics. This hypothesis has been tested in studies on human volunteers, as follows.

Effects of anticholinesterase on anesthesia-induced loss of consciousness in human volunteers

The series of experiments described below tested the hypothesis that the loss of consciousness produced by anesthetic drugs involves depression, directly or indirectly, of central cholinergic transmission. We investigated whether physostigmine, a centrally-acting anticholinesterase, reverses the loss of consciousness produced by continuous administration of anesthetic drugs. In addition, we determined whether the reversal was blocked by scopolamine, a non-selective centrally-acting muscarinic antagonist. We chose three pharmacologically distinct anesthetics for investigation: (1) propofol, a substituted isopropyl-phenol administered intravenously; (2) remifentanil, a synthetic ultra-short acting mu opiod given intravenously; (3) sevoflurane, an inhaled volatile fluorinated methyl isopropyl ether. Studies were done in a functioning operating room at the Royal Victoria Hospital, and subjects were treated exactly as if they were to undergo outpatient surgery. The operating room was equipped with a pre-checked anesthesia machine, wall suction, and airway equipment and drugs for emergency resuscitation. A catheter was inserted into a forearm vein for drug infusion, and into a radial artery for monitoring of systemic arterial pressure and blood sampling. End-tidal CO_2, EKG, noninvasive blood pressure, pulse oximetry, and respiratory rate were monitored continuously. Oxygen was administered via nasal prongs, and a certified anesthesiologist continuously monitored the subjects for the duration of the experiment.

The dose of physostigmine (28 μg kg^{-1}) was that which has been used clinically to reverse the sedative effects of anesthesia and the central anticholinergic syndrome. Physostigmine was always coadministered with glycopyrrolate (0.3 mg), a nonselective muscarinic antagonist that does not cross the blood brain barrier, to block the peripheral muscarinic side effects of physostigmine (salivation, nausea, vomiting and bradycardia). The dose of scopolamine (8.6 μg kg^{-1}) was that which is use clinically to produce sedation.

Propofol-induced loss of consciousness (Meuret et al., 2000)

Propofol was administered intravenously to 17 volunteers using a computer controlled infusion pump to achieve stable plasma and CNS effector site concentration of drug. The infusion rate was determined using population-based pharmacokinetic data to achieve predictable drug concentration, and arterial blood samples were obtained for *post hoc* plasma concentration analysis. The rate of infusion was increased stepwise until loss of consciousness was achieved and subjects were motionless. Loss of consciousness was defined as the inability to respond to simple verbal commands ('open your eyes'), and gentle tactile stimulation (shoulder prodding). With loss of consciousness maintained by the continuous administration of propofol, subjects received an injection of saline (control) or physostigmine in a randomized, double-blind, crossover fashion. In six additional subjects, the effect of scopolamine pretreatment on the physostigmine-induced return to consciousness was assessed. In this series of

experiments, the protocol was similar to that described above except that following loss of consciousness, scopolamine was administered, followed by a 1 h delay before administration of physostigmine. The administration of scopolamine 1 h before physostigmine is based on the latency for its maximal effect to produce amnesia.

In addition to assessing the level of consciousness clinically (response to 'open your eyes' command), responsiveness was assessed by the electro-oculogram (EOG). CNS function was also assessed by measurement of the auditory steady state response (ASSR) and bispectral index (BIS). The ASSR is a sinusoidal response of the brain to periodically presented auditory stimuli. It appears when the rate of stimulus delivery is sufficiently rapid to produce overlapping of the responses to individual stimuli, and the response is most prominent for stimulus rates near 40 Hz. The ASSR is a steady-state equivalent of the transient middle latency response that demonstrates graded, dose-related amplitude reductions with anesthetic drugs (Plourde and Picton, 1990). The BIS is a complex parameter composed of a combination of time domain, frequency domain and high order spectral sub-parameters of ongoing EEG activity. It integrates several disparate descriptors of the EEG into a single variable that correlates well with behavioral assessment of anesthetic-induced sedation and hypnosis based on a large volume of clinical data (Rampil, 1998).

Loss of consciousness was produced by propofol at a target plasma concentration of 3.1 ± 0.6 (sd) μg ml^{-1} which was similar to the measured plasma concentration of 3.2 ± 0.8 μg ml^{-1}. Loss of consciousness was associated with the inability to respond to the command 'open your eyes' and the abolition of the EOG response (baseline awake: latency 576.3 ± 297.8 ms; amplitude 250.2 ± 138.1 μv). Loss of consciousness was also associated with reductions in ASSR (awake 0.32 ± 0.18 μv; unconscious 0.1 ± 0.08 μv, $P < 0.001$) and BIS (awake 92.4 ± 3.9; unconscious 55.7 ± 8.8, $P < 0.001$). Physostigmine restored consciousness unequivocally in 9 of 11 subjects, while saline was without effect. Return to consciousness was accompanied by the reappearance of the EOG response (latency 687.5 ± 526.7 ms; amplitude 60.1 ± 114.4 μv), and increases in the ASSR (0.38 ± 0.17 μv, $P < 0.01$) and

BIS (75.3 ± 8.3, $P < 0.001$). Of the remaining 2 of 11 subjects in whom physostigmine did not fully reverse the loss of consciousness, one remained fully unresponsive and the ASSR and BIS remained depressed. The other subject demonstrated wakefulness (commands were followed in an equivocal manner) and increases in ASSR and BIS were observed. In all six subjects, scopolamine pretreatment blocked the physostigmine-induced reversal of unconsciousness and the ASSR and BIS remained depressed (unconscious ASSR: 0.09 ± 0.09 μv, BIS: 58.2 ± 7.5; post physostigmine ASSR: 0.08 ± 0.06 μv, BIS 56.8 ± 6.7). Plasma propofol concentrations remained stable throughout the different experimental protocols (data not shown).

Remifentanil-induced loss of consciousness (Talbot et al., 2000)

In this series of experiments on 13 volunteers, remifentanil was infused intravenously using a computer-controlled infusion pump to achieve stable plasma and CNS effector site concentration of drug. The study protocol was similar to that described above for propofol, whereby drug concentration was increased in a step-wise fashion until consciousness was lost. At each target plasma concentration of remifentanil, the ASSR and BIS were measured. Following loss of consciousness, physostigmine or saline was administered in a randomized, double-blind, crossover fashion while maintaining infusion of remifentanil.

Loss of consciousness was achieved in 8/13 volunteers at a target plasma concentration of 28.7 ± 3.5 ng ml^{-1}. Larger doses of drug, which would have produced unconsciousness in the remaining subjects, were not used for reasons of safety (hypotension, prolonged respiratory depression). Cessation of spontaneous respiration was observed in 8/8 unconscious subjects, and respiration was manually assisted with a tight-fitting facemask to maintain end-tidal CO_2 at 40–50 mmHg. Unconsciousness was accompanied by reductions in ASSR (awake 0.55 ± 0.27; unconscious 0.32 ± 0.08 μv, $P < 0.05$) and BIS (awake 96.9 ± 3.5; unconscious 56.0 ± 26.6, $P < 0.05$). Physostigmine administration reversed unconsciousness in all 8 subjects, and was accompanied by an increase in ASSR (0.56 ± 0.15 μv, $P < 0.05$) and BIS (83.6 ± 12.5, $P < 0.05$).

Sevoflurane-induced loss of consciousness (Plourde et al., 2001)

Sevoflurane was administered to nine volunteers using a tightly fitting face mask connected to an anesthesia machine. Sevoflurane (carrier gas 100% O_2) was slowly titrated to produce unconsciousness. ASSR and BIS were recorded to assess the hypnotic effects. Physostigmine or saline was administered sequentially in a randomized, double-blind crossover manner while the end-tidal concentration of sevoflurane (measured by a Datex Ultima^{TM} gas analyzer) was kept constant.

Loss of consciousness was produced at a sevoflurane end-tidal concentration of 1.06 ± 0.37 vol% and was accompanied by reductions in ASSR (awake: 0.40 ± 0.16, unconscious: 0.08 ± 0.05 μv, $P < 0.001$) and BIS (awake 94 ± 9; unconscious: 66 ± 17, $P < 0.01$). Following administration of physostigmine, one subject regained consciousness and three regained wakefulness (spontaneous eye opening without clear response to commands). In the one subject who regained consciousness, ASSR and BIS increased (ASSR: 0.08–0.20 μv; BIS: 66–83). In the remaining five subjects who appeared to be unaffected by physostigmine, ASSR (0.08 ± 0.06 μv) and BIS (62 ± 13) did not change.

Conclusion: is central cholinergic transmission a final common pathway affected by anesthetics?

Although current knowledge implicates altered central cholinergic transmission as a mediator of anesthetic-induced altered states of consciousness, our human data must be interpreted with caution. In volunteers, physostigmine clearly reversed the loss of consciousness produced by propofol and remifentanil. The reversal was blocked by pretreatment with scopolamine (propofol study). Physostigmine did not reliably reverse the unconsciousness produced by sevoflurane, as it did with propofol and remifentanil, although evidence for awakening was indeed observed. This suggests that impairment of central cholinergic transmission may not play a pivotal role in sevoflurane-induced unconsciousness. Alternatively, central cholinergic transmission may have been so severely impaired that the inhibition of central cholinesterase activity was not sufficient to reverse it. In this regard, it may be significant that nAChRs appear to be exquisitely sensitive to the inhibitory effect of volatile drugs (see above). When the findings are considered as a whole, physostigmine's reversal of the sedative effects produced by disparate anesthetic drugs raises the possibility that depression of central cholinergic transmission is a final common pathway mediating anesthesia-induced loss of consciousness.

We cannot discount the possibility that physostigmine produced a return to consciousness by altering transmission in a pathway not affected by the anesthetic. For example, it is known that noxious stimulation may improve the level of consciousness in a sedated patient, yet this cannot be interpreted as proof that the somnolence was produced by block of sensory input. Likewise, we cannot discount the possibility that scopolamine alters neuronal activity in a pathway not affected by anesthetics. Conceivably, the anesthetic could mediate its effect on consciousness by interfering with nicotinic transmission and the administration of scopolamine, via inhibition of central muscarinic transmission, simply augments the depth of anesthesia such that physostigmine no longer reverses unconsciousness.

The experimental design of studies on humans is constrained by safety issues related to the dose of drugs that can be administered. The dose of anesthetic drug was titrated to produce loss of consciousness and larger doses were not administered to avoid excessive depression of the CNS. This had practical relevance in that subjects were able to maintain spontaneous respiration (propofol and sevoflurane) despite unconsciousness. This may have theoretical relevance because it is anticipated that with any drug there is a greater degree of both selective and nonselective effect as the dose is increased. The dose of physostigmine was also limited for reasons of safety. The effect of anticholinesterase on central cholinergic drive will depend on the dose-dependant magnitude of cholinesterase inhibition on the one hand, and on the level of endogenous central cholinergic activity on the other.

Given these caveats, it is fortuitous that the dose of physostigmine was of sufficient magnitude such that it produced the intended effect (propofol and remifentanil studies in particular), indicating that the dose-effect relationship between the relevant endogenous ligands and receptors were amenable to these pharmacological manipulations.

The ability to titrate anesthetic drugs safely to produce a rapid and reversible behavioral response is a powerful research tool. When this is technique is combined with brain imaging (Fiset et al., 1999) and ligand displacement binding studies (Alkire and Haier, 2001), important information may be obtained concerning the role of chemical neurotransmitter systems mediating altered states of consciousness. The administration of sedating doses of anesthetics may prove useful in studies concerned with the neural substrate underlying cognitive function such as memory, attention and learning. It may also serve as a useful model to investigate pathological conditions such as Alzheimer's disease. In particular, the ability to reverse the sedative effects of anesthesia by inhibition of central cholinesterase activity may prove useful for studying the efficacy of various anticholinesterase drugs to treat this condition.

References

Alkire, M.T. and Haier, R.J. (2001) Correlating *in vivo* anaesthetic effects with *ex vivo* receptor density data supports a GABAergic mechanism of action for propofol, but not for isoflurane. Br. J. Anaesth., 86: 618–626.

Angel, A. (1993) Central neuronal pathways and the process of anaesthesia. Br. J. Anaesth., 71: 148–163.

Anthony, B.L., Dennison, R.L., Narayanan, T.K. and Aronstam, R.S. (1988) Diethyl ether effects on muscarinic acetylcholine complexes in rat brainstem. Biochem. Pharmacol., 37: 4041–4046.

Anthony, B.L., Dennison, R.L. and Aronstam, R.S. (1989) Disruption of muscarinic receptor-G protein coupling is a general property of liquid volatile anesthetics. Neurosci. Lett., 99: 191–196.

Artru, A.A. and Hui, G.S. (1986) Physostigmine reversal of general anesthesia for intraoperative neurological testing: Associated EEG changes. Anesth. Analg., 65: 1059–1062.

Baghdoyan, H.A., Spotts, J.L. and Snyder, S.G. (1993) Sleep cycle alterations following simultaneous pontine and forebrain injections of carbachol. J. Neurosci., 3: 227–240.

Caldwell, C.B. and Gross, J.B. (1982) Physostigmine reversal of midazolam-induced sedation. Anesthesiology, 57: 125–127.

Cardoso, R.A., Yakamura, T., Brozowski, S.J., Chavez-Noriega, L.E. and Harris, R.A. (1999) Human neuronal nicotinic acetylcholine receptors expressed in *Xenopus* oocytes predict efficacy of halogenated compounds that disobey the Meyer-Overton rule. Anesthesiology, 91: 1370–1377.

Celesia, G.G. and Jasper, H.H. (1966) Acetylcholine released from cerebral cortex in relation to state of activation. Neurology, 16: 1053–1064.

Churchland, P.M. (1996) The Hornswoggle Problem. J. Consciousness Studies, 5–6: 402–408.

Coates, K.M., Mather, L.E., Johnson, R. and Flood, P. (2001) Thiopental is a competitive inhibitor at the human [alpha]7 nicotine acetylcholine receptor. Anesth. Analg., 92: 930–933.

Dennison, R.L., Anthony, B.L., Narayanan, T.K. and Aronstam, R.S. (1987) Effects of halothane on high affinity agonist binding and guanine nucleotide sensitivity of muscarinic acetylcholine receptors from brainstem of rats. Neuropharmacology, 26: 1201–1205.

Do, S-H., Kamatchi, G. and Durieux, M.E. (2001) The effect of isoflurane on native and chimeric muscarinic acetylcholine receptors: the role of protein kinase C. Anesth. Analg., 93: 375–381.

Downie, D.L., Franks, N.P. and Lieb, W. (2000) Effects of thiopental and its optical isomers on nicotinic acetylcholine receptors. Anesthesiology, 93: 774–783.

Durieux, M.E. (1995) Halothane inhibits signalling though m1 muscarinic receptors expressed in *Xenopus* oocytes. Anesthesiology, 82: 174–182.

Durieux, M.E. (1996) Muscarinic signalling in the central nervous system. Recent developments and anesthetic implications. Anesthesiology, 84: 173–189.

Durieux, M.E. and Gregor, N. (1997) Synergistic inhibition of muscarinic signalling by ketamine stereoisomers and the preservative benzethonium chloride. Anesthesiology, 86: 1326–1333.

Eger, E.I., Saidman, L.J. and Brandstater, B. (1995) Minimum alveolar anesthetic concentration: A standard of anesthetic potency. Anesthesiology, 26: 756–763.

Fassoulaki, A., Sarantopouilos, C. and Derveniotis, Ch. (1997) Physostigmine increases the dose of propofol required to induce anaesthesia. Can. J. Anaesth., 44: 1148–1151.

Firestone, L.L., Alifmoff, J.K. and Miller, K.W. (1994) Does general anesthetic-induced desensitization of the Torpedo acetylcholine receptor correlate with lipid disordering? Mol. Pharmacol., 46: 508–515.

Fiset, P., Paus, T. and Daloze, T. (1999) Brain mechanisms of propofol-induced loss of consciousness in humans: A positron emission tomographic study. J. Neuroscience, 19: 5506–5513.

Flood, P. and Krasowski, M.D. (2000) Intravenous anesthetics differentially modulate ligand-gated ion channels. Anesthesiology, 92: 1418–1425.

Flood, P., Ramirez-Latorre, J. and Role, L. (1997) Alpha 4 beta 2 neuronal nicotinic acetylcholine receptors in the central nervous system are inhibited by isoflurane and propofol, but alpha 7-type nicotinic acetylcholine receptors are unaffected. Anesthesiology, 86: 859–865.

Flood, P., Sonner, J.M., Gong, D. and Coates, K. (2002) Heteromeric nicotinic inhibition by isoflurane does not mediate MAC or loss of righting reflex. Anesthesiology, 97: 902–905.

Franks, N.P. and Lieb, WR. (1991) Stereospecific effects of inhalational general anesthetic optical isomers on nerve ion channels. Science, 254: 427–430.

Franks, N.P. and Lieb, W.R. (1993) Selective actions of volatile general anaesthetics at molecular and cellular levels. Br. J. Anaesth., 71: 65–76.

Franks, N.P. and Lieb, W.R. (1994) Molecular and central mechanisms of general anaesthesia. Nature, 367: 607–614.

Friederich, P., Dybek, A. and Urband, B. (2000) Stereospecific interaction of ketamine with nicotinic acetylcholine receptors in human sympathetic ganglion-like SH-SY5Y cells. Anesthesiology, 93: 818–824.

Furuya, R., Oka, K., Watanabe, I., Kamiya, Y., Itoh, H. and Andoh, T. (1999) The effects of ketamine and propofol on neuronal nicotinic acetylcholine receptors and P_{2x} purino-ceptors in PC12 cells. Anesth. Analg., 88: 174–180.

Granacher, R.P. and Baldessarini, RJ. (1975) Physostigmine: Its use in acute anticholinergic syndrome with antidepressant and antiparkinson drugs. Arch. Gen. Psychiatry, 32: 375–380.

Guedel, A.E. (1937). Inhalation Anesthesia, A Fundamental Guide. Macmillan, New York.

Hill, G.E., Stanley, T.H. and Sentker, C.R. (1977) Physostigmine reversal of postoperative somnolence. Can. Anaesth. Soc. J., 24: 707–711.

Honda, T. and Semba, K. (1994) Serotonergic synaptic input to cholinergic neurons in the rat mesopontine tegmentum. Brain Res., 647: 299–306.

Jones, B.E. and Cuello, A.C. (1989) Afferents to the basal forebrain cholinergic cells area from pontomesencephalic-catecholamine, serotonin, and acetylcholine neurons. Neuroscience, 31: 37–61.

Kamiya, Y., Tomio, A., Itaru, W., Higashi, T. and Itoh, H. (2001) Inhibitory effects of barbiturates on nicotinic acetylcholine receptors in rat central nervous system neurons. Anesthesiology, 94: 694–704.

Keifer, J.C., Baghdoyan, H.A., Becker, L. and Lydic, R. (1994) Halothane decreases pontine acetylcholine release and increases EEG spindles. Neuro. Report, 5: 577–580.

Keifer, J.C., Baghdoyan, H.A. and Lydic, R. (1996) Pontine cholinergic mechanisms modulate the cortical EEG spindles of halothane anesthesia. Anesthesiology, 84: 945–954.

Kesecioglu, J., Rupreht, J., Telci, L., Dzoljic, M. and Erdmann, W. (1991) Effect of aminophylline or physostigmine on recovery from nitrous oxide-enflurane anaesthesia. Acta Anaesthesiol. Scand., 35: 616–620.

Kikuchi, T., Wang, Y., Sato, K. and Okumura, F. (1998) *In vivo* effects of propofol on acetylcholine release from the frontal cortex, hippocampus and striatum studied by intracerebral microdialysis in freely moving rats. Br. J. Anaesth., 80: 644–648.

Koblin, D.D., Chortkoff, B.S., Laster, M.J., Eger, E.I. II., Halsey, M.J. and Ionescu, P. (1994) Polyhalogenated and perfluorinated compounds that disobey the Meyer-Overton hypothesis. Anesth. Analg., 79: 1043–1048.

Lin, L.H., Leonard, S. and Harris, R.A. (1993) Enflurane inhibits the function of mouse and human brain phosphatidylionositol-linked acetylcholine and serotonin receptors expressed in Xenopus oocytes. Mol. Pharmacol., 43: 941–948.

Luekbe, J.L., Greene, R.W., Semba, K., Kamondi, A., McCarley, R.W. and Reiner, P.B. (1992) Serotonin hyperpolarizes cholinergic low threshold burst neurons in the rat laterodorsal tegmental nucleus in vitro. Proc. Natl. Acad. Sci. USA, 89: 743–747.

Lydic, R. and Baghdoyan, H.A. (1997). Cholinergic contributions to the control of consciousness. In: Yaksh T.L., Lynch C.III, Zapol W.M., Maze M., Biebuyck J.F. and Saidman L.J. (Eds.), Anesthesia: Biologic Foundations. Lippincott-Raven, Philadelphia, pp. 433–450.

Lydic, R., Baghdoyan, H.A., Hibbard, L., Bonyak, E.V., DeJoseph, M.R. and Hawkins, R.A. (1991a) Regional brain glucose metabolism is altered during rapid eye movement sleep in the cat. J. Comp. Neurol., 304: 517–529.

Lydic, R., Baghdoyan, H.A. and Lorinc, Z. (1991b) Microdialysis of cat pons reveals enhanced acetylcholine release during state-dependent respiratory depression. Am. J. Physiol., 261: R766–R770.

Lydic, R., Keifer, J.C., Baghdoyan, H.A. and Becker, L. (1993) Microdialysis of the pontine reticular formation reveals inhibition of acetylcholine release by morphine. Anesthesiology, 79: 1003–1012.

Lysko, G.S., Robinson, J.L., Casto, R. and Ferrone, R.A. (1994) The stereospecific effects of isoflurane isomers in vivo. Eur. J. Pharmacol., 263: 25–29.

MacDermott, A.B., Role, L.W. and Seigelbaum, S.A. (1999) Presynaptic ionotropic receptors and the control of transmitter release. Ann. Rev. Neurosci., 22: 443–485.

Magyar, J. and Szabo, G. (1996) Effects of volatile anesthetics on the G protein-regulated muscarinic potassium channel. Mol. Pharmacol., 50: 1520–1528.

Matsuzaki, M., Okada, Y. and Shuto, S. (1968) Cholinergic agents related to para-sleep in acute brain stem preparations. Brain Res., 9: 253–267.

Meuret, P., Backman, S.B., Bonhomme, V., Plourde, G. and Fiset, P. (2000) Physostigmine reverses propofol-induced unconsciousness and attenuation of the auditory steady state response and bispectral index in human volunteers. Anesthesiology, 93: 708–717.

Minami, K., Vanderah, T.W., Minami, M. and Harris, R.A. (1997) Inhibitory effects of anesthetics and ethanol of muscarinic receptors expressed in Xenopus oocytes. Eur. J. Pharm., 339: 237–244.

Mitchell, J.F. (1963) The spontaneous and evoked release of acetylcholine from the cerebral cortex. J. Physiol., 165: 98–116.

Mortazavi, S., Thompson, J., Baghdoyan, H. and Lydic, R. (1999) Fentanyl and morphine, but not remifentanil, inhibit acetylcholine release in pontine regions modulating arousal. Anesthesiology, 90: 1070–1077.

Ngai, S.H., Cheney, D.L. and Finck, A.D. (1978) Acetylcholine concentrations and turnover in rat brain structures during anesthesia with halothane, enflurane and ketamine. Anesthesiology, 48: 4–10.

Nietgen, G.W., Honemann, C.W. and Chan, C.K. (1998) Volatile anesthetics have differential effects on recombinant m1 and m3 muscarinic acetylcholine receptor function. Br. J. Anaesth., 81: 569–577.

Paraskeva, A., Papilas, K., Fassoulaki, A., Melemeni, A. and Papadopoulos, G. (2002) Physostigmine does not antagonize sevoflurane anesthesia assessed by bispectral index or enhances recovery. Anesth. Analg., 94: 569–572.

Plourde, G., Fiset, P., Chartrand, D. and Backman, S.B. (2001) Physostigmine does not reliably reverse unconsciousness induced by sevoflurane. Anesth. Analg., 92: S302.

Plourde, G. and Picton, T.W. (1990) Human auditory steady-state response during general anesthesia. Anesth. Analg., 71: 460–468.

Raines, D.E. (1996) Anesthetic and nonanesthetic halogenated volatile compounds have dissimilar activities on nicotinic acetylcholine receptor desensitization kinetics. Anesthesiology, 84: 663–671.

Richter, D. and Crossland, J. (1949) Variation in acetylcholine content of the brain with physiological state. Am. J. Physiol., 159: 247–255.

Rampil, I.J. (1998) A primer for EEG signal processing in anesthesia. Anesthesiology, 89: 980–1002.

Rossi, M., Chan, C.K., Christensen, J.D., DeGuzman, E.J. and Durieux, M.E. (1996) Interactions between propofol and lipid mediator receptors: inhibition of lysophosphatidate signalling. Anesth. Analg., 83: 1090–1096.

Roy, R.C. and Stullken, E.H. (1981) Electroencephalographic evidence of arousal in dogs from halothane after doxapram, physostigmine, or naloxone. Anesthesiology, 55: 392–397.

Rumack, BH. (1973) Anticholinergic poisoning: Treatment with physostigmine. Pediatrics, 52: 449–451.

Sagales, T., Erill, S. and Somino, E.F. (1969) Differential effects of scopolamine and chlorpromaxine on REM and NREM sleep in normal state subjects. Clin. Pharmacol. Ther., 10: 522–529.

Sarlord, F., Keita, H., Lecharny, J-B., Henzel, D., Desmonts, J.-M. and Mantz, J. (1997) Halothane and isoflurane differentially affect the regulation of dopamine and gamma-aminobutyric acid release mediated by presynaptic acetylcholine receptors in the rat striatum. Anesthesiology, 86: 632–641.

Sasaki, T., Andoh, T. and Watanabe, I. (2000) Nonstereoselective inhibition of neuronal nicotinic acetylcholine receptors by ketamine isomers. Anesth. Analg., 91: 741–748.

Sitaram, N., Wyatt, R.J., Dawson, S. and Gillin, J.C. (1976) REM sleep induction by physostigmine infusion during sleep. Science, 191: 1281–1283.

206

Smith, D.B., Clark, R.B., Stephens, S.R., Sherman, R.L. and Hyde, M.L. (1976) Physostigmine reversal of sedation in parturients. Anesth. Analg., 55: 478–480.

Snir-Mor, I., Weinstock, M., Davidson, J.T. and Bahar, M. (1983) Physostigmine anagonizes morphine-induced respiratory depression in human subjects. Anesthesiology, 59: 6–9.

Steriade, M. (1999). Cellular substrates of oscillations in corticothalamic systems during states of vigilance. In: Lydic R. and Baghdoyan H.A. (Eds.), Handbook of Behavioral State Control, Cellular and Molecular Mechanisms. CRC, New York, pp. 327–348.

Talbot, M., Fiset, P., Backman, S.B., Plourde, G. and Chartrand, D. (2000) Physostigmine reverses unconsciousness produced by remifentanil in human healthy volunteers. Am. Soc. Anesthesiol., 93: A389.

Tassonyi, E., Charpentier, E., Muller, D., Dumont, L. and Bertrand, D. (2002) The role of nicotinic acetylcholine receptors in the mechanisms of anesthesia. Brain Res. Bull., 57: 133–150.

Toro-Matos, A., Rendon-Platas, A.M., Avil-Valdez, E. and Villarreal-Guzman, R.A. (1980) Physostigmine antognizes ketamine. Anesth. Analg., 59: 764–767.

Yamakura, T., Ghavez-Noriega, L.E. and Harris, R.A. (2000) Subunit dependent inhibition of human neuronal nicotinic acetylcholine receptors and other ligand-gated ion channels by dissociative anesthetics ketamine and dizoclipine. Anesthesiology, 92: 1144–1153.

Yamamoto, S., Kawana, S., Miyamoto, A., Oshika, H. and Namika, A. (1999) Propofol-induced depression of cultured rat ventricular myocytes is related to the M2-acetylcholine receptor-NO-cGMP signalling pathway. Anesthesiology, 91: 1712–1725.

Záborsky, L., Heimer, L., Eckenstein, F. and Leranth, C. (1986) GABAergic input to cholinergic forebrain neurons: an ultrastructural study using retrograde tracing of HRP and double immuno-labeling. J. Comp. Neurol., 250: 282–295.

Záborsky, L., Cullinan, W.E. and Luine, V.N. (1993) Catecholamine-cholinergic interaction in the basal forebrain. Prog. Brain. Res., 98: 31–49.

Zucker, J. (1991) Central cholinergic depression reduces MAC for isoflurane in rats. Anesth. Analg., 72: 790–795.

Progress in Brain Research, Vol. 145
ISSN 0079-6123

CHAPTER 15

High acetylcholine levels set circuit dynamics for attention and encoding and low acetylcholine levels set dynamics for consolidation

Michael E. Hasselmo* and Jill McGaughy

Department of Psychology, Center for Memory and Brain and Program in Neuroscience, Boston University, 2 Cummington St., Boston, MA 02215, USA

Introduction

Extensive physiological research has demonstrated a number of effects of acetylcholine within the hippocampus, piriform cortex, neocortex, and thalamus (Krnjević and Phillis, 1963; Krnjević et al., 1971; see review in Hasselmo, 1995). Here the review will focus on data regarding cholinergic modulation in the hippocampus and piriform cortex, but data from the neocortex suggests similar principles apply in other cortical structures.

At times the effects of acetylcholine on specific neuron types and synaptic pathways in cortical structures appear paradoxical and inconsistent. For example, why should acetylcholine simultaneously enhance pyramidal cell spiking through depolarization (Krnjević et al., 1971; Cole and Nicoll, 1984), while suppressing excitatory glutamatergic synaptic transmission at intrinsic synapses in the hippocampus (Hounsgaard, 1978; Valentino and Dingledine, 1981; Dutar and Nicoll, 1988; Hasselmo and Bower, 1992; Hasselmo and Schnell, 1994) and neocortex (Brocher et al., 1992; Gil et al., 1997; Hsieh et al., 2000)? Similarly, why should acetylcholine in the hippocampus simultaneously depolarize interneurons (Frazier et al., 1998; McQuiston and Madison 1999a,b;

Alkondon and Albuquerque, 2001) while suppressing hippocampal inhibitory synaptic transmission (Pitler and Alger, 1992; Patil and Hasselmo, 1999)? A unifying theoretical framework is required for understanding these disparate physiological effects.

Computational modeling offers a unifying theoretical framework for understanding the functional properties of acetylcholine within cortical structures (Hasselmo, 1995, 1999). This chapter will provide a description of how the different physiological effects of acetylcholine could interact to alter specific functional properties of the cortex. In particular, acetylcholine enhances the response to afferent sensory input while decreasing the internal processing based on previously formed cortical representations. These same circuit level effects can be categorized with different colloquial terms at a behavioral level, sometimes being interpreted as an enhancement of attention, sometimes as an enhancement of memory encoding. But the same change in circuit level dynamics could underlie all these behavioral effects. In this chapter, we will present the basic theoretical framework of enhanced response to input, with reduced feedback processing. We will then discuss individual physiological effects of acetylcholine in the context of this framework. Finally, we will discuss how the loss of cholinergic modulation will shift network dynamics toward those appropriate for the consolidation of previously encoded information.

*Corresponding author: Tel.: (617) 353-1397; Supported by NIH grants MH60013, MH61492, MH60450 and DA16454
Fax: (617) 353-1431; E-mail: hasselmo@bu.edu

DOI: 10.1016/S0079-6123(03)45015-2

Evidence for diffuse modulatory state changes caused by acetylcholine

The computational models described here assume that acetylcholine causes diffuse modulatory state changes within cortical structures. This assumption is based on the following evidence: (1) microdialysis studies show dramatic changes in acetylcholine level in cortex during different stages of waking and sleep; (2) anatomical studies of cholinergic fibers suggest diffuse modulatory influences on cortical function; and (3) the slow transition between different states is supported by data showing a relatively slow time course of changes in physiological effects of acetylcholine.

Microdialysis studies of acetylcholine

The physiological effects of acetylcholine on cortical circuits will change in magnitude as the extracellular concentration of acetylcholine changes within cortical circuits. Microdialysis studies demonstrate striking changes in the levels of acetylcholine within the cortex during different behaviors (Giovannini et al., 2001) and during different stages of waking and sleep (Jasper and Tessier, 1971; Kametani and Kawamura, 1990; Marrosu et al., 1995). In particular, acetylcholine levels are higher during active waking in freely moving rats (Kametani and Kawamura, 1990) and cats (Marrosu et al., 1995), as summarized in Fig. 1. In experiments on rats, active waking is defined as periods of time during which the rat is actively exploring the environment, scurrying along the walls or across the floor, sniffing novel objects and rearing up extensively. In EEG recordings from hippocampus and entorhinal cortex, this period is characterized by large amplitude oscillations in the theta frequency range (Buzsaki, 1989; Bland and Colom, 1993; Chrobak and Buzsaki, 1994), whereas the neocortex displays high frequency, low-amplitude activity with local synchronization (Steriade et al., 1996) and some periods of theta in certain regions (Maloney et al., 1997). The increase in acetylcholine during waking is particularly strong when a rat is initially exposed to a novel environment, apparently in association with both the increase in fear elicited by such an environment, as well as the increase in attention to

stimuli within the environment (Giovannini et al., 2001). Cortical acetylcholine levels rise dramatically during performance of tasks requiring sustained attention for detection of a stimulus (Himmelheber et al., 2000; Arnold et al., 2002).

In contrast, acetylcholine levels decrease during periods of 'quiet waking' during which animals are immobile or performing consummatory behaviors such as eating or grooming (Marrosu et al., 1995). Recordings of the EEG in this phase of behavior shows irregular EEG activity with periodic appearance of brief, large amplitude events termed sharp waves (Buzsaki, 1986; Chrobak and Buzsaki, 1994). Acetylcholine levels decrease even more dramatically during slow wave sleep, to levels less than one third of those observed during active waking (Jasper and Tessier, 1971; Kametani and Kawamura, 1990; Marrosu et al., 1995). Slow wave sleep is defined by the characteristic EEG phenomena occurring during this phase of sleep, particularly the large amplitude, low frequency oscillations found in neocortical structures and commonly termed slow waves (Steriade, 1994, 2001). Thus, there are striking changes in acetylcholine levels within cortical circuits which are correlated with striking changes in behavior and electroencephalographic dynamics within these structures. Computational modeling can help to elucidate how the changes in acetylcholine

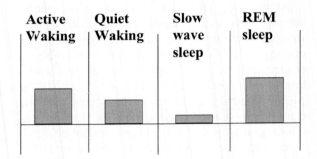

Fig. 1. Schematic representation of the microdialysis data showing changes in acetylcholine levels during different stages of waking and sleep. During active waking (exploration), animals have higher levels of acetylcholine than during quiet waking (immobility, eating, grooming). Acetylcholine levels fall to 1/3 of waking levels during slow wave sleep, but rise to levels above active waking during REM sleep (based on Jasper and Tessier, 1971; Kametani and Kawamura, 1990; Marrosu et al., 1995).

concentration could contribute to the change in EEG dynamics and functional properties of cortical circuits during these different periods.

Anatomical support for volume transmission

In the computational modeling work presented here and previously (Hasselmo et al., 1992; Hasselmo and Schnell, 1994; Patil and Hasselmo, 1999; Linster and Hasselmo, 2001), we model acetylcholine effects as being diffuse and relatively homogeneous within cortical circuits. That is, we assume that volume transmission provides a general activation of cholinergic receptors at a number of receptor sites, rather than focused effects localized at individual synaptic contacts (Descarries et al., 1997). Anatomical evidence supports this concept of volume transmission for acetylcholine. In particular, the axonal varicosities on cholinergic fibers predominantly are not accompanied by specific postsynaptic densities (Umbriaco et al., 1994, 1995), suggesting that the release sites of acetylcholine are not associated with specific clusters of cholinergic receptors. For example, in the hippocampus, only 7% of the axonal varicosities on cholinergic fibers are associated with junctional specialization, whereas all GABAergic varicosities showed synaptic specializations (Umbriaco et al., 1995). In the parietal cortex, less than 15% of cholinergic varicosities were associated with post-synaptic junctions (Umbriaco et al., 1994). This data supports the concept of volume transmission for acetylcholine within the hippocampus and neocortex.

In addition to this anatomical data, functional considerations support the concept of volume transmission. The number of cholinergic neurons within the basal forebrain is relatively small, on the order of 10^5 in the rat (Mesulam et al., 1983a,b). These cholinergic neurons innervate cortical structures containing many orders of magnitude more pyramidal cells and interneurons. These structures mediate encoding of a large number of different patterns of activity in semantic memory. This large discrepancy in number of neurons raises serious doubts about the ability of the basal forebrain to selectively regulate activity associated with specific memory patterns in cortical structures. Even if each stored pattern were being regulated by only a single basal forebrain neuron, the number of stored patterns in semantic memory would far exceed the capacity of the basal forebrain for selective regulation. In contrast, this relatively small number of neurons could work in a more cohesive manner to set different functional properties during longer functional stages of waking and sleep. There are sufficient numbers of cholinergic neurons to regulate transitions between active waking and quiet waking, and the numbers are sufficient to allow these functional stages to be regulated separately for larger cortical regions.

Slow time course of modulatory changes caused by acetylcholine

The large scale regulation of functional state via volume transmission is also supported by the relatively long time course of cholinergic modulatory effects. Experimental data demonstrates that activation of muscarinic cholinergic receptors causes physiological changes which take several seconds to reach their maximum, and persist for a minimum of 10–20 s, both for measurements of membrane potential depolarization (Krnjević et al., 1971; Cole and Nicoll, 1984) and for cholinergic modulation of excitatory synaptic transmission (Hasselmo and Fehlau, 2001). This slow time course means that even weak, temporally variable diffusion of acetylcholine within the extracellular space will build up over many seconds to cause strong and tonic changes in functional state within broad cortical regions. Thus, these considerations support the modeling of acetylcholine as a diffuse and relatively homogeneous regulation of circuit properties.

Acetylcholine enhances input relative to feedback

This chapter focuses on a single general framework for interpreting effects of acetylcholine within cortical structures. As summarized in Fig. 2, acetylcholine appears to enhance the strength of input relative to feedback in the cortex. The physiological effects of acetylcholine serve to enhance the influence of feedforward afferent input to the cortex, while decreasing background activity due to spontaneous

210

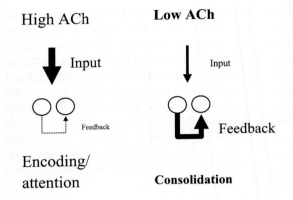

High ACh　　Low ACh

Input　　　　Input

Feedback　　Feedback

Encoding/
attention　　Consolidation

Fig. 2. General theory of acetylcholine effects in cortical circuits. This circuit diagram summarizes the predominant effect of acetylcholine within cortical circuits. High Ach levels cause an increase in the influence of afferent input relative to internal processing. Low Ach levels allow strong feedback relative to weaker afferent input. This is due to three sets of effects. (1) Intrinsic properties. Acetylcholine causes depolarization of pyramidal cells, and reduction in spike-frequency accommodation, allowing pyramidal cells to respond more robustly to external afferent input. (2) Modulation of inhibition. Acetylcholine depolarizes inhibitory interneurons, decreasing background spiking activizty, while suppressing inhibitory synaptic transmission, allowing a stronger response to afferent input due to reduced inhibitory feedback. (3) Modulation of excitatory synaptic transmission. When acetylcholine is present, activation of nicotinic receptors enhances thalamic afferent input, while muscarinic suppression reduces excitatory recurrent processing in cortex.

spiking and the spread of activity via excitatory feedback connections within cortical circuits. Thus, by enhancing the response to sensory input, high levels of acetylcholine enhance attention to the environment, making cortical circuits more responsive to specific features of sensory stimuli. Likewise, by enhancing the response to external input, high levels of acetylcholine enhance the encoding of memory for specific stimuli, by making cortical circuits respond to the specific features of sensory stimuli, allowing more effective and accurate encoding of sensory events.

This basic theoretical framework will be applied in discussing a number of effects of acetylcholine within cortical structures. The change in dynamics results from three primary sets of effects on a physiological level: (i) modulation of intrinsic properties of pyramidal cells; (ii) modulation of inhibitory neuron depolarization and synaptic transmission; and

(iii) selective modulation of excitatory synaptic transmission; The physiological effects of acetylcholine will be described within this functional framework in the following sections.

Acetylcholine enhances spiking response to afferent input

The physiological effects of acetylcholine on cortical pyramidal cells act to enhance the spiking response to excitatory afferent input, consistent with the enhanced response to input summarized in Fig. 2. Early studies using single unit recordings from the neocortex showed that application of cholinergic agonists to the cortex would cause a strong increase in firing rate of neurons (Krnjević and Phillis, 1963; Krnjević et al., 1971; Krnjević, 1984). This increase in firing rate was demonstrated to result from a slow depolarization of cortical pyramidal cells due to blockade of a potassium current, causing the membrane potential to move away from the reversal potential of potassium (Krnjević et al., 1971). Subsequent research in brain slice preparations of cortical structures have consistently demonstrated depolarization of pyramidal cells with application of cholinergic agonists (Benardo and Prince, 1982; Cole and Nicoll, 1984; Barkai and Hasselmo, 1994). This slow depolarization of membrane potential will enhance the spiking response to excitatory synaptic input (see Fig. 3).

Another physiological effect which enhances the spiking response to afferent input is the suppression of spike frequency accommodation, also shown in Fig. 3. Pyramidal cells normally respond to a sustained current injection with a high initial firing rate which gradually slows down. This results from activation of a voltage-sensitive potassium current (the M current) and also from voltage-dependent calcium influx causing activation of calcium-sensitive potassium currents (the AHP current) (Constanti and Galvan, 1983; Madison and Nicoll, 1984; Constanti and Sim, 1987; Madison et al., 1987; Schwindt et al., 1988). Activation of muscarinic receptors decreases activation of both of these potassium currents, allowing neurons to fire in a more sustained manner in response to input. This reduction in spike frequency accommodation appears in neocortical

A. Membrane potential depolarization

-71mV

-76mV

B. Reduced spike frequency accomodation

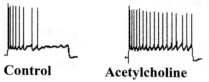

Control   **Acetylcholine**

Fig. 3. Intrinsic effects which enhance spiking response to afferent input. (A) Acetylcholine causes direct depolarization of pyramidal cell membrane potential (Krnjević et al., 1971; Krnjević, 1984; Cole and Nicoll, 1984; Barkai and Hasselmo, 1994), making cells more likely to generate spikes. (B) In addition, acetylcholine allows cells to generate spikes more persistently, due to cholinergic reduction in spike frequency accommodation (Madison and Nicoll, 1984; Tseng and Haberly, 1989; Barkai and Hasselmo, 1994).

structures (McCormick and Prince, 1986) as well as the piriform cortex (Tseng and Haberly, 1989; Barkai and Hasselmo, 1994) and hippocampal region CA1 (Madison and Nicoll, 1984). This effect allows neurons to continue to generate spiking responses to sustained afferent input, which could be very important for maintaining responsiveness to sensory input in attentional tasks. In particular, behavioral accuracy in continuous performance tasks requires neurons to remain responsive to subtle sensory input over extended periods of time. Suppression of spike frequency accommodation could prevent spontaneous background activity from reducing the sensitivity of cortical pyramidal cells to afferent input. Similarly, more sustained spiking activity would be important for maintaining responses necessary for encoding new memories.

Cholinergic modulation of inhibitory interneurons suppresses background activity while enhancing response to input

Acetylcholine also regulates the functional properties of cortical circuits through modulation of inhibitory interneurons. On first inspection, some of the data on these modulatory effects of acetylcholine appear contradictory and paradoxical, but they make sense when analyzed computationally, as shown in Fig. 4. Experimental data in the hippocampus demonstrates that acetylcholine simultaneously depolarizes inhibitory interneurons, while suppressing the evoked release of GABA at inhibitory synaptic terminals (Pitler and Alger, 1992; Behrends and ten Bruggencate, 1993). Cholinergic agonists have been shown to suppress inhibitory synaptic potentials in the hippocampal formation. In whole cell clamp recordings, the cholinergic agonist carbachol suppresses spontaneous GABA$_A$ inhibitory synaptic potentials, suggesting a direct suppression of the release of synaptic vesicles containing GABA (Pitler and Alger, 1992). However, carbachol also increases the number of miniature synaptic potentials presumed to result from the spontaneous spiking of inhibitory interneurons (Pitler and Alger, 1992). This coincides with other evidence showing that application of acetylcholine causes direct increases in spiking activity of inhibitory interneurons (McCormick and Prince, 1986). This evidence includes recordings from inhibitory interneurons in cortical structures which demonstrate direct depolarization of inhibitory interneurons by activation of cholinergic receptors in the hippocampus (Frazier et al., 1998; McQuiston and Madison, 1999a,b; Chapman and Lacaille, 1999). The activation of muscarinic receptors causes depolarization in many individual interneurons (McQuiston and Madison, 1999b). Similarly activation of nicotinic receptors depolarizes interneurons with different receptor properties in different neurons. In the hippocampus, neurons which responded with both fast nicotinic alpha-7 receptor responses and slow non-alpha-7 responses had cell bodies in oriens and projected to lacunosum-moleculare, while another set of neurons were depolarized by only alpha-7 receptors and appeared spread through many layers (McQuiston and Madison, 1999a). The direct depolarization of interneurons is consistent with the fact that nicotinic receptor activation causes an increase in GABA currents in hippocampal pyramidal cells and interneurons (Alkondon and Albuquerque, 2001). Similarly, in the neocortex, there is a selective nicotinic depolarization of specific interneurons, including neurons containing

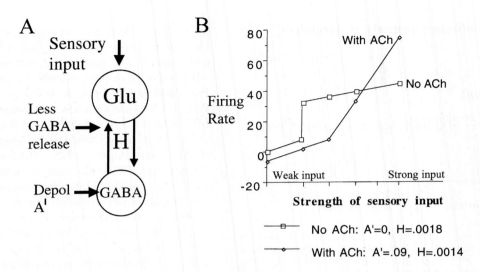

Fig. 4. Schematic diagram of cholinergic modulation of inhibition. (A) Basic circuit diagram depicting feedback inhibition. A population of excitatory neurons (labeled by Glu) receives depolarizing afferent input A. These neurons send excitatory output to inhibitory interneurons (labeled GABA). Modulatory effects of acetylcholine include direct suppression of inhibitory synaptic transmission (H) and direct depolarization of inhibitory interneuron membrane potential (represented by depolarizing input A′). (B) Analysis of this circuit demonstrates that these effects of acetylcholine decrease background activity while enhancing the response to strong afferent input. The equilibrium (steady state) of the network is plotted for different levels of afferent input A. When acetylcholine is not present (A′ = 0, H = 0.0018), the network responds to weak afferent input and only shows slight increases as afferent input increases. When acetylcholine is present, causing depolarization of interneurons (A′ = 0.09) and suppression of inhibitory transmission (H = 0.0014), the network shows little response to weak afferent input, but an enhanced response to strong afferent input (modified from Patil and Hasselmo, 1999).

vasoactive intestinal protein (Porter et al., 1999). Depending on the magnitude of this direct depolarizing effect on interneurons, it can result in periods of apparent enhanced inhibition after cholinergic application in slices (McCormick and Prince, 1986) or the enhancement of inhibition upon awakening (Steriade and Deschenes, 1974).

Thus, acetylcholine appears to increase spiking activity in inhibitory interneurons, while decreasing synaptic transmission from these neurons. These effects appear somewhat paradoxical, but as demonstrated in Fig. 4, computational modeling provides a framework for understanding such a combination. The influence of these two effects was analyzed in a circuit model evaluating the steady-state response to different levels of afferent input A. The cholinergic depolarization of interneurons has the effect of reducing the background firing rate of pyramidal cells during weak afferent input. In contrast, the cholinergic suppression of GABAergic transmission has the effect of enhancing the steady-state response

to strong afferent input. Thus, these cholinergic effects reduce background activity, but heighten the response to suprathreshold stimuli. Overall, these modulation effects on inhibition could enhance the sensitivity of cortical circuits to specific sensory input, important for performance in attention tasks as well as encoding tasks. The reduction in background firing could enhance the detection of subtle sensory input (assuming that the input strongly activates a subset of cells). In addition, this reduction in background firing activity would prevent activation of the calcium-dependent potassium currents underlying spike frequency accommodation. Thus, depolarization of interneurons could enhance the ability of cortical circuits to maintain responses to sensory input over extended periods, while the suppression of inhibitory transmission would reduce inhibition when a signal activates the cortex.

Evidence for selective cholinergic regulation of cortical circuitry has also been demonstrated in rat visual neocortex circuits. In particular, nicotinic depolarization of low threshold spiking interneurons could inhibit activity in upper layers (and dendritic inputs to layer V pyramidal cells), while muscarinic hyperpolarization of fast spiking interneurons could release inhibition at the soma of layer V pyramidal cells and increase spiking activity (Xiang et al., 1998, 2002). Acetylcholine also causes suppression of inhibitory synaptic potentials in visual cortex, while simultaneously depolarizing pyramidal cells (Murakoshi, 1995). These combined effects could play a similar role in reducing background activity while enhancing responses to suprathreshold stimulation. Acetylcholine has selective effects on inhibitory synaptic transmission within other structures as well. For example, stimulation of mesopontine cholinergic nuclei causes suppression of inhibitory potentials elicited in the anterior thalamus by cortical stimulation, while enhancing inhibitory potentials elicited by mammillary body stimulation (Curro Dossi et al., 1992). These studies emphasize the potential importance of selective modulation of different network properties.

Acetylcholine selectively suppresses excitatory feedback but does not suppress afferent input

Acetylcholine appears to reduce the internal processing of information by cortical structures, due to suppression of excitatory synaptic transmission at excitatory feedback connections within cortical circuits. This suppression of excitatory glutamatergic transmission contrasts with the depolarization of excitatory pyramidal cells in the same manner that the suppression of inhibitory GABAergic transmission contrasts with the depolarization of inhibitory interneurons. In the framework of computational modeling, these competing effects of acetylcholine make sense when it is emphasized that the cholinergic suppression of excitatory transmission is selective for the excitatory feedback connections. As described below, acetylcholine allows afferent input to maintain a strong influence on cortical circuits.

Selective suppression of intrinsic but not afferent fiber transmission in olfactory cortex. The selective cholinergic suppression of excitatory feedback, but not afferent input to the cortex has been demonstrated in a number of different regions, using a

number of different techniques. This differential cholinergic modulation was first demonstrated in slice preparations of the piriform cortex (Hasselmo and Bower, 1992). Earlier studies had demonstrated cholinergic suppression of excitatory synaptic transmission in tangential slices of the piriform cortex (Williams and Constanti, 1988), but those slices did not allow comparison of different synapses in different layers. As shown in Fig. 5, the use of transverse slices allowed direct comparison of the cholinergic effects on excitatory afferent input synapses from the olfactory bulb in layer Ia of piriform cortex, versus excitatory feedback connections in layer Ib (Hasselmo and Bower, 1992). As summarized in Fig. 5, acetylcholine and muscarine caused selective suppression of excitatory feedback potentials in layer Ib, while having a much weaker effect on afferent input in layer Ia (Hasselmo and

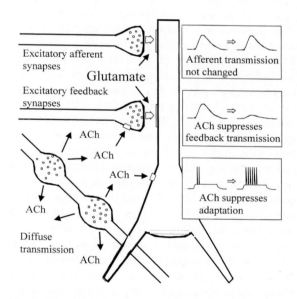

Fig. 5. Acetylcholine selectively suppresses excitatory synaptic transmission at feedback synapses, but not afferent input and feedforward connections. This diagram shows synaptic transmission in the piriform cortex at afferent fiber synapses on distal dendrites (top) and excitatory feedback synapses on proximal dendrites (middle). Activation of cholinergic receptors selectively suppresses glutamatergic transmission (Hasselmo and Bower, 1992; Linster et al., 1999) at excitatory feedback synapses (middle), but not excitatory afferent fibers (top). Acetylcholine simultaneously enhances the spiking response of neurons to current injection (shown) or synaptic input (modified from Patil and Hasselmo, 1999; Linster et al., 1999).

Bower, 1992). This effect has been confirmed in anesthetized preparations *in vivo*, in which direct stimulation of the cholinergic innervation of piriform cortex suppresses feedback from posterior piriform cortex to anterior piriform cortex, but does not influence the afferent input to piriform cortex from the lateral olfactory tract (Linster et al., 1999; Linster and Hasselmo, 2001).

Selective suppression of intrinsic but not afferent fiber transmission in neocortex. Subsequently, a similar selectivity of cholinergic suppression has been demonstrated in the neocortex. In particular, in slice preparations of somatosensory cortex, cholinergic modulation causes suppression of excitatory transmission at feedback connections from higher order somatosensory cortex, while having less effect on synaptic potentials elicited in layer IV by stimulation of subcortical white matter (Hasselmo and Cekic, 1996). The differential regulation of different pathways was demonstrated further in thalamocortical slice preparations (Gil et al., 1997), where activation of nicotinic receptors enhanced thalamic input to the neocortex, while that of muscarinic receptors suppressed both intracortical and thalamocortical synaptic transmission. Similarly, in the primary auditory cortex, acetylcholine suppressed intracortical synaptic potentials (Hsieh et al., 2000) while having no effect on or enhancing thalamocortical connections (Metherate and Ashe, 1993; Hsieh et al., 2000). Cholinergic modulation has also been demonstrated to suppress intracortical synaptic transmission in primary visual cortex (Brocher et al., 1992) and frontal cortex (Vidal and Changeux, 1993). In studies of visual cortex using optical imaging, cholinergic modulation appeared to regulate the intracortical spread of activity, while having a much weaker effect on thalamic input (Kimura et al., 1999; Kimura, 2000). The nicotinic enhancement of thalamic input to the neocortex has also been demonstrated for the medial dorsal thalamic input to prefrontal cortex (Vidal and Changeux, 1993; Gioanni et al., 1999). Thus, the differential suppression of intracortical feedback connections with sparing of afferent input connections which was demonstrated in piriform cortex appears to generalize to most neocortical regions, supporting the functional framework illustrated in Fig. 2.

Selective suppression of intrinsic but not afferent fiber transmission in hippocampus. The selective

suppression of excitatory synaptic transmission at feedback but not feedforward synapses has also been demonstrated at connections within the hippocampal formation, as described in a recent review (Hasselmo, 1999). The suppression of excitatory synaptic transmission by acetylcholine was described in a number of early studies in slice preparations of the hippocampal formation. Muscarinic suppression of excitatatory transmission was reported at connections including the medial entorhinal input to the middle molecular layer of the dentate gyrus (Yamamoto and Kawai, 1967; Kahle and Cotman, 1989) as well as the Schaffer collaterals projecting from hippocampal region CA3 to region CA1 (Hounsgaard, 1978; Valentino and Dingledine, 1981). Similar suppression of Schaffer collateral synaptic potentials was obtained in anesthetized animals with microintophoretic application of acetylcholine (Rovira et al., 1982), with stimulation of the medial septum (Rovira et al., 1983), and with sensory stimulation which activates hippocampal theta rhythm (Herreras et al., 1988a,b). Cholinergic modulation has also been demonstrated to suppress excitatory synaptic transmission in the amygdala (Yajeya et al., 2000).

The differential modulation of excitatory synaptic transmission has been explicitly demonstrated in slice preparations of the hippocampal formation. Within the dentate gyrus, the synaptic inputs to the outer molecular layer from lateral entorhinal cortex show little decrease in the presence of cholinergic agonists (Yamamoto and Kawai, 1967; Kahle and Cotman, 1989). Within hippocampal region CA1, there is strong suppression of excitatory transmission in the stratum radiatum, where Schaffer collateral inputs terminate (Hounsgaard, 1978; Valentino and Dingledine, 1981; Hasselmo and Schnell, 1994), but there is a much weaker suppression of synaptic potentials in the stratum lacunosum-moleculare, where input from entorhinal cortex layer III terminates (Hasselmo and Schnell, 1994). In hippocampal region CA3, excitatory recurrent connections in stratum radiatum are strongly suppressed (Hasselmo et al., 1995; Vogt and Regehr, 2001), whereas the effect at synapses from dentate gyrus terminating in stratum lucidum appears much weaker (Hasselmo et al., 1995). There was some evidence for suppression of the mossy fiber input from dentate gyrus, but this appeared to be due to secondary cholinergic activation of inhibitory interneurons, as there was no cholinergic suppression of mossy fiber input in the presence of GABAB receptor blockers (Vogt and Regehr, 2001).

The selectivity of the cholinergic suppression of synaptic transmission is summarized in Fig. 5. This general suppression of excitatory feedback would act to reduce feedback within the hippocampus and from hippocampus to other cortical areas during high acetylcholine levels. But none of these papers show total suppression—the influence of hippocampus is reduced but not removed; there is still sufficient feedback to allow retrieval of relevant stored information.

These specific data are consistent with recordings showing changes in feedback transmission in awake, behaving animals. Recordings from the entorhinal cortex suggest that during active waking the influence of hippocampus on entorhinal cortex is weak, as determined by the low rates of spiking activity in deep layers of entorhinal cortex, which receive output from the hippocampus, in contrast to the higher rates of activity in the superficial layers of entorhinal cortex which send input to the hippocampus (Chrobak and Buzsaki, 1994). In contrast to this weak hippocampal feedback during active waking, the spiking activity in deep entorhinal layers is much higher during quiet waking and slow wave sleep (Chrobak and Buzsaki, 1994). In addition, stimulation of entorhinal input pathways causes very small amplitude evoked potentials in the entorhinal cortex during active waking, whereas the same stimulation amplitude will evoke much larger evoked potentials in the entorhinal cortex during slow wave sleep (Winson and Abzug, 1978; Buzsaki, 1986). A weaker cholinergic influence on perforant path input to the dentate gyrus is supported by the fact that EPSPs evoked in the dentate gyrus by angular bundle stimulation are larger during the high acetylcholine levels of active waking than during the lower acetylcholine levels of slow wave sleep (Winson and Abzug, 1978). In anesthetized animals, sensory stimulation activating theta rhythm appears to increase the sensory response within the dentate gyrus (Herreras et al., 1988b), though this may be due to a cholinergic increase in postsynaptic excitability, because stimulation of the medial septum causes

increases in population spike activity in the dentate gyrus, but does not have a systematic effect on EPSPs (Mizumori et al., 1989). Cholinergic modulation influences synaptic transmission in the thalamus as well. In particular, stimulation of the cholinergic laterodorsal nucleus enhances excitatory synaptic potentials in the anterior thalamus induced by mammillary body stimulation (Pare et al., 1990) as well as ventrolateral thalamic EPSPs induced by cerebellar stimulation (Timofeev et al., 1996). It is not clear if these effects can be described within the same functional framework presented here.

Functional data

The theoretical framework presented in Fig. 2 raises the question: What is the functional purpose of the alteration in circuit dynamics induced by cholinergic modulation? Why would it be necessary to selectively enhance the afferent input relative to feedback excitation? This section will review behavioral evidence demonstrating the potential role of this change in circuit dynamics, showing how acetylcholine effects may enhance attention to external stimuli, and may enhance encoding of new input. This raises another question: If this enhancement of attention and encoding is important, then why should circuit dynamics not remain in this state at all times? The existence of a selective modulator for changing these dynamics suggests that the absence of these dynamics are important for some function. The last section here will review how the low acetylcholine state may be important for a separate function. In this state, excitatory feedback is strong, whereas afferent input is relatively weak. Additional physiological and behavioral data suggests that low levels of acetylcholine may set appropriate dynamics for the consolidation of previously encoded information.

Acetylcholine and the enhancement of attention

Behavioral data supports the functional framework presented here for the role of acetylcholine within cortical structures. In particular, the enhancement of afferent input relative to internal processing could enhance performance in attention tasks. The performance in attention tasks often depends on the

Suppression of Feedback Excitation (60%) and Inhibition (40%)

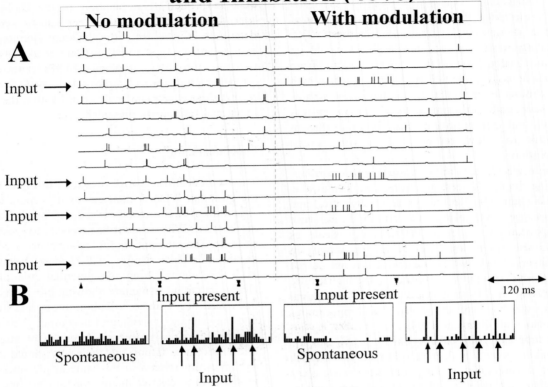

Fig. 6. Suppression of synaptic transmission at excitatory feedback synapses and inhibitory feedback synapses enhances signal to noise ratio in a simulation containing multiple integrate-and-fire neurons (modified from Hasselmo et al., 1997; Linster and Hasselmo, 2001). On the left, in the absence of modulation, there is broadly distributed background spiking activity, which makes it difficult to distinguish the response to direct afferent sensory input (arrows). The histogram on the bottom shows the number of spikes fired by individual neurons (arrows indicate afferent input). On the right, modulation of synaptic transmission enhances signal to noise, increasing the number of spikes generated by neurons receiving afferent input (arrows), while reducing the spontaneous spiking activity of other neurons in the network. (Modified from Hasselmo et al. 1997)

sensitivity to specific weak stimuli over an extended period of time. Performance in these tasks will be enhanced if the neuronal response to the stimulus is strong (allowing rapid and accurate behavioral responses), whereas the background noise in neural systems should be relatively weak (thereby preventing generation of incorrect responses at inappropriate times in the task).

The cholinergic influence on circuit dynamics illustrated in Fig. 2 has the net effect of enhancing the response to external sensory stimuli relative to background noise. This can be seen in Fig. 6, which shows the results of a spiking network simulation of piriform cortex (Linster and Hasselmo, 2001). This model implements selective suppression of excitatory feedback synaptic transmission, as well as modulation which reduces feedback inhibition (analogous to the suppression of GABAergic transmission shown experimentally). These effects serve to enhance the response of the network to the pattern of input, while reducing the amount of background spiking activity (Linster and Hasselmo, 2001). These effects are analogous to what has been reported in single unit recordings from neurons in the primary visual cortex, which shows that local application of acetylcholine enhances the response of neurons to visual input (Sato et al., 1987) and enhances the direction selectivity of individual neurons (Murphy and Sillito, 1991). Thus, the change in network dynamics enhances response to sensory stimulation relative to background noise.

This functional interpretation is supported by extensive data indicating a role for acetylcholine in attention processes at a behavioral level. Neuropsychological data from humans subjects has consistently suggested that neuropathology in the ascending cholinergic system underlies deficits in cognition and specifically attention (Parasuraman et al., 1992; Parasuraman and Haxby, 1993; Greenwood et al., 1997; Calderon et al., 2001). Cholinergic effects at nicotinic receptors enhance performance in attention tasks (Wesnes and Revell, 1984; Wesnes and Warburton, 1984), whereas scopolamine impairs performance (Wesnes and Warburton, 1983). These findings have prompted neuroscientific investigations to better understand the precise role of the cholinergic system in attention.

A great deal of research has been directed at understanding the role of the cholinergic system in attention using either cholinergic drugs or a variety of lesioning techniques. These studies have been critical to the formulation of current hypotheses regarding the role of this system in attention, but the conclusions drawn were necessarily limited by the lack of selectivity of previous methods for acetylcholine (for review see McGaughy et al., 2000). Recently, the immunotoxin 192 IgG-saporin (SAP), which couples the ribosome inactivating toxin saporin to an antibody that recognizes the low affinity nerve growth factor (NGF) receptor, p75, has been developed to allow selective destruction of cortical cholinergic afferents (Wiley et al., 1991; Book et al., 1994; Wiley et al., 1995). 192 IgG-saporin also spares non-cholinergic (GABAergic and glutamatergic) neurons in the basal forebrain and permits a direct investigation of the effects of acetylcholine on cognition not previously possible using excitotoxins (Dunnett et al., 1991; Muir et al., 1994, 1995, 1996).

Most cholinergic cell bodies throughout the basal forebrain, including the nucleus basalis of Meynert, medial septum (ms), vertical and horizontal limbs of the diagonal band, bear these p75 receptors and undergo apoptotic cell death following infusion of this toxin (Wiley et al., 1991; Book et al., 1994; Heckers et al., 1994; Wiley et al., 1995). In addition, this receptor is colocalized on the terminal fields of cholinergic projections in the neocortex and hippocampus, which allows neuroanatomically restricted deafferentation following retrograde transport of the toxin to the cell bodies (Wiley et al., 1991; Book et al., 1994; Heckers et al., 1994; Wiley et al., 1995). Cholinergic projections from the pedunculopontine and laterodorsal tegmental nucleus, which innervate the thalamic nuclei and provide a negligible contribution to the neocortex as well as those from the nucleus basalis of Meynert to the amygdala, lack the p75 receptor and are unaffected by the toxin (Book et al., 1994; Heckers et al., 1994).

Behavioral studies using this immunotoxin have provided consistent evidence of the role of acetylcholine in attention (Chiba et al., 1995; McGaughy et al., 1996; Turchi and Sarter, 1997; Bucci et al., 1998; McGaughy and Sarter, 1998; McGaughy et al., 1999). Intrabasalis infusions of the immunotoxin severely impaired sustained attention (McGaughy et al., 1996). Rats were trained to detect rare and temporally unpredictable visual stimuli and report the presence or absence of these signals via two distinct response levers (McGaughy and Sarter, 1995). The detection of signals was robustly and persistently impaired following the lesions, while the ability to respond correctly to non-signals was spared. Additionally, this impairment in signal detection was exacerbated over the course of the testing session (McGaughy et al., 1996). The maintenance of prolonged attentional performance throughout a testing session has been previously shown to increase attentional demands (Parasuraman and Giambra, 1991; McGaughy and Sarter, 1995) and was predicted to augment the attentional impairments produced by lesioning the nucleus basalis of Meynert. The continued ability of rats to respond to non-signal events after lesioning suggests that impairments in task performance were not confounded by a loss of memory for the rules of the task (McGaughy et al., 1996). Multiple cortical

218

infusions of the toxin (McGaughy and Sarter, 1998) or smaller doses of the toxin into the nucleus basalis of Meynert (McGaughy et al., 1999) produced qualitatively similar though less severe impairment in sustained attention. In all studies, the degree of attentional impairment was correlated with the extent of cortical deafferentation (McGaughy et al., 1996; McGaughy and Sarter, 1998; McGaughy et al., 1999).

The framework illustrated in Figs. 2, 4, 5 and 6 can be used to account for the importance of cholinergic modulation in sustained attention performance as described below (McGaughy and Sarter, 1995; Arnold et al., 2002). Effective performance in that task requires discrimination of neural activity induced by a brief light stimulus, in contrast with the activity associated with the absence of a stimulus. Cholinergic modulation of cellular and synaptic properties could contribute to this task in two ways. First, the depolarization of interneurons (Fig. 4) and the suppression of excitatory feedback transmission (Figs. 5 and 6) could reduce spontaneous background activity as well as any response to distractor stimuli— such as a flashing house light (McGaughy and Sarter, 1995), thereby reducing false alarms. Second, the depolarization of pyramidal cells and the suppression of GABAergic transmission could enhance the magnitude of response to sensory stimuli, which will increase the propensity of the target light stimulus for pushing network activity over threshold, resulting in correct detection of the target light stimulus (increasing hits relative to misses). Excessive cholinergic modulation could cause too much depolarization of pyramidal cells, and too strong a response to afferent input, resulting in generation of false alarms. This effect has been observed in behavioral experiments in which cholinergic modulation was enhanced by infusion of benzodiazepine inverse agonists into the basal forebrain (which reduces inhibition of cholinergic innervation, thereby enhancing acetylcholine release). This manipulation enhances the number of false alarms (Holley et al., 1995).

Cholinergic deafferentation was also found to impair divided attention (Turchi and Sarter, 1997). Rats were trained in a conditional discrimination with different response rules dependent upon the modality (auditory or visual) of the stimuli presented.

Testing sessions consisted of blocks of modality-certain trials (only auditory or visual trials) followed by blocks of modality-uncertain trials (all stimulus types randomly interspersed) (McGaughy et al., 1994). When compared to sham-lesioned rats, nucleus basalis of Meynert-lesioned rats showed greater increases in the response latencies during modality-uncertain blocks but no difference in response latencies during modality certain blocks. These data suggested that cholinergic deafferentation decreased processing capacity and produced impairments in performance only when the concurrent maintenance of both sets of response rules was required (modality uncertainty) (Turchi and Sarter, 1997).

In vivo microdialysis in rats performing various tests of attention has repeatedly demonstrated an increase in acetylcholine (ACh) efflux in the area of the frontoparietal (Himmelheber et al., 1997, 2000, 2001) or medial prefrontal cortex (mPFC) (Passetti et al., 2000; Dalley et al., 2001; McGaughy et al., 2002). In rats trained in a 5 choice serial reaction time task (5CSRTT), a test of sustained attention, intra-nucleus basalis of Meynert infusions of high doses of 192 IgG-saporin (SAP-HIGH) produced robust decreases in ACh efflux in the mPFC. This comprised efflux correlated with extensive cortical cholinergic deafferentation and severe attentional impairments (McGaughy et al., 2002). Rats infused with lower doses of the toxin (SAP-LOW), tested under standard conditions, showed no significant differences in either ACh efflux or attentional performance between lesioned rats and sham-lesioned rats. This similarity in attentional performance was hypothesized to result from neurochemical compensation in the cholinergic system following the smaller lesions, as histological analyses later confirmed that SAP-LOW rats had significant cholinergic cell loss in the nucleus basalis of Meynert (McGaughy et al., 2002). Though no differences existed under baseline conditions, attentional performance of SAP-LOW rats was more vulnerable than in sham-lesioned rats to increases in attentional demands produced by prolonged time on task or an increased event rate. Data from sham-lesioned, SAP-HIGH and SAP-LOW rats showed that cortical ACh correlated with accuracy in the task and residual cholinergic neurons in the nucleus basalis of Meynert (McGaughy et al., 2002). These data provided strong evidence that cholinergic efflux in the mPFC was dependent upon nucleus basalis of Meynert (but not vertical limb of the diagonal band of Broca) cholinergic neurons and this pathway was critical to the maintenance of sustained attention performance.

Cholinergic input to the neocortex from the nucleus basalis of Meynert is hypothesized by Sarter and colleagues to play a crucial part in activating 'top–down' control of various forms of attention via the convergence of inputs from the nucleus basalis of Meynert and other sensory and associational cortices in the neocortex (Sarter and Bruno, 2000; Sarter et al., 2001). Intra-nucleus basalis of Meynert infusions of SAP impaired the ability of rats to 'increment attention' in response to changes in the predictive relationship of a stimulus to a reward (Chiba et al., 1995). These attentional deficits were replicated when the immunotoxin was infused directly into the area of the posterior parietal cortex, highlighting the specific importance of nucleus basalis of Meynert projections to this area in mediating the effect (Bucci et al., 1998). 'Decremental attention', as required to disregard previously irrelevant stimuli, was not affected by nucleus basalis of Meynert lesions (Chiba et al., 1995), but was impaired by lesions of medial septum/vertical limb of the diagonal band (Baxter et al., 1997).

Gill et al. (2000) have used electrophysiological methods to determine precisely what aspects of attentional performance are mediated by cortical, cholinergic afferents. Single unit recordings of neuronal activity in the area of the mPFC of well-trained rats performing in a test of sustained attention (McGaughy and Sarter, 1995) showed units correlated with correct responses, trial type (signal or non-signal) or the expectation of reinforcement (Gill et al., 2000). Presentation of a distractor stimulus, hypothesized to increase attentional demands, coincided with an increase in behaviorally correlated unit activity (Gill et al., 2000). This increase in unit activity was hypothesized to provide 'top–down' excitation of the nucleus basalis of Meynert via glutamatergic afferents from the mPFC, thereby facilitating the processing of sensory information during conditions of heightened attentional demand (Sarter et al., 2001). Unilateral, cholinergic deafferentation produced by cortical

infusions of SAP in these same animals severely attenuated event-related behavioral activity during baseline conditions, but did not impair attentional performance (this requires bilateral deafferentation). Furthermore, increases in mPFC unit activity previously associated with the introduction of a visual distractor were absent in the lesioned hemisphere (Gill et al., 2000). In summary, converging neuroscientific data continue to provide support for the hypothesis that corticopetal cholinergic afferents originating in the nucleus basalis of Meynert are critical to attentional processing.

Moreover, the role of cholinergic afferents in attention can be dissociated from that of noradrenergic afferents by both lesions and in vivo microdialysis in behaving rats. In contrast to cholinergic lesions, large noradrenergic lesions (>90% loss) produced no impairment in attentional performance in a sustained attention task, even in the presence of a visual distractor stimulus (McGaughy et al., 1997). In vivo microdialysis of the prefrontal cortex of rats performing the 5 CSRTT have shown that an unexpected change in reinforcement contingencies selectively increases noradrenergic but not cholinergic efflux. This change in efflux is found in the first but not the second exposure to these conditions (Dalley et al., 2001). In general, it has been suggested that stimuli that are sufficiently novel engage the noradrenergic system to produce a broadening of attention to context (Sarter and Bruno, 2000) and may be seen as a kind of sentry for potentially threatening stimuli.

Acetylcholine and the enhancement of encoding

As noted earlier, the enhanced response to afferent input with the reduction of feedback can play a role in enhancing performance in attention tasks. But this same change in dynamics could also be important for the encoding of new information in memory. Traditionally, researchers have attempted to distinguish and differentiate the role of acetylcholine in attention from the role in encoding. However, in this section we will review how these may not be separable functions. The same enhancement of afferent input relative to feedback excitation may be interpreted as enhancing attention when it occurs in neocortical structures, whereas this same effect in medial temporal structures such as the hippocampus could serve to enhance the encoding of new memories.

Numerous human memory studies demonstrate that blockade of muscarinic acetylcholine receptors by systemic administration of the drug scopolamine interferes with the encoding of new verbal information, while having little effect on retrieval of previously stored information (Crow and Grove-White, 1973; Drachman and Leavitt, 1974; Ghonheim and Mewaldt, 1975; Peterson, 1977; Beatty et al., 1986; Sherman et al., 2003; see Hasselmo, 1995; Hasselmo and Wyble, 1997 for review). Scopolamine appears to primarily affect episodic memory, while sparing semantic and procedural memory (Caine et al., 1981; Broks et al., 1988) and short-term memory phenomena such as the recency component of a serial position curve (Crow and Grove-White, 1973) and digit span (Drachman and Leavitt, 1974; Beatty et al., 1986).

A selective impairment of encoding has also been demonstrated in experiments testing memory function in animals. In particular, one series of experiments used a task with an encoding phase during which a monkey viewed a series of visual objects, followed by a later recognition phase during which they were tested for their recognition of these items (Aigner and Mishkin, 1986; Aigner et al., 1991). In these experiments, systemic injections of scopolamine impaired the encoding of new objects, while having little effect when administered during the recognition phase for objects encoded without scoplamine. These encoding effects appeared to be focused in the parahippocampal regions, because encoding of stimuli in this task was impaired by local infusions of scopolamine into the perirhinal (and entorhinal) regions but not by infusions into the dentate gyrus or inferotemporal cortex (Tang et al., 1997). Microdialysis showed a 41% increase in acetylcholine levels in perirhinal cortex during performance of this visual recognition task (Tang and Aigner, 1996).

In rats, effects of muscarinic receptor blockade by scopolamine have been observed in tasks where the rat must encode episodic memories—events occurring at a specific place and time. For example, injections of scopolamine impaired the encoding of platform location in a task in which the platform was moved on a day by day basis (Whishaw, 1985; Buresova et al., 1986). In the 8-arm radial maze task, the encoding of previously visited arms appeared to be impaired in a similar manner by systemic scopolamine injections, as well as by lesions of the fornix which destroys the cholinergic innervation of the hippocampus (Cassel and Kelche, 1989). The effects of scopolamine were stronger when a delay was interposed between response and test (Bolhuis et al., 1988). Scopolamine had an effect when it was present during encoding, for example the first four arm visits before a delay, but did not affect retrieval when injected during the delay (Buresova et al., 1986).

Scopolamine has also been shown to impair learning in various conditioning tasks. In eye blink conditioning tasks, the rate of learning was significantly slowed by electrolytic lesions of the medial septum (Berry and Thompson, 1979) as well as by ibotenic acid lesions of the medial septum (Allen et al., 2002). Systemic scopolamine also slowed the learning of classical conditioning in eye-blink conditioning tasks in rabbits (Solomon et al., 1983; Salvatierra and Berry, 1989) and in humans (Solomon et al., 1993). These effects have been modeled in a hippocampal simulation (Myers et al., 1998). The slowing of eye blink conditioning showed a direct correlation with reductions in hippocampal theta rhythm (Berry and Thompson, 1979; Salvatierra and Berry, 1989). The hippocampal theta rhythm appearing during alert immobility in these types of experiments was sensitive to cholinergic blockade (Kramis et al., 1975). A similar correlation with theta rhythm has been shown in experiments where presentation of the stimulus during periods of theta rhythm enhances the rate of learning (Berry and Seager, 2001; Seager et al., 2002). Scopolamine also impairs appetitive jaw movement conditioning (Seager et al., 1999), and scopolamine has also been shown to impair encoding of fear conditioning (Anagnostaras et al., 1995; Young et al., 1995; Anagnostaras et al., 1999), even when injected directly into the hippocampus (Gale et al., 2001), but appears to enhance consolidation of fear conditioning when injected after training (Young et al., 1995). This enhancement of consolidation by cholinergic blockade supports the hypothesis presented below that low levels of acetylcholine are important for consolidation.

Acetylcholine also appears important for encoding of sensory representations in neocortical structures such as the auditory cortex. Experimental recordings from individual neurons in auditory cortex showed a tuning curve focused on specific frequencies. These tuning curves were altered by conditioning and could also be altered by auditory stimulation combined with microiontophoretic application of acetylcholine (Metherate and Weinberger, 1990; Metherate et al., 1990). Cholinergic modulation has also been shown to cause long-term alterations in responses to somatosensory stimulation (Tremblay et al., 1990; Dykes, 1997).

Cholinergic suppression of feedback may prevent interference

The cholinergic suppression of excitatory transmission might appear somewhat paradoxical when considering encoding. Why would a substance that is important for learning cause suppression of excitatory transmission? As noted earlier, it is important to emphasize the selectivity of this suppression for intrinsic but not afferent fibers. The importance of this selective suppression of transmission has been analyzed in computational models of associative memory function (Hasselmo et al., 1992; Hasselmo and Bower, 1993; DeRosa and Hasselmo, 2000; Linster and Hasselmo, 2001). These models demonstrate that cholinergic suppression of transmission prevents retrieval of previously encoded associations from interfering with the encoding of new associations. For example, if an association A–B has been encoded, then subsequent presentation of an association A–C could cause retrieval of the A–B association. This would cause the new association A–C to suffer from interference from the A–B association and cause associations between C and B. Recent experiments have tested behavioral predictions of these computational models (DeRosa and Hasselmo, 2000; DeRosa et al., 2001). In one experiment, rats were initially trained to respond to odor A when presented with the odor pair A–B. Then in a separate phase of the experiment, the rat had to learn to respond to odor C when presented with odor pair A–C, and during the same period had to learn to respond to odor D when presented with odor pair

D–E. In a counterbalanced design, rats received injections of scopolamine, methylscopolamine or saline after learning of A–B and before learning of A–C and D–E. Injections of scopolamine during encoding caused a greater impairment in the learning of overlapping odor pairs (A–C) than non-over-lapping odor pairs (D–E). Thus, scopolamine appears to enhance proactive interference, consistent with its blockade of the cholinergic suppression of exci-tatory synaptic transmission at intrinsic synapses in the piriform cortex. 192 IgG, Saporin lesions of the limb of the diagonal band, resulting in cholinergic denervation of piriform cortex, heightened the sensitivity to scopolamine in this paradigm (DeRosa et al., 2001). This model is further supported by experimental data showing that electrical stimulation of the olfactory cortex can modulate the activity of neurons in the HDB, thus providing a pathway for regulation of cholinergic activity (Linster and Hasselmo, 2000). Cholinergic blockade also increases the generalization between similar odorants seen in an odor guided digging tasks (Linster et al., 2001). Similar effects have been obtained in an experiment performed in human subjects, in which scopolamine caused greater impairments in the encoding of overlapping versus non-overlapping word pairs (Kirchhoff et al., 2000; Atri et al., 2003). These data suggest that interference effects may underlie the impairments caused by lesions of the medial septum, including impairments of reversal learning (M'Harzi et al., 1987), and delayed alternation (Numan et al., 1995). Interference effects could also contribute to the impairment of 8-arm radial maze performance caused by scopolamine (Buresova et al., 1986; Bolhuis et al., 1988).

Acetylcholine enhances long-term potentiation

Activation of acetylcholine receptors also enhances synaptic modification in long-term potentiation experiments. This enhancement would naturally be important for the encoding of new information. Physiological experiments in brain slice preparations have demonstrated enhancement of LTP by choli-nergic agonists at a number of different synaptic pathways, including the perforant path input to the dentate gyrus (Burgard and Sarvey, 1990), the

Schaffer collateral input to region CA1 (Blitzer et al., 1990; Huerta and Lisman, 1993), excitatory synaptic connections in primary visual cortex (Brocher et al., 1992) and association fiber connections in the piriform cortex (Hasselmo and Barkai 1995; Patil et al., 1998). In slice studies of region CA1, it has been shown that LTP is most strongly enhanced by stimulation in phase with spontaneous oscillatory activity (Huerta and Lisman, 1993). Drugs which block these neuromodulatory effects on LTP appear to impair memory function: muscarinic receptor antagonists such as scopolamine block the cholinergic enhancement of LTP (Burgard and Sarvey 1990; Huerta and Lisman, 1993; Patil et al., 1998) and these antagonists also impair encoding as described above. These results demonstrate that modulators could contribute to encoding of new information through enhancement of long-term potentiation.

Acetylcholine enhances sustained spiking activity

Acetylcholine also appears to influence the firing activity of cortical circuits by enhancing intrinsic mechanisms for sustained spiking activity in individual neurons. Data from slice preparations of entorhinal cortex demonstrate this cellular mechanism for sustained spiking activity. In physiological recordings from non-stellate cells in slice preparations (Klink and Alonso, 1997a,b), application of the cholinergic agonist carbachol causes long-term depolarizations, which have been termed plateau potentials. If the cells generate an action potential during cholinergic modulation, either due to cholinergic depolarization or current injection, these neurons show sustained spiking activity. This sustained spiking activity appears to result from activation of a non-specific cation current (termed I_{NCM}) which is sensitive to muscarinic receptor activation, as well as the intracellular concentration of calcium (Shalinsky et al., 2002). This intrinsic capacity for self-sustained spiking activity of individual neurons could underlie the sustained spiking activity observed during performance of delayed nonmatch and delayed match to sample tasks in the entorhinal cortex of rats (Young et al., 1997) and monkeys (Suzuki et al., 1997). Simulations demonstrate how these phenomena could directly arise from

I_{NCM} in individual entorhinal neurons (Fransen et al., 2002). These phenomena include stimulus selective spiking activity during the delay period, as well as enhancement of spiking response to stimuli, which match the previously presented sample stimulus. In addition, incorporation of these effects in a network of excitatory and inhibitory neurons can create other phenomena such as match suppression, and non-match enhancement and suppression (Suzuki et al., 1997). Recent data demonstrates that cholinergic modulation of neurons in entorhinal cortex layer V causes activation of a current capable of maintaining graded levels of spiking activity, potentially relevant to the maintenance of analog representations of external stimuli (Egorov et al., 2002).

If cholinergic activation of I_{NCM} is important to provide intrinsic mechanisms for self-sustained spiking activity, then blockade of this cholinergic activation should prevent sustained spiking activity during the delay period and match enhancement. This effect of muscarinic antagonists could underlie the behavioral impairments in delayed matching tasks seen with systemic injections of muscarinic antagonists (Bartus and Johnson, 1976; Penetar and McDonough, 1983). In addition to this role in short-term memory function, sustained activity in entorhinal cortex could also be very important for effective encoding of long-term representations through synaptic modification in the hippocampal formation. Thus, the blockade of sustained spiking activity in entorhinal cortex could contribute to the encoding impairment caused by injections of scopolamine (Aigner and Mishkin, 1986; Buresova et al., 1986; Aigner et al., 1991; Anagnostaras et al., 1995; Tang et al., 1997; Anagnostaras et al., 1999).

Low levels of acetylcholine set appropriate dynamics for consolidation

If acetylcholine plays such an important role in attention and encoding, then why is it not present at high concentrations at all times? Or why could the modulatory effects of acetylcholine not be maintained as the baseline parameters for cortical circuits? The ability of acetylcholine to selectively regulate these parameters suggests that the low acetylcholine state has functional importance. In this section, we review the hypothesis that low levels of acetylcholine are important for the consolidation of previously encoded information (see Hasselmo, 1999 for more detailed review). This consolidation would take place during quiet waking and slow wave sleep, when levels of acetylcholine are low and cortical network dynamics include EEG phenomena such as slow waves (Steriade, 1994, 2001) and sharp waves (Buzsaki, 1989).

The hypothesis that consolidation occurs during slow wave sleep and quiet waking has been discussed for many years (Buzsaki, 1989; Wilson and McNaughton, 1994). This hypothesis proposes that initial encoding of memories occurs within the hippocampal formation during active waking. Subsequently, during quiet waking or slow wave sleep, in the absence of specific sensory input, random activity in the hippocampus causes reactivation of memory representations. Some research has focused on recordings of hippocampal place cells, showing that when pairs of place cells code adjacent positions during a period of active waking, these neurons show greater correlations of firing during subsequent slow-wave sleep, as compared to slow-wave sleep preceding the training session (Wilson and McNaughton, 1994). Another set of experiments have focused on the predominant flow of activity during different behavioral states (as symbolized by the arrows in Fig. 7). These experiments demonstrate that during active waking, when theta rhythm is present in the hippocampus, there is extensive neuronal activity in the layer of entorhinal cortex which provides input to the hippocampus (layer II), but not in the entorhinal layers receiving output from the hippocampus (layers V and VI) (Chrobak and Buzsaki, 1994). In contrast, during quiet waking and slow wave sleep, EEG phenomena termed sharp waves originate among the strong excitatory recurrent collaterals in hippocampal region CA3 and spread back through region CA1 to deep, output layers of entorhinal cortex (Chrobak and Buzsaki, 1994). This suggests that hippocampus could be inducing coactivation of neurons in neocortical regions which could form new cross-modal associations. During this stage, slow waves are prominent in the cortical EEG arising from neocortical and thalamocortical circuits (Steriade, 2001). These slow waves include both delta waves and lower frequency oscillations which with sleep spindles

are also postulated to contribute to the consolidation of memory traces acquired during the state of wakefulness (Steriade, 2001).

The sharp waves observed in the hippocampus during quiet waking and slow wave sleep could directly arise from the change in dynamics caused by a drop in acetylcholine levels during these behavioral states. As summarized in Fig. 7, lower levels of acetylcholine would release glutamatergic feedback synapses from the cholinergic suppression described above (Hounsgaard, 1978; Valentino and Dingledine, 1981; Rovira et al., 1982; Hasselmo and Schnell,

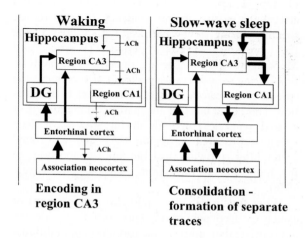

Fig. 7. Schematic of cholinergic modulation of hippocampal dynamics during active waking and slow wave sleep. Left: During active waking, high levels of acetylcholine set appropriate dynamics for encoding. Sensory information from neocortical structures flows through the entorhinal cortex and dentate gyrus (DG) into hippocampal region CA3, where cholinergic enhancement of synaptic modification helps in formation of an intermediate term representation binding together different elements of an episodic memory. Feedback connections to region CA1, entorhinal cortex and association cortex are strong enough to mediate immediate retrieval, but cholinergic suppression of these connections (ACh) prevents them from dominating over the feedforward connectivity. Right: During quiet waking or slow wave sleep, much lower levels of acetylcholine release the suppression of excitatory feedback. This strong excitatory feedback mediates reactivation of memories stored in region CA3 during EEG phenomena termed sharp waves. These waves of activity flow back through region CA1 to entorhinal cortex. This will enable the slow consolidation of long-term episodic memory in hippocampal region CA1, entorhinal cortex and association neocortex, and may underlie modification of semantic memory within circuits of association neocortex.

1994), resulting in strong excitatory feedback. Slow-wave sleep would be characterized by a great increase in the effect of excitatory recurrent connections in region CA3 and excitatory feedback connections from CA3 to CA1 and entorhinal cortex. This drop in cholinergic modulation could thereby underlie the increase in sharp wave activity observed during slow wave sleep (Buzsaki, 1989; Chrobak and Buzsaki, 1994). In fact, muscarinic antagonists such as atropine put the hippocampus into a sharp wave state (Buzsaki, 1986). In addition, this spread of activity should be influenced by synaptic modification during the previous waking period. Thus, the release of suppression of excitatory transmission could contribute to the greater tendency of cells to fire together during slow-wave sleep if they fired during the previous waking period (Wilson and McNaughton, 1994). The loss of cholinergic modulation during slow-wave sleep should also enhance the spread of excitatory activity in response to stimulation. This could underlie the increase in magnitude of evoked synaptic potentials during slow-wave sleep which is observed in region CA1 and entorhinal cortex after stimulation of the input connections to the hippocampal formation (Winson and Abzug, 1978).

What functional role could this enhancement of excitatory feedback have? This would provide the appropriate dynamics for the formation of additional traces within region CA3 and region CA1, and could allow the hippocampus to further strengthen internal connections and 'train' the entorhinal cortex or association neocortex on the basis of previously encoded associations (Buzsaki, 1989; Wilson and McNaughton, 1994; Hasselmo et al., 1996; Hasselmo, 1999). As shown in Fig. 7, the spontaneous reactivation of neurons coding an association in the hippocampus would then be able to drive cells in entorhinal cortex and neocortex without any assistance from sensory input. The reduction of cholinergic suppression might provide the opportunity for this strong feedback influence. The physiological activity during slow wave sleep has been proposed to be appropriate for modification of synaptic components (Trepel and Racine, 1998). Behavioral data suggests that slow-wave sleep may be important for the declarative component of behavioral tasks which correspond most closely to episodic memories.

Subjects are better at retrieval of word lists if they learn the list before falling asleep and are tested on retrieval after being awakened in the middle of the night, than if they learn the list after many hours sleep and are tested in the morning (Stickgold, 1998).

What is the functional purpose of suppressing feedback to entorhinal cortex during active waking? To begin with, this suppression should not be total, as recently stored memories from the hippocampus are still accessible for retrieval. But the strength of connections necessary for the strong transmission of stored memories back to region CA1 and entorhinal cortex would allow them to dominate over afferent input. This could distort the initial perception of sensory information, causing interference during learning in temporal structures—and if the retrieval activity is sufficiently dominant—causing hallucinations such as those observed under the influence of cholinergic antagonists at high doses (Perry and Perry, 1995). Thus, partial cholinergic suppression of excitatory feedback might allow cued retrieval without hallucinatory retrieval.

Concluding remarks

Acetylcholine has a number of different physiological effects on cortical circuits which often appear inconsistent. Computational modeling provides a unifying theoretical framework for understanding these different physiological effects, as summarized in this paper. Modeling demonstrates that the combined physiological effects of acetylcholine serve to enhance the influence of afferent input on neuronal spiking activity, while reducing the influence of internal and feedback processing. Computational models demonstrate how these network properties can be interpreted functionally as both enhancing attention to sensory stimuli and enhancing the encoding of new memories. The levels of acetylcholine in the hippocampus and neocortex change dramatically during different stages of waking and sleep. High levels of acetylcholine during active waking may set appropriate dynamics for attention to sensory input or encoding of new information. At the same time, the cholinergic suppression of excitatory feedback connections prevents interference from internal processing of previously stored information. Lower levels of acetylcholine during quiet waking and slow wave sleep may provide a release from this suppression of excitatory feedback, allowing stronger spread of activity within the hippocampus and from hippocampus to entorhinal cortex, thereby facilitating the process of consolidation of separate memory traces.

References

Aigner, T.G. and Mishkin, M. (1986) The effects of physostigmine and scopolamine on recognition memory in monkeys. Behav. Neurosci., 45: 81–87.

Aigner, T.G., Walker, D.L. and Mishkin, M. (1991) Comparison of the effects of scopolamine administered before and after acquisition in a test of visual recognition memory in monkeys. Behav. Neural Biol., 55: 61–67.

Alkondon, M. and Albuquerque, E.X. (2001) Nicotinic acetylcholine receptor alpha7 and alpha4beta2 subtypes differentially control GABAergic input to CA1 neurons in rat hippocampus. J. Neurophysiol., 86: 3043–3055.

Alkondon, M. and Albuquerque, E.X. (2002) A non-alpha7 nicotinic acetylcholine receptor modulates excitatory input to hippocampal CA1 interneurons. J. Neurophysiol., 87: 1651–1654.

Albuquerque, E.X., Pereira, E.F., Mike, A., Eisenberg, H.M., Maelicke, A. and Alkondon, M. (2000) Neuronal nicotinic receptors in synaptic functions in humans and rats: physiological and clinical relevance. Behav. Brain Res., 113: 131–141.

Allen, M.T., Padilla, Y. and Gluck, M.A. (2002) Ibotenic acid lesions of the medial septum retard delay eyeblink conditioning in rabbits (Oryctolagus cuniculus). Behav. Neurosci., 116: 733–738.

Anagnostaras, S.G., Maren, S. and Fanselow, M.S. (1995) Scopolamine selectively disrupts the acquisition of contextual fear conditioning in rats. Neurobiol Learn Mem., 64: 191–194.

Anagnostaras, S.G., Maren, S., Sage, J.R., Goodrich, S. and Fanselow, M.S. (1999) Scopolamine and Pavlovian fear conditioning in rats: dose-effect analysis. Neuropsychopharmacology, 21: 731–744.

Arnold, H.M., Burk, J.A., Hodgson, E.M., Sarter, M. and Bruno, J.P. (2002) Differential cortical acetylcholine release in rats performing a sustained attention task versus behavioral control tasks that do not explicitly tax attention. Neurosci., 114: 451–460.

Atri, A., Sherman, S.J., Norman, K.A., Kirchhoff, B.A., Nicolas, M.M., Greicius, M.D., Cramer, S.C., Breiter, H.C., Hasselmo, M.E. and Stern, C.E. (2003) Blockade of central cholinergic receptors impairs new learning and increases proactive interference in a word paired-associate memory task. Behavioral Neurosci., submitted.

Barkai, E. and Hasselmo, M.E. (1994) Modulation of the input/output function of rat piriform cortex pyramidal cells. J. Neurophysiol., 72: 644–658.

Bartus, R.T. and Johnson, H.R. (1976) Short term memory in the rhesus monkey: disruption from the anticholinergic scopolamine. Pharmacol. Biochem. Behav., 5: 39–40.

Baxter, M.G., Holland, P.C. and Gallagher, M. (1997) Disruption of decrements in conditioned stimulus processing by selective removal of hippocampal cholinergic input. J. Neurosci., 17: 5230–5236.

Beatty, W.W., Butters, N. and Janowsky, D.S. (1986) Patterns of memory failure after scopolamine treatment: Implications for cholinergic hypotheses of dementia. Behav. Neural Biol., 45: 196–211.

Behrends, J.C. and ten Bruggencate, G. (1993) Cholinergic modulation of synaptic inhibition in the guinea pig hippocampus in vitro: excitation of GABAergic interneurons and inhibition of GABA release. J. Neurophysiol., 69: 626–629.

Benardo, L.S. and Prince, D.A. (1982) Ionic mechanisms of cholinergic excitation in mammalian hippocampal pyramidal cells. Brain Res., 249: 333–344.

Berry, S.D. and Seager, M.A. (2001) Hippocampal theta oscillations and classical conditioning. Neurobiol. Learn Mem., 76: 298–313.

Berry, S.D. and Thompson, R.F. (1979) Medial septal lesions retard classical conditioning of the nicitating membrane response in rabbits. Science, 205: 209–211.

Bland, B.H. and Colom, L.V. (1993) Extrinsic and intrinsic properties underlying oscillation and synchrony in limbic cortex. Prog. Neurobiol., 41: 157–208.

Blitzer, R.D., Gil, O. and Landau, E.M. (1990) Cholinergic stimulation enhances long-term potentiation in the CA1 region of rat hippocampus. Neurosci. Lett., 119: 207–210.

Bolhuis, J.J., Strijkstra, A.M. and Kramers, R.J. (1988) Effects of scopolamine on performance of rats in a delayed-response radial maze task. Physiol. Behav., 43: 403–409.

Book, A.A., Wiley, R.G. and Schweitzer, J.B. (1994) 192 IgG-saporin: I. specific lethality for cholinergic neurons in the basal forebrain of the rat. J. Neuropathol. Exp. Neurol., 53: 95–102.

Brocher, S., Artola, A. and Singer, W. (1992) Agonists of cholinergic and noradrenergic receptors facilitate synergistically the induction of long-term potentiation in slices of rat visual cortex. Brain Res., 573: 27–36.

Broks, P., Preston, G.C., Traub, M., Poppleton, P., Ward, C. and Stahl, S.M. (1988) Modelling dementia: Effects of scopolamine on memory and attention. Neuropsychologia, 26: 685–700.

Bucci, D.J., Holland, P.C. and Gallagher, M. (1998) Removal of cholinergic input to rat posterior parietal cortex disrupts incremental processing of conditioned stimuli. J. Neurosci., 18: 8038–8046.

Buresova, O., Bolhuis, J.J. and Bures, J. (1986) Differential effects of cholinergic blockade on performance of rats in the water tank navigation task and in a radial water maze. Behav. Neurosci., 100: 476–482.

Burgard, E.C. and Sarvey, J.M. (1990) Muscarinic receptor activation facilitates the induction of long-term potentiation (LTP) in the rat dentate gyrus. Neurosci. Lett., 116: 34–39.

Buzsaki, G. (1986) Hippocampal sharp waves: their origin and significance. Brain Res., 398: 242–252.

Buzsaki, G. (1989) Two-stage model of memory trace formation: a role for 'noisy' brain states. Neuroscience, 31: 551–570.

Caine, E.D., Weingartner, H., Ludlow, C.L., Cudahy, E.A. and Wehry, S. (1981) Qualitative analysis of scopolamine-induced amnesia. Psychopharm., 74: 74–80.

Calderon, J., Perry, R.J., Erzinclioglu, S.W., Berrios, G.E., Dening, T.R. and Hodges, J.R. (2001) Perception, attention and working memory are disproportionately impaired in dementia with Lewy bodies compared with Alzheimer's disease. J. Neurol. Neurosurg. Psychiatry, 70: 157–164.

Cassel, J.C. and Kelche, C. (1989) Scopolamine treatment and fimbria-fornix lesions: mimetic effects on radial maze performance. Physiol Behav., 46: 347–353.

Chapman, C.A. and Lacaille, J.C. (1999) Intrinsic theta-frequency membrane potential oscillations in hippocampal CA1 interneurons of stratum lacunosum-moleculare. J. Neurophysiol., 81: 1296–1307.

Chiba, A.A., Bucci, D.J., Holland, P.C. and Gallagher, M. (1995) Basal forebrain cholinergic lesions disrupt increments but not decrements in conditioned stimulus processing, J. Neurosci., 15: 7315–7322.

Chrobak, J.J. and Buzsaki, G. (1994) Selective activation of deep layer (V-VI) retrohippocampal cortical neurons during hippocampal sharp waves in the behaving rat. J. Neurosci, 14: 6160–6170.

Cole, A.E. and Nicoll, R.A. (1984) Characterization of a slow cholinergic postsynaptic potential recorded in vitro from rat hippocampal pyramidal cells. J. Physiol. (London), 352: 173–188.

Constanti, A. and Galvan, M. (1983) M-current in voltage clamped olfactory cortex neurones. Neurosci. Lett., 39: 65–70.

Constanti, A. and Sim, J.A. (1987) Calcium-dependent potassium conductance in guinea-pig olfactory cortex neurons in vitro. J. Physiol., 387: 173–194.

Crow, T.J. and Grove-White, I.G. (1973) An analysis of the learning deficit following hyoscine administration to man. Br. J. Pharmacol., 49: 322–327.

Curro Dossi, R., Pare, D. and Steriade, M. (1992) Various types of inhibitory postsynaptic potentials in anterior thalamic cells are differentially altered by stimulation of laterodorsal tegmental cholinergic nucleus. Neuroscience, 47: 279–289.

Dalley, J.W., McGaughy, J., O'Connell, M.T., Cardinal, R.N., Levita, L. and Robbins, T.W. (2001) Distinct changes in cortical acetylcholine and noradrenaline efflux during contingent and non-contingent performance of a visual attentional task. J. Neurosci., 21: 4908–4914.

DeRosa, E. and Hasselmo, M.E. (2000) Muscarinic cholinergic neuromodulation reduces proactive interference between stored odor memories during associative learning in rats. Behav. Neurosci., 114: 32–41.

DeRosa, E., Hasselmo, M.E. and Baxter, M.G. (2001) Contribution of the cholinergic basal forebrain to proactive interference from stored odor memories during associative learning in rats. Behav. Neurosci., 115: 314–327.

Descarries, L., Gisiger, V. and Steriade, M. (1997) Diffuse transmission by acetylcholine in the CNS. Prog. Neurobiol., 53: 603–625.

Drachman, D.A. and Leavitt, J. (1974) Human memory and the cholinergic system. Arch. Neurol., 30: 113–121.

Dunnett, S.B., Everitt, B.J. and Robbins, T.W. (1991) The basal forebrain-cortical cholinergic system: interpreting the functional consequences of excitotoxic lesions. Trends Neurosci., 14: 494–501.

Dutar, P. and Nicoll, R.A. (1988) Classification of muscarinic responses in hippocampus in terms of receptor subtypes and 2nd-messenger systems—electrophysiological studies in vitro. J. Neurosci., 8: 4214–4224.

Dykes, R.W. (1997) Mechanisms controlling neuronal plasticity in somatosensory cortex. Can. J. Physiol. Pharmacol., 75: 535–545.

Egorov, A.V., Hamam, B.N., Fransen, E., Hasselmo, M.E. and Alonso, A.A. (2002) Graded persistent activity in entorhinal cortex neurons. Nature, 420: 173–178.

Fransen, E., Alonso, A.A. and Hasselmo, M.E. (2002) Simulations of the role of the muscarinic-activated calcium-sensitive non-specific cation current I(NCM) in entorhinal neuronal activity during delayed matching tasks. J. Neurosci., 22: 1081–1097.

Frazier, C.J., Rollins, Y.D., Breese, C.R., Leonard, S., Freedman, R. and Dunwiddie, T.V. (1998) Acetylcholine activates an alpha-bungarotoxin-sensitive nicotinic current in rat hippocampal interneurons, but not pyramidal cells. J. Neurosci., 18: 1187–1195.

Gale, G.D., Anagnostaras, S.G. and Fanselow, M.S. (2001) Cholinergic modulation of pavlovian fear conditioning: effects of intrahippocampal scopolamine infusion. Hippocampus, 11: 371–376.

Ghonheim, M.M. and Mewaldt, S.P. (1975) Effects of diazepan and scopolamine on storage, retrieval, and organisational processes in memory. Psychopharm., 44: 257–262.

Ghonheim, M.M. and Mewaldt, S.P. (1977) Studies on human memory: The interactions of diazepam, scopolamine and physostigmine. Psychopharmacol., 52: 1–6.

Gil, Z., Connors, B.W. and Amitai, Y. (1997) Differential regulation of neocortical synapses by neuromodulators and activity. Neuron, 19: 679–686.

Gill, T.M., Sarter, M. and Givens, B. (2000) Sustained visual attention performance-associated prefrontal neuronal activity evidence for cholinergic modulation. J. Neurosci., 20: 4745–4757.

Gioanni, Y., Rougeot, C., Clarke, P.B., Lepouse, C., Thierry, A.M. and Vidal, C. (1999) Nicotinic receptors in the rat prefrontal cortex: increase in glutamate release and facilitation of mediodorsal thalamo-cortical transmission. Eur. J. Neurosci., 11: 18–30.

Giovannini, M.G., Rakovska, A., Benton, R.S., Pazzagli, M., Bianchi, L. and Pepeu, G. (2001) Effects of novelty and habituation on acetylcholine, GABA, and glutamate release from the frontal cortex and hippocampus of freely moving rats. Neuroscience, 106: 43–53.

Greenwood, P.M., Parasuraman, R. and Alexander, G.E. (1997) Controlling the focus of spatial attention during visual search: effects of advanced aging and Alzheimer's disease. Neuropsychology, 11: 3–12.

Hagan, J.J. and Morris, R.G.M. (1989). The cholinergic hypothesis of memory: A review of animal experiments. In: Iversen L.L., Iversen S.D. and Snyder S.H. (Eds.), Psychopharmacology of the Aging Nervous System. Plenum Press, New York, pp. 237–324.

Hasselmo, M.E. (1993) Acetylcholine and learning in a cortical associative memory. Neural Comp., 5: 32–44.

Hasselmo, M.E. (1995) Neuromodulation and cortical function: Modeling the physiological basis of behavior. Behav. Brain Res., 67: 1–27.

Hasselmo, M.E. (1999) Neuromodulation: Acetylcholine and memory consolidation. Trends Cognitive Sciences, 3: 351–359.

Hasselmo, M.E. and Barkai, E. (1995) Cholinergic modulation of activity-dependent synaptic plasticity in rat piriform cortex. J. Neurosci., 15: 6592–6604.

Hasselmo, M.E. and Bower, J.M. (1992) Cholinergic suppression specific to intrinsic not afferent fiber synapses in rat piriform (olfactory) cortex. J. Neurophysiol., 67: 1222–1229.

Hasselmo, M.E. and Bower, J.M. (1993) Acetylcholine and memory. Trends Neurosci., 16: 218–222.

Hasselmo, M.E. and Cekic, M. (1996) Suppression of synaptic transmission may allow combination of associative feedback and self-organizing feedforward connections in the neocortex. Behav. Brain Res., 79: 153–161.

Hasselmo, M.E. and Fehlau, B.P. (2001) Differences in time course of cholinergic and GABAergic modulation of excitatory synaptic potentials in rat hippocampal slice preparations. J. Neurophysiol., 86: 1792–1802.

Hasselmo, M.E. and Schnell, E. (1994) Laminar selectivity of the cholinergic suppression of synaptic transmission in rat hippocampal region CA1: computational modeling and brain slice physiology. J. Neurosci., 14: 3898–3914.

Hasselmo, M.E. and Wyble, B.P. (1997) Free recall and recognition in a network model of the hippocampus: simulating effects of scopolamine on human memory function. Behav. Brain Res., 89: 1–34.

Hasselmo, M.E., Anderson, B.P. and Bower, J.M. (1992) Cholinergic modulation of cortical associative memory function. J. Neurophysiol., 67: 1230–1246.

Hasselmo, M.E., Schnell, E. and Barkai, E. (1995) Dynamics of learning and recall at excitatory recurrent synapses and cholinergic modulation in hippocampal region CA3. J. Neurosci., 15: 5249–5262.

Hasselmo, M.E., Wyble, B.P. and Wallenstein, G.V. (1996) Encoding and retrieval of episodic memories: Role of cholinergic and GABAergic modulation in the hippocampus. Hippocampus, 6: 693–708.

Heckers, S., Ohtake, T., Wiley, R.G., Lappi, D.A., Geula, C. and Mesulam, M.M. (1994) Complete and selective cholinergic denervation of the rat neocortex and hippocampus but not amygdala by an immunotoxin against the p75 receptor. J. Neurosci., 14: 1271–1289.

Herreras, O., Solis, J.M., Munoz, M.D., Martin del Rio, R. and Lerma, J. (1988a) Sensory modulation of hippocampal transmission. I. Opposite effects on CA1 and dentate gyrus synapsis. Brain Res., 461: 290–302.

Herreras, O., Solis, J.M., Herranz, A.S., Martin del Rio, R. and Lerma, J. (1988b) Sensory modulation of hippocampal transmission. II. Evidence for a cholinergic locus of inhibition in the Schaffer-CA1 synapse. Brain Res., 461: 303–313.

Himmelheber, A.M., Sarter, M. and Bruno, J.P. (1997) Operant performance and cortical acetylcholine release: role of response rate, reward density, and non-contingent stimuli. Cognitive Brain Res., 6: 23–36.

Himmelheber, A., Sarter, M. and Bruno, J. (2000) Increases in cortical acetylcholine release during sustained attention performance in rats. Cognitive Brain Research, 9: 313–325.

Himmelheber, A.M., Sarter, M. and Bruno, J.P. (2001) The effects of manipulations of attentional demand on cortical acetylcholine release. Cognitive Brain Res., 12: 353–370.

Holley, L.A., Turchi, J., Apple, C. and Sarter, M. (1995) Dissociation between the attentional effects of infusions of a benzodiazepine receptor agonist and an inverse agonist into the basal forebrain. Psychopharmacology, 120: 99–108.

Hounsgaard, J. (1978) Presynaptic inhibitory action of acetylcholine in area CA1 of the hippocampus. Exp. Neurol., 62: 787–797.

Hsieh, C.Y., Cruikshank, S.J. and Metherate, R. (2000) Differential modulation of auditory thalamocortical and intracortical synaptic transmission by cholinergic agonist. Brain Res., 880: 51–64.

Huerta, P.T. and Lisman, J.E. (1993) Heightened synaptic plasticity of hippocampal CA1 neurons during a cholinergically induced rhythmic state. Nature, 364: 723–725.

Jasper, H.H. and Tessier, J. (1971) Acetylcholine liberation from cerebral cortex during paradoxical (REM) sleep. Science, 172: 601–602.

Kahle, J.S. and Cotman, C.W. (1989) Carbachol depresses the synaptic responses in the medial but not the lateral perforant path. Brain Res., 482: 159–163.

Kametani, H. and Kawamura, H. (1990) Alterations in acetylcholine release in the rat hippocampus during sleep-wakefulness detected by intracerebral dialysis. Life Sci., 47: 421–426.

Kimura, F. (2000) Cholinergic modulation of cortical function: a hypothetical role in shifting the dynamics in cortical network. Neurosci Res., 38(1): 19–26.

Kimura, F., Fukuda, M. and Tsumoto, T. (1999) Acetylcholine suppresses the spread of excitation in the visual cortex revealed by optical recording: possible differential effect depending on the source of input. Eur. J. Neurosci., 11(10): 3597–3609.

Kirchhoff, B.A., Hasselmo, M.E., Norman, K.A., Nicolas, M.M., Greicius, M.D., Breiter, H.C. and Stern, C.E. (2000) Effect of cholinergic blockade on paired associate learning in humans. Soc. Neurosci. Abstr., 26: 263.18.

228

Klink, R. and Alonso, A. (1997a) Muscarinic modulation of the oscillatory and repetitive firing properties of entorhinal cortex layer II neurons. J. Neurophysiol., 77: 1813–1828.

Klink, R. and Alonso, A. (1997b) Ionic mechanisms of muscarinic depolarization in entorhinal cortex layer II neurons. J. Neurophysiol., 77: 1829–1843.

Kramis, R., Vanderwolf, C.H. and Bland, B.H. (1975) Two types of hippocampal rhythmical slow activity in both the rabbit and the rat: relations to behavior and effects of atropine, diethyl ether, urethane, and pentobarbital. Exp. Neurol., 49(1 Pt 1): 58–85.

Krnjević, K. (1984). Neurotransmitters in cerebral cortex: A general account. In: E.G. Jones and A. Peters (Eds.), Cerebral Cortex. New York, Plenum, pp. 39–61.

Krnjević, K. and Phillis, J.W. (1963) Acetylcholine-sensitive cells in the cerebral cortex. J. Physiol., 166: 296–327.

Krnjević, K., Pumain, R. and Renaud, L. (1971) The mechanism of excitation by acetylcholine in the cerebral cortex. J. Physiol., 215: 247–268.

Linster, C. and Hasselmo, M.E. (2000) Neural activity in the horizontal limb of the diagonal band of broca can be modulated by electrical stimulation of the olfactory bulb and cortex in rats. Neurosci Lett., 282: 157–160.

Linster, C. and Hasselmo, M.E. (2001) Neuromodulation and the functional dynamics of piriform cortex. Chem. Senses, 26: 585–594.

Linster, C., Garcia, P.A., Hasselmo, M.E. and Baxter, M.G. (2001) Selective loss of cholinergic neurons projecting to the olfactory system increases perceptual generalization between similar, but not dissimilar, odorants. Behav Neurosci., 115: 826–833.

Linster, C., Wyble, B.P. and Hasselmo, M.E. (1999) Electrical stimulation of the horizontal limb of the diagonal band of Broca modulates population EPSP's in piriform cortex. J. Neurophysiol., 81: 2737–2742.

Madison, D.V., Lancaster, B. and Nicoll, R.A. (1987) Voltage clamp analysis of cholinergic action in the hippocampus. J. Neurosci., 7: 733–741.

Madison, D.V. and Nicoll, R.A. (1984) Control of the repetitive discharge of rat CA1 pyramidal neurones in vitro. J. Physiol., 354: 319–331.

Maloney, K.J., Cape, E.G., Gotman, J. and Jones, B.E. (1997) High frequency gamma electroencephalogram activity in association with sleep-wake states and spontaneous behaviors in the rat. Neuroscience, 76: 541–555.

Marrosu, F., Portas, C., Mascia, M.S., Casu, M.A., Fa, M., Giagheddu, M., Imperato, A. and Gessa, G.L. (1995) Microdialysis measurement of cortical and hippocampal acetylcholine release during sleep-wake cycle in freely moving cats. Brain Res., 671: 329–332.

McCormick, D.A. and Prince, D.A. (1986) Mechanisms of action of acetylcholine in the guinea-pig cerebral cortex in vitro. J. Physiol., 375: 169–194.

McGaugh, J.L. (1989) Involvement of hormonal and neuromodulatory systems in the regulation of memory storage. Ann. Rev. Neurosci., 12: 255–287.

McGaughy, J. and Sarter, M. (1995) Behavioral vigilance in rats: task validation and effects of age, amphetamine and benzodiazepine receptor ligands. Psychopharmacology, 117: 340–357.

McGaughy, J. and Sarter, M. (1998) Sustained attention performance in rats with intracortical infusions of 192 IgG-saporin-induced cortical cholinergic deafferentation: effects of physostigmine and FG 7142. Behavioral Neurosci., 110: 247–265.

McGaughy, J., Turchi, J. and Sarter, M. (1994) Crossmodal divided attention in rats: effects of chlordiazepoxide and scopolamine. Psychopharmacology, 115: 213–220.

McGaughy, J., Kaiser, T. and Sarter, M. (1996) Behavioral vigilance following infusions of 192 IgG-saporin into the basal forebrain: selectivity of the behavioral impairment and relation to cortical AChE-positive fiber density. Behav. Neurosci., 110: 247–265.

McGaughy, J., Sandstrom, M., Ruland, S., Bruno, J.P. and Sarter, M. (1997) Lack of effects of lesions of the dorsal noradrenergic bundle on behavioral vigilance. Behavioral Neuroscience, 111: 646–652.

McGaughy, J., Decker, M.W. and Sarter, M. (1999) Enhancement of sustained attention performance by the nicotinic receptor agonist ABT-418 in intact but not basal forebrain-lesioned rats. Psychopharmacology, 144: 175–182.

McGaughy, J., Everitt, B.J., Robbins, T.W. and Sarter, M. (2000) The role of cholinergic afferent projections in cognition: impact of new selective immunotoxins. Behav. Brain Res., 115: 251–263.

McGaughy, J., Dalley, J.W., Morrison, C.H., Everitt, B.J. and Robbins, T.W. (2002) Selective behavioral and neurochemical effects of cholinergic lesions produced by intrabasalis infusions of 192 IgG-saporin on attentional performance in a 5 choice serial reaction time task. J. Neurosci., 22: 1905–1913.

McQuiston, A.R. and Madison, D.V. (1999a) Nicotinic receptor activation excites distinct subtypes of interneurons in the rat hippocampus. J. Neurosci., 19: 2887–2896.

McQuiston, A.R. and Madison, D.V. (1999b) Muscarinic receptor activity has multiple effects on the resting membrane potentials of CA1 hippocampal interneurons. J. Neurosci., 19: 5693–5702.

Mesulam, M.-M., Mufson, E.J., Wainer, B.H. and Levey, A.I. (1983a) Central cholinergic pathways in the rat: An overview based on an alternative nomenclature (Ch1-Ch6). Neurosci., 10: 1185–1201.

Mesulam, M.M., Mufson, E.J., Levey, A.I. and Wainer, B.H. (1983b) Cholinergic innervation of cortex by the basal forebrain: cytochemistry and cortical connections of the septal area, diagonal band nuclei, nucleus basalis (substantia innominata and hypothalamus) in the rhesus monkey. J. Comp. Neurol., 214: 170–197.

Metherate, R. and Ashe, J.H. (1993) Nucleus basalis stimulation facilitates thalamocortical synaptic transmission in the rat auditory cortex. Synapse, 14(2): 132–143.

Metherate, R. and Weinberger, N.M. (1990) Cholinergic modulation of responses to single tones produces tone-specific receptive-field alterations in cat auditory-cortex. Synapse, 6: 133–145.

Metherate, R., Ashe, J.H. and Weinberger, N.M. (1990) Acetylcholine modifies neuronal acoustic rate level functions in guinea pig auditory cortex by an action at muscarinic receptors. Synapse, 6: 364–368.

M'Harzi, M., Palacios, A., Monmaur, P., Willig, F., Houcine, O. and Delacour, J. (1987) Effects of selective lesions of fimbria-fornix on learning set in the rat. Physiol. Behav., 40: 181–188.

Mizumori, S.J., McNaughton, B.L. and Barnes, C.A. (1989) A comparison of supramammillary and medial septal influences on hippocampal field potentials and single-unit activity. J. Neurophysiol., 61(1): 15–31.

Muir, J.L., Everitt, B.J. and Robbins, T.W. (1994) AMPA-induced excitotoxic lesions of the basal forebrain: a significant role for the cortical cholinergic system in attentional function. J. Neurosci., 14: 2313–2326.

Muir, J.L., Everitt, B.J. and Robbins, T.W. (1995) Reversal of visual attentional dysfunction following lesions of the cholinergic basal forebrain by physostigmine and nicotine and not by the 5-HT3 receptor antagonist, ondansetron. Psychopharmacology, 118: 82–92.

Muir, J.L., Bussey, T.J., Everitt, B.J. and Robbins, T.W. (1996) Dissociable effects of AMPA-induced lesions of the vertical limb diagonal band of Broca on performance of the 5-Choice serial reaction time task and on acquisition of a conditional visual discrimination. Behav. Brain Res., 82: 31–44.

Murakoshi, T. (1995) Cholinergic modulation of synaptic transmission in the rat visual cortex in vitro. Vision Res., 35(1): 25–35.

Murphy, P.C. and Sillito, A.M. (1991) Cholinergic enhancement of direction selectivity in the visual cortex of the cat. Neuroscience, 40(1): 13–20.

Myers, C.E., Ermita, B.R., Hasselmo, M. and Gluck, M.A. (1998) Further implications of a computational model of septohippocampal cholinergic modulation in eyeblink conditioning. Psychobiol., 26: 1–20.

Numan, R., Feloney, M.P., Pham, K.H. and Tieber, L.M. (1995) Effects of medial septal lesions on an operant go/no-go delayed response alternation task in rats. Physiol. Behav., 58: 1263–1271.

Parasuraman, R. and Giambra, L. (1991) Skill development in vigilance effects of event rate and age. Psychol. Aging, 6: 155–169.

Parasuraman, R. and Haxby, J.V. (1993) Attention and brain function in Alzheimer's disease: a review. Neuropsychology, 7: 242–272.

Parasuraman, R., Greenwood, P.M., Haxby, J.V. and Grady, C.L. (1992) Visuospatial attention in dementia of the Alzheimer's type. Brain, 115: 711–733.

Pare, D., Steriade, M., Deschenes, M. and Bouhassira, D. (1990) Prolonged enhancement of anterior thalamic synaptic responsiveness by stimulation of a brain-stem cholinergic group. J. Neurosci., 10: 20–33.

Passetti, F., Dalley, J.W., O'Connell, M.T., Everitt, B.J. and Robbins, T.W. (2000) Increased acetylcholine release in the rat medial prefrontal cortex during performance of a visual attentional task. Eur. J. Neurosci., 12: 3051–3058.

Patil, M.M. and Hasselmo, M.E. (1999) Modulation of inhibitory synaptic potentials in the piriform cortex. J. Neurophysiol., 81: 2103–2118.

Patil, M.M., Linster, C., Lubenov, E. and Hasselmo, M.E. (1998) Cholinergic agonist carbachol enables associative long-term potentiation in piriform cortex slices. J. Neurophysiol., 80: 2467–2474.

Penetar, D.M. and McDonough, J.H. (1983) Effects of cholinergic drugs on delayed match to sample performance of Rhesus monkeys. Pharmacol. Biochem. Behav., 19: 963–967.

Perry, E.K. and Perry, R.H. (1995) Acetylcholine and hallucinations: disease-related compared to drug-induced alterations in human consciousness. Brain Cogn., 28: 240–258.

Peterson, R.C. (1977) Scopolamine-induced learning failures in man. Psychopharmacologia, 52: 283–289.

Pitler, T.A. and Alger, B.E. (1992) Cholinergic excitation of GABAergic interneurons in the rat hippocampal slice. J. Physiol., 450: 127–142.

Porter, J.T., Cauli, B., Tsuzuki, K., Lambolez, B., Rossier, J. and Audinat, E. (1999) Selective excitation of subtypes of neocortical interneurons by nicotinic receptors. J. Neurosci., 19: 5228–5235.

Rovira, C., Cherubini, E. and Ben-Ari, Y. (1982) Opposite actions of muscarinic and nicotinic agents on hippocampal dendritic negative fields recorded in rats. Neuropharmacology, 21(9): 933–936.

Rovira, C., Ben-Ari, Y., Cherubini, E., Krnjević, K. and Ropert, N. (1983) Pharmacology of the dendritic action of acetylcholine and further observations on the somatic disinhibition in the rat hippocampus in situ. Neuroscience, 8: 97–106.

Salvatierra, A.T. and Berry, S.D. (1989) Scopolamine disruption of septo-hippocampal activity and classical conditioning. Behav Neurosci., 103(4): 715–721.

Sarter, M. and Bruno, J.P. (2000) Cortical cholinergic inputs mediating arousal, attentional processing and dreaming: differential afferent regulation of the basal forebrain by telecephalic and brainstem afferents. Neuroscience, 95: 933–952.

Sarter, M., Givens, B. and Bruno, J.P. (2001) The cognitive neuroscience of sustained attention: where top-down meets bottom-up. Brain Res. Rev., 35: 146–160.

Sato, H., Hata, Y., Masui, H. and Tsumoto, T. (1987) A functional role of cholinergic innervation to neurons in the cat visual cortex. J. Neurophysiol., 58: 765–780.

Schwindt, P.C., Spain, W.J., Foehring, R.C., Stafstrom, C.E., Chubb, M.C. and Crill, W.E. (1988) Slow conductances in neurons from cat sensorimotor cortex and their role in slow excitability changes. J. Neurophysiol., 59: 450–467.

Seager, M.A., Asaka, Y. and Berry, S.D. (1999) Scopolamine disruption of behavioral and hippocampal responses in appetitive trace classical conditioning. Behav. Brain Res., 100: 143–151.

230

Seager, M.A., Johnson, L.D., Chabot, E.S., Asaka, Y. and Berry, S.D. (2002) Oscillatory brain states and learning: Impact of hippocampal theta-contingent training. Proc. Natl. Acad. Sci. USA, 99: 1616–1620.

Shalinsky, M.H., Magistretti, J., Ma, L. and Alonso, A.A. (2002) Muscarinic activation of a cation current and associated current noise in entorhinal-cortex layer-II neurons. J. Neurophysiol., 88(3): 1197–1211.

Sherman, S.J., Atri, A., Hasselmo, M.E., Stern, C.E. and Howard, M.W. (2003) Scopolamine impairs human recognition memory: Data and modeling, Behavioral Neurosci., 117(3): 526–539.

Sillito, A.M. and Kemp, J.A. (1983) Cholinergic modulation of the functional organization of the cat visual cortex. Brain Res., 289: 143–155.

Solomon, P.R., Solomon, S.D., Schaaf, E.V. and Perry, H.E. (1983) Altered activity in the hippocampus is more detrimental to classical conditioning than removing the structure. Science, 220(4594): 329–331.

Solomon, P.R., Groccia-Ellison, M.E., Flynn, D., Mirak, J., Edwards, K.R., Dunehew, A. and Stanton, M.E. (1993) Disruption of human eyeblink conditioning after central cholinergic blockade with scopolamine. Behav. Neurosci., 107(2): 271–279.

Steriade, M. (1994) Sleep oscillations and their blockage by activating systems. J. Psychiatry Neurosci., 19: 354–358.

Steriade, M. (2001) Impact of network activities on neuronal properties in corticothalamic systems. J. Neurophysiol., 86: 1–39.

Steriade, M., Amzica, F. and Contreras, D. (1996) Synchronization of fast (30–40 Hz) spontaneous cortical rhythms during brain activation,. J. Neurosci., 16: 392–417.

Steriade, M. and Deschenes, M. (1974) Inhibitory processes and interneuronal apparatus in motor cortex during sleep and waking. II. Recurrent and afferent inhibition of pyramidal tract neurons. J. Neurophysiol., 37: 1093–1113.

Stickgold, R. (1998) Sleep: Off-line memory reprocessing. Trends Cognit. Sci., 2: 484–492.

Sutherland, R.J., Wishaw, I.Q. and Regehr, J.C. (1982) Cholinergic receptor blockade impairs spatial localization by use of distal cues in the rat. J. Comp. Physiol. Psychol., 96: 563–573.

Suzuki, W.A., Miller, E.K. and Desimone, R. (1997) Object and place memory in the macaque entorhinal cortex. J. Neurophysiol., 78: 1062–1081.

Tang, Y. and Aigner, T.G. (1996) Release of cerebral acetylcholine increases during visually mediated behavior in monkeys. NeuroReport, 7: 2231–2235.

Tang, Y., Mishkin, M. and Aigner, T.G. (1997) Effects of muscarinic blockade in perirhinal cortex during visual recognition. Proc. Natl. Acad. Sci. USA, 94: 12667–12669.

Timofeev, I., Contreras, D. and Steriade, M. (1996) Synaptic responsiveness of cortical and thalamic neurones during various phases of slow sleep oscillation in cat. J. Physiol., 494: 265–278.

Tremblay, N., Warren, R.A. and Dykes, R.W. (1990) Electrophysiological studies of acetylcholine and the role of the basal forebrain in the somatosensory cortex of the cat. II. Cortical neurons excited by somatic stimuli. J. Neurophysiol., 64(4): 1212–1222.

Trepel, C. and Racine, R.J. (1998) Long-term potentiation in the neocortex of the adult, freely moving rat. Cereb. Cortex, 8: 719–729.

Tseng, G.-F. and Haberly, L.B. (1989) Deep neurons in piriform cortex. II. Membrane properties that underlie unusual synaptic responses. J. Neurophysiol., 62(2): 386–400.

Turchi, J. and Sarter, M. (1997) Cortical acetylcholine and processing capacity: effects of cortical cholinergic differentiation on crossmodal divided attention in rats. Cognitive Brain Res., 6: 147–158.

Umbriaco, D., Watkins, K.C., Descarries, L., Cozzari, C. and Hartman, B.K. (1994) Ultrastructural and morphometric features of the acetylcholine innervation in adult rat parietal cortex: an electron microscopic study in serial sections. J. Comp. Neurol., 348: 351–373.

Umbriaco, D., Garcia, S., Beaulieu, C. and Descarries, L. (1995) Relational features of acetylcholine, noradrenaline, serotonin and GABA axon terminals in the stratum radiatum of adult rat hippocampus (CA1). Hippocampus, 5(6): 605–620.

Valentino, R.J. and Dingledine, R. (1981) Presynaptic inhibitory effect of acetylcholine in the hippocampus. J. Neurosci., 1: 784–792.

Vidal, C. and Changeux, J.-P. (1993) Nicotinic and muscarinic modulations of excitatory synaptic transmission in the rat prefrontal cortex in vitro. Neuroscience, 56: 23–32.

Vogt, K.E. and Regehr, W.G. (2001) Cholinergic modulation of excitatory synaptic transmission in the CA3 area of the hippocampus. J. Neurosci., 21(1): 75–83.

Wesnes, K. and Revell, A. (1984) The separate and combined effects of scopolamine and nicotine on human information processing. Psychopharmology, 84: 5–11.

Wesnes, K. and Warburton, D.M. (1983) Effects of scopolamine on stimulus sensitivity and response bias in a visual vigilance task. Neuropsychobiology, 9: 154–157.

Wesnes, K. and Warburton, D.M. (1984) Effects of scopolamine and nicotine on human rapid information processing performance. Psychopharmacology, 82: 147–150.

Whishaw, I.Q. (1985) Cholinergic receptor blockade in the rat impairs locale but not taxon strategies for place navigation in a swimming pool. Behav. Neurosci., 99: 979–1005.

Wiley, R., Berbos, T.G., Deckwerth, T., Johnson, E.M. and Lappi, D.A. (1995) Destruction of the cholinergic basal forebrain using immunotoxin to rat NGF receptor: modeling the cholinergic degeneration in Alzheimer's disease. J. Neurol. Sci., 128: 157–166.

Wiley, R., Oeltmann, T. and Lappi, D. (1991) Immuonlesioning: selective destruction of neurons using immunotinx to rat NGF receptor. Brain Res., 562: 149–153.

Williams, S.H. and Constanti, A. (1988) Quantitative effects of some muscarinic agonists on evoked surface-negative field potentials recorded from the guinea-pig olfactory cortex slice. Br. J. Pharmacol., 93: 846–854.

Wilson, M.A. and McNaughton, B.L. (1994) Reactivation of hippocampal ensemble memories during sleep. Science, 265: 676–679.

Winson, J. and Abzug, C. (1978) Dependence upon behavior of neuronal transmission from perforant pathway through entorhinal cortex. Brain Res., 147: 422–427.

Xiang, Z., Huguenard, J.R. and Prince, D.A. (1998) Cholinergic switching within neocortical inhibitory networks. Science, 281: 985–988.

Xiang, Z., Huguenard, J.R. and Prince, D.A. (2002) Synaptic inhibition of pyramidal cells evoked by different interneuronal subtypes in layer v of rat visual cortex. J. Neurophysiol., 88: 740–750.

Yajeya, J., De La Fuente, A., Criado, J.M., Bajo, V., Sanchez-Riolobos, A. and Heredia, M. (2000) Muscarinic agonist carbachol depresses excitatory synaptic transmission in the rat basolateral amygdala in vitro. Synapse, 38(2): 151–160.

Yamamoto, C. and Kawai, N. (1967) Presynaptic action of acetylcholine in thin sections from the guinea-pig dentate gyrus in vitro. Exp. Neurol., 19: 176–187.

Young, S.L., Bohenek, D.L. and Fanselow, M.S. (1995) Scopolamine impairs acquisition and facilitates consolidation of fear conditioning: differential effects for tone vs context conditioning. Neurobiol. Learn Mem., 63: 174–180.

Young, B.J., Otto, T., Fox, G.D. and Eichenbaum, H. (1997) Memory representation within the parahippocampal region. J. Neurosci., 17(13): 5183–5195.

Clinical, pathological and therapeutic implications

Progress in Brain Research, Vol. 145
ISSN 0079-6123

CHAPTER 16

Knockout and knockin mice to investigate the role of nicotinic receptors in the central nervous system

Nicolas Champtiaux and Jean-Pierre Changeux*

Laboratoire de Neurobiologie Moléculaire, Centre National de la Recherche scientifique, Unité de Recherche Associée 2182 "Récepteurs et Cognition", Institut Pasteur, 75724 Paris Cedex 15, France

Abstract: The recent use of genetically engineered knockout (Ko) and knockin (Kin) animals for neurotransmitter receptor genes, in particular, nicotinic acetylcholine receptors (nAChRs) in the brain, has provided a powerful alternative to the classical pharmacological approach. These animal models are not only useful in order to reexamine and refine the results derived from pharmacological studies, but they do also provide a unique opportunity to determine the subunit composition of the nicotinic receptors which modulate various brain functions. Ultimately, this knowledge will be valuable in the process of designing new drugs that will mimic the effects of nicotine on several important pathologies or on smoking cessation therapies. In this review, we present recent data obtained from the studies of mutant animals that contributed to our understanding of the role and composition of nAChRs in the central nervous system (CNS). The advantages and pitfalls of Ko animal models will also be discussed.

Keywords: addiction; aging; Alzheimer's disease; analgesia; attention; cognition; development; dopaminergic; knockout; knockin; neurodegeneration; nicotinic receptors; reinforcement; schizophrenia

Introduction

Acetylcholine, acting on central nAChRs, is involved in a wide range of brain activities including attentional processes, learning, memory, and even brain development and degeneration (Levin, 1992; Broide and Leslie, 1999). In animals or humans, nicotine exerts diverse cellular and behavioral effects with potential therapeutic applications. Indeed, drugs affecting nicotinic transmission might be beneficial for the treatment of neurodegenerative disorders, schizophrenia and drug addiction. A prerequisite for the development of new drugs with high specificity and limited side effects is the identification of the

different nAChR subtypes that mediate particular effects of nicotine.

nAChRs are cationic ligand-gated ion channels expressed throughout the nervous system as well as in sensory organs, and in muscles (reviewed in Role and Berg, 1996; Corringer et al., 2000). They are pentameric proteins organized around an axis of quasi-symmetry delineating the pore of the channel, with at least two ligand-binding sites situated at the interface between the subunits. Upon application of nicotinic agonists, nAChRs undergo fast activation leading to an open-channel state, and slow desensitization leading to a closed-channel state refractory to activation. Activation and desensitization correspond to transitions between a small number of discrete structural states that potentially exist in the absence of effectors. Nicotinic ligands promote the transitions between allosteric states by

*Corresponding author: Tel.: +33-01-45688805;
Fax: +33-01-45688836; E-mail: changeux@pasteur.fr

DOI: 10.1016/S0079-6123(03)45016-4

236

selectively stabilizing those they bind with high affinity. The pharmacological, physiological and kinetic properties of these allosteric states depend on the subunit composition of the receptor (references in Galzi and Changeux, 1995; Edelstein and Changeux, 1998).

Among the 16 genes encoding nAChR subunits that have been identified and cloned in mammals, 9 (α2–α7 and β2–β4) are expressed in the CNS. These subunits can coassemble to form functional pentameric receptors. When expressed in *Xenopus* oocytes, α7 forms homopentamers, while other subunits (α2–4, α6, β2 and β4) coassemble to form heteromeric receptors with a putative 2α. 3β stoichiometry (reviewed in McGehee and Role, 1995). Finally, the structural subunits, β3 and α5, which lack critical amino acids necessary to contribute to the ligand binding site, are incorporated into heteromeric receptors with other α and β subunits (Ramirez-Latorre et al., 1996; Groot-Kormelink et al., 1998). Although largely established in heterologous expression systems, these association rules suggest the existence of a wide variety of native nAChR subtypes with different pharmacological and electrophysiological properties. In particular, neurons in the visual system and in the brain catecholaminergic system express most, if not all of the neuronal nicotinic subunits (Le Novère et al., 1996; Vailati et al., 2000; Klink et al., 2001). More generally, the existence of overlapping patterns of expression of nAChR subunits and the lack of specific pharmacological agents have greatly hampered the identification of the native nAChR subtypes present in the nervous system.

What is the functional relevance of the diversity of nAChR subtypes and of their individual expression profiles? What is the subunit composition of nAChRs involved in the various functions of nicotinic transmission? Is there a relationship between the allosteric properties of nAChRs and their role in the brain? In the present review, we present evidence that recombinant DNA technology provides a valuable complementary approach to classical pharmacological approaches to answer these questions. Knockout mice lacking specific nAChR subunits (α3 (Xu et al., 1999a), α4 (Marubio et al., 1999; Ross et al., 2000), α5 (Orr-Urtreger et al., 2000), α6 (Champtiaux et al., 2002), α7 (Orr-Urtreger et al., 1997), α9 (Vetter et al.,

1999), β2 (Picciotto et al., 1995), β3 (Booker et al., 1999), or β4 (Xu et al., 1999b)) have been generated, and characterization of their phenotype has allowed a successful evaluation of the contribution of particular subunits to specific aspects of nicotinic transmission. More recently, Kin techniques that permit the introduction of point mutations into a single gene were used to generate mice expressing mutant forms of the α4 (Labarca et al., 2001) and α7 (Orr-Urtreger et al., 2000) subunits. The generalization of this kind of studies will help to understand the relationships between the structural characteristics of individual receptors and their function in vivo.

The role of nAChRs in normal and pathological behavioral and cognitive states

Nicotinic transmission has been proposed to participate in many cognitive processes. Not only can nicotine improve performance in various tasks involving spatial and associative learning, working memory and attention, but mecamylamine, a general nicotinic antagonist, has been demonstrated to impair memory performance (Levin, 1992). Furthermore, lesions of the cholinergic forebrain system result in cognitive deficits reverted by nicotine (Tilson et al., 1988; Decker et al., 1992; but see Gallagher and Colombo, 1995; Chappell et al., 1998). These findings, together with the observation that cholinergic function is impaired in the brains of Alzheimer's disease patients have contributed to the formulation of a cholinergic hypothesis to explain the cognitive impairments observed in Alzheimer's disease (Robbins et al., 1997). Prompted by epidemiological studies suggesting that the incidence of Alzheimer's and Parkinson's disease is significantly reduced in smokers (reviewed in Fratiglioni and Wang, 2000), the role of nAChRs in these disorders has been examined and a therapeutic approach using nicotinic agents has been proposed (Newhouse et al., 1997). These treatments might help to improve cognitive performance, protect against neurodegeneration and alleviate the psychotic symptoms observed in dementia. The following sections discuss the role of Ko animal studies in the examination of these different points.

Examination of behavioral impairments in β2 and α7 Ko animals

In a passive avoidance test, β2 Ko animals exhibited longer latencies than their wild-type (Wt) controls to enter a dark (preferred) compartment previously paired with an aversive stimulus. While nicotine reliably improved performance in Wt animals, it did not influence β2 Ko mice behavior (Picciotto et al., 1995). This result demonstrates the involvement of endogenous nicotinic transmission in the regulation of the cognitive mechanisms recruited in the passive avoidance paradigm. It further establishes that nicotinic agents modulate passive avoidance performance through β2-containing (β2*) nAChRs. On the other hand, young adult β2 as well as α7 Ko animals, performed normally in spatial learning (Morris Water Maze) and pavlovian fear conditioning tasks (Picciotto et al., 1995; Paylor et al., 1998; Caldarone et al., 2000).

The interpretation of these data requires a clear distinction between the cognitive enhancing properties of nicotine, which seem to be dependent on β2* nAChR activation, and the physiological role of nicotinic transmission in normal cognitive function, which has been more difficult to demonstrate. Apart from functional compensations, which could explain these findings, it is striking to note that highly specific lesions of basal forebrain cholinergic neurons fail similarly to produce overt behavioral deficits (Gallagher and Colombo, 1995; Chappell et al., 1998). In both cases, it is likely that in the absence of any other lesion, redundant neurotransmitter systems are sufficient to maintain high levels of performance in simple behavioral tasks. Examining the performance of Ko animals during aging (see below) or in more difficult tasks might be necessary to reveal a deficit. In fact, the physiological involvement of nAChRs in cognitive function seems particularly prominent in cognitive processes with higher demands on attention and working memory (Levin, 1992). However, the performance of Ko animals in these tasks has not been extensively examined at this point. For example, it could be very interesting, although difficult in mice, to study the performance of Ko animals in delayed matching or nonmatching to sample tasks, which are known to be sensitive to the administration of nicotinic agonist and antagonists in rats (Hironaka et al., 1992; Granon et al., 1995). On the

other hand, recent quantitative analysis of the spontaneous motor behavior of β2 Ko and WT mice significant deficits in executive functions and social behavior (Granon et al., 2003).

Attentional deficits in α7 and β2 Ko mice?

The impairment of attentional processes is a common trait in many neuropsychiatric disorders including dementias and schizophrenia. In both types of disease, these deficits may cause complex psychotic symptoms such as hallucinations or delusions (Sarter, 1994). More specifically, schizophrenic patients exhibit impaired sensory gating (i.e. the selective filtering of irrelevant stimuli) as assessed by the measure of sensory inhibition and latent inhibition (Light and Braff, 1999; Moser et al., 2000). There are several lines of evidence that link the nAChRs with the sensory deficits observed in schizophrenic patients. Deficits in sensory inhibition can be revealed by measuring prepulse inhibition of the startle response (PPI), or by recording evoked potentials following the paired presentation of an auditory stimulus. In these paradigms, nicotinic agonists have been demonstrated to increase the sensory inhibition in both schizophrenic patients and in normal subjects (review in Braff et al., 2001). Moreover, in humans (Freedman et al., 1997; Riley et al., 2000; but see Neves-Pereira et al., 1998; Curtis et al., 1999) or in mice, genetic studies have established a link between a chromosomal region containing the α7 nAChR gene and schizophrenia or auditory gating deficits. Finally, in various strains of mice, a positive correlation between the level of auditory gating and the level of αBgtX binding has been found (Stevens et al., 1996).

In order to determine the contribution of α7-containing (α7*) nAChRs to sensory inhibition, the performance of animals lacking the α7 subunit was evaluated in the PPI paradigm. In contrast to what had been expected, α7 Ko animals exhibited normal levels of PPI of the acoustic or tactile startle response (Paylor et al., 1998). This observation, which seems to contradict the wealth of evidence supporting the role of α7* nAChRs in attentional processes, should be tempered by the following remarks. Using the PPI paradigm, pharmacological studies performed in humans and animals have been contradictory and failed to establish the role of $\alpha7^*$ nAChRs in sensory inhibition (Acri et al., 1994; Curzon et al., 1994; but see Decker et al., 1997; Faraday et al., 1998, 1999; Olivier et al., 2001). On the other hand, studies evaluating the extent of sensory inhibition by recording auditory evoked potentials have consistently reported an effect of $\alpha7$ agonists or antagonists (Luntz-Leybman et al., 1992; Stevens and Wear, 1997; Stevens et al., 1998; Simosky et al., 2001). As a result, the lack of data using such electrophysiological measures in $\alpha7$ Ko animals precludes the exact evaluation of the role of $\alpha7^*$ nAChRs in sensory inhibition.

Another way to reveal deficits in attentional processes is to measure the extent of latent inhibition. This term refers to a phenomenon by which noncontingent preexposure to a stimulus retards subsequent conditioning to that stimulus. In schizophrenic patients, latent inhibition is greatly impaired, but only a few studies have addressed the influence of nicotine on this process in humans (Thornton et al., 1996; Della Casa et al., 1999). In $\beta2$ Ko mice, latent inhibition levels measured in a fear conditioning paradigm were found to be normal (Caldarone et al., 2000). In rats however, nicotine has been demonstrated to either increase or decrease latent inhibition depending on the preexposure phase parameters (number and duration of the stimulus presentations) (Rochford et al., 1996). Therefore, the data on $\beta2$ Ko animals, that were generated using a single set of preconditioning parameters, are still too incomplete to be conclusive. However, in the study on rats, the potency of cytisine, a poor agonist at $\beta2^*$ nAChRs, seems to support the hypothesis that nicotinic effects on latent inhibition do not depend on the activation of $\beta2^*$ nAChRs.

Overall, the studies on Ko animals to bring only limited or no evidence that endogenous cholinergic activation of $\beta2^*$ or $\alpha7^*$ nAChRs is necessary for attentional processing of information, possibly because of methodological limitations. Moreover, these studies did not investigate the influence on sensory gating of nicotinic agents that could be used in the treatment of psychotic symptoms. Further experiments on Ko animals are thus required to address these issues.

Contribution of Ko mice to the study of nAChRs in brain reinforcement systems

Recent theoretical models for cognitive learning have underlined the importance of reinforcement mechanisms in the organization and selection of brain representations (Dehaene and Changeux, 2000). Although the precise role of the ascending DA system is still debated (Koob, 1996; Schultz, 1997; Berridge and Robinson, 1998; Di Chiara, 1998), most authors agree that its integrity is crucial to the expression of motivated behaviors. By modulating the activity of the main reward pathway, nicotinic transmission is thought to play a role in reinforcement. First, nicotine acts as a positive reinforcer, as shown by place preference and self-administration studies (see Di Chiara, 2000 and references there in). Second, lesions of the pedunculopontine nucleus, the major cholinergic input to mesencephalic DA neurons, have been reported to impair learning and expression of behaviors reinforced by non-nicotinic addictive drugs (Olmstead and Franklin, 1994), brain stimulation (Lepore and Franklin, 1996), or natural rewards (Stefurak and van der Kooy, 1994; but see Olmstead et al., 1999). The subunit composition of the nAChRs in DA neurons and their contribution to these effects have been examined using Ko animals.

Subunit composition of nAChRs involved in nicotine addiction

As many drugs of abuse, nicotine is thought to exert its addictive properties through stimulation of DA release in the nucleus accumbens (Nac) (Pontieri et al., 1996; Di Chiara, 2000). This effect of nicotine has been attributed, at least in part, to the direct activation of nAChRs located on DA neurons of the ventral tegmental area (VTA). Indeed, nicotine is known to increase the firing rate of these neurons, both in vitro (Pidoplichko et al., 1997) and in vivo (Grenhoff et al., 1986). In addition, nicotinic antagonists infused into the VTA prevent nicotine-elicited DA release in the Nac and disrupt systemic nicotine self-administration in rats (Corrigall et al., 1994; Nisell et al., 1994). Since nAChR subtypes mediating the reinforcing properties of nicotine should be valuable targets of new pharmacological agents for the treatment of nicotine abuse, much research has been devoted to identifying their subunit composition.

For instance, electrophysiological experiments in $\beta2$ Ko animals demonstrated that somatic $\beta2*$ nAChRs are required for the effect of nicotine on the firing rate of DA neurons in vitro (Picciotto et al., 1998). Also, in synaptosomes prepared from the striatum of $\beta2$ Ko animals, nicotine failed to stimulate DA release by a direct action on nAChRs in DA terminals (Grady et al., 2001). Last but not least, in vivo, $\beta2*$ nAChRs are also required for the stimulation of striatal DA release by systemic nicotine, and for the maintenance of nicotine self-administration (Picciotto et al., 1998).

While these studies have demonstrated that nicotine reinforcement is dependent on the activation of $\beta2*$ nAChRs, other studies have tried to determine the nature of the subunits associated with $\beta2$. Relying on pharmacological evidence, some authors have proposed the $\alpha4$ and $\alpha3$ subunits as the likely partners of $\beta2$ in DA neurons (Sharples et al., 2000). This hypothesis, however, has only been partially confirmed by Ko animal studies. Patch-clamp recording of mesencephalic neurons from $\alpha4$ Ko mice show that the majority of slowly desensitizing nicotine-induced currents disappear in these mice (Klink et al., 2001), demonstrating the existence of $\alpha4*$ nAChRs in these neurons. Moreover, the release of DA induced by systemic injections of nicotine is abolished, in vivo, in $\alpha4$ Ko mice (Marubio et al., 2003). On the other hand, the role of the $\alpha3$ subunit in nicotine reinforcement has been challenged. Indeed, recent studies on $\alpha3$ and $\alpha6$ Ko animals suggest a novel interpretation of available pharmacological data. The contribution of the $\alpha3$ subunit had been initially proposed in view of the selectivity of αCtxMII for recombinant $\alpha3\beta2*$ nAChRs expressed in oocytes and the discovery that this toxin partially inhibits nicotine-induced DA release in striatal synaptosomes (Kulak et al., 1997; Kaiser et al., 1998). However, this selectivity has been questioned by the observation that high-affinity αCtxMII binding sites are preserved in $\alpha3$ Ko animals but completely disappear in $\alpha6$ Ko mice (Fig. 1) (Champtiaux et al., 2002; Whiteaker et al., 2002). Thus, in DA neurons, $\alpha6$ rather than $\alpha3$ is likely to associate with the $\beta2$ subunit to form

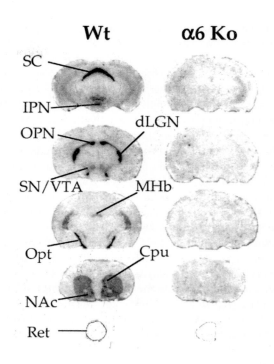

Fig. 1. $\alpha6$ Ko animals reveal the selectivity of αCtxMII for native $\alpha6*$ receptors. Autoradiographic representation of $[^{125}I]\alpha$CtxMII (0.5 nM) binding in mouse brain. The abbreviations used to identify brain regions are: Cpu, caudate putamen; dLGN, dorso-lateral geniculate nucleus; IPN, interpeduncular nucleus; MHb, medial habenula; Nac, nucleus accumbens; OPN, olivary pretectal nucleus; Opt, optic tract; Ret, Retina; SC, superior colliculus; SN, Substantia nigra, pars compacta; VTA, ventral tegmental area. In Wt animals, labeling is found in three nuclei (MHb, SN/VTA and Ret) in which $\alpha6$ mRNA has also been detected, and in their projections areas. High-affinity αCtXMII binding completely disappears in the brain of $\alpha6$ Ko animals, demonstrating that, on native receptors, this toxin is specific for $\alpha6*$ nAChRs. (From Champtiaux et al., reproduced with permission from Bentham Science Publishers Ltd.)

functional nAChRs. Altogether, these results demonstrate that the main heteromeric nAChRs subtypes expressed in DA neurons are the $\alpha4\beta2*$ and $\alpha6\beta2*$ subtypes. A contribution of the $\alpha5$ and $\beta3$ structural subunits, abundantly expressed in these neurons, remains to be established. Ultimately, the contribution of individual nAChR subtypes to the reinforcing properties of nicotine remains to be assessed more directly. One possible way to do this would be by studying nicotine self-administration and nicotine-induced place preference in Ko animals.

It has also been suggested that the reinforcing properties of nicotine might be attributed to an indirect stimulation of DA neurons (Nomikos et al., 2000; Corrigall et al., 2001). Indeed, nicotine is known to modify the firing pattern of DA neurons from a regular spiking to a burst-firing mode, through a mechanism which involves the activation of NMDA receptors. Relying on pharmacological evidence, it has been proposed that the activation of presynaptic $\alpha 7^*$ nAChRs located on cortical glutamatergic afferents to the VTA stimulates glutamate release, which in turn promotes a switch in the firing mode of DA neurons and causes DA release in the Nac (reviewed in Nomikos et al., 2000). However, the finding that nicotine fails to stimulate DA release in $\alpha 4$ and $\beta 2$ Ko animals as well as the recent observation that methyllycaconitine, a 'selective' $\alpha 7^*$ antagonist, can block some non-$\alpha 7^*$ nAChRs in DA neurons (Klink et al., 2001), in contrast with this hypothesis. Further experiments on $\alpha 7$ Ko animals will help to clarify this issue.

Role of β2 nAChRs in general mechanisms of reinforcement

Ko animals have also been used to investigate the role of nAChRs in the reinforcing properties of other addictive drugs, such as cocaine or morphine (Zachariou et al., 2001). Notably, $\beta 2$ Ko animals fail to develop a place preference to a low concentration of cocaine (5 mg/kg), which normally is effective in Wt animals. Yet, place preference to higher cocaine doses (10 mg/kg) or to morphine is preserved in these mice.

This finding is unlikely to be due to developmental compensations since mecamylamine treatment similarly disrupts cocaine-induced place preference in Wt animals. Rather, it has been proposed that in Wt mice, endogenous nicotinic transmission normally potentiates cocaine reinforcement through a tonic control of DA release in Nac. Unlike those of morphine, the reinforcing effects of cocaine are in part mediated through a direct blockade of DA reuptake. Cocaine reinforcement is thus critically dependent on the spontaneous release of DA. There are at least two ways by which endogenous nicotinic transmission could modulate spontaneous DA

release. First, since mecamylamine can decrease the firing rate of DA neurons in the VTA, these neurons are thought to be tonically stimulated by cholinergic projections arising from the pedunculopontine nucleus (Grenhoff et al., 1986). Second, at the terminal level, in striatal slices, endogenous acetylcholine released by cholinergic interneurons potentiates the spontaneous or electrically evoked release of DA. This latter effect is absent in mice lacking the $\beta 2$ subunit, which thus exhibit reduced basal levels of DA release (Zhou et al., 2001).

The observed decrease in sensitivity for cocaine in $\beta 2$ Ko animals is in agreement with the observation that cocaine self-administration is reduced by mecamylamine treatment in rats (Levin et al., 2000). Moreover, in drug addicts, mecamylamine has been reported to reduce cue-induced craving for cocaine (Reid et al., 1999). Overall, these results demonstrate that cholinergic control of the DA system contributes to general mechanisms of reinforcement, and that $\beta 2^*$ nAChRs play a pivotal role in this control. Finally, these observations open up new therapeutic avenues for the treatment of drug addiction.

Role of nAChRs in nicotine-elicited analgesia

Nicotinic agonists have long been known to exert analgesic effects. However, the high doses required to produce analgesia, close to those inducing seizures, have limited their therapeutic use. On the other hand, the serious side effects of morphine administration in pain alleviation have pin-pointed the necessity to develop non opiod analgesic drugs. Interestingly, a new nicotinic agonist, ABT594, has recently been shown to exert analgesic effects comparable to those of morphine at doses that do not produce peripheral side effects (Bannon et al., 1998).

In an attempt to understand nicotine-induced analgesia, the antinociceptive properties of nicotine were studied in mice lacking the $\alpha 4$ and $\beta 2$ nAChR subunits (Marubio et al., 1999). In the hot plate test, a model of the supraspinal response to acute thermal nociception, nicotine or epibatidine failed to produce analgesia in both lines of mutant mice. Also, patch-clamp recording from serotoninergic neurons in the raphe magnus showed a loss of nicotine-elicited currents in $\alpha4$ and $\beta2$ Ko animals. These findings are consistent with the notion that supraspinal analgesic actions of nicotine are due to the activation of neurons in this nucleus. Finally, Kin mice expressing a mutant form of the $\alpha4$ nAChR subunit with increased affinity for agonists are hypersensitive to the analgesic effects of nicotine in the hot plate test (ED50 was 5.3 times lower in Kin animals compared to their Wt controls; Fonck et al., 2001). All in all, these results demonstrate that the activation of $\alpha4\beta2^*$ nAChRs is required for supraspinal analgesic effects of nicotine.

A somewhat different picture emerges from the study of nicotine-induced analgesia in the tail-flick response, a spinal reflex to acute thermal nociception. Nicotine is still able to produce dose-dependent analgesia in mice lacking the $\beta2$ or $\alpha4$ nAChR subunits, but the dose–response curve is shifted to the right. The subunit composition of the nAChRs involved in this residual response to nicotine is unknown. However, these receptors are likely to be located in the spinal cord or in peripheral ganglia. Actually, nicotine is still able to evoke currents in neurons from the dorsal horn of the spinal cord in $\beta2$ or $\alpha4$ Ko animals. Furthermore, the presynaptic modulation of serotonin release in the spinal cord is controlled by nAChRs that do not contain the $\alpha4$ or $\beta2$ subunits (Cordero-Erausquin and Changeux, 2001). The precise nature of these unidentified nAChRs needs further examination.

The role of nAChRs in CNS development

Several lines of evidence suggest that nAChRs contribute to the development of the CNS. nAChRs are expressed very early in the developing CNS, as demonstrated by in situ hybridization as well as ligand binding studies (Zoli et al., 1995; Conroy and Berg, 1998). Moreover, their level of expression is dynamically regulated during development (Fuchs, 1989; Zoli et al., 1995; Zhang et al., 1998). In vitro, nAChR activation has been shown to regulate neurite outgrowth and axonal growth cone guidance through a mechanism which implicates Ca^{2+} influx (Chan and Quik, 1993; Pugh and Berg, 1994; Zheng et al., 1994). In Ko animals, the developmental

contribution of nAChRs has been examined in the visual and somatosensory systems.

A role for β2-containing (β2*) receptors in the maturation of retinothalamic projections

In the visual system of mammals, the refinement of retinothalamic projections occurs during early post-natal development (P4–P9 in mice) (So et al., 1990). Initially intermixed in their thalamic target, the dorso-lateral geniculate nucleus, retinal ganglion cell (RGC) projections from each eye segregate into eye specific layers during that period (Fig. 2a, b). Segregation of retinothalamic projections is a competitive process

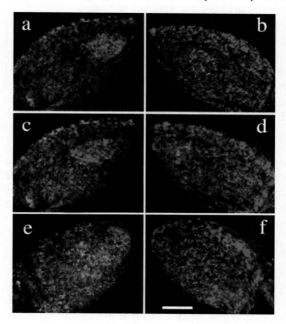

Fig. 2. Abnormal development of retinogeniculate projections in adult $\beta2$ Ko animals. Representative coronal sections of left (a, c, e) and right (b, d, f) dorso-lateral geniculate nuclei in adult Wt (a, b), $\alpha6$ Ko (c, d) and $\beta2$ Ko animals (e, f). Both eyes were injected with cholera toxin B fragment coupled to two different fluorophors (Alexa Fluor 594, red, in the left eye, and Alexa Fluor 488, green, in the right eye). Note the absence of overlap in projections from the two eyes, characteristic of correct fiber segregation in Wt and $\alpha6$ Ko animals. On the other hand, in $\beta2$ Ko animals, retinothamamic projections from both eyes do not segregate into eye specific layers. Scale bar 200 μm. (From Rossi et al., reproduced with permission from Bentham Science Publishers Ltd.)

controlled by spontaneous RGC activity (Penn et al., 1998). The RGCs fire periodic bursts of action potentials that sweep across the retina in a wave-like manner (see reviews of Wong, 1999; Roerig and Feller, 2000). Although the mechanisms of wave initiation and propagation are not fully understood, it has been established that nAChR blockade in the retina prevents the generation of waves and inhibits the segregation process (Feller et al., 1996; Penn et al., 1998).

In agreement with these pharmacological observations, it was recently demonstrated that during the first week of postnatal development, spontaneous retinal activity is absent in mice lacking the $\beta2$ subunit (Bansal et al., 2000). At P4 (before the normal onset of binocular segregation), the pattern of retinogeniculate projections is normal in $\beta2$ Ko animals. In contrast, adult mutant mice exhibit anomalous overlap of retinogeniculate projections, comparable to the overlap observed in the immature geniculate body of Wt neonates (Rossi et al., 2001) (Fig. 2e, f). In addition, mutant mice exhibit reduced visual acuity as well as functional expansion of the binocular visual cortex (i.e., binocular responses to light stimulation were recorded even at cortical locations normally responsive only to the contralateral eye). Although the absence of nAChRs in the adult brain can contribute to these deficits, the anatomical abnormalities observed in the geniculate body may be sufficient to explain the functional impairment.

While $\beta2*$ nAChRs, presumably located on RGCs, are necessary for normal development of the visual system in mice, the α subunits which coassemble with $\beta2$ to form the nAChRs mediating these developmental effects remain to be identified. In mice lacking $\alpha4$ or $\alpha6$ nAChR subunits, the pattern of retinothalamic projections is normal (Rossi et al., 2001; Champtiaux et al., 2002) (Fig. 2c, d). Conversely, in $\alpha3$ Ko mice, spontaneous retinal activity is not abolished. However, the spatiotemporal characteristics of the waves are slightly altered (Bansal et al., 2000). To explain these findings, we propose that the population of $\beta2*$ nAChRs contributing to the development of the visual system in mice is not homogeneous. Interestingly, αConotoxinMII (αCtxMII) completely prevents the generation of spontaneous retinal waves (Bansal et al., 2000). Since

this toxin is known to block $\alpha3\beta2*$ as well as $\alpha6\beta2*$ nAChRs (Cartier et al., 1996; Champtiaux et al., 2002), and since $\alpha3$ inactivation does not abolish spontaneous retinal activity as previously mentioned, we suggest that both subtypes play an equivalent role in the retina. Inactivating only one of these receptors subtype is not sufficient to produce detectable alterations of the segregation process, or to disrupt retinal activity. This hypothesis could be tested in $\alpha3$-$\alpha3\alpha6$ double mutant mice.

A role for $\alpha7$ in the development of the sensory cortex?

$\alpha7*$ nAChRs have been hypothesized to play a role in brain development and plasticity (Broide and Leslie, 1999). In situ hybridization, as well as αBungarotoxin (αBgtX, a competitive blocker of $\alpha7*$ nAChRs) binding experiments, have demonstrated a transitory increase of $\alpha7$ expression in the thalamus and neocortex during prenatal and early postnatal development, starting from E13 (Fuchs, 1989; Bina et al., 1995; Broide et al., 1995). In the primary sensory cortex, this period corresponds to the formation of barrel structures arising from whisker projections. Since αBgtX binding outlines the barrels at the time of their appearance, it has been proposed that $\alpha7*$ nAChRs might play a critical role in barrel formation.

In contrast to this hypothesis, $\alpha7$ Ko mice do not exhibit gross morphological abnormalities in the brain (Orr-Urtreger et al., 1997). Close examination of the neocortex and hippocampus, where $\alpha7$ is abundantly expressed in Wt animals, reveals a normal organization of these regions (including normal cell density, layering, and barrel formation in the cortex). However, subtle ultrastructural abnormalities might have been overlooked in $\alpha7$ Ko mice. Moreover, this particular study focused on morphological, rather than functional, aspects of development. For example, the synaptic plasticity that takes place in cortical barrels following whiskers stimulation is known to be modulated by basal forebrain cholinergic projections to the somatosensory cortex (Sachdev et al., 1998; Ego-Stengel et al., 2001). It would be interesting to examine the consequences of $\alpha7$ inactivation on this plasticity.

α7 activation and cell death

Another mechanism through which α7* nAChRs activation may influence development is through stimulation of apoptosis. Indeed, in the developing chick embryo, αBgtX infusions rescued motoneurons from naturally occurring cell death (Renshaw et al., 1993). Recent findings also suggest that α7* nAChRs may be recruited naturally in pathological conditions (Laudenbach et al., 2002). In an experimental model of neonatal anoxia, i.e., ibotenate-induced cortical lesion in newborns, the size of the lesion was significantly smaller in α7 Ko mice than their Wt controls. This observation is in agreement with pharmacological data. DMXB, a partial α7 nAChR agonist, and methyllycaconitine, a potent α7 nAChR antagonist, respectively increased and reduced the extent of the ibotenate-induced lesion in Wt animals. These findings suggest that endogenous acetylcholine, acting at α7* receptors, potentiates the excitotoxic effects of ibotenate in Wt animals.

In an attempt to explain the deleterious effects of α7* activation, Berger and colleagues (Berger et al., 1998) demonstrated that the cytotoxic effects of α7* activation in hippocampal progenitor cells is attributable to an excessive influx of calcium. Indeed, hippocampal progenitor cells are sensitive to α7* stimulation because their calcium buffering mechanisms are not established before differentiation. In the case of Laudenbach et al., (2002) study, the deleterious effects of α7* activation were observed in young animals (P4–5), when α7 expression in the cortex is maximal (Bina et al., 1995; Broide et al., 1995). Moreover, in cortical cultures from E14.5 animals, when α7 expression is low, DMXB had neuroprotective effects. Although in vitro and in vivo studies are difficult to compare, it is possible that excessive Ca^{2+} influx through α7* receptors is responsible for the potentiation of ibotenate-induced lesions.

The idea that excessive stimulation of nAChRs can trigger neuronal cell death is also supported by studies of Kin animals expressing mutant forms of the α4 or α7 subunits. In these mice, the gain of function mutation (α4Leu-9′Ser or α7L250T) causes the selective degeneration of neuronal populations in which the α4 or α7 subunits are normally expressed at very high levels (Orr-Urtreger et al., 2000; Broide et al., 2001; Labarca et al., 2001).

Aging reveals cognitive impairment in β2 Ko animals

In Alzheimer's disease, a selective decrease in the number of telencephalic nAChRs has been observed (Aubert et al., 1992; Nordberg, 2001), and the severity of the cognitive deficits has been correlated with the loss of cholinergic function (Mega, 2000). Therefore, the impact of nicotinic dysfunction in mutant mice might be aggravated by other biological mechanisms occurring in the course of normal aging. To investigate this question, cognitive tests were performed in old β2 Ko mice. The performance of aged Ko animals was found to be significantly impaired, when compared to age-matched Wt animals, in the Morris Water Maze and a fear conditioning task (Fig. 3) (Zoli et al., 1999; Caldarone et al., 2000). These deficits can be mainly attributed to increased neurodegeneration during aging in β2 Ko animals (Zoli et al., 1999). While a decrease in neocortical thickness and a loss of pyramidal hippocampal neurons were also observed in aged Wt animals, the neuronal atrophy was accelerated in β2 mutant mice. Additionally, the authors observed an increase of cortico-hippocampal astro- and microgliosis, a prominent characteristic of pathological aging (Selkoe, 1994).

Two nonmutually exclusive hypotheses can explain these observations. On the one hand, the impairment of nicotinic transmission itself might cause neurodegeneration. Indeed, corticosterone levels are increased in aged β2 Ko mice (Zoli et al., 1999), and it is known that chronic elevation of corticosterone levels causes a marked loss of hippocampal neurons (McEwen and Sapolsky, 1995; Sapolsky, 1996). On the other hand, it is possible that endogenous acetylcholine, acting at β2* nAChRs, normally exerts neuroprotective effects against repeated exposure to normal cellular stress accumulated during aging. Indeed, as discussed in the following section, there is much evidence for a neuroprotective effect of nAChR activation.

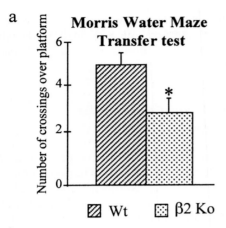

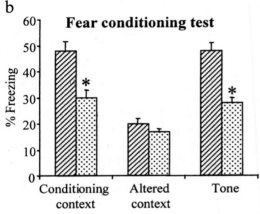

Fig. 3. Impaired cognitive performance in aged $\beta2$ Ko animals. (a) Transfer test performance for 22- to 24-month old $\beta2$ Ko mice and their control siblings in the Morris Water Maze. Performance on the visible platform test was identical between $\beta2$ Ko and Wt animals, demonstrating that the observed impairment was not due to sensory or motor deficits ($*p < 0.05$ by Mann-Whitney U-test). (b) Conditioned fear in aged (9- to 20-months) $\beta2$ Ko mice and their control siblings. Compared to Wt mice, mutant animals exhibited less freezing to the tone and in the conditioning context, but not in the altered context. $*p < 0.05$ by one way ANOVA. (From Zoli et al., 1999, Caldarone et al., 2000. Reproduced with permission from Bentham Science Publishers Ltd.).

Work on Ko animals has significantly contributed to research on the neuroprotective action of nicotine in various models of neurodegeneration, both in vitro and in vivo (see references in Belluardo et al., 2000). In an experimental model of parkinsonism (Ryan et al., 2001), acute nicotine treatment protected against metamphetamine-induced neurodegeneration

of dopaminergic (DA) terminals in Wt mice. In mice lacking the $\alpha4$ subunit, the extent of the lesion was similar to what was observed in Wt controls, and it could not be prevented by nicotine. The role of $\beta2^*$ nAChR in nicotine's neuroprotective actions has also been examined in an experimental model of excitotoxic brain injury, in neonates (Laudenbach et al., 2002). In this study, nicotine protected against the toxic effects of intracerebral ibotenate injections in Wt but not in $\beta2$ Ko animals. It should be noted, however, that in the absence of pharmacological treatment, the size of the lesion was significantly smaller in $\beta2$ Ko animals compared to their Wt controls. This finding is puzzling since the nicotinic antagonists dihydro-beta-erythroidine and mecamylamine did not exhibit a neuroprotective effect in Wt animals. The current interpretation is that compensatory changes might have occurred in systems that influence the outcome of ibotenate-induced lesions in $\beta2$ Ko animals. Despite this last caveat, these studies firmly established the role of $\alpha4^*$ and $\beta2^*$ nAChRs in the neuroprotective effects of nicotine both in vitro and in vivo. Therefore, nicotinic therapies aimed at $\alpha4\beta2^*$ nAChRs in Alzheimer patients could combine the benefits of functional compensation of the impact of cholinergic loss on cognitive function with neuroprotective effects.

New perspectives from the study of Kin mice

The recent development of Kin techniques to introduce point mutations into nAChR subunits allows the examination of the relationships between structural properties of nAChRs and their function in vivo. Novel impetus to address such a fundamental question was recently given by the finding that some forms of familial epilepsies (autosomal dominant nocturnal frontal lobe epilepsies) were linked to single point mutations in the $\alpha4$ or $\beta2$ nAChR subunits (for review, see Sutor and Zolles, 2001). In oocytes, these mutations have been reported to increase the apparent affinity for acetylcholine, to alter Ca^{2+} permeability, and/or to modify the desensitization rate of $\alpha4\beta2$ nAChRs (Bertrand, 1999; Phillips et al., 2001). It is not known, however, how these functional changes cause the transient alteration in consciousness that characterizes the pathology. Introducing these mutations in Kin mice would greatly help to address this issue and would provide a solid animal model of the disease.

The Kin approach may also be used to investigate issues such as the role of the desensitization and calcium permeability characteristics of nAChRs. Extensive research over the past twenty years has led to the identification of amino acids crucially involved in several functional properties of the nAChR (reviewed in Corringer et al., 2000). Recently, two groups generated Kin mice expressing 'gain of function' mutant forms of the $\alpha4$ or $\alpha7$ subunits (Labarca et al., 2001; Orr-Urtreger et al., 2000; Broide et al., 2001). In oocytes, these mutations ($\alpha7$L250T and $\alpha4$Leu-9'Ser) increase the affinity for the agonists and decrease the rate of desensitization of $\alpha7^*$ or $\alpha4\beta2^*$ nAChRs (Bertrand et al., 1992). As previously mentioned, expression of these mutant nAChR subunits in Kin mice causes neurodegeneration in DA neurons for $\alpha4$Leu-9'Ser or in the somatosensory cortex for $\alpha7$L250T. Moreover, $\alpha4$Leu-9'Ser Kin mice exhibit increased levels of anxiety and impaired motor learning. However, the severe developmental abnormalities in the DA system observed in these animals, hampers the interpretation of the behavioral data.

In the future, more subtle mutations of the nAChRs subunits will help to address specific questions. For example, the finding that $\alpha7^*$ nAChRs exhibit very high calcium permeability raises the question of the role of this property in signal transduction following $\alpha7^*$ activation (Castro and Albuquerque, 1995). This issue could be investigated by generating Kin animals expressing mutant $\alpha7^*$ nAChRs impermeable to calcium (Bertrand et al., 1993). Using such mice, the role of calcium influx through $\alpha7^*$ nAChRs in neurodegeneration or in presynaptic facilitation of neurotransmitter release could be further examined.

Another important question that could be addressed using a Kin approach is the role of upregulation of high affinity nicotine binding sites following chronic nicotine treatment. Some authors have postulated that upregulation, as observed in vitro and in vivo (see references in Wonnacott, 1990; Ochoa, 1994; Dani and Heinemann, 1996), represents an homeostatic adaptation to chronic desensitization

of the nAChRs and that it plays a fundamental role in nicotine addiction (Dani and Heinemann, 1996). By using mutant $\alpha4\beta2$ nAChRs displaying increased or decreased sensitivity to nicotine-mediated upregulation (Fenster et al., 1999) the validity of this model could be examined in Kin mice.

These are only examples among many possibilities, and we assume that the development of these techniques will provide new insights into the role of nAChRs in brain function.

Advantages and limitations of the Ko approach

Throughout this chapter, we have illustrated the advantages of using genetically modified mice to investigate the function and composition of native nAChRs. As shown, one cannot rely exclusively on pharmacological data derived from in vitro studies to establish the contribution of a particular nAChR subtype to an observed phenotype. This is especially important when complex behaviors are explored, and when the concentrations of agonists or antagonists used in vivo are higher than those required to ensure specificity.

While the Ko approach overcomes these drawbacks, it has its own limitations. One major difficulty of Ko studies is the occurrence of developmental modifications as a result of gene inactivation that interferes with the interpretation of the observed phenotype. Although these alterations can actually reveal the contribution of the inactivated gene to normal development, they can mask the role of the protein in an adult animal. For instance, some alterations have been found in the development of the visual system of $\beta2$ Ko mice, which lead to impairment of binocular vision (Rossi et al., 2001).

The existence of functional compensatory mechanisms in Ko animals is the second major difficulty of the molecular approach. For instance, the expression level of many proteins, including nAChR subunits, muscarinic receptors, acetylcholine biosynthetic enzymes or even DA markers, has been found to be normal in various lines of nAChR subunit Ko mice (Picciotto et al., 1995, 1998; Marubio et al., 1999; Zoli et al., 1999; Ross et al., 2000; Champtiaux et al., 2002). Yet, we recently found a significant upregulation of

αCtxMII-resistant [^{3}H]epibatidine binding sites in the striatum of α6 Ko animals (Champtiaux et al., 2002). Of course, more complex compensations than upregulation of other nAChR subunits might exist and will be more difficult to demonstrate.

In many instances, however, discrepancies between pharmacological and molecular data have been reported which do not seem to result from compensatory mechanisms or developmental effects. The genetic background is known to have a strong influence on the expression of a given mutation. The same mutation in different mouse strains can have rather different phenotypic consequences, due to the existence of modifier genes (for review, see Nadeau, 2001). Interestingly, most nAChR subunit mutations have been studied after backcrossing to the C57Bl6 strain (α3, α4, α6, α7, β2 and β4). Backcrossing these mutations onto other genetic backgrounds might reveal so far unobserved phenotypes.

Moreover, it is also likely that different nAChR subtypes may contribute to the same function. Inactivating only one subtype might not be sufficient to produce a deficit. This possibility, proposed to explain the lack of developmental alterations in the visual system of α3, α4 or α6 Ko mice, was recently demonstrated in the autonomic nervous system. The major nAChR subtype expressed in autonomic ganglia is α3β4* (Vernallis et al., 1993; Conroy and Berg, 1995). However, major autonomic dysfunction is only observed in α3 or β2xβ4 Ko animals (Xu et al., 1999a,b), whereas mice lacking only one β subunit (β2 or β4) exhibit a normal autonomic phenotype. A close examination of the physiology of the superior cervical ganglion in β4 Ko animals reveals that nicotine elicited currents are actually reduced by 98% in ganglion cells, demonstrating that β4* nAChRs are not replaced by new β2* receptors. Rather, it can be assumed that residual nicotinic transmission through α3β2* nAChRs is sufficient to maintain a normal phenotype.

We tentatively expand this idea of redundancy between different nAChR subtypes at the level of neurotransmitter systems. Selective impairment of nicotinic transmission through β2 subunit gene inactivation may fail to produce dramatic behavioral deficits because other neurotransmitter systems are sufficient to maintain alterations of behavioral performance below detection threshold. Indeed, it

has already been reported that either serotoninergic or cholinergic denervation of the forebrain alone fails to produce significant impairment in the Morris Water maze task, whereas combined lesions produce severe deficits (Nilsson et al., 1988). The same idea may also explain the lack of cognitive impairment following selective immunotoxin-induced lesions of forebrain cholinergic neurons (Gallagher and Colombo, 1995).

Conclusion

Ko studies have confirmed that the inactivation of a single nAChR subunit gene abolishes the pharmacological response to specific nicotinic agents in the CNS, and has allowed the identification of particular nAChR subtypes involved in nicotine's neuroprotective, analgesic or rewarding effects. Moreover, these mice have revealed the importance of nAChR activation in normal development and aging. Further studies are required to investigate the role of nAChRs in cognitive function, setting up the need for refined tools strategies. In the future, development of tissue-specific or inducible Ko strategies and the use of Kin animals will help to overcome the present limitations of the classical Ko approach and improve our understanding of the role of nAChRs in the brain.

Acknowledgments

This work was supported by the Collège de France, the Centre National de la Recherche Scientifique, the Institut Pasteur, the AFM, ARC, and CEE contracts no. 097038 & 097038. We thank Naguib Mechawar, Clément Léna, Michele Zoli, Morten Sunesen and Marina Picciotto for critical reading of this review.

Abbreviations

αBgtX	alpha-bungarotoxin
αCtxMII	alpha-conotoxin MII
CNS	central nervous system
DA	dopamine/dopaminergic
Kin	knockin
Ko	knockout

Nac   nucleus accumbens
nAChR   nicotinic acetylcholine receptor
PPI   prepulse inhibition
RGC   retinal ganglion cells
VTA   ventral tegmental area
Wt   wild-type
$\alpha 7^*$ $\alpha 7$-containing receptor
$\beta 2^*$ $\beta 2$-containing receptor

References

Acri, J.B., Morse, D.E., Popke, E.J. and Grunberg, N.E. (1994) Nicotine increases sensory gating measured as inhibition of the acoustic startle reflex in rats. Psychopharmacology (Berl.), 114: 369–374.

Aubert, I., Araujo, D.M., Cecyre, D., Robitaille, Y., Gauthier, S. and Quirion, R. (1992) Comparative alterations of nicotinic and muscarinic binding sites in Alzheimer's and Parkinson's diseases. J. Neurochem., 58: 529–541.

Bannon, A.W., Decker, M.W., Holladay, M.W., Curzon, P., Donnelly-Roberts, D., Puttfarcken, P.S., Bitner, R.S., Diaz, A., Dickenson, A.H., Porsolt, R.D., Williams, M., Arneric, S.P. (1998) Broad-spectrum, non-opioid analgesic activity by selective modulation of neuronal nicotinic acetylcholine receptors. Science, 279: 77–81.

Bansal, A., Singer, J.H., Hwang, B.J., Xu, W., Beaudet, A. and Feller, M.B. (2000) Mice lacking specific nicotinic acetylcholine receptor subunits exhibit dramatically altered spontaneous activity patterns and reveal a limited role for retinal waves in forming ON and OFF circuits in the inner retina. J. Neurosci., 20: 7672–7681.

Belluardo, N., Mudo, G., Blum, M., Amato, G. and Fuxe, K. (2000) Neurotrophic effects of central nicotinic receptor activation. J. Neural. Transm. Suppl.(60): 227–245.

Berger, F., Gage, F.H. and Vijayaraghavan, S. (1998) Nicotinic receptor-induced apoptotic cell death of hippocampal progenitor cells. J. Neurosci., 18: 6871–6881.

Berridge, K.C. and Robinson, T.E. (1998) What is the role of dopamine in reward: hedonic impact, reward learning, or incentive salience?. Brain Res. Brain Res. Rev., 28: 309–369.

Bertrand, D. (1999) Neuronal nicotinic acetylcholine receptors: their properties and alterations in autosomal dominant nocturnal frontal lobe epilepsy. Rev. Neurol. (Paris), 155: 457–462.

Bertrand, D., Devillers-Thiery, A., Revah, F., Galzi, J.L., Hussy, N., Mulle, C., Bertrand, S., Ballivet, M. and Changeux, J.P. (1992) Unconventional pharmacology of a neuronal nicotinic receptor mutated in the channel domain. Proc. Natl. Acad. Sci. USA, 89: 1261–1265.

Bertrand, D., Galzi, J.L., Devillers-Thiery, A., Bertrand, S. and Changeux, J.P. (1993) Mutations at two distinct sites within the channel domain M2 alter calcium permeability of neuronal alpha 7 nicotinic receptor. Proc. Natl. Acad. Sci. USA, 90: 6971–6975.

Bina, K.G., Guzman, P., Broide, R.S., Leslie, F.M., Smith, M.A. and O'Dowd, D.K. (1995) Localization of alpha 7 nicotinic receptor subunit mRNA and alpha-bungarotoxin binding sites in developing mouse somatosensory thalamocortical system. J. Comp. Neurol., 363: 321–332.

Booker, T.K., Allen, R.S., Marks, M.J., Grady, S.R., Whiteaker, P., Smith, K;W., Collins, A.C. and Heinemann, S.F. (1999) In: Neuronal Nicotinic Receptors: From Structure to Therapeutics, Venice, p. 22.

Braff, D.L., Geyer, M.A. and Swerdlow, N.R. (2001) Human studies of prepulse inhibition of startle: normal subjects, patient groups, and pharmacological studies. Psychopharmacology (Berl.), 156: 234–258.

Broide, R.S. and Leslie, F.M. (1999) The alpha7 nicotinic acetylcholine receptor in neuronal plasticity. Mol. Neurobiol., 20: 1–16.

Broide, R.S., O'Connor, L.T., Smith, M.A., Smith, J.A. and Leslie, F.M. (1995) Developmental expression of alpha 7 neuronal nicotinic receptor messenger RNA in rat sensory cortex and thalamus. Neuroscience, 67: 83–94.

Broide, R.S., Orr-Urtreger, A. and Patrick, J.W. (2001) Normal apoptosis levels in mice expressing one alpha7 nicotinic receptor null and one L250T mutant allele. Neuroreport, 12: 1643–1648.

Caldarone, B.J., Duman, C.H. and Picciotto, M.R. (2000) Fear conditioning and latent inhibition in mice lacking the high affinity subclass of nicotinic acetylcholine receptors in the brain. Neuropharmacology, 39: 2779–2784.

Cartier, G.E., Yoshikami, D., Gray, W.R., Luo, S., Olivera, B.M.and and McIntosh, J.M. (1996) A new alpha-conotoxin which targets alpha3beta2 nicotinic acetylcholine receptors. J. Biol. Chem., 271: 7522–7528.

Castro, N.G. and Albuquerque, E.X. (1995) alpha-Bungarotoxin-sensitive hippocampal nicotinic receptor channel has a high calcium permeability. Biophys. J., 68: 516–524.

Champtiaux, N., Han, Z.Y., Bessis, A., Rossi, F.M., Zoli, M., Marubio, L., McIntosh, J.M. and Changeux, J.P. (2002) Distribution and pharmacology of alpha 6-containing nicotinic acetylcholine receptors analyzed with mutant mice. J. Neurosci., 2: 1208–1217.

Chan, J. and Quik, M. (1993) A role for the nicotinic alpha-bungarotoxin receptor in neurite outgrowth in PC12 cells. Neuroscience, 56: 441–451.

Chappell, J., McMahan, R., Chiba, A. and Gallagher, M. (1998) A re-examination of the role of basal forebrain cholinergic neurons in spatial working memory. Neuropharmacology, 37: 481–487.

Conroy, W.G. and Berg, D.K. (1995) Neurons can maintain multiple classes of nicotinic acetylcholine receptors distinguished by different subunit compositions. J. Biol. Chem., 270: 4424–4431.

Conroy, W.G. and Berg, D.K. (1998) Nicotinic receptor subtypes in the developing chick brain: appearance of a species containing the alpha4, beta2, and alpha5 gene products. Mol. Pharmacol., 53: 392–401.

248

Cordero-Erausquin, M. and Changeux, J.P. (2001) Tonic nicotinic modulation of serotoninergic transmission in the spinal cord. Proc. Natl. Acad. Sci. USA, 98: 2803–2807.

Corrigall, W.A., Coen, K.M. and Adamson, K.L. (1994) Self-administered nicotine activates the mesolimbic dopamine system through the ventral tegmental area. Brain Res., 653: 278–284.

Corrigall, W.A., Coen, K.M., Zhang, J. and Adamson, K.L. (2001) GABA mechanisms in the pedunculopontine tegmental nucleus influence particular aspects of nicotine self-administration selectively in the rat. Psychopharmacology (Berl.), 158: 190–197.

Corringer, P.J., Le Novere, N. and Changeux, J.P. (2000) Nicotinic receptors at the amino acid level. Annu. Rev. Pharmacol. Toxicol., 40: 431–458.

Curtis, L., Blouin, J.L., Radhakrishna, U., Gehrig, C., Lasseter, V.K., Wolyniec, P., Nestadt, G., Dombroski, B., Kazazian, H.H., Pulver, A.E., Housman, D., Bertrand, D., Antonarakis, S.E. (1999) No evidence for linkage between schizophrenia and markers at chromosome 15q13-14. Am. J. Med. Genet., 88: 109–112.

Curzon, P., Kim, D.J. and Decker, M.W. (1994) Effect of nicotine, lobeline, and mecamylamine on sensory gating in the rat. Pharmacol. Biochem. Behav., 49: 877–882.

Dani, J.A. and Heinemann, S. (1996) Molecular and cellular aspects of nicotine abuse. Neuron, 16: 905–908.

Decker, M.W., Bannon, A.W., Curzon, P., Gunther, K.L., Brioni, J.D., Holladay, M.W., Lin, N.H., Li, Y., Daanen, J.F., Buccafusco, J.J., Prendergast, M.A., Jackson, W.J., Arneric, S.P. (1997) ABT-089 [2-methyl-3-(2-(S)-pyrrolidinyl-methoxy)pyridine dihydrochloride]: II. A novel cholinergic channel modulator with effects on cognitive performance in rats and monkeys. J. Pharmacol. Exp. Ther., 283: 247–258.

Decker, M.W., Majchrzak, M.J. and Anderson, D.J. (1992) Effects of nicotine on spatial memory deficits in rats with septal lesions. Brain Res., 572: 281–285.

Dehaene, S. and Changeux, J.P. (2000) Reward-dependent learning in neuronal networks for planning and decision making. Prog. Brain Res., 126: 217–229.

Della Casa, V., Hofer, I. and Feldon, J. (1999) Latent inhibition in smokers vs. nonsmokers: interaction with number or intensity of preexposures?. Pharmacol. Biochem. Behav., 62: 353–359.

Di Chiara, G. (1998) A motivational learning hypothesis of the role of mesolimbic dopamine in compulsive drug use. J. Psychopharmacol., 12: 54–67.

Di Chiara, G. (2000) Role of dopamine in the behavioural actions of nicotine related to addiction. Eur. J. Pharmacol., 393: 295–314.

Edelstein, S.J. and Changeux, J.P. (1998) Allosteric transitions of the acetylcholine receptor. Adv. Protein Chem., 51: 121–184.

Ego-Stengel, V., Shulz, D.E., Haidarliu, S., Sosnik, R. and Ahissar, E. (2001) Acetylcholine-dependent induction and expression of functional plasticity in the barrel cortex of the adult rat. J. Neurophysiol., 86: 422–437.

Faraday, M.M., O'Donoghue, V.A. and Grunberg, N.E. (1999) Effects of nicotine and stress on startle amplitude and sensory gating depend on rat strain and sex. Pharmacol. Biochem. Behav., 62: 273–284.

Faraday, M.M., Rahman, M.A., Scheufele, P.M. and Grunberg, N.E. (1998) Nicotine administration impairs sensory gating in Long-Evans rats. Pharmacol. Biochem. Behav., 61: 281–289.

Feller, M.B., Wellis, D.P., Stellwagen, D., Werblin, F.S. and Shatz, C.J. (1996) Requirement for cholinergic synaptic transmission in the propagation of spontaneous retinal waves. Science, 272: 1182–1187.

Fenster, C.P., Whitworth, T.L., Sheffield, E.B., Quick, M.W. and Lester, R.A. (1999) Upregulation of surface alpha4beta2 nicotinic receptors is initiated by receptor desensitization after chronic exposure to nicotine. J. Neurosci., 19: 4804–4814.

Fonck, C.I., Damaj, M.I., Deshpande, P., Bowers, B.J., Wehner, J.M., Nashmi, R., Khakh, B.S., Schwarz, J., Lester, H.A., Labarca, C. (2001) Knock-in mice carrying hypersensitive alpha4 nicotinic receptors: nicotine responses. Soc. Neurosci. Abstr., 27: 2135.

Fratiglioni, L. and Wang, H.X. (2000) Smoking and Parkinson's and Alzheimer's disease: review of the epidemiological studies. Behav. Brain Res., 113: 117–120.

Freedman, R., Coon, H., Myles-Worsley, M., Orr-Urtreger, A., Olincy, A., Davis, A., Polymeropoulos, M., Holik, J., Hopkins, J., Hoff, M., Rosenthal, J., Waldo, M.C., Reimherr, F., Wender, P., Yaw, J., Young, D.A., Breese, C.R., Adams, C., Patterson, D., Adler, L.E., Kruglyak, L., Leonard, S. and Byerley, W. (1997) Linkage of a neurophysiological deficit in schizophrenia to a chromosome 15 locus. Proc. Natl. Acad. Sci. USA, 94: 587–592.

Fuchs, J.L. (1989) [125I]alpha-bungarotoxin binding marks primary sensory area developing rat neocortex. Brain Res., 501: 223–234.

Gallagher, M. and Colombo, P.J. (1995) Ageing: the cholinergic hypothesis of cognitive decline. Curr. Opin. Neurobiol., 5: 161–168.

Galzi, J.L. and Changeux, J.P. (1995) Neuronal nicotinic receptors: molecular organization and regulations. Neuropharmacology, 34: 563–582.

Grady, S.R., Meinerz, N.M., Cao, J., Reynolds, A.M., Picciotto, M.R., Changeux, J.P., McIntosh, J.M., Marks, M.J. and Collins, A.C. (2001) Nicotinic agonists stimulate acetylcholine release from mouse interpeduncular nucleus: a function mediated by a different nAChR than dopamine release from striatum. J. Neurochem., 76: 258–268.

Granon, S., Poucet, B., Thinus-Blanc, C., Changeux, J.P. and Vidal, C. (1995) Nicotinic and muscarinic receptors in the rat prefrontal cortex: differential roles in working memory, response selection and effortful processing. Psychopharmacology (Berl.), 119: 139–144.

Granon, S., Faure, P. and Changeux, J. P. (2003) Executive and social behaviors under nicotinic receptor regulation. Proc. Natl. Acad. Sci., USA, (in press).

Grenhoff, J., Aston-Jones, G. and Svensson, T.H. (1986) Nicotinic effects on the firing pattern of midbrain dopamine neurons. Acta Physiol. Scand., 128: 351–358.

Groot-Kormelink, P.J., Luyten, W.H., Colquhoun, D. and Sivilotti, L.G. (1998) A reporter mutation approach shows incorporation of the 'orphan' subunit beta3 into a functional nicotinic receptor. J. Biol. Chem., 273: 15317–15320.

Hironaka, N., Miyata, H. and Ando, K. (1992) Effects of psychoactive drugs on short-term memory in rats and rhesus monkeys. Jpn. J. Pharmacol., 59: 113–120.

Kaiser, S.A., Soliakov, L., Harvey, S.C., Luetje, C.W. and Wonnacott, S. (1998) Differential inhibition by alpha-conotoxin-MII of the nicotinic stimulation of [3H]dopamine release from rat striatal synaptosomes and slices. J. Neurochem., 70: 1069–1076.

Klink, R., de Kerchove d'Exaerde, A., Zoli, M. and Changeux, J.P. (2001) Molecular and physiological diversity of nicotinic acetylcholine receptors in the midbrain dopaminergic nuclei. J. Neurosci., 21: 1452–1463.

Koob, G.F. (1996) Hedonic valence, dopamine and motivation. Mol. Psychiatry, 1: 186–189.

Kulak, J.M., Nguyen, T.A., Olivera, B.M. and McIntosh, J.M. (1997) Alpha-conotoxin MII blocks nicotine-stimulated dopamine release in rat striatal synaptosomes. J. Neurosci., 17: 5263–5270.

Labarca, C., Schwarz, J., Deshpande, P., Schwarz, S., Nowak, M.W., Fonck, C., Nashmi, R., Kofuji, P., Dang, H., Shi, W., Fidan, M., Khakh, B.S., Chen, Z., Bowers, B.J., Boulter, J., Wehner, J.M., Lester, H.A. (2001) Point mutant mice with hypersensitive alpha 4 nicotinic receptors show dopaminergic deficits and increased anxiety. Proc. Natl Acad. Sci. USA, 98: 2786–2791.

Laudenbach, V., Medja, F., Zoli, M., Rossi, F.M., Evrard, P., Changeux, J.P. and Gressens, P. (2002) Selective activation of central subtypes of the nicotinic acetylcholine receptor has opposite effects on neonatal excitotoxic brain injuries. Faseb J., 16: 423–425.

Le Novère, N., Zoli, M. and Changeux, J.P. (1996) Neuronal nicotinic receptor alpha 6 subunit mRNA is selectively concentrated in catecholaminergic nuclei of the rat brain. Eur. J. Neurosci., 8: 2428–2439.

Lepore, M. and Franklin, K.B. (1996) N-methyl-D-aspartate lesions of the pedunculopontine nucleus block acquisition and impair maintenance of responding reinforced with brain stimulation. Neuroscience, 71: 147–155.

Levin, E.D. (1992) Nicotinic systems and cognitive function. Psychopharmacology (Berl.), 108: 417–431.

Levin, E.D., Mead, T., Rezvani, A.H., Rose, J.E., Gallivan, C. and Gross, R. (2000) The nicotinic antagonist mecamylamine preferentially inhibits cocaine vs. food self-administration in rats. Physiol. Behav., 71: 565–570.

Light, G.A. and Braff, D.L. (1999) Is there a critical developmental 'window' for isolation rearing-induced changes in prepulse inhibition of the acoustic startle response? Curr. Psychiatry Rep., 1: 31–40.

Luntz-Leybman, V., Bickford, P.C. and Freedman, R. (1992) Cholinergic gating of response to auditory stimuli in rat hippocampus. Brain Res., 587: 130–136.

Marubio, L.M., del Mar Arroyo-Jimenez, M., Cordero-Erausquin, M., Lena, C., Le Novere, N., de Kerchove d'Exaerde, A., Huchet, M., Damaj, M.I. and Changeux, J.P. (1999) Reduced antinociception in mice lacking neuronal nicotinic receptor subunits. Nature, 398: 805–810.

Marubio, L.M., Gardier A.M. Durier, S. David, D., Klink, R. del Mar Arroyo-Jimenez, M. McIntosh, J.M., Rossi, F.M.R., Champtiaux, N., Zoli, M. and Changeux, J.P. (2003) Effects of nicotine in the dopaminergic system of mice lacking the alpha4 subunit of neuronal nicotinic acetylcholine receptors. Eur. J. Neurosci., 17: 1329–1337.

McGehee, D.S. and Role, L.W. (1995) Physiological diversity of nicotinic acetylcholine receptors expressed by vertebrate neurons. Annu. Rev. Physiol., 57: 521–546.

McEwen, B.S. and Sapolsky, R.M. (1995) Stress and cognitive function. Curr. Opin. Neurobiol., 5: 205–216.

Mega, M.S. (2000) The cholinergic deficit in Alzheimer's disease: impact on cognition, behaviour and function. Int. J. Neuropsychopharmacol., 3: 3–12.

Moser, P.C., Hitchcock, J.M., Lister, S. and Moran, P.M. (2000) The pharmacology of latent inhibition as an animal model of schizophrenia. Brain Res. Brain Res. Rev., 33: 275–307.

Nadeau, J.H. (2001) Modifier genes in mice and humans. Nat. Rev. Genet., 2: 165–174.

Neves-Pereira, M., Bassett, A.S., Honer, W.G., Lang, D., King, N.A. and Kennedy, J.L. (1998) No evidence for linkage of the CHRNA7 gene region in Canadian schizophrenia families. Am. J. Med. Genet., 81: 361–363.

Newhouse, P.A., Potter, A. and Levin, E.D. (1997) Nicotinic system involvement in Alzheimer's and Parkinson's diseases. Implications for therapeutics. Drugs Aging, 11: 206–228.

Nilsson, O.G., Strecker, R.E., Daszuta, A. and Bjorklund, A. (1988) Combined cholinergic and serotonergic denervation of the forebrain produces severe deficits in a spatial learning task in the rat. Brain Res., 453: 235–240.

Nisell, M., Nomikos, G.G. and Svensson, T.H. (1994) Systemic nicotine-induced dopamine release in the rat nucleus accumbens is regulated by nicotinic receptors in the ventral tegmental area. Synapse, 16: 36–44.

Nomikos, G.G., Schilstrom, B., Hildebrand, B.E., Panagis, G., Grenhoff, J. and Svensson, T.H. (2000) Role of alpha7 nicotinic receptors in nicotine dependence and implications for psychiatric illness. Behav. Brain Res., 113: 97–103.

Nordberg, A. (2001) Nicotinic receptor abnormalities of Alzheimer's disease: therapeutic implications. Biol. Psychiatry, 49: 200–210.

Ochoa, E.L. (1994) Nicotine-related brain disorders: the neurobiological basis of nicotine dependence. Cell Mol. Neurobiol., 14: 195–225.

Olivier, B., Leahy, C., Mullen, T., Paylor, R., Groppi, V.E., Sarnyai, Z. and Brunner, D. (2001) The DBA/2J strain and prepulse inhibition of startle: a model system to test antipsychotics?. Psychopharmacology (Berl.), 156: 284–290.

Olmstead, M.C. and Franklin, K.B. (1994) Lesions of the pedunculopontine tegmental nucleus block drug-induced reinforcement but not amphetamine-induced locomotion. Brain Res., 638: 29–35.

Olmstead, M.C., Inglis, W.L., Bordeaux, C.P., Clarke, E.J., Wallum, N.P., Everitt, B.J. and Robbins, T.W. (1999) Lesions of the pedunculopontine tegmental nucleus increase sucrose consumption but do not affect discrimination or contrast effects. Behav. Neurosci., 113: 732–743.

Orr-Urtreger, A., Broide, R.S., Kasten, M.R., Dang, H., Dani, J.A., Beaudet, A.L. and Patrick, J.W. (2000) Mice homozygous for the L250T mutation in the alpha7 nicotinic acetylcholine receptor show increased neuronal apoptosis and die within 1 day of birth. J. Neurochem., 74: 2154–2166.

Orr-Urtreger, A., Goldner, F.M., Saeki, M., Lorenzo, I., Goldberg, L., De Biasi, M., Dani, J.A., Patrick, J.W. and Beaudet, A.L. (1997) Mice deficient in the alpha7 neuronal nicotinic acetylcholine receptor lack alpha-bungarotoxin binding sites and hippocampal fast nicotinic currents. J. Neurosci., 17: 9165–9171.

Paylor, R., Nguyen, M., Crawley, J.N., Patrick, J., Beaudet, A. and Orr-Urtreger, A. (1998) Alpha7 nicotinic receptor subunits are not necessary for hippocampal-dependent learning or sensorimotor gating: a behavioral characterization of Acra7-deficient mice. Learn. Mem., 5: 302–316.

Penn, A.A., Riquelme, P.A., Feller, M.B. and Shatz, C.J. (1998) Competition in retinogeniculate patterning driven by spontaneous activity. Science, 279: 2108–2112.

Phillips, H.A., Favre, I., Kirkpatrick, M., Zuberi, S.M., Goudie, D., Heron, S.E., Scheffer, I.E., Sutherland, G.R., Berkovic, S.F., Bertrand, D., Mulley, J.C. (2001) CHRNB2 is the second acetylcholine receptor subunit associated with autosomal dominant nocturnal frontal lobe epilepsy. Am. J. Hum. Genet., 68: 225–231.

Picciotto, M.R., Zoli, M., Lena, C., Bessis, A., Lallemand, Y., LeNovere, N., Vincent, P., Pich, E.M., Brulet, P., Changeux, J.P. (1995) Abnormal avoidance learning in mice lacking functional high-affinity nicotine receptor in the brain. Nature, 374: 65–67.

Picciotto, M.R., Zoli, M., Rimondini, R., Lena, C., Marubio, L.M., Pich, E.M., Fuxe, K. and Changeux, J.P. (1998) Acetylcholine receptors containing the beta2 subunit are involved in the reinforcing properties of nicotine. Nature, 391: 173–177.

Pidoplichko, V.I., DeBiasi, M., Williams, J.T. and Dani, J.A. (1997) Nicotine activates and desensitizes midbrain dopamine neurons. Nature, 390: 401–404.

Pontieri, F.E., Tanda, G., Orzi, F. and Di Chiara, G. (1996) Effects of nicotine on the nucleus accumbens and similarity to those of addictive drugs. Nature, 382: 255–257.

Pugh, P.C. and Berg, D.K. (1994) Neuronal acetylcholine receptors that bind alpha-bungarotoxin mediate neurite retraction in a calcium-dependent manner. J. Neurosci., 14: 889–896.

Ramirez-Latorre, J., Yu, C.R., Qu, X., Perin, F., Karlin, A. and Role, L. (1996) Functional contributions of alpha5 subunit

to neuronal acetylcholine receptor channels. Nature, 380: 347–351.

Reid, M.S., Mickalian, J.D., Delucchi, K.L. and Berger, S.P. (1999) A nicotine antagonist, mecamylamine, reduces cue-induced cocaine craving in cocaine-dependent subjects. Neuropsychopharmacology, 20: 297–307.

Renshaw, G., Rigby, P., Self, G., Lamb, A. and Goldie, R. (1993) Exogenously administered alpha-bungarotoxin binds to embryonic chick spinal cord: implications for the toxin-induced arrest of naturally motoneuron death. Neuroscience, 53: 1163–1172.

Riley, B.P., Makoff, A., Mogudi-Carter, M., Jenkins, T., Williamson, R., Collier, D. and Murray, R. (2000) Haplotype transmission disequilibrium and evidence for linkage of the CHRNA7 gene region to schizophrenia in Southern African Bantu families. Am. J. Med. Genet., 96: 196–201.

Rochford, J., Sen, A.P. and Quirion, R. (1996) Effect of nicotine and nicotinic receptor agonists on latent inhibition in the rat. J. Pharmacol. Exp. Ther., 277: 1267–1275.

Robbins, T.W., McAlonan, G., Muir, J.L. and Everitt, B.J. (1997) Cognitive enhancers in theory and practice: studies of the cholinergic hypothesis of cognitive deficits in Alzheimer's disease. Behav. Brain Res., 83: 15–23.

Roerig, B. and Feller, M.B. (2000) Neurotransmitters and gap junctions in developing neural circuits. Brain Res. Brain Res. Rev., 32: 86–114.

Role, L.W. and Berg, D.K. (1996) Nicotinic receptors in the development and modulation of CNS synapses. Neuron, 16: 1077–1085.

Ross, S.A., Wong, J.Y., Clifford, J.J., Kinsella, A., Massalas, J.S., Horne, M.K., Scheffer, I.E., Kola, I., Waddington, J.L., Berkovic, S.F., Drago, J. (2000) Phenotypic characterization of an alpha 4 neuronal nicotinic acetylcholine receptor subunit knock-out mouse. J. Neurosci., 20: 6431–6441.

Rossi, F.M., Pizzorusso, T., Porciatti, V., Marubio, L.M., Maffei, L. and Changeux, J.P. (2001) Requirement of the nicotinic acetylcholine receptor beta 2 subunit for the anatomical and functional development of the visual system. Proc. Natl Acad. Sci. USA, 98: 6453–6458.

Ryan, R.E., Ross, S.A., Drago, J. and Loiacono, R.E. (2001) Dose-related neuroprotective effects of chronic nicotine in 6-hydroxydopamine treated rats, and loss of neuroprotection in alpha4 nicotinic receptor subunit knockout mice. Br. J. Pharmacol., 132: 1650–1656.

Sachdev, R.N., Lu, S.M., Wiley, R.G. and Ebner, F.F. (1998) Role of the basal forebrain cholinergic projection in somatosensory cortical plasticity. J. Neurophysiol., 79: 3216–3228.

Sapolsky, R.M. (1996) Why stress is bad for your brain. Science, 273: 749–750.

Sarter, M. (1994) Neuronal mechanisms of the attentional dysfunctions in senile dementia and schizophrenia: two sides of the same coin?. Psychopharmacology (Berl.), 114: 539–550.

Schultz, W. (1997) Dopamine neurons and their role in reward mechanisms. Curr. Opin. Neurobiol., 7: 191–197.

Selkoe, D.J. (1994) Cell biology of the amyloid beta-protein precursor and the mechanism of Alzheimer's disease. Annu. Rev. Cell Biol., 10: 373–403.

Sharples, C.G., Kaiser, S., Soliakov, L., Marks, M.J., Collins, A.C., Washburn, M., Wright, E., Spencer, J.A., Gallagher, T., Whiteaker, P., Wonnacott, S. (2000) UB-165: a novel nicotinic agonist with subtype selectivity implicates the alpha4beta2* subtype in the modulation of dopamine release from rat striatal synaptosomes. J. Neurosci., 20: 2783–2791.

Simosky, J.K., Stevens, K.E., Kem, W.R. and Freedman, R. (2001) Intragastric DMXB-A, an alpha7 nicotinic agonist, improves deficient sensory inhibition in DBA/2 mice. Biol. Psychiatry, 50: 493–500.

So, K.F., Campbell, G. and Lieberman, A.R. (1990) Development of the mammalian retinogeniculate pathway: target finding, transient synapses and binocular segregation. J. Exp. Biol., 153: 85–104.

Stefurak, T.L. and van der Kooy, D. (1994) Tegmental pedunculopontine lesions in rats decrease saccharin's rewarding effects but not its memory-improving effect. Behav. Neurosci., 108: 972–980.

Stevens, K.E. and Wear, K.D. (1997) Normalizing effects of nicotine and a novel nicotinic agonist on hippocampal auditory gating in two animal models. Pharmacol. Biochem. Behav., 57: 869–874.

Stevens, K.E., Freedman, R., Collins, A.C., Hall, M., Leonard, S., Marks, M.J. and Rose, G.M. (1996) Genetic correlation of inhibitory gating of hippocampal auditory evoked response and alpha-bungarotoxin-binding nicotinic cholinergic receptors inbred mouse strains. Neuropsychopharmacology, 15: 152–162.

Stevens, K.E., Kem, W.R., Mahnir, V.M. and Freedman, R. (1998) Selective alpha7-nicotinic agonists normalize inhibition of auditory response in DBA mice. Psychopharmacology (Berl.), 136: 320–327.

Sutor, B. and Zolles, G. (2001) Neuronal nicotinic acetylcholine receptors and autosomal dominant nocturnal frontal lobe epilepsy: a critical review. Pflugers Arch., 442: 642–651.

Thornton, J.C., Dawe, S., Lee, C., Capstick, C., Corr, P.J., Cotter, P., Frangou, S., Gray, N.S., Russell, M.A., Gray, J.A. (1996) Effects of nicotine and amphetamine on latent inhibition in human subjects. Psychopharmacology (Berl.), 127: 164–173.

Tilson, H.A., McLamb, R.L., Shaw, S., Rogers, B.C., Pediaditakis, P. and Cook, L. (1988) Radial-arm maze deficits produced by colchicine administered into the area of the nucleus basalis are ameliorated by cholinergic agents. Brain Res., 438: 83–94.

Vailati, S., Moretti, M., Balestra, B., McIntosh, M., Clementi, F. and Gotti, C. (2000) beta3 subunit is present in different nicotinic receptor subtypes in chick retina. Eur. J. Pharmacol., 393: 23–30.

Vernallis, A.B., Conroy, W.G. and Berg, D.K. (1993) Neurons assemble acetylcholine receptors with as many as three kinds of subunits while maintaining subunit segregation among receptor subtypes. Neuron, 10: 451–464.

Vetter, D.E., Liberman, M.C., Mann, J., Barhanin, J., Boulter, J., Brown, M.C., Saffiote-Kolman, J., Heinemann, S.F. and Elgoyhen, A.B. (1999) Role of alpha9 nicotinic ACh receptor subunits in the development and function of cochlear efferent innervation. Neuron, 23: 93–103.

Whiteaker, P., Peterson, C.G., Xu, W., McIntosh, J.M., Paylor, R., Beaudet, A.L., Collins, A.C. and Marks, M.J. (2002) Involvement of the alpha3 subunit in central nicotinic binding populations. J. Neurosci., 22: 2522–2529.

Wong, R.O. (1999) Retinal waves and visual system development. Annu. Rev. Neurosci., 22: 29–47.

Wonnacott, S. (1990) The paradox of nicotinic acetylcholine receptor upregulation by nicotine. Trends Pharmacol. Sci., 11: 216–219.

Xu, W., Gelber, S., Orr-Urtreger, A., Armstrong, D., Lewis, R.A., Ou, C.N., Patrick, J., Role, L., De Biasi, M., Beaudet, A.L. (1999) Megacystis, mydriasis, and ion channel defect in mice lacking the alpha3 neuronal nicotinic acetylcholine receptor. Proc. Natl. Acad. Sci. USA, 96: 5746–5751.

Xu, W., Orr-Urtreger, A., Nigro, F., Gelber, S., Sutcliffe, C.B., Armstrong, D., Patrick, J.W., Role, L.W., Beaudet, A.L., De Biasi, M. (1999) Multiorgan autonomic dysfunction in mice lacking the beta2 and the beta4 subunits of neuronal nicotinic acetylcholine receptors. J. Neurosci., 19: 9298–9305.

Zachariou, V., Caldarone, B.J., Weathers-Lowin, A., George, T.P., Elsworth, J.D., Roth, R.H., Changeux, J.P. and Picciotto, M.R. (2001) Nicotine receptor inactivation decreases sensitivity to cocaine. Neuropsychopharmacology, 24: 576–589.

Zhang, X., Liu, C., Miao, H., Gong, Z.H. and Nordberg, A. (1998) Postnatal changes of nicotinic acetylcholine receptor alpha 2, alpha 3, alpha 4, alpha 7 and beta 2 subunits genes expression in rat brain. Int. J. Dev. Neurosci., 16: 507–518.

Zheng, J.Q., Felder, M., Connor, J.A. and Poo, M.M. (1994) Turning of nerve growth cones induced by neurotransmitters. Nature, 368: 140–144.

Zhou, F.M., Liang, Y. and Dani, J.A. (2001) Endogenous nicotinic cholinergic activity regulates dopamine release in the striatum. Nat. Neurosci., 4: 1224–1229.

Zoli, M., Le Novere, N., Hill, J.A., Jr. and Changeux, J.P. (1995) Developmental regulation of nicotinic ACh receptor subunit mRNAs in the rat central and peripheral nervous systems. J. Neurosci., 15: 1912–1939.

Zoli, M., Picciotto, M.R., Ferrari, R., Cocchi, D. and Changeux, J.P. (1999) Increased neurodegeneration during ageing in mice lacking high-affinity nicotine receptors. EMBO J., 18: 1235–1244.

Progress in Brain Research, Vol. 145
ISSN 0079-6123

CHAPTER 17

Nicotinic modulation of thalamocortical neurotransmission

Paul B.S. Clarke*

*Department of Pharmacology and Therapeutics, McGill University, 3655 Promenade Sir William Osler,
Montreal, QC H3G 1Y6, Canada*

Introduction

Nicotinic acetylcholine receptors (nAChRs) are widely distributed in the brain (Wada et al., 1989; Clarke, 1999; Paterson and Nordberg, 2000). Nicotinic AChR expression appears largely neuronal (Clarke, 1999), although in vitro evidence suggests that some may also be glial (Sharma and Vijayaraghavan, 2001). Acetylcholine (ACh) is implicated in numerous brain functions, and the relative contributions of nicotinic and muscarinic AChRs are under active investigation. Nicotinic AChRs also play a crucial role in tobacco smoking, and are thought to mediate most and perhaps all of the numerous psychopharmacological effects of nicotine (Benowitz, 1996); progress is being made in identifying the nAChR subtypes and sites responsible (Jones et al., 1999; Cordero-Erausquin et al., 2000; Di Chiara, 2000). In particular, it has been widely concluded from animal studies that the reinforcing effect of nicotine occurs mainly through activation of the mesolimbic dopamine (DA) system. However, this conclusion has recently been challenged (Laviolette and van der Kooy, 2003), and other mechanisms should be considered. This review focuses on thalamocortical neurons, which likely represent a major target for both ACh and nicotine.

*Corresponding author: Tel.: + 514-398-3616;
Fax: + 514-398-6690; E-mail: pclarke@pharma.mcgill.ca

DOI: 10.1016/S0079-6123(03)45017-6

Nicotinic acetylcholine receptor diversity

Nicotinic AChRs form a heterogeneous family of pentameric proteins (Lukas et al., 1999). Most neuronal nAChRs comprise α and β subunits in combination, and at least six α and three β subunits are expressed in the brain. Different subunit combinations give rise to different nAChR subtypes (Lukas et al., 1999). The theoretical number of subtypes is vast, and several pharmacologically distinct subtypes have already been reported in vivo or after heterologous expression. It is difficult to gauge the relative abundance of different nAChR subtypes, since available probes (e.g., antibodies) tend to target particular subunits rather than particular subtypes. Nevertheless, based on relative transcript abundance and other convergent evidence, it appears that two nAChR subtypes are particularly prominent in mammalian brain. One of these comprises $\alpha4$ and $\beta2$ subunits, and is preferentially labeled by high-affinity agonists such as [^{3}H]nicotine (Whiting and Lindstrom, 1987, 1988; Flores et al., 1992; Marubio et al., 1999). The other is an $\alpha7$ homooligomer and is labeled by [^{125}I]α-bungarotoxin and [^{3}H]methyllycaconitine (Couturier et al., 1990; Schoepfer et al., 1990; Seguela et al., 1993; Davies et al., 1999). Although in the context of smoking, most attention has been paid to $\alpha4\beta2$ receptors, recent evidence suggests that $\alpha7$ receptors should not be discounted (Papke and Porter Papke, 2002).

Nicotinic receptor expression in thalamic nuclei

In adult rats and mice, cortically-projecting thalamic nuclei (i.e., both principal and association nuclei) express abundant $\alpha4\beta2$ receptors. Transcript levels for $\alpha4$ and $\beta2$ subunits are very high relative to other subunits and compared to most other brain areas (Wada et al., 1989; Marks et al., 1992). In addition, labeling by [3H]nicotine and other high-affinity agonists is dense (Clarke et al., 1985; Marks et al., 1992; Happe et al., 1994). In contrast, $\alpha7$ receptor expression in these nuclei tends to be low (Clarke et al., 1985; Seguela et al., 1993; Marks et al., 1996).

Several early studies using single unit recording in cortically projecting thalamic nuclei, performed in cats, revealed that some neurons were excited by local application of ACh or nicotine. Both thalamocortical projection neurons and local circuit neurons were shown to be responsive (Andersen and Curtis, 1964). Analogous nicotinic excitations have also been reported in rats and mice, and shown to be due to a direct agonist action (Lena and Changeux, 1997; Zhu and Uhlrich, 1997; Cox and Sherman, 2000). Since thalamic interneurons are inhibitory (GABAergic), the overall effect of nAChR stimulation on thalamo-cortical impulse flow may be complex; in those thalamic nuclei that possess a significant population of interneurons. In this context, it should be recalled that in the rat and mouse thalamus, GABAergic interneurons are far from ubiquitous; for example, they are sparse or absent in the ventrobasal complex and the anterior and medial nuclear groups (Ottersen and Storm-Mathisen, 1984; Price, 1995). A further source of GABAergic inhibition is provided by the reticular thalamic nucleus, which innervates both projection neurons and local circuit neurons (Price, 1995). This nucleus strongly expresses $\alpha4\beta2$ nAChRs in rats (Happe et al., 1994) and is therefore a likely target for nicotine.

Nicotinic receptor expression in cerebral cortex

Both $\alpha4\beta2$ and $\alpha7$ receptors appear widely expressed in rat cerebral cortex, and as elsewhere in the brain, [3H]nicotine and [125I]α-bungarotoxin reveal different patterns of binding (Clarke et al., 1985). In adult rat neocortex, binding of high-affinity agonists such as [3H]nicotine and [3H]cytisine is particularly abundant in layers 1 and 3/4 (Clarke et al., 1985; Happe et al., 1994). In contrast, [125I]α-bungarotoxin labeling is densest in layers 5 and 6, as well as in layer 1 (Clarke et al., 1985).

Several nAChR subunit transcripts are expressed in rat cerebral cortex, including $\alpha4$, $\alpha7$ and $\beta2$ (Wada et al., 1989; Seguela et al., 1993). In addition, numerous neocortical neurons are immunostained by nAChR subunit-selective antibodies (Hill et al., 1993; Britto et al., 1994; Nakayama et al., 1995). Thus, some neocortical nAChRs are probably expressed by intrinsic neurons. However, other evidence suggests that a portion of cortical nAChRs are located on cortical afferents. In this context, two initial observations supported the possibility of nAChRs on basalocortical cholinergic afferents. First, nicotinic agonists were shown to promote [3H]ACh release from synaptosomes prepared from mouse cerebral cortex (Rowell and Winkler, 1984). Second, [3H]nicotine binding was decreased in post mortem Alzheimer brain, concomitant with the loss of cholinergic afferents (Whitehouse et al., 1986). These observations have been repeated and extended (e.g., Araujo et al., 1988; Court et al., 2001). However, neither of these sets of observations is definitive; not all rat cortical ACh is extrinsic, and cholinergic denervation is not the only pathological change in Alzheimer disease. The existence of nAChRs on basalocortical cholinergic afferents has also been sought by examining the effects of basal forebrain lesions on high-affinity [3H]agonist binding in rat neocortex. The results have been almost uniformly negative (e.g., Schwartz et al., 1984; Rossner et al., 1995). This suggests that in rats, few if any $\alpha4\beta2$ nAChRs are located on cholinergic afferents from basal forebrain.

Presynaptic nicotinic receptors on thalamocortical afferents

Autoradiographic studies

Convergent evidence suggests that a portion of cere-bral cortical nAChRs are located on thalamocortical terminals. The first evidence for this notion was correlative. In particular, dense [³H]nicotine labeling was found not only in rat thalamic relay nuclei, but also in neocortical layers that receive a major thalamocortical input (Clarke et al., 1984). Thalamo-cortical deafferentation studies have since provided firmer evidence for such an arrangement. The first such studies were performed in adult cats. In these species, a pronounced [³H]nicotine binding pattern is observed in the thalamocortical terminal zone of cat primary visual cortex (Prusky et al., 1987). This band was shown to be extrinsic, since it was little affected by local excitotoxic lesions, but virtually eliminated by surgical removal of subcortical afferents (Prusky et al., 1987). The probable source of these afferents was subsequently shown to be the lateral geniculate nucleus (i.e., the thalamic visual relay nucleus). Thus, focal excitotoxic lesions of this structure resulted in an anatomically discrete loss of [³H]nicotine binding in visual cortex. An analogous loss was seen after electrolytic lesions of the lateral geniculate (Prusky et al., 1987; Parkinson et al., 1988); this approach provided useful confirmation, given the possibility of remote damage after excitotoxin infusion.

These findings have been confirmed and extended in rats. In an initial study, it was shown that excitoxic or electrolytic lesions centered on the ventrobasal thalamic complex (which includes the main somato-sensory relay nucleus) led to a reduction in fronto-parietal [³H]nicotine binding (Sahin et al., 1992). However, the excitotoxic lesions encompassed much more than the target nucleus and radioligand binding was assessed only semi-quantitatively.

We have extended these findings to five thalamic projections. These pathways originate either from association nuclei (anterior nuclear group, medio-dorsal nucleus) or from relay nuclei (ventral nuclear group, medial and dorsal lateral geniculate nuclei)(Lavine et al., 1997; Gioanni et al., 1999). To achieve the desired anatomical resolution, the excitotoxin N-methyl-D-aspartate (NMDA) was stereotaxically infused in small volumes and in doses individualized for each thalamic target. Destruction of a given thalamic target area led to a characteristic decline in [³H]nicotine labeling in the cortical areas and layers. In each case, this loss corresponded to the termination zone of the particular thalamocortical projection. In each

thalamocortical projection examined, lesions reduced cortical binding by 35–45%. As in earlier studies, direct intracortical infusion of excitoxin into cerebral cortex itself produced little change in [³H]nicotine binding despite massive neuronal loss in the vicinity (Lavine et al., 1997).

In all the above autoradiographic studies, a single, non-saturating concentration of radioligand was used, making it impossible to distinguish changes in receptor density vs. affinity. However, a saturation binding analysis in neocortical homogenates indi-cated that the decline in binding reflected a reduction in receptor density not affinity. Thus, based on lesion studies, it appears that a substantial proportion of [³H]nicotine binding sites in cerebral cortex are located on thalamocortical afferents. This arrange-ment would seem to allow nAChRs to modulate transmission not only in sensory projections to neocortex, but also in the limbic projections from association nuclei that innervate allocortex.

Electrophysiological evidence

Electrophysiological experiments provide further support for this arrangement. In rat prelimbic cortex (i.e., part of medial prefrontal cortex), local applica-tion of nicotine was found to enhance field potentials that were recorded in vivo in response to adjacent electrical stimulation (Vidal and Changeux, 1989). With extracellular recording, possible pre- and postsynaptic modulatory effects of nicotine could not be distinguished. Subsequently, intracellular electrodes were used in the same preparation to examine electrically evoked synaptic excitation of pyramidal cells (Vidal and Changeux, 1993). This excitation was mediated by non-NMDA glutamater-gic receptors and the nicotinic enhancement was shown to be presynaptic in origin. Microdialysis experiments have shown that nicotine can increase extracellular glutamate concentrations in this cortical area in freely moving rats (Gioanni et al., 1999). Whether this finding reflects a nicotinic facilitation of glutamate release from thalamocortical terminals remains to be determined.

Lesion experiments therefore suggest that thala-mocortical afferents possess nAChRs, and functional studies suggest that presynaptic (or preterminal) nAChRs can increase glutamate release in at least one cortical area. Two studies suggest that nAChRs located on thalamocortical projections are indeed functional, at least in a pharmacological sense (Gil et al., 1997; Gioanni et al., 1999). The first study used a rodent brain slice preparation that contains an intact thalamocortical projection to somatosensory cortex (Gil et al., 1997). By electrically stimulating the appropriate thalamic nucleus, it was possible to record monosynaptic excitatory postsynaptic potentials in layer 3 pyramidal cells. These potentials, presumably glutamatergic, were enlarged by addition of nicotine or ACh to the superfusion solution. Although not definitive, antagonist experiments indicated that $\alpha 4\beta 2$ nAChRs may have been responsible. The nicotinic enhancement was selective, in that corticocortical transmission was not affected and nicotinic agonists did not affect pyramidal cell activity directly.

Additional evidence for nicotinic enhancement of thalamocortical transmission has emerged from electrophysiological recordings in vivo (Gioanni et al., 1999). In anesthetized rats, it was possible to evoke monosynaptic, AMPA receptor-mediated responses in prelimbic cortex by electrical stimulation of the mediodorsal thalamus. These excitatory responses were enhanced by microiontophoretic administration of nicotine and of the nicotinic agonist DMPP; neither agonist directly affected spontaneous cell firing, pointing to a presynaptic mechanism. As in the earlier in vivo study, pharmacological tests indicated that the nAChRs responsible are possibly of the $\alpha 4\beta 2$ subtype, consistent with the expression profile of thalamocortical neurons.

Species differences

Nicotinic modulation of thalamocortical activity has been studied most extensively in the rat. Evidence from other species is at best fragmentary. In the mouse, thalamic expression patterns of nAChR subunit transcripts and [³H]nicotine binding are highly reminiscent of the rat (Marks et al., 1992; Zoli et al., 1998). In addition, electrophysiological recordings have shown that mouse thalamic neurons express functional $\alpha 4\beta 2$ nAChRs at the somatodendritic level (Lena and Changeux, 1997; Zoli et al., 1998; Marubio et al., 1999). However, in contrast to the rat, cortical [³H]nicotine labeling is not noticeably more dense in layers 3/4 (Pauly et al., 1989; Rogers et al., 1998), and whether nAChRs reside on thalamocortical terminals is unknown. In the cat, [³H]nicotine labeling is clearly concentrated in the lateral geniculate nucleus and in its terminal zone in visual cortex, and lesion studies have indicated the likely presence of presynaptic nAChRs (see above). In primates (including humans), available evidence indicates high thalamic [³H]nicotine labeling, but cortical binding patterns appear markedly species-dependent, with no obvious enrichment in layers receiving thalamocortical projections (Adem et al., 1988; Perry et al., 1992; Spurden et al., 1997; Sihver et al., 1998).

Nicotinic cholinergic transmission

Although cholinergic nerve fibers and nAChRs are widely distributed in the brain, only a few sites of nicotinic cholinergic transmission have so far been proposed. In this regard, the thalamus is a prime candidate. As discussed, it possesses a very high abundance of nAChR-related transcripts and [³H]agonist binding sites, and many of its neurons are excited by local application of nicotine. It also enjoys a rich cholinergic innervation, derived principally from Ch5 and Ch6 cholinergic cell groups (Mesulam et al., 1983). At the electron microscopic level, cholinergic synapses are formed mainly with the dendrites of principal cells and interneurons (Hallanger et al., 1990; Beaulieu and Cynader, 1992). It is not known whether thalamic nAChRs are located at synaptic junctions.

Steriade and coworkers have provided functional evidence for nicotinic cholinergic transmission in several thalamic nuclei (Hu et al., 1988, 1989; Curró Dossi et al., 1991). In essence, these experiments showed that electrical stimulation of brainstem sites, intended to excite ascending cholinergic projections, led to a rapid excitation of thalamic relay neurons; this excitation was blocked by a nicotinic antagonist given intravenously or locally by iontophoresis. These experiments were performed in cats; it will be interesting to see whether the findings can be repeated in rodents. Given that nAChRs and cholinergic markers coexist at high levels in almost all thalamic nuclei, it seems likely that nicotinic cholinergic transmission is a widespread phenomenon in this structure.

The situation in the cerebral cortex is less clear. Cholinergic innervation pervades most cortical regions and has been extensively characterized in many respects (Woolf, 1991). As noted above, nAChRs are present in all cortical areas and layers, and at least some of these receptors are pharmacologically active. However, the extent to which cholinergic transmission is synaptic vs. paracrine is currently a matter of debate (Descarries et al., 1997; Turrini et al., 2001). Essentially, several ultrastructural studies have provided evidence that cholinergic fibers form mainly synaptic contacts, whereas other studies have reached the opposite conclusion. Synaptic structures, where seen, were typically axo-dendritic or axo-somatic; it is tempting to speculate that nAChRs located on presynaptic terminals, if targeted by endogenous ACh, subserve a paracrine role.

Functional evidence for nicotinic cholinergic transmission in cerebral cortex is limited at present to two electrophysiological studies, both using slice preparations prepared from rat (Chu et al., 2000) or ferret (Roerig et al., 1997). These studies provide important evidence for nicotinic cholinergic transmission. However, in these studies electrical stimulation was used in order to evoke ACh release; a question that remains to be addressed is how significant spontaneous ACh release is in vivo.

Broader significance

Acetylcholine is a widely distributed neurotransmitter, and contributes to a plethora of behavioral and physiological functions. Accordingly, nicotinic and muscarinic receptors are also widely expressed throughout the brain. At the receptor level, nicotine mimics certain actions of ACh. However, as a self-administered drug, nicotine exerts multiple effects that are complicated by numerous factors. These factors include the existence of nAChR subtypes differing in agonist sensitivity and desensitization kinetics, and in other, more persistent forms of neuroregulation. An additional and potentially important consideration is that nicotine may be able to activate nAChRs that are not cholinergically innervated. For example, it is likely that thalamocortical neurons receive a nicotinic cholinergic input at the somatodendritic level, but whether the same is true of thalamocortical terminals is an open question.

Nicotine acts in a highly pervasive manner in the brain. It is also important, therefore, to understand how this drug effects the functioning of multiple interacting neuronal systems. For example, moderate to high nAChR subunit expression occurs in cortical layer 6, suggesting that *corticothalamic* neurons may also express nAChRs (Wada et al., 1989; Seguela et al., 1993). If this is indeed the case, activity in thalamocortical loops may be affected in complex ways.

Acknowledgments

The author gratefully acknowledges the contributions of colleagues (N. Lavine, M. Reuben) and collaborators (C. Vidal, A.-M. Thierry et al.) to some of the published work described. The author's own work has been supported by the Canadian Institutes of Health Research, Natural Sciences and Engineering Research Council of Canada, and the Fonds de la Recherche en Santé du Québec.

References

Adem, A., Jossan, S.S., d'Argy, R., Brandt, I., Winblad, B. and Nordberg, A. (1988) Distribution of nicotinic receptors in human thalamus as visualized by 3H-nicotine and 3H-acetylcholine receptor autoradiography. J. Neural Transm., 73: 77–83.

Andersen, P. and Curtis, D.R. (1964) The excitation of thalamic neurones by acetylcholine. Acta Physiol. Scand., 61: 85–99.

Araujo, D.M., Lapchak, P.A., Collier, B. and Quirion, R. (1988) Characterization of N-[3H]methylcarbamylcholine binding sites and effect of N-methylcarbamylcholine on acetylcholine release in rat brain. J. Neurochem., 51: 292–299.

Beaulieu, C. and Cynader, M. (1992) Preferential innervation of immunoreactive choline acetyltransferase synapses on relay cells of the cat's lateral geniculate nucleus: A double-labelling study. Neuroscience, 47: 33–44.

Benowitz, N.L. (1996) Pharmacology of nicotine: Addiction and therapeutics. Annu. Rev. Pharmacol. Toxicol., 36: 597–613.

258

Britto, L.R.G., Torrao, A.S., Hamassaki-Britto, D.E., Mpodozis, J., Keyser, K.T., Lindstrom, J.M. and Karten, H.J. (1994) Effects of retinal lesions upon the distribution of nicotinic acetylcholine receptor subunits in the chick visual system. J. Comp. Neurol., 350: 473–484.

Chu, Z.G., Zhou, F.M. and Hablitz, J.J. (2000) Nicotinic acetylcholine receptor-mediated synaptic potentials in rat neocortex. Brain Res., 887: 399–405.

Clarke, P.B.S. (1999) Functional anatomy of nicotinic cholinoceptors in mammalian brain. In: S.P. Arneric and J.D. Brioni (Eds.), Neuronal Nicotinic Receptors: Pharmacology and Therapeutic Opportunities. John Wiley, New York, pp. 127–139.

Clarke, P.B.S., Pert, C.B. and Pert, A. (1984) Autoradiographic distribution of nicotine receptors in rat brain. Brain Res., 323: 390–395.

Clarke, P.B.S., Schwartz, R.D., Paul, S.M., Pert, C.B. and Pert, A. (1985) Nicotinic binding in rat brain: autoradiographic comparison of ^{3}H-acetylcholine, ^{3}H-nicotine, and ^{125}I-alpha-bungarotoxin. J. Neurosci., 5: 1307–1315.

Cordero-Erausquin, M., Marubio, L.M., Klink, R. and Changeux, J.P. (2000) Nicotinic receptor function: new perspectives from knockout mice. Trends Pharmacol. Sci., 21: 211–217.

Court, J., Martin-Ruiz, C., Piggott, M., Spurden, D., Griffiths, M. and Perry, E.K. (2001) Nicotinic receptor abnormalities in Alzheimer's disease. Biol. Psychiatry, 49: 175–184.

Couturier, S., Bertrand, D., Matter, J.M., Hernandez, M.C., Bertrand, S., Millar, N., Valera, S., Barkas, T. and Ballivet, M. (1990) A neuronal nicotinic acetylcholine receptor subunit (alpha 7) is developmentally regulated and forms a homo-oligomeric channel blocked by alpha-BTX. Neuron, 5: 847–856.

Cox, C.L. and Sherman, S.M. (2000) Control of dendritic outputs of inhibitory interneurons in the lateral geniculate nucleus. Neuron, 27: 597–610.

Curró Dossi, R., Paré, D. and Steriade, M. (1991) Short-lasting nicotinic and long-lasting muscarinic depolarizing responses of thalamocortical neurons to stimulation of mesopontine cholinergic nuclei. J. Neurophysiol., 65: 393–405.

Davies, A.R.L., Hardick, D.J., Blagbrough, I.S., Potter, B.V.L., Wolstenholme, A.J. and Wonnacott, S. (1999) Characterisation of the binding of [^{3}H]methyllycaconitine: a new radioligand for labelling α7-type neuronal nicotinic acetylcholine receptors. Neuropharmacology, 38: 679–690.

Descarries, L., Gisiger, V. and Steriade, M. (1997) Diffuse transmission by acetylcholine in the CNS. Prog. Neurobiol., 53: 603–625.

Di Chiara, G. (2000) Behavioural pharmacology and neurobiology of nicotine reward and dependence. In: F. Clementi, D. Fornasari and C. Gotti (Eds.), Neuronal Nicotinic Receptors, Handbook of Experimental Pharmacology, Vol. 144, Springer, Berlin, pp. 603–750.

Flores, C.M., Rogers, S.W., Pabreza, L.A., Wolfe, B.B. and Kellar, K.J. (1992) A subtype of nicotinic cholinergic receptor in rat brain is composed of alpha4 and beta2

subunits and is up-regulated by chronic nicotine treatment. Mol. Pharmacol., 41: 31–37.

Gil, Z., Connors, B.W. and Amitai, Y. (1997) Differential regulation of neocortical synapses by neuromodulators and activity. Neuron, 19: 679–686.

Gioanni, Y., Rougeot, C., Clarke, P.B.S., Lepousé, C., Thierry, A.M. and Vidal, C. (1999) Nicotinic receptors in the rat prefrontal cortex: increase in glutamate release and facilitation of mediodorsal thalamo-cortical transmission. Eur. J. Neurosci., 11: 18–30.

Hallanger, A.E., Price, S.D., Lee, H.J., Steininger, T.L. and Wainer, B.H. (1990) Ultrastructure of cholinergic synaptic terminals in the thalamic anteroventral, ventroposterior, and dorsal lateral geniculate nuclei of the rat. J. Comp. Neurol., 299: 482–492.

Happe, H.K., Peters, J.L., Bergman, D.A. and Murrin, L.C. (1994) Localization of nicotinic cholinergic receptors in rat brain: autoradiographic studies with [^{3}H]cytisine. Neuroscience, 62: 929–944.

Hill, J.A., Zoli, M., Bourgeois, J.-P. and Changeux, J.P. (1993) Immunocytochemical localization of a neuronal nicotinic receptor: The beta2-subunit. J. Neurosci., 13: 1551–1568.

Hu, B., Bouhassira, D., Steriade, M. and Deschenes, M. (1988) The blockage of ponto-occipital waves in the cat lateral geniculate nucleus by nicotinic antagonists. Brain Res., 473: 394–397.

Hu, B., Steriade, M. and Deschenes, M. (1989) The effects of brainstem peribrachial stimulation on neurons of the lateral geniculate nucleus. Neuroscience, 31: 13–24.

Jones, S., Sudweeks, S. and Yakel, J.L. (1999) Nicotinic receptors in the brain: correlating physiology with function. Trends Neurosci., 22: 555–561.

Lavine, N., Reuben, M. and Clarke, P.B.S. (1997) A population of nicotinic receptors is associated with thalamocortical afferents in the adult rat: laminal and areal analysis. J. Comp. Neurol., 380: 175–190.

Laviolette, S.R. and van der Kooy, D. (2003) Blockade of mesolimbic dopamine transmission dramatically increases sensitivity to the rewarding effects of nicotine in the ventral tegmental area. Mol. Psychiat., 8: 50–59.

Lena, C. and Changeux, J.P. (1997) Role of Ca^{2+} ions in nicotinic facilitation of GABA release in mouse thalamus. J. Neurosci., 17: 576–585.

Lukas, R.J., Changeux, J.P., Le Novère, N., Albuquerque, E.X., Balfour, D.J.K., Berg, D.K., Bertrand, D., Chiappinelli, V.A., Clarke, P.B.S., Collins, A.C., Dani, J.A., Grady, S.R., Kellar, K.J., Lindstrom, J.M., Marks, M.J., Quik, M., Taylor, P.W. and Wonnacott, S. (1999) International Union of Pharmacology. XX. Current status of the nomenclature for nicotinic acetylcholine receptors and their subunits. Pharmacol. Rev., 51: 397–401.

Marks, M.J., Pauly, J.R., Gross, S.D., Deneris, E.S., Hermans-Borgmeyer, I., Heinemann, S.F. and Collins, A.C. (1992) Nicotine binding and nicotinic receptor subunit RNA after chronic nicotine treatment. J. Neurosci., 12: 2765–2784.

Marks, M.J., Pauly, J.R., Grun, E.U. and Collins, A.C. (1996) ST/b and DBA/2 mice differ in brain α-bungarotoxin binding and α7 nicotinic receptor subunit mRNA levels: A quantitative autoradiographic analysis. Mol. Brain Res., 39: 207–222.

Marubio, L.M., Arroyo-Jimenez, M.D., Cordero-Erausquin, M., Lena, C., Le Novère, N., D'Exaerde, A.D., Huchet, M., Damaj, M.I. and Changeux, J.P. (1999) Reduced antinociception in mice lacking neuronal nicotinic receptor subunits. Nature, 398: 805–810.

Mesulam, M.M., Mufson, E.J., Wainer, B.H. and Levey, A.I. (1983) Central cholinergic pathways in the rat: an overview based on an alternative nomenclature (Ch1–Ch6). Neuroscience, 10: 1185–1201.

Nakayama, H., Shioda, S., Okuda, H., Nakashima, T. and Nakai, Y. (1995) Immunocytochemical localization of nicotinic acetylcholine receptor in rat cerebral cortex. Mol. Brain Res., 32: 321–328.

Ottersen, O.P. and Storm-Mathisen, J. (1984) GABA-containing neurons in the thalamus and pretectum of the rodent. An immunocytochemical study. Anat. Embryol. (Berl), 170: 197–207.

Papke, R.L. and Porter Papke, J.K. (2002) Comparative pharmacology of rat and human alpha7 nAChR conducted with net charge analysis. Br. J. Pharmacol., 137: 49–61.

Parkinson, D., Kratz, K.E. and Daw, N.W. (1988) Evidence for a nicotinic component to the actions of acetylcholine in cat visual cortex. Exp. Brain Res., 73: 553–568.

Paterson, D. and Nordberg, A. (2000) Neuronal nicotinic receptors in the human brain. Prog. Neurobiol., 61: 75–111.

Pauly, J.R., Stitzel, J.A., Marks, M.J. and Collins, A.C. (1989) An autoradiographic analysis of cholinergic receptors in mouse brain. Brain Res. Bull., 22: 453–459.

Perry, E.K., Court, J.A., Johnson, M., Piggott, M.A. and Perry, R.H. (1992) Autoradiographic distribution of [³H]nicotine binding in human cortex: Relative abundance in subicular complex. J. Chem. Neuroanat., 5: 399–405.

Price, J.L. (1995) Thalamus. In: G. Paxinos (Ed.), The Rat Nervous System. Academic Press, New York, pp. 629–648.

Prusky, G.T., Shaw, C. and Cynader, M.S. (1987) Nicotine receptors are located on lateral geniculate nucleus terminals in cat visual cortex. Brain Res., 412: 131–138.

Roerig, B., Nelson, D.A. and Katz, L.C. (1997) Fast synaptic signaling by nicotinic acetylcholine and serotonin 5-HT3 receptors in developing visual cortex. J. Neurosci., 17: 8353–8362.

Rogers, S.W., Gahring, L.C., Collins, A.C. and Marks, M. (1998) Age-related changes in neuronal nicotinic acetylcholine receptor subunit α4 expression are modified by long-term nicotine administration. J. Neurosci., 18: 4825–4832.

Rossner, S., Schliebs, R., Perez-Polo, J.R., Wiley, R.G. and Bigl, V. (1995) Differential changes in cholinergic markers from selected brain regions after specific immunolesion of the rat cholinergic basal forebrain system. J. Neurosci. Res., 40: 31–43.

Rowell, P.P. and Winkler, D.L. (1984) Nicotinic stimulation of [³H]acetylcholine release from mouse cerebral cortical synaptosomes. J. Neurochem., 43: 1593–1598.

Sahin, M., Bowen, W.D. and Donoghue, J.P. (1992) Location of nicotinic and muscarinic cholinergic and μ-opiate receptors in rat cerebral neocortex: Evidence from thalamic and cortical lesions. Brain Res., 579: 135–147.

Schoepfer, R., Conroy, W.G., Whiting, P., Gore, M. and Lindstrom, J. (1990) Brain alpha-bungarotoxin binding protein cDNAs and MAbs reveal subtypes of this branch of the ligand-gated ion channel gene superfamily. Neuron, 5: 35–48.

Schwartz, R.D., Lehmann, J. and Kellar, K.J. (1984) Presynaptic nicotinic cholinergic receptors labeled by ³H-acetylcholine on catecholamine and serotonin axons in brain. J. Neurochem., 42: 1495–1498.

Seguela, P., Wadiche, J., Dineley-Miller, K., Dani, J.A. and Patrick, J.W. (1993) Molecular cloning, functional properties, and distribution of rat brain α7: A nicotinic cation channel highly permeable to calcium. J. Neurosci., 13: 596–604.

Sharma, G. and Vijayaraghavan, S. (2001) Nicotinic cholinergic signaling in hippocampal astrocytes involves calcium-induced calcium release from intracellular stores. Proc. Natl. Acad. Sci. USA, 98: 4148–4153.

Sihver, W., Gillberg, P.G. and Nordberg, A. (1998) Laminar distribution of nicotinic receptor subtypes in human cerebral cortex as determined by [³H](−)nicotine, [³H]cytisine and [³H]epibatidine in vitro autoradiography. Neuroscience, 85: 1121–1133.

Spurden, D.P., Court, J.A., Lloyd, S., Oakley, A., Perry, R., Pearson, C., Pullen, R.G.L. and Perry, E.K. (1997) Nicotinic receptor distribution in the human thalamus: autoradiographical localization of [³H]nicotine and [¹²⁵I]α-bungarotoxin binding. J. Chem. Neuroanat., 13: 105–113.

Turrini, P., Casu, M.A., Wong, T.P., De Koninck, Y., Ribeiro-da-Silva, A. and Cuello, A.C. (2001) Cholinergic nerve terminals establish classical synapses in the rat cerebral cortex: synaptic pattern and age-related atrophy. Neuroscience, 105: 277–285.

Vidal, C. and Changeux, J.P. (1989) Pharmacological profile of nicotinic acetylcholine receptors in the rat prefrontal cortex: an electrophysiological study in a slice preparation. Neuroscience, 29: 261–270.

Vidal, C. and Changeux, J.P. (1993) Nicotinic and muscarinic modulations of excitatory synaptic transmission in the rat prefrontal cortex in vitro. Neuroscience, 56: 23–32.

Wada, E., Wada, K., Boulter, J., Deneris, E., Heinemann, S., Patrick, J. and Swanson, L.W. (1989) Distribution of alpha 2, alpha 3, alpha 4, and beta 2 neuronal nicotinic receptor subunit mRNAs in the central nervous system: a hybridization histochemical study in the rat. J. Comp. Neurol., 284: 314–335.

Whitehouse, P.J., Martino, A.M., Antuono, P.G., Lowenstein, P.R., Coyle, J.T., Price, D.L. and Kellar, K.J. (1986) Nicotinic acetylcholine binding sites in Alzheimer's disease. Brain Res., 371: 146–151.

Whiting, P. and Lindstrom, J. (1987) Purification and characterization of a nicotinic acetylcholine receptor from rat brain. Proc. Natl. Acad. Sci. USA, 84: 595–599.

Whiting, P.J. and Lindstrom, J.M. (1988) Characterization of bovine and human neuronal nicotinic acetylcholine receptors using monoclonal antibodies. J. Neurosci., 8: 3395–3404.

Woolf, N.J. (1991) Cholinergic systems in mammalian brain and spinal cord. Prog. Neurobiol., 37: 475–524.

Zhu, J.J. and Uhlrich, D.J. (1997) Nicotinic receptor-mediated responses in relay cells and interneurons in the rat lateral geniculate nucleus. Neuroscience, 80: 191–202.

Zoli, M., Lena, C., Picciotto, M.R. and Changeux, J.P. (1998) Identification of four classes of brain nicotinic receptors using $\beta2$ mutant mice. J. Neurosci., 18: 4461–4472.

Progress in Brain Research, Vol. 145
ISSN 0079-6123

CHAPTER 18

Amyloid β peptides and central cholinergic neurons: functional interrelationship and relevance to Alzheimer's disease pathology

Satyabrata Kar and Rémi Quirion*

Douglas Hospital Research Center, Department of Psychiatry, McGill University, 6875 La Salle Blvd., Verdun, Montreal, QC H4H 1R3, Canada

Introduction

Alzheimer's disease (AD) is the most common form of late-life dementia and a leading cause of death in the developed world. It is characterized clinically not only by an impairment in cognition but also by a decline in global function, a deterioration in the ability to perform activities in daily living, and the appearance of behavioral disturbances. The average course of AD is approximately a decade, but the rate of progression is variable. Approximately 8–10% of all persons older than 65 years have AD and the prevalence of the disease doubles every 5 years thereafter. At present, there is no remission in the progression of the disease nor is there any truly effective treatment. This fact, together with a steady increase in life expectancy, particularly in industrialized nations, makes AD one of the most serious health problems of this century (Katzman and Kawas, 1994; Irizarry and Hyman, 2001).

AD appears to have a heterogeneous etiology and can be caused by mutations in the β-amyloid precursor (APP) gene on chromosome 21, the presenilin 1 (PS1) gene on chromosome 14 and the presenilin 2 (PS2) gene on chromosome 1. However, these mutations account for only a small percentage of AD patients with early age of onset and a family

history of the disease. The vast majority of AD patients do not have a familial history of the disease and do not show mutations in these genes (Mullan and Crawford, 1993; Levy-Lahad et al., 1995; Sherrington et al., 1995; Hardy, 1997; St George-Hyslop, 1999). The factors, secondary to causative genes, which play an important role in the pathogenesis of AD, include age, genetic predisposition (e.g. apolipoprotein E, ApoE genotypes) and possibly certain environmental denominators such as head injury and stress (see Poirier et al., 1993; Strittmatter et al., 1993; Muller-Spahn and Hock, 1999; Heininger, 2000; Holmes, 2002).

Neuropathological features of AD

The major neuropathological changes in the brains of AD patients are synaptic loss and neuronal death, particularly in regions related to memory and cognition, and the presence of abnormal intra- and extracellular protein aggregates in the form of neurofibrillary tangles and amyloid plaques, respectively (see Goedert, 1993; Mullan and Crawford, 1993; Cordell, 1994; Lee, 1995; Hardy, 1997; Price and Sisodia, 1998; Selkoe, 2001). Structurally, tangles are paired helical filaments (PHF) composed of largely abnormally hyperphosphorylated microtubule associated protein tau. Under normal condition, tau binds and stabilizes microtubule by reversible

*Corresponding author: Tel.: + 514-762-3048 ext. 2934;
Fax: + 514-762-3034; E-mail: quirem@douglas.mcgill.ca

DOI: 10.1016/S0079-6123(03)45018-8

phosphorylation and dephosphorylation processes mediated via protein kinases and phosphatases, respectively. Phosphorylated tau is unable to bind microtubules, undergoes polymerization into straight filaments and is then cross-linked by glycosylation to form PHF-tau (Goedert, 1993; Lee, 1995; Iqbal et al., 1998; Brion et al., 2001). It is suggested that formation of PHF-tau reduces the ability of tau to stabilize microtubules, leading to disruption of neuronal transport and eventual death of the affected neurons (Johnson and Jenkins, 1996; Lee, 1996; Billingsley and Kincaid, 1997; Iqbal et al., 1998; Brion et al., 2001). The extent of neurofibrillary pathology, particularly the number of cortical neurofibrillary tangles, correlates positively with the severity of dementia (Arriagada et al., 1992; Bierer et al., 1995; Lee, 1995; Iqbal et al., 1998; Selkoe, 2001).

Neuritic plaques are spherical, multicellular lesions containing a compact deposit of amyloid peptides surrounded by dystrophic neurites, activated microglia and reactive astrocytes. The major component of the amyloid deposits is an intermix of β-amyloid$_{1-42}$ (Aβ_{1-42}) and Aβ_{1-40}, the peptides which are generated by proteolytic cleavage of APP. Recent studies have shown that in AD brain Aβ_{1-42} is deposited first and is the predominant form in senile plaques, whereas Aβ_{1-40} is deposited later in the disease process. The neuritic plaques are most prevalent in areas that show extensive neurodegeneration such as entorhinal cortex, hippocampus and association cortices (see Dickson, 1997; Wisniewski et al., 1997; Clippingdale et al., 2001; Selkoe, 2001). At present, the mechanisms which trigger Aβ deposition or their subsequent transformation to neuritic plaques in selected brain regions remain unclear. However, several lines of experimental approaches including (i) the association of some AD cases to inherited APP mutations (see Goedert, 1993; Mullan and Crawford, 1993; Cordell, 1994; Lee, 1995; Hardy, 1997; Selkoe, 2001), (ii) the observed deposition of Aβ peptides preceding other lesions in AD and Down's syndrome brains (Giaccone et al., 1989; Tanzi, 1996), and (iii) the in vitro neurotoxic potential of fibrillar Aβ peptides (Busciglio et al., 1992; Pike et al., 1993; Yankner, 1996; Selkoe, 2001), indicate that assembly of Aβ peptide in the brain may over time initiate and/or

contribute to AD pathogenesis. Given that Aβ peptides are produced constitutively by normal cells in culture and detected in the plasma and cerebrospinal fluid of healthy humans and other mammals, it is possible that overproduction and/or a lack of degradation or clearance mechanisms, may lead to amyloid aggregation which could in turn, contribute to neuronal degeneration and development of AD pathology (Haass et al., 1992; Seubert et al., 1992; Shoji et al., 1992; Selkoe, 2001). There is also evidence to suggest that soluble Aβ peptide can mediate neurotoxicity under in vitro parameters (Clippingdale et al., 2001; Ditaranto et al., 2001; Selkoe, 2001). Recent data from transgenic mice (Games et al., 1995; Hsiao et al., 1996; Irizarry et al., 1997; Calhoun et al., 1998) as well as some in vivo studies (Nitta et al., 1994; Giovannelli et al., 1995; Harkany et al., 1995; Itoh et al., 1996; Maurice et al., 1996, Geula et al., 1998) have reinforced the notion that overexpression of Aβ peptide or injection of aggregated Aβ induces subcellular alterations or neuronal loss in selected brain regions. Although these results implicate a role for Aβ peptides in the process of neurodegeneration, at present neither the significance of Aβ in the normal brain nor the mechanisms by which it may trigger the loss of neurons in selected regions of AD brains are clearly understood.

A well studied component of AD pathology is the loss of synapses and neurons in discrete brain regions, i.e., hippocampus, entorhinal cortex, neocortex and some subcortical areas. The associated neurotransmitters/modulators such as acetylcholine (ACh), noradrenaline, glutamate, neuropeptide Y and somatostatin have also been reported to be differentially altered in AD brains. However, among all the afflicted regions/neurotransmitters, the basal forebrain cholinergic neurons providing major input to the hippocampus and neocortex are the ones found to be most consistently and severely affected in AD pathology (Whitehouse et al., 1982; Hyman et al., 1990; Braak and Braak, 1994; Geula and Mesulam, 1994; DeKosky et al., 1996; Ladner and Lee, 1998). There are also profound decreases in the activity of the acetylcholine synthesizing enzyme, choline acetyltransferase (ChAT), high-affinity uptake of choline and ACh levels in the cortex and hippocampus, which correlate with the number of plaques and tangles, as well as with the clinical severity of the disease (Davies and Maloney, 1976; Whitehouse et al., 1982; Quirion, 1993; Geula and Mesulam, 1994; Kasa et al., 1997; Ladner and Lee, 1998; Blusztajn and Berse, 2000; Auld et al., 2002; Hartig et al., 2002). Interestingly, other cholinergic neurons which are located in the brainstem or in the striatum are found to be relatively spared or affected during late stages of the disease (see Geula and Mesulam, 1994; Muir, 1997; Ladner and Lee, 1998; Francis et al., 1999). These findings, together with the evidence that cholinergic blockade in normal humans has negative effects on cognition, led to the development of the 'cholinergic hypothesis' of AD. According to this hypothesis, the deterioration of cognitive function associated with AD is attributable to the degeneration of cholinergic neurons in the basal forebrain and the associated loss of cholinergic neurotransmission in the cerebral cortex and other areas (Drachman and Leavitt, 1974; Perry et al., 1978; Bartus et al., 1982; Quirion, 1993; Kasa et al., 1997; Muir, 1997; Ladner and Lee, 1998; Francis et al., 1999; Blusztajn and Berse, 2000). Support for the 'cholinergic hypothesis' has also emerged from the findings that drugs which potentiate central cholinergic function have thus far proven to be the most effective forms of therapeutic treatment in AD.

In view of the loss of basal forebrain cholinergic neurons, ACh receptors have also been studied extensively in AD brains (see Nordberg et al., 1992; Quirion, 1993; Geula and Mesulam, 1994; Kasa et al., 1997; Ladner and Lee, 1998; Schroder and Wevers, 1998; Francis et al., 1999; Blusztajn and Berse, 2000). It is generally believed that M2 receptors, most of which are located on presynaptic cholinergic terminals, are significantly reduced in AD brains (Aubert et al., 1992; Nordberg et al., 1992; Quirion, 1993; Ladner and Lee, 1998). The density of postsynaptically located M1 receptors remains unaltered, but there is some evidence for a disruption of the coupling between the receptors, their G-proteins and second messengers (Nordberg et al., 1992; Quirion, 1993; Warpman et al., 1993). As for nicotinic receptors, that are mostly located on cholinergic terminals, the majority of studies have shown that high-affinity nicotinic binding sites are markedly reduced in the hippocampus and neocortex of the AD brain both in vitro, with postmortem tissue, and in

vivo, with positron emission tomography (Aubert et al., 1992; Quirion, 1993; Nordberg et al., 1995; Schroder and Wevers, 1998; Francis et al., 1999). Some recent studies also indicate a significant decrease in α-bungarotoxin (BgTx) binding sites in the hippocampus of AD brains (Schroder and Wevers, 1998). Despite a large body of literature on the cholinergic system in AD brains, there is little information on how the central cholinergic system is influenced by other key components of AD pathology such as, hyperphosphorylated tau protein and $A\beta$-related peptides. This review will define the interactions between $A\beta$ peptides, tau phosphorylation and cholinergic neurons, and will discuss the potential relevance of these relationships to the cholinergic deficits observed in AD brains.

Modulation of cholinergic functions by $A\beta$ peptides

Effects of $A\beta$ peptide on cholinergic neurons under acute conditions

Several lines of evidence suggest that pM-nM $A\beta$ peptides can negatively regulate various steps of ACh synthesis and release, without inducing apparent neurotoxicity. The highly potent and reversible nature of this effect, together with the fact that $A\beta$ peptides are produced constitutively in normal brain cells, suggest that $A\beta$-related peptides can act as physiological active modulator of cholinergic function (see Auld et al., 1998; Mattson and Pedersen, 1998; Blusztajn and Berse, 2000; Hellstrom-Lindahl, 2000; Jhamandas et al., 2001; Auld et al., 2002). We have earlier shown that pM-nM concentrations of $A\beta$ peptides, under acute conditions can potently inhibit high K$^+$—as well as veratridine-evoked ACh release from rat hippocampal and cortical slices. This effect is insensitive to the sodium channel blocker tetrodotoxin, suggesting that $A\beta$ peptide may act at the level of cholinergic terminals (Kar et al., 1996, 1998). Given the evidence that several $A\beta$ fragments, including $A\beta_{1-42}$, $A\beta_{1-40}$, $A\beta_{1-28}$ and $A\beta_{25-35}$ (but not random sequences, all D-isomers or reverse sequence control peptides) can similarly inhibit ACh release from rat hippocampal slices, it is likely that the sequence $A\beta_{25-28}$ (GSNK; the C-terminal domain of the $A\beta_{1-28}$ fragment) is crucial for this effect. It is rather intriguing that striatal ACh release, in contrast to hippocampal or cortical release, is not significantly altered by $A\beta$ peptide at the concentration tested (Kar et al., 1996). The basis for this regional selectivity is currently unclear, but could relate to the distinct cellular characteristics of basal forebrain cholinergic projection neurons versus striatal interneurons (e.g. distribution of $A\beta$ 'receptor' or other intrinsic differences). These results, however, suggest that most vulnerable brain regions in AD, which include the neocortex and the hippocampus, are sensitive to $A\beta$ peptide in the rat brain, whereas striatal cholinergic interneurons are neither as affected in AD nor sensitive to $A\beta$ neuromodulation.

The inhibitory influence of $A\beta$ peptide on ACh release has recently been demonstrated in guinea pig cortical synaptosomes (Wang et al., 1999), as well as in the pure population of cholinergic synaptosomes prepared from the electric organ of the electric ray *Narke japonica* (Satoh et al., 2001). It is also reported that nM $A\beta$ peptide can act synergistically with the neuropeptide galanin to attenuate ACh release from cortical synaptosomal preparations (Wang et al., 1999). More recently, we have shown that $A\beta_{1-40}$ levels in the hippocampus are significantly higher in aged rats compared to young adult rats and that the cholinergic neurons of aged cognitively impaired rats, as evident from $A\beta$-mediated inhibition of hippocampal ACh release, are possibly more sensitive than either aged cognitively unimpaired or young adult rats (Vaucher et al., 2001). Interestingly, using a protocol similar to ours, Lee et al. (2001) reported that the inhibition of ACh release by $A\beta_{25-35}$ could be reversed by certain ginseng saponins (i.e., Rb1), at concentrations which did not by themselves alter ACh release. This effect was not blocked by tetrodotoxin, suggesting a direct interaction at the level of the cholinergic synapse. Given the evidence that ginseng extracts can ameliorate age-related spatial memory impairments in rats, it would be of interest to determine whether the beneficial effects of ginseng involve protection of the cholinergic system against the inhibitory influence of $A\beta$ peptide or its ability to promote ACh release.

The cellular mechanisms by which $A\beta$-related peptides can regulate endogenous ACh release from selected brain regions under acute conditions remain unclear. However, given the nature and potency of the effect, it is possible that several processes important for ACh synthesis and release could be impaired by $A\beta$ peptides. Earlier studies reported that the turnover of ACh in the cholinergic terminals is regulated such that increased transmitter release is associated with an increased rate of synthesis (Tucek, 1985; Collier, 1988). When brain slices are exposed to submaximal concentration of depolarizing agents such K^+ or veratridine, ongoing synthesis of ACh keeps pace with the release from the terminals (Collier, 1988; Wecker, 1991). ACh synthesis under these conditions depends on the high-affinity uptake of choline from extracellular sources and intracellular acetyl CoA and ChAT. The availability of choline, but not ChAT activity, is considered to serve as a rate-limiting step in ACh synthesis (Tucek, 1985; Wecker, 1991). Our results suggest that soluble pM-nM $A\beta$-related peptides, under acute conditions, do not alter ChAT activity in tissue homogenates or in slice preparations from hippocampus, cortex or striatum. Interestingly, temperature-dependent high-affinity [^{3}H]choline uptake is found to be significantly decreased, in a concentration-dependent manner, following a 20 min preincubation with $A\beta$-related peptides, particularly in the hippocampus and cortex, mirroring the observed effect of $A\beta$ peptide on ACh release (Kar et al., 1998). Recently, it was reported that incubation of hippocampal synaptosomes with low nM $A\beta_{1-40}$ for 10 min suppressed depolarization-induced high-affinity choline uptake as well as [^{3}H]hemicholinium-3 ([^{3}H]HC-3) binding. Detailed analysis of these data suggests that changes in the transport are predominantly due to an alteration of V_{max}, whereas the changes in the specific binding probably depend on alterations of both B_{max} and K_D. Higher μM $A\beta_{1-40}$ decreased the high-affinity choline uptake and [^{3}H]HC-3 binding under basal conditions in a time-dependent manner (Kristofikova et al., 2001). These results, taken together, suggest that $A\beta$-related peptides, under acute conditions, can regulate ACh release, at least in part by influencing high-affinity choline uptake. However, their possible involvement in other processes such as intracellular transport or vesicular uptake and release of ACh remains to be established.

Although a variety of receptors [e.g., receptors for advanced glycation end products (RAGE), class A scavenger receptor (SR), the 75kD-neurotrophin receptor (p75^{NTR}), serpin-enzyme complex receptors and nicotinic receptor] have been reported to interact with Aβ peptides (Joslin et al., 1991; El Khoury et al., 1996; Yan et al., 1996, 1997; Kuner et al., 1998; Wang et al., 2000a,b), it is not known whether the effect of Aβ on ACh release is mediated via any of these receptor subtypes. Earlier studies have shown that inhibition of ACh release by Aβ peptide is pertussis toxin-insensitive and is not therefore possibly considered to be mediated via a G-protein coupled receptor sensitive to Gi or Go protein (Wang et al., 1999). More recently, it has been reported that Aβ_{1-42} can be coimmunoprecipitated with α_7 nicotinic receptors and can bind with high affinity to α_7/α-BgTx and with lower affinity to $\alpha_4\beta_2$/cytisine nicotinic (but not muscarinic) receptors in the rat and guinea pig hippocampus and cerebral cortex (Wang et al., 2000a,b; albeit these data need to be replicated by other groups). This is supported, in part, by whole-cell patch-clamp studies which showed that nM Aβ_{1-40}/Aβ_{1-42} can specifically and reversibly block the α_7 nicotinic receptor current in cultured primary hippocampal neurons. The impairment of this current is non-competitive, voltage-independent and is mediated through the extracellular N-terminal length of the α_7 subunit (Liu et al., 2001). Interestingly, Pettit et al. (2001), using hippocampal slices from 13- to 18-day-old rats, found that Aβ_{1-42} can also reversibly inhibit carbachol-induced whole cell nicotinic currents. This study, in contrast to the earlier report, indicates that Aβ peptide can inhibit both α_7 and non-α_7 nicotinic receptor currents—a discrepancy which could relate to the concentration of Aβ peptide and/or the use of cultured neurons vs slice preparations. Collectively, these results provide support for a functional interrelationship between Aβ peptide and nicotinic receptors, but its significance in relation to the modulation of neurotransmitter release remains unclear.

Solubilized Aβ peptide, apart from interacting with nicotinic receptors, also disrupts transduction of the muscarinic M1-like receptor signal. A 4 h exposure to nM–μM Aβ_{25-35} reduces carbachol-induced GTPase activity in rat cortical cultured neurons without affecting muscarinic receptor ligand binding parameters. At higher concentrations, similar treatment with Aβ peptide reduced the intracellular

consequences of muscarinic M1 receptor signaling, by decreasing intracellular Ca^{2+} and the accumulation of Ins(1)P, Ins(1,4)P$_2$, Ins(1,4,5)P$_3$ and Ins(1,3,4,5)P$_4$ (Kelly et al., 1996). More recently, it was shown that exposure of rat cortical cultured neurons to nM Aβ_{1-42}/Aβ_{25-35} inhibited carbachol, but not glutamate, and induced an increase in intracellular Ca^{2+} and Ins(1,4,5)P$_3$ (Huang et al., 2000), thus indicating that a rather selective disruption of muscarinic receptor signaling is another mechanism through which Aβ peptide might affect cholinergic neurotransmission.

It is of interest to note that application of 1 μM Aβ_{25-35} or Aβ_{1-42} to acutely dissociated rat neurons from the diagonal band of Broca decreases whole cell voltage-sensitive currents in cholinergic neurons identified by single cell RT-PCR. This reduction is associated with changes in several K$^+$ currents, including the Ca^{2+}-activated K$^+$ current, the delayed rectifier current and transient outward K$^+$ conductances, but not calcium or sodium currents. These effects are blocked by tyrosine kinase inhibitors, suggesting the possible involvement of tyrosine phoshorylation in mediating the response. Under current-clamp conditions, Aβ evoked an increase in excitability and a loss of accommodation in cholinergic neurons (Jhamandas et al., 2001). These results suggest that Aβ peptide, in contrast to its effect on the hippocampus/cortical cholinergic terminals, can act directly at the level of cell bodies to increase neuronal excitability entailing a higher frequency of action potential generation.

Effects of prolonged exposure of Aβ peptide on cholinergic neurons

Prolonged exposure to pM-nM Aβ-related peptides (i.e., Aβ_{1-42}, Aβ_{1-28}, Aβ_{25-35} and Aβ_{25-28}) has been reported to decrease the intracellular concentration of ACh in the hybrid SN56 cell line (mouse septal neurons x neuroblastoma) expressing the cholinergic phenotype, without any indication of toxicity. This inhibitory action is less potent with Aβ_{25-28} than that seen with larger peptides, suggesting that surrounding amino acids are necessary to obtain a maximal effect. Concomitant with this ACh decrease, there is a reduction in ChAT activity, but not acetylcholinesterase (AChE) activity. The reduction in ACh content could be prevented by a cotreatment with all-trans retinoic acid, a compound previously shown to increase ChAT mRNA expression in SN56 cells, and by the tyrosine kinase inhibitors, genistein and tyrphostin A25 (Pedersen et al., 1996; Pedersen and Blusztajn, 1997; Blusztajn and Berse, 2000). Interestingly, nM concentrations of $A\beta_{1-42}$ have been reported to decrease ACh production in primary cultures of rat septal neurons with concomitant reductions in the activity of pyruvate dehydrogenase (PDH), an enzyme that generates acetyl-CoA from pyruvate, but without affecting ChAT activity or neuronal survival. The decreased PDH activity likely results from $A\beta$ activation of the τ protein kinase I/glycogen synthase kinase-3β (GSK-3β), which is considered to phosphorylate and inactivate PDH (Hoshi et al., 1997). This study indicated that chronic exposure to $A\beta$ peptide might impair ACh processing by an additional mechanism i.e., by undermining the availability of acetyl CoA.

A number of in vitro studies have shown that chronic exposure to $A\beta$ peptides can be toxic in a variety of cell lines, as well as in primary cultures of rat or human neurons. The toxic potency of the peptide, unlike its neuromodulatory effects, is related to its ability to form insoluble aggregates, which depends not only on the presence of the 25–35 decapeptide sequence but also on the concentration of $A\beta$ used in the study (Busciglio et al., 1992; Pike and Cotman, 1993; Yankner, 1996). It is noteworthy that some neuronal phenotypes, such as GABAergic and serotonergic neurons, are apparently resistant to $A\beta$ toxicity, while various cell lines differ in their degree of sensitivity (Pike and Cotman, 1993; Olesen et al., 1998; Gschwind and Huber, 1995). It is also reported that differentiation of RN46A cell line into cholinergic phenotype by ciliary neurotrophic factor resulted in a cell population that is highly sensitive to $A\beta$ peptide, whereas differentiation of RN46A cell line into serotonergic phenotype by brain derived growth factor yielded a cell population that was unaffected by $A\beta$ treatment (Olesen et al., 1998). Our recent data indicate that chronic exposure of rat primary septal culture neurons to μM $A\beta$ peptides can induce toxicity with a concomitant decrease in ChAT activity (Zheng et al., 2002). These results suggest that μM concentrations of $A\beta$ peptide can

induce toxicity in neurons expressing the cholinergic phenotype, whereas pM–nM concentrations of the peptide can reduce cholinergic markers without affecting neuronal survival.

The mechanisms by which $A\beta$ peptides are toxic to cholinergic neurons are still unclear, but may involve, as indicated from other studies, alterations in intracellular calcium and/or the production of free radicals (Behl et al., 1992; Mattson et al., 1992; Hensley et al., 1994; Mattson and Pedersen, 1998). A variety of experimental data from different cell lines and primary cultured neurons suggest that $A\beta$ toxicity might be mediated either by interaction with a hydroxysteroid dehydrogenase enzyme or by the plasma membrane RAGE, SR, p75[NTR] and/or nicotinic receptors (El Khoury et al., 1996; Yan et al., 1996, 1997; Kuner et al., 1998; Wang et al., 2000a,b). Whether $A\beta$ peptides induce toxicity in cholinergic neurons by interacting with any of these receptor subtypes remain to be established. However, it has been demonstrated that 24 h exposure to μM $A\beta_{1-40}$ can increase choline fluxes from PC12 cells (Allen et al., 1997). If μM $A\beta$ peptide does so in cholinergic neurons, it is likely that the observed vulnerability could partly be the consequence of a concomitant decrease in the uptake and an increase leakage of choline from the neurons. More recently, it was demonstrated that aggregated μM $A\beta$ peptides can induce the phosphorylation of tau protein in SN56 cholinergic cell lines, which may contribute to the vulnerability of these neurons by destabilizing microtubules and impaired axonal transport (Le et al., 1997). Our results from primary rat septal cultured neurons also indicate that aggregated $A\beta$ peptides, apart from being toxic, can significantly increase the phosphorylation of tau protein in a time- and concentration-dependent manner. At the cellular level, phosphorylated tau immunoreactivity in control cultures was detected primarily in the distal axons, whereas staining in $A\beta$-treated cultures was not only confined to axons, but also appeared in the soma, as well as dendrites of the neurons (Zheng et al., 2002). Interestingly, the levels of total tau were also found to be slightly increased in $A\beta$-treated cultures, which may account, at least in part, for the observed increase in phospho-tau levels (Busciglio et al., 1995; Le et al., 1997; Shea et al., 1997; Alvarez et al., 1999). Given the evidence that phosphorylated tau protein can lead to the death of neurons, possibly via disruption of the neuronal network, in a variety of cell lines as well as in primary cultured neurons (see Billingsley and Kincaid, 1997; Iqbal et al., 1998; Brion et al., 2001), it is likely that increased phosphorylation of tau protein induced by $A\beta$ peptides may underlie the cause of neurotoxicity in septal cultured neurons including those expressing the cholinergic phenotype.

The mechanisms by which $A\beta$-related peptides induce the phosphorylation of the tau protein remain to be identified. It has been proposed that reactive oxygen species and the lipid peroxidation product 4-hydroxynonenal may be involved in $A\beta$-neurotoxicity and the cross-linking of tau proteins (Mark et al., 1997). However, a growing body of evidence suggests that the effects of $A\beta$ on tau phosphorylation could be the direct consequence of increased kinase or decreased phosphatase activity (Busciglio et al., 1995; Greenberg and Kosik, 1995; Le et al., 1997; Takashima et al., 1998; Alvarez et al., 2001). Our recent experiments with cultured rat septal neurons showed that chronic exposure to μM $A\beta$ peptide, in addition to inducing tau phosphorylation, significantly increases the levels of mitogen activated protein (MAP) kinase and GSK-3β, whereas treatment with specific inhibitors of MAP kinase (i.e., PD98059) and GSK-3β (i.e., lithium chloride) attenuates $A\beta$-induced tau phosphorylation. These data indicate a role for MAP kinase and GSK-3β in $A\beta$-mediated tau phosphorylation in rat septal cultured neurons (Zheng et al., 2002). Earlier studies have shown that the activation of GSK-3β by $A\beta$-related peptides induces tau protein phosphorylation and cell death in cortical and hippocampal cultures, whereas blockade of GSK-3β expression or activity, either by antisense oligonucleotides or by lithium, prevented $A\beta$-induced neurodegeneration (Hong et al., 1997; Takashima et al., 1998; Alvarez et al., 1999). MAP kinase activation has also been implicated in tau phosphorylation and/or degeneration of neurites/neurons in hippocampal, cortical and PC12 cell cultures (Greenberg et al., 1994; Ekinci et al., 1999; Rapoport and Ferreira, 2000). Given the evidence that various kinases including MAP kinase and GSK-3β phosphorylate tau at some, but not all sites, it is likely that the phosphorylation of tau protein in rat septal cultured neurons could result

from the activation of multiple kinases. It is therefore of interest to determine not only the possible involvement of other tau kinases, such as cyclin-dependent kinase 5, PKC and calcium-calmodulin kinase (Billingsley and Kincaid, 1997; Brion et al., 2001), in $A\beta$-induced phosphorylation of tau protein, but also to examine the association, if any, between specific kinase(s) and neurons expressing the cholinergic phenotype.

Several attempts have also been made to measure the impact of $A\beta$ peptide on cholinergic system under in vivo conditions. Despite some inconsistent data, which may relate to the experimental paradigms or to differences in delivering aggregated peptide to the brain, it has been reported that $A\beta$ peptides can induce cholinergic hypofunction when administered to the brain (Auld et al., 1998; Blusztajn and Berse, 2000). Injection of $A\beta$ peptides into the rat medial septum was shown to cause a reduction in ACh release from the hippocampus in the absence of toxicity (Abe et al., 1994). Using a similar approach, Harkany et al. (1995) demonstrated that $A\beta_{1-42}$ exhibits toxic effects on cholinergic neurons, as indicated by a decrease in ChAT-immunoreactive neurons within basal forebrain, and by a reduction in ChAT immunoreactive fibers in the cerebral cortex. Additional studies have shown that infusion of $A\beta$ into the lateral ventricles of adult rats results in impaired performance on learning and memory tasks which correlate with deficits in cholinergic transmission (Nitta et al., 1994; Giovannelli et al., 1995; Itoh et al., 1996; Maurice et al., 1996). Whether the effect of $A\beta$ on the cholinergic system in these in vivo studies occurs through similar mechanisms as observed in in vitro studies remains to be established.

In the last few years, the structural integrity of the central cholinergic system has been studied in mutant APP, PS1 or APP/PS1 transgenic mice with elevated $A\beta$ levels. Human APP Swedish mutant (hAPPswe) transgenic mice at 12 months of age were shown to display dystrophic AChE fibers in the vicinity of neuritic plaques and cell losses in the CA1 region of the hippocampus (Sturchler-Pierrat et al., 1997; Calhoun et al., 1998). Similarly, in 17- to 22-month hAPP London mutant transgenic mice, AChE-positive fibers are reorganized in the hippocampus, and amyloid plaques in the cortex are surrounded by dystrophic AChE-positive fibers (Bronfman et al.,

2000). In 8-month-old hAPPswe/PS1$_{M146L}$ double transgenic mice, which display extensive amyloid plaques in the brain, the density and size of cholinergic synapses are decreased in the frontal cortex and hippocampus. By contrast, PS1$_{M146L}$ single transgenic mice demonstrated no apparent changes in either the size or density of cholinergic synapses, whereas APPswe mutant mice showed an upregulation in the density of cholinergic synapses in the frontal and parietal cortices and in the hippocampus (Wong et al., 1999). Interestingly, a stereological study demonstrated a selective increase in p^{75NTR}- (a putative marker of basal forebrain cholinergic neurons) positive neurons in the medial septum of 12-month-old hAPPswe and PS1$_{M146L}$ single transgenic mice, but not in hAPPswe/PS1$_{M146L}$ double transgenic mice. Furthermore, p^{75NTR}-immunoreactive fiber staining in hippocampus was reported to be more robust in both single transgenic mice as compared to the non-transgenic controls, while double transgenic mice displayed the least amount of p^{75NTR} fiber staining (Jaffar et al., 2001). More recently, a longitudinal (i.e., 14, 18, and 23 months of age) study of APPswe transgenic mice revealed no significant differences in either ChAT activity, AChE activity, vesicular ACh transporter binding or high-affinity choline uptake sites in any brain regions (i.e., cortex, hippocampus, striatum and cerebellum) as compared to wild-type control mice (Gau et al., 2002). Whether these differences relate to the type of mutation, the level of transgene expression, age or to the experimental paradigm remain unclear. With regards to cholinergic receptors, we failed to observe significant differences in the level of M1/[^{3}H]pirenzepine, M2/[^{3}H]AF-DX 384 or α_7 nicotinic/[^{125}I]α-BgTx receptor binding sites in any brain regions among mutant PS1$_{L286V}$ transgenic, wild-type PS1 transgenic and littermate non-transgenic controls (Vaucher et al., 2002). However, a significant increase in hippocampal α_7 nicotinic receptor levels has recently been demonstrated in hAPPswe single as well as hAPPswe/PS1$_{A246E}$ double transgenic mice as compared to age-matched controls (Dineley et al., 2001, 2002). Taken together, these findings suggest that increased levels of Aβ peptides, as observed in mutant APP, PS1 or APP/PS1 transgenic mice, may result in subtle alterations in the basal forebrain cholinergic system.

Implications of Aβ peptide and cholinergic neuron interactions

Aβ related peptides are produced constitutively by brain cells and are found in pM-nM range in the CSF of normal individuals (Haass et al., 1992; Seubert et al., 1992; Shoji et al., 1992; Selkoe, 2001). Physiological concentrations of Aβ-related peptides, under acute conditions, have been shown to regulate negatively various steps in the synthesis and release of ACh, thus suggesting a possible neuromodulatory role for Aβ peptide in the regulation of normal cholinergic functions. On the contrary, it is reported that lesioning of the basal forebrain cholinergic neurons or a transient inhibition of cortical ACh release elevates the synthesis of APP in the cerebral cortex (Iverfeldt et al., 1993; Wallace et al., 1993; Roberson and Harrell, 1997; Lin et al., 1999). Furthermore, the agonist-induced activation of muscarinic m$_1$ and m$_3$ receptor subtypes has also been shown to increase the secretion of soluble APP derivatives and to reduce the production of amyloidogenic Aβ peptides (Buxbaum et al., 1992; Nitsch et al., 1992; Pittel et al., 1996; Roberson and Harrell, 1997; Mills and Reiner, 1999; Hellstrom-Lindahl, 2000; Auld et al., 2002). Thus, it appears that the normal cholinergic innervation participates in the nonamyloidogenic maturation of APP, whereas Aβ peptides seem to be involved in the regulation of ACh release from selected brain regions. These results, taken together, suggest the existence of a reciprocal control mechanism between Aβ related peptides and the functioning of long projection cholinergic neurons originating from the basal forebrain. If this is indeed the case, then it is of interest to determine how the modulatory role of Aβ peptides is being compromised, leading to Aβ deposition and/or the degeneration of the neurons seen in AD brains. It is possible that an overproduction or lack of degradation/clearance mechanism either by mutations in specific genes (i.e., APP, PS1 or PS2) and/or insults to the cholinergic neurons increases the levels of Aβ peptides. This alteration, together with the possible changes in the local microenvironment, can subsequently enhance the vulnerability of neurons to direct toxicity seen at high concentrations of Aβ peptides (Pike et al., 1993; Yankner, 1996; Selkoe, 2001), and/ or by short supply of choline (Wurtman, 1992; Allen et al., 1997; Auld et al., 1998; Kar et al., 1998; Kristofikova et al., 2001). Since cholinergic neurons can utilize choline from membrane phosphatidylcholine to synthesize ACh, it is also possible that $A\beta$-induced alteration in intracellular choline levels could trigger membrane disruption to sustain neurotransmission which, in turn, could lead to degeneration of the neurons by increased $A\beta$ production from APP and/or autocannibalism (Wurtman, 1992). Furthermore, given the evidence that $A\beta$ deposits precede any other lesions in AD brains (Giaccone et al., 1989; Tanzi, 1996), it is likely that amyloid-induced tau phosphorylation plays a critical role in neuronal losses observed in the AD brain. This is supported by some in vivo studies which showed that injection or infusion of $A\beta$ peptides directly into the brain or overexpression of $A\beta$ peptide in APP/PS1 double transgenic mice can induce loss of neurons and/or presynaptic cholinergic markers in selected brain regions (Nitta et al., 1994; Giovannelli et al., 1995; Harkany et al., 1995; Itoh et al., 1996; Maurice et al., 1996). These data together with the observed selective effects of $A\beta$ peptides on certain cholinergic neurons provide a basis to suggest that the vulnerability of basal forebrain cholinergic neurons and their projection in AD could relate, at least in part, to their sensitivity to $A\beta$ peptides.

Acknowledgments

The work was supported by grants from the Alzheimer Society of Canada and Canadian Institute of Health Research to S.K. and R.Q. S.K. is recipient of a Chercheur-Boursier Senior award from the 'Fonds de la Recherche en Santé du Québec'.

Abbreviations

$A\beta$	beta amyloid
ACh	acetylcholine
AD	Alzheimer disease
AChE	acetylcholinesterase
APP	amyloid precursor protein
α-BgTx	α-bungarotoxin
ChAT	choline acetyltransferase
DS	Down's syndrome
GSK	glycogen synthase kinase
MAPK	mitogen activated protein kinase
PHF	paired helical filaments
PDH	pyruvate dehydrogenase
PKC	protein kinase C
PS	presenilin
RAGE	receptors for advanced glycation end products
p^{75NTR}	75kD-neurotrophin receptor

References

Abe, E., Casamenti, F., Giovannelli, L., Scali, C. and Pepeu, G. (1994) Administration of amyloid β-peptides into the medial septum of rats decreases acetylcholine release from hippocampus in vivo. Brain Res. 636: 162–164.

Allen, D.D., Galdzicki, Z., Brining, S.K., Fukuyama, R., Rapoport, S.I. and Smith, Q.R. (1997) Beta-amyloid induced increase in choline flux across PC12 cell membranes. Neurosci. Lett. 234: 71–73.

Alvarez, G., Munoz-Montano, J.R., Satrustegui, J., Avila, J., Bogonez, E. and Diaz-Nido, J. (1999) Lithium protects cultured neurons against β-amyloid-induced neurodegeneration. FEBS Lett. 453: 260–264.

Alvarez, A., Toro, R., Caceres, A. and Maccioni, R.B. (2001) Inhibition of tau phosphorylating protein kinase cdk5 prevents β-amyloid-induced neuronal death. FEBS Lett. 459: 421–426.

Arriagada, P.V., Growdon, J.H., Hedley-Whyte, E.T. and Hyman, B.T. (1992) Neurofibrillary tangles but not senile plaques parallel duration and severity of Alzheimer's disease. Neurology 42: 631–639.

Aubert, I., Araujo, D.M., Cecyre, D., Robitaille, Y., Gauthier, S. and Quirion, R. (1992) Comparative alterations of nicotinic and muscarinic receptor binding sites in Alzheimer's and Parkinson's diseases. J. Neurochem. 58: 529–541.

Auld, D.S., Kar, S. and Quirion, R. (1998) β-amyloid peptides as direct cholinergic neuromodulators: a missing link. Trends Neurosci. 21: 408–417.

Auld, D.S., Kornecook, T.J., Bastianetto, S. and Quirion, R. (2002) Alzheimer's disease and the basal forebrain cholinergic system: relations to beta-amyloid peptides, cognition, and treatment strategies. Prog. Neurobiol. 68: 209–245.

Bartus, R.T., Dean, R.L., III, Beer, B. and Lipa, A.S. (1982) The cholinergic hypothesis of geriatric memory dysfunction. Science 217: 408–417.

Behl, C., Cole, G.M. and Schubert, D. (1992) Vitamin E protects nerve cells from amyloid β protein toxicity. Biochem. Biophys. Res. Commun. 186: 944–950.

Bierer, L.M., Hof, P.R., Purohit, D.P., Carlin, L., Schmeidler, J., Davis, K.L. and Perl, D.P. (1995)

Neocortical neurofibrillary tangles correlate with dementia severity in Alzheimer's disease. Arch. Neurol. 52: 81–88.

Billingsley, M.L. and Kincaid, R.L. (1997) Regulated phosphorylation and dephosphorylation of tau protein: effects on microtubule interaction, intracellular trafficking and neurodegeneration. Biochem. J. 323: 577–591.

Blusztajn, J.K. and Berse, B. (2000) The cholinergic neuronal phenotype in Alzheimer's disease. Met. Brain Res. 15: 45–64.

Braak, H. and Braak, E. (1994) Pathology of Alzheimer's disease. In: D.B. Calne (Ed.), Neurodegenerative Diseases, Saunders, Philadelphia, pp. 585–613.

Brion, J.P., Anderton, B.H., Authelet, M., Dayanandan, R., Leroy, K., Lovestone, S., Octave, J.N., Pradier, L., Touchet, N. and Tremp, G. (2001) Neurofibrillary tangles and tau phosphorylation. Biochem. Soc. Symp. 67: 81–88.

Bronfman, F.C., Moechars, D. and Van Leuven, F. (2000) Acetylcholinesterase-positive fiber deafferentation and cell shrinkage in the septphippocampal pathway of aged amyloid precursor protein london mutant transgenic mice. Neurobiol. Dis. 7: 152–168.

Busciglio, J.A., Lorenzo, A. and Yankner, B.A. (1992) Methodological variables in the assessment of β amyloid neurotoxicity. Neurobiol. Aging 13: 609–612.

Busciglio, J., Lorenzo, A., Yeh, J. and Yankner, B.A. (1995) β-amyloid fibrils induce tau phosphorylation and loss of microtubule binding. Neuron 14: 879–888.

Buxbaum, J.D., Oishi, M., Chen, H.I., Pinkas-Kramarski, R., Jaffe, E.A., Gandy, S.E. and Greengard, P. (1992) Cholinergic agonists and interleukin 1 regulate processing and secretion of the (β/A$_4$ amyloid precursor protein. Proc. Natl. Acad. Sci. USA 89: 10075–10078.

Calhoun, M., Wiederhold, K., Abramowski, D., Phinney, A.L., Sturchler-Pierrat, C., Staufenbiel, M., Sommer, B. and Jucker, M. (1998) Neuron loss in APP transgenic mice. Nature 395: 755–756.

Clippingdale, A.B., Wade, J.D. and Barrow, C.J. (2001) The amyloid-β peptide and its role in Alzheimer's disease. J. Peptide Sci. 7: 227–249.

Collier, B. (1988) The synthesis and storage of acetylcholine in mammalian cholinergic nerve terminals. In: M. Avoli, T.A. Reader, R.W. Dykes and P.P. Gloor (Eds.), Neurotransmitters and Cortical Function: From Molecules to Mind, Plenum Press, New York, pp. 261–276.

Cordell, B. (1994) β-amyloid formation as a potential therapeutic target for Alzheimer's disease. Annu. Rev. Pharmacol. Toxicol. 34: 69–89.

Davies, P. and Maloney, A.J.F. (1976) Selective loss of central cholinergic neurons in Alzheimer's disease. Lancet 2: 1403.

DeKosky, S.T., Scheff, S.W. and Styren, S.D. (1996) Structural correlates of cognition in dementia: quantification and assessment of synapse change. Neurodegeneration 5: 417–421.

Dickson, D.W. (1997) The pathogenesis of senile plaques. J. Neuropathol. Exp. Neurol. 56: 321–339.

Dineley, K.T., Xia, X., Bui, D., Sweatt, J.D. and Zheng, H. (2002) Accelerated plaque accumulation, associative learning deficits and upregulation of α_7 nicotinic receptor protein in transgenic mice co-expressing mutant human presenilin 1 and amyloid precursor proteins. J. Biol. Chem. 227: 22768–22780.

Dineley, K.T., Westerman, M., Bui, D., Bell, K., Ashe, K.H. and Sweatt, J.D. (2001) β-amyloid activates the mitogen-activated protein kinase cascade via hippocampal α_7 nicotinic acetylcholine receptors: in vitro and in vivo mechanisms related to Alzheimer's disease. J. Neurosci. 21: 4125–4133.

Ditaranto, L., Tekirian, T.L. and Yang, A.L. (2001) Lysosomal membrane damage in soluble Abeta-mediated cell death in Alzheimer's disease. Neurobiol. Dis. 8: 19–31.

Drachman, D.A. and Leavitt, J. (1974) Human memory and the cholinergic system: a relationship to aging? Arch. Neurol. 3: 113–121.

Ekinci, E.J., Malik, K.U. and Shea, T.B. (1999) Activation of the voltage-sensitive calcium channel by mitogen-activated protein (MAP) kinase following exposure of neuronal cells to β-amyloid. J. Biol. Chem. 274: 30322–30327.

El Khoury, J., Hickman, S.E., Thomas, C.A., Cao, L., Silverstein, S.C. and Loike, J.D. (1996) Scavenger receptor-mediated adhesion of microglia to β-amyloid fibrils. Nature 382: 716–719.

Francis, P.T., Palmer, A.M., Snape, M. and Wilcock, G.K. (1999) The cholinergic hypothesis of Alzheimer's disease: a review of progress. J. Neurol. Neurosurg. Psychiatry 66: 137–147.

Games, D., Adams, D., Alessandrini, R., Barbour, R., Berthelette, P., Blackwell, C., Carr, T., Clemens, J., Donaldson, T., Gillespie, F., Guido, T., Hagopian, S., Johnson-Wood, K., Khan, K., Lee, M., Leibowitz, P., Lieberburg, I., Little, S., Masliah, E., McConlogue, L., Montoya-Zavala, M., Mucke, L., Paganini, L., Penniman, E., Power, M., Schenk, D., Seubert, P., Snyder, B., Soriano, F., Tan, H., Vitale, J., Wadsworth, S., Wolozin, B. and Zhao, J. (1995) Alzheimer type neuropathology in transgenic mice overexpressing V717F β-amyloid precursor protein. Nature 373: 523–527.

Gau, J.T., Steinhilb, M.L., Kao, T.C., D'Amato, C.J., Gaut, J.R., Frey, K.A. and Turner, R.S. (2002) Stable β-secretase activity and presynaptic cholinergic markers during progressive central nervous system amyloidogenesis in Tg2576 mice. Am. J. Pathol. 160: 731–738.

Geula, C. and Mesulam, M.M. (1994) Cholinergic system and related neuropathological predilection patterns in Alzheimer's disease. In: R.D. Terry, R. Katzman and K.L. Bick (Eds.), Alzheimer Disease, Raven Press, New York, pp. 263–291.

Geula, C., Wu, C.K., Saroff, D., Lorenzo, A., Yuan, M. and Yanker, B.A. (1998) Aging renders the brain vulnerable to amyloid β-protein neurotoxicity. Nat. Med. 4: 827–831.

Giaccone, G., Tagliavini, F., Linoli, G., Bouras, C., Frigerio, L., Frangione, B. and Bugiani, O. (1989) Down patients: extracellular preamyloid deposits precede neuritic degeneration and senile plaques. Neurosci. Lett. 97: 232–238.

Giovannelli, L., Casamenti, F., Scali, C., Bartolini, L. and Pepeu, G. (1995) Differential effects of amyloid peptides β-(1–40) and β-(25–35) injections into rat nucleus basalis. Neuroscience 66: 781–792.

Goedert, M. (1993) Tau protein and the neurofibrillary pathology of Alzheimer's disease. Trends Neurosci. 16: 460–465.

Greenberg, S.M. and Kosik, K.S. (1995) Secreted β-APP stimulates MAP kinase and phosphorylation of tau in neurons. Neurobiol. Aging 16: 403–408.

Greenberg, S.M., Koo, E.H., Selkoe, D.J., Qiu, W.Q. and Kosik, K.S. (1994) Secreted β-amyloid precursor protein stimulates mitogen-activated protein kinase and enhances phosphorylation. Proc. Natl. Acad. Sci. USA 91: 7104–7108.

Gschwind, M. and Huber, S. (1995) Apoptotic cell death induced by β-amyloid1-42 is cell type dependent. J. Neurochem. 65: 292–300.

Haass, C., Schlossmacher, M.G., Hung, A.Y., Vigo-Pelfrey, C., Mellon, A., Ostaszewski, B.L., Lieberburg, I., Koo, E.H., Schenk, D., Teplow, D.B. and Selkoe, D.J. (1992) Amyloid β-peptide is produced by cultured cells during normal metabolism. Nature 359: 322–325.

Hardy, J. (1997) Amyloid, the presenilins and Alzheimer's disease. Trends Neurosci. 20: 154–159.

Harkany, T., Lengyel, Z., Soos, K., Penke, B., Luiten, P.G.M. and Gulya, K. (1995) Cholinotoxic effects of β-amyloid1-42 peptide on cortical projections of the rat nucleus basalis magnocellularis. Brain Res. 695: 71–75.

Hartig, W., Bauer, A., Brauer, K., Grosche, J., Hortobagyi, T., Penke, B., Schliebs, R. and Harkany, T. (2002) Functional recovery of cholinergic basal forebrain neurons under disease conditions: old problems, new solutions? Rev. Neurosci. 13: 95–165.

Heininger, K. (2000) A unifying hypothesis of Alzheimer's disease. IV. causation and sequence of events. Rev. Neurosci. 11: 213–328.

Hellstrom-Lindahl, E. (2000) Modulation of β-amyloid precursor protein processing and tau phosphorylation by acetylcholine receptors. Eur. J. Pharmacol. 393: 255–263.

Hensley, K., Carney, J.M., Mattson, M.P., Aksenova, M., Harris, M., Wu, J.F., Floyd, R.A. and Butterfield, D.A. (1994) A model for β-amyloid aggregation and neurotoxicity based on free radical generation by the peptide: relevance to Alzheimer disease. Proc. Natl. Acad. Sci. USA 91: 3270–3274.

Holmes, C. (2002) Genotype and phenotype in Alzheimer's disease. Brit. J. Psychiatry 180: 131–134.

Hong, M., Chen, D.C.R., Klein, P.S. and Lee, V.M.Y. (1997) Lithium reduces tau phosphorylation by inhibition of glycogen synthase kinase-3. J. Biol. Chem. 40: 25326–25332.

Hoshi, M., Takashima, A., Murayama, M., Yasutake, K., Yoshida, N., Ishiguro, K., Hoshino, T. and Imahori, K. (1997) Nontoxic amyloid β peptide$_{1-42}$ supresses acetylcholine synthesis. J. Biol. Chem. 272: 2038–2041.

Hsiao, K., Chapman, P., Nilsen, S., Eckman, C., Harigaya, Y., Younkin, S., Yang, F. and Cole, G. (1996) Correlative memory deficits, Aβ elevation, and amyloid plaques in transgenic mice. Science 274: 99–102.

Huang, H.M., Ou, H.C. and Hsieh, S.J. (2000) Amyloid beta peptide impaired carbachol but not glutamate-mediated phosphoinositide pathways in cultured rat cortical neurons. Neurochem. Res. 25: 303–312.

Hyman, B.T., Van Hoesen, G.W. and Damasio, A.R. (1990) Memory-related neuronal systems in Alzheimer's disease: an anatomic study. Neurology 40: 55–66.

Iqbal, K., Alonso, A.C., Gong, C.X., Khatoon, S., Pei, J.J., Wang, J.Z. and Grundke-Iqbal, I. (1998) Mechanisms of neurofibrillary degeneration and the formation of neurofibrillary tangles, J. Neural Transm. (suppl) 53: 169–180.

Irizarry, M. and Hyman, B. (2001) Alzheimer's disease. In: T. Batchlor and M. Cudkowicz (Eds.), Principles of Neuroepidemiology, Butterworth-Heinemann, Boston, pp. 69–98.

Irizarry, M.C., Soriano, F., McNamara, M., Page, K.J., Schenk, D., Games, D. and Hyman, B.T. (1997) Aβ deposition is associated with neuropil changes, but not with overt neuronal loss in the human amyloid precursor protein V717 (PDAPP) transgenic mouse. J. Neurosci. 17: 7053–7059.

Itoh, A., Nitta, A., Nadai, M., Nishimura, K., Hirose, M., Hasegawa, T. and Nabeshima, T. (1996) Dysfunction of cholinergic and dopaminergic neuronal systems in β-amyloid protein-infused rats. J. Neurochem. 66: 1113–1117.

Iverfeldt, K., Walaas, S.I. and Greengard, P. (1993) Altered processing of Alzheimer amyloid precursor protein in response to neuronal degeneration. Proc. Natl. Acad. Sci. USA 90: 4146–4150.

Jaffar, S., Counts, S.E., Ma, S.Y., Dadko, E., Gordon, M.N., Morgan, D. and Mufson, E.J. (2001) Neuropathology of mice carrying mutant APPswe and/or PS1$_{M146L}$ transgenes: alterations in the p^{75NTR} cholinergic basal forebrain septo-hippocampal pathway. Exp. Neurol. 170: 227–243.

Jhamandas, J.H., Cho, C., Jassar, B., Harris, K., MacTavish, D. and Easaw, J. (2001) Cellular mechanisms for amyloid b-protein activation of rat basal forebrain neurons. J. Neurophysiol. 86: 1312–1320.

Johnson, G.V.W. and Jenkins, S.M. (1996) Tau protein in normal and Alzheimer's disease brain. Alzheimer's Dis. Rev. 1: 38–54.

Joslin, G., Krause, J.E., Hershey, A.D., Adams, S.P., Fallon, R.J. and Perlmutter, D.H. (1991) Amyloid-β peptide, substance, and bombesin bind to the serpin-enzyme complex receptor. J. Biol. Chem. 266: 21897–21902.

Kar, S., Seto, D., Gaudreau, P. and Quirion, R. (1996) β-amyloid-related peptides inhibit potassium-evoked acetylcholine release from rat hippocampal formation. J. Neurosci. 16: 1034–1040.

Kar, S., Issa, A.M., Seto, D., Auld, D.S., Collier, B. and Quirion, R. (1998) Amyloid β-peptide inhibits high-affinity choline uptake and acetylcholine release in rat hippocampal slices. J. Neurochem. 70: 2179–2187.

Kasa, P., Rakonczay, Z. and Gulya, K. (1997) The cholinergic system in Alzheimer's disease. Prog Neurobiol. 52: 511–535.

Katzman, R. and Kawas, C.H. (1994) The epidemiology of dementia and Alzheimer's disease. In: R.D. Terry, R. Katzman and K.L. Bick (Eds.), Alzheimer Disease, Raven Press, New York, pp. 105–122.

Kelly, J.F., Furukawa, K., Barger, S.W., Rengen, M.R., Mark, R.J., Blanc, E.M., Roth, G.S. and Mattson, M.P. (1996) Amyloid β-peptide disrupts carbachol-induced muscarinic cholinergic signal transduction in cortical neurons. Proc. Natl. Acad. Sci. USA 93: 6753–6758.

Kristofikova, Z., Tekalova, H. and Klaschka, J. (2001) Amyloid beta peptide1-40 and the function of rat hippocampal hemicholinium-3 sensitive choline carriers: effects of a proteolytic degradation *in vitro*. Neurochem. Res. 26: 203–212.

Kuner, P., Schubenel, R. and Hertel, C. (1998) β-amyloid binds to p75^{NTR} and activates NFκ B in human neuroblastoma cells. J. Neurosci. Res. 54: 798–804.

Ladner, C.J. and Lee, J.M. (1998) Pharmacological drug treatment of Alzheimer disease: the cholinergic hypothesis revisited. J. Neuropathol. Exp. Neurol. 57: 719–731.

Le, W., Xie, W.J., Kong, R. and Appel, S.H. (1997) β-amyloid-induced neurotoxicity of a hybrid septal cell line associated with increased tau phosphorylation and expression β-amyoid precursor protein. J. Neurochem. 69: 978–985.

Lee, T.F., Shiao, Y.J., Chen, C.F. and Wang, L.C. (2001) Effect of ginseng saponins on beta-amyloid-suppressed acetylcholine release from rat hippocampal slices. Planta Med. 67: 634–637.

Lee, V.M. (1995) Disruption of the cytoskeleton in Alzheimer's disease. Curr. Opin. Neurobiol. 5: 663–668.

Lee, V.M. (1996) Regulation of tau phosphorylation in Alzheimer's disease. Ann. N.Y. Acad. Sci. 293: 1446–1447.

Levy-Lahad, E., Wasco, W., Pookraj, P., Romano, D.M., Oshima, J., Pettingell, W.H., Yu, C., Jondro, P.D., Schmidt, S.D., Wang, K., Crowley, A.C., Fu, Y.H., Guenette, S.Y., Galas, D., Nemens, E., Wijsman, E.M., Bird, T.D., Schellenberg, G.D. and Tanzi, R.E. (1995) Candidate gene for the chromosome 1 familial Alzheimer's disease locus. Science 269: 973–977.

Lin, L., Georgievska, B., Mattsson, A. and Isacson, O. (1999) Cognitive changes and modified processing of amyloid precursor protein in the cortical and hippocampal system after cholinergic synapse loss and muscarinic receptor activation. Proc. Natl. Acad. Sci. USA 96: 12108–12113.

Liu, Q., Kawai, H. and Berg, D.K. (2001) β-amyloid peptide blocks the response of α_7-containing nicotinic receptors on hippocampal neurons. Proc. Natl. Acad. Sci. USA 98: 4734–4739.

Mark, R.J., Lovell, M.A., Markesbery, W.R., Uchida, K. and Mattson, M.P. (1997) A role for 4-hydroxynonenal, an aldehydic product of lipid peroxidation, in disruption of ion homeostasis and neuronal death induced by amyloid β-peptide. J. Neurochem. 68: 255–264.

Mattson, M.P., Cheng, B., Davis, D., Bryant, K., Lieberburg, I. and Rydel, R.E. (1992) β-amyloid peptides destabilize calcium homeostasis and render human cortical neurons vulnerable to excitotoxicity. J. Neurosci. 12: 376–389.

Mattson, M.P. and Pedersen, W.A. (1998) Effects of amyloid precursor protein derivatives and oxidative stress on basal forebrain cholinergic systems in Alzheimer's disease. Int. J. Dev. Neurosci. 16: 737–753.

Maurice, T., Lockhart, B.P. and Privat, A. (1996) Amnesia induced in mice by centrally administered β-amyloid peptides involves cholinergic dysfunction. Brain Res. 706: 181–193.

Mills, J. and Reiner, P.B. (1999) Regulation of amyloid precursor protein cleavage. J. Neurochem. 72: 443–460.

Mullan, M. and Crawford, F. (1993) Genetic and molecular advances in Alzheimer's disease. Trends Neurosci. 16: 398–402.

Muller-Spahn, F. and Hock, C. (1999) Risk factors and differential diagnosis of Alzheimer's disease. Eur. Arch. Psychiatry Clin. Neurosci. 249(suppl. 3): III/37–III/42.

Muir, J.L. (1997) Acetylcholine, aging and Alzheimer's disease. Pharmacol. Biochem. Behav. 56: 687–696.

Nitsch, R.M., Slack, B.E., Wurtman, R.J. and Growdon, J.H. (1992) Release of Alzheimer amyloid precursor derivatives stimulated by activation of muscarinic cholinergic receptor. Science 258: 304–307.

Nitta, A., Itoh, A., Hasegawa, T. and Nabeshima, T. (1994) β-amyloid protein-induced Alzheimer's disease animal model. Neurosci. Lett. 170: 63–66.

Nordberg, A., Alafuzoff, I. and Winbald, B. (1992) Nicotinic and muscarinic receptor subtypes in the human brain: changes with aging and dementia. J. Neurosci. Res. 31: 103–111.

Nordberg, A., Lundqvist, H., Hartvig, P., Lilja, A. and Langstrom, B. (1995) Kinetic analysis of regional $(S)(-)^{11}C$-nicotine binding in normal and Alzheimer brains-*in vivo* assessment using positron emission tomography. Alzheimer Dis. Assoc. Disord. 9: 21–27.

Olesen, O.F., Dago, L. and Mikkelsen, J.D. (1998) Amyloid β neurotoxicity in the cholinergic but not in the serotonergic phenotype of RN46A cells. Mol. Brain Res. 57: 266–274.

Pedersen, W.A. and Blusztajn, J.K. (1997) Characterization of the acetylcholine reducing effect of the amyloid-beta peptide in mouse SN56 cells. Neurosci. Lett. 239: 77–80.

Pedersen, W.A., Kloczewiak, M.A. and Blusztajn, J.K. (1996) Amyloid β-protein reduces acetylcholine synthesis in a cell line derived from cholinergic neurones of the basal forebrain. Proc. Natl. Acad. Sci. USA 93: 8068–8071.

Perry, E.K., Tomlinson, B.E., Blessed, G., Bergman, K., Gibson, P.H. and Perry, R.H. (1978) Correlation of cholinergic abnormalities with senile plaques and mental test scores in senile dementia. Br. Med. J. 2: 1457–1459.

Pettit, D.L., Shao, Z. and Yakel, J.L. (2001) β-myloid$_{1-42}$ peptide directly modulates nicotinic receptors in the rat hippocampal slice. J. Neurosci. 21: RC120–124.

Pike, C.J. and Cotman, C.W. (1993) Cultured GABA-immunoreactive neurons are resistant to toxicity induced by β-amyloid. Neuroscience 56: 269–274.

Pike, C.J., Burdick, D., Walencewicz, A.J., Glabe, C.G. and Cotman, C.W. (1993) Neurodegeneration induced by β-amyloid peptides *in vitro*: the role of peptide assembly state. J. Neurosci. 13: 1676–1687.

Pittel, Z., Heldman, E., Barg, J., Haring, R. and Fisher, A. (1996) Muscarinic control of amyloid precursor protein secretion in rat cerebral cortex and cerebellum. Brain Res. 742: 299–304.

Poirier, J., Davignon, J., Bouthillier, D., Kogan, S., Bertrand, P. and Gauthier, S. (1993) Apolipoprotein E polymorphism and Alzheimer's disease. Lancet 342: 697–699.

Price, D.L. and Sisodia, S.S. (1998) Mutant genes in familial Alzheimer's disease and transgenic models. Annu. Rev. Neurosci. 21: 479–505.

Quirion, R. (1993) Cholinergic markers in Alzheimer's disease and the autoregulation of acetylcholine release. J. Psychiatry Neurosci. 18: 226–234.

Rapoport, M. and Ferreira, A. (2000) PD98059 prevents neurite degeneration induced by fibrillar β-amyloid in mature hippocampal neurons. J. Neurochem. 74: 125–133.

Roberson, M.R. and Harrell, L.E. (1997) Cholinergic and amyloid precursor protein metabolism. Brain Res. Rev. 25: 50–69.

Satoh, Y., Hirakura, Y., Shibayama, S., Hirashima, N., Suzuki, T. and Kirino, Y. (2001) Beta-amyloid peptides inhibit acetylcholine release from cholinergic nerve endings isolated from an electric ray. Neurosci. Letts. 302: 97–100.

Schroder, H. and Wevers, A. (1998) Nicotinic acetylcholine receptors in Alzheimer's disease. Alz. Disease Rev. 3: 20–27.

Selkoe, D.J. (2001) Alzheimer's disease: genes, proteins and therapy. Physiol. Rev. 81: 741–766.

Seubert, P., Vigo-Pelfrey, C., Esch, F., Lee, M., Dovey, H., Davis, D., Sinha, S., Schlossmacher, M., Whaley, J., Swindlehurst, C., McCormack, R., Wolfert, R., Selkoe, D., Lieberburg, I. and Schenk, D. (1992) Isolation and quanti-fication of soluble Alzheimer β-peptide from biological fluids. Nature 359: 325–327.

Shea, T.B., Prabhakar, S. and Ekinci, F.J. (1997) β-amyloid and ionophore A23187 evoke tau hyper-phosphorylation by distinct intracellular pathways: differential involvement of the calpain/protein kinase C system. J. Neurosci. Res. 49: 759–768.

Sherrington, R., Rogaev, E.I., Liang, Y., Rogaeva, E.A., Levesque, G., Ikeda, M., Chi, H., Lin, C., Li, G., Holman, K., Tsuda, T., Mar, L., Foncin, J.F., Bruni, A.C., Montesi, M.P., Sorbi, S., Rainero, I., Pinessi, L., Nee, L., Chumakov, I., Pollen, D., Brookes, A., Sanseau, P., Polinsky, R.J., Wasco, W., DaSilva, H.A.R., Haines, J.L., Pericak-Vance, M.A., Tanzi, R.E., Roses, A.D., Fraser, P.E., Rommens, J.M. and St George-Hyslop, P.H. (1995) Cloning of a gene bearing missense mutations in early-onset familial Alzheimer's disease. Nature 375: 754–760.

Shoji, M., Golde, T.E., Ghiso, J., Cheung, T.T., Estus, S., Shaffer, L.M., Cai, X.D., Makay, D.M., Tintner, R., Frangione, B. and Younkin, S.G. (1992) Production of Alzheimer's β protein by normal proteolytic processing. Science 258: 126–129.

St George-Hyslop, P.H. (1999) Molecular genetics of Alzheimer disease. Semin Neurol. 19: 371–383.

Strittmatter, W.J., Saunders, A.M., Schmeckel, D., Pericak-Vance, M., Enghild, J., Salvesen, G.S. and Roses, A.D. (1993) Apolipoprotein E: High-avidity binding to β-amyloid and increased frequency of type 4 allele in late-onset familial Alzheimer disease. Proc. Natl. Acad. Sci. USA 90: 1977–1981.

Sturchler-Pierrat, C., Abramowski, D., Duke, M., Wiederhold, K.H., Mistl, C., Rothacher, S., Ledermann, B., Burki, K., Frey, P., Paganetti, P.A., Waridel, C., Calhoun, M.E., Jucker, M., Probst, A., Staufenbiel, M. and Sommer, B. (1997) Two amyloid precursor protein transgenic mouse models with Alzheimer's disease-like pathology. Proc. Natl. Acad. Sci. USA 94: 13287–13292.

Takashima, A., Honda, T., Yasutake, K., Michel, G., Murayama, O., Murayama, M., Ishiguro, K. and Yamaguchi, H. (1998) Activation of tau protein kinase I/glycogen synthase kinase-3 beta by amyloid beta peptide (25–35) enhances phosphorylation of tau in hippo-campal neurons. Neurosci. Res. 4: 317–323.

Tanzi, R.E. (1996) Neuropathology in the Down's syndrome brain. Nat. Med. 2: 31–32.

Tucek, S. (1985) Regulation of acetylcholine synthesis in the brain. J. Neurochem. 44: 11–24.

Vaucher, E., Amount, N., Rowe, W., Pearson, D., Poirier, J. and Kar, S. (2001) Amyloid β peptide levels and its effects on hippocampal acetylcholine release in aged, cognitively-impaired and -unimpaired rats. J. Chem. Neuroanat. 21: 323–329.

Vaucher, E., Fluit, P., Chishti, M.A., Westaway, D., Mount, H.T.J. and Kar, S. (2002) Alteration in working memory but not cholinergic receptor binding sites in transgenic mice expressing human presenilin 1 transgenes. Exp. Neurol. 175: 398–406.

Wallace, W., Ahlers, S.T., Gotlib, J., Bragin, V., Sugar, J., Gluck, R., Shea, P.A., Davis, K.L. and Haroutunian, V. (1993) Amyloid precursor protein in the cerebral cortex is rapidly and persistently induced by loss of subcortical innervation. Proc. Natl. Acad. Sci. USA 90: 8712–8716.

Wang, H.Y., Wild, K.D., Shank, R.P. and Lee, D.H.S. (1999) Galanin inhibits acetylcholine release from rat cerebral cortex *via* a pertussis toxin-sensitive G_i protein. Neuropeptides 33: 197–205.

Wang, H.Y., Lee, D.H.S., Davis, C.B. and Shank, R.P. (2000a) Amyloid peptide $A\beta$1-42 binds selectively and with pico-molar affinity to α_7 nicotinic acetylcholine receptors. J. Neurochem. 75: 1155–1161.

Wang, H.Y., Lee, D.H.S., D'Andrea, M.R., Peterson, P.A., Shank, R.P. and Reitz, A.B. (2000b) β-amyloid1-42 binds α_7 nicotinic acetylcholine receptor with high affinity: implica-tions for Alzheimer's disease pathology. J. Biol. Chem. 275: 5626–5632.

Warpman, U., Alafuzoff, I. and Nordberg, A. (1993) Coupling of muscarinic receptors to GTP proteins in postmortem human brain—alterations in Alzheimer's disease. Neurosci. Lett. 150: 39–43.

Wecker, L. (1991) The synthesis and release of acetylcholine by depolarized hippocampal slices is increased by increased choline available *in vitro* prior to stimulation. J. Neurochem. 57: 119–1127.

Whitehouse, P.J., Price, D.L., Struble, R.G., Clark, A.W., Coyle, J.T. and Delon, M.R. (1982) Alzheimer's disease and senile dementia: loss of neurons in the basal forebrain. Science 215: 1237–1239.

274

Wisniewski, T., Ghiso, J. and Frangione, B. (1997) Biology of Aβ amyloid in Alzheimer's disease. Neurobiol. Dis. 4: 313–328.

Wong, T.P., Debeir, T., Duff, K. and Cuello, A.C. (1999) Reorganization of cholinergic terminals in the cerebral cortex and hippocampus in transgenic mice carrying mutated presenilin-1 and amyloid precursor protein transgenes. J. Neurosci. 19: 2706–2716.

Wurtman, R. (1992) Choline metabolism as a basis for the selective vulnerability of cholinergic neurones. Trends Neurosci. 15: 117–122.

Yan, S.D., Chen, X., Fu, J., Chen, M., Zhu, H., Roher, A., Slattery, T., Zhao, L., Nagashima, M., Morser, J., Migheli, A., Nawroth, P., Stern, D. and Schmidt, A.M. (1996) RAGE and amyloid-β peptide neurotoxicity in Alzheimer's disease. Nature 382: 685–691.

Yan, S.D., Fu, J., Soto, C., Chen, X., Zhu, H., Al-Mohanna, F., Collison, K., Zhu, A., Stern, E., Saido, T., Tohyama, M., Ogawa, S., Roher, A., Stern, D. (1997) An intracellular protein that binds amyloid-β peptide and mediates neurotoxicity in Alzheimer's disease. Nature 389: 689–695.

Yankner, B.A. (1996) Mechanisms of neuronal degeneration in Alzheimer's disease. Neuron 16: 921–932.

Zheng, W.H., Bastianetto, S., Mennicken, F., Ma, W. and Kar, S. (2002) Amyloid β peptide induces tau phosphorylation and neuronal degeneration in rat primary septal cultured neurons. Neuroscience 115: 201–211.

Progress in Brain Research, Vol. 145
ISSN 0079-6123

CHAPTER 19

Nicotinic receptor mutations in human epilepsy

Ortrud K. Steinlein*

*Institute of Human Genetics, University Hospital Bonn, Friedrich-Wilhelms-University,
Wilhelmstr. 31, D-53111 Bonn, Germany*

Inheritance and mutations in idiopathic epilepsy

Familial occurrence of idiopathic epilepsies has long been recognized, but progress in identifying genes has been rather slow. This is especially true for the more common forms of the disease, like juvenile myoclonic epilepsy, juvenile or childhood absence epilepsies, or *grand mal*-epilepsy on awakening. A genetic component to aetiology is well established in these syndromes, but the mechanism of inheritance and the genes involved are unknown. Molecular genetic approaches have been more successful regarding the rare idiopathic epilepsies with single gene inheritance. The genes that have been identified so far are associated with different forms of idiopathic epilepsies. Mutations within the genes coding for the α4- and β2-subunits (CHRNA4 and CHRNB2) of the neuronal nicotinic acetylcholine receptor (nAChR) are found in families with autosomal dominant nocturnal frontal lobe epilepsy (ADNFLE) (Steinlein et al., 1995; Fusco et al., 2000; Phillips et al., 2001). Two of the voltage gated potassium channel subunits forming the muscarinic-regulated potassium current (M-current), KCNQ2 and KCNQ3, are associated with benign familial neonatal convulsions (BFNC) (Biervert et al., 1998; Charlier et al., 1998; Singh et al., 1998). Mutations in different subunits of the family of voltage gated sodium channels as well as in the $GABA_A$-receptor subunit GABRG2, have been described in a syndrome named 'generalized epilepsy

*Tel.: +49-228-287-2644; Fax: +49-228-287-2380;
E-mail: ortrud.steinlein@ukb.uni-bonn.de

DOI: 10.1016/S0079-6123(03)45019-X

with febrile seizure plus' (GEFS[+]) (Wallace et al., 1998; Escayg et al., 2000; Baulac et al., 2001; Sugawara et al., 2001; Wallace et al., 2001). Most recently, the first gene which is not apparently coding for an ion channel was found to be associated with familial lateral temporal lobe epilepsy (Kalachikov et al., 2002), also named autosomal dominant partial epilepsy with auditory features (Ottman et al., 1995).

Autosomal dominant nocturnal frontal lobe epilepsy

ADNFLE was the first partial epilepsy found to be segregating as a single gene disorder (Scheffer et al., 1994). Although several monogenic generalized epilepsies had already been described, partial epilepsies were mostly regarded as symptomatic disorders caused by trauma or infections, for example. Since 1994 additional forms of partial epilepsies with monogenic inheritance have been identified (Ottman et al., 1995; Giupponi et al., 1997; Szepetowski et al., 1997; Neubauer et al., 1998; Scheffer et al., 1998; Xiong et al., 1999; Malacarne et al., 2001). These reports confirm that mutations which have been passed through the germline, and are therefore present in the patients every cell, can express themselves functionally only in circumscribed parts of the brain.

ADNFLE is characterized by nocturnal motor seizures, which tend to cluster, and often occur several times a night. Seizures are mainly occurring during nonREM sleep, either shortly after falling asleep, or in the early morning hours. Some patients also report about seizures during daytime naps. The main age of onset is within the first or second decade of life, but a much later onset is not unusual. Even within the same family, the variation in seizure onset as well as in severity of the disease can be considerable (Scheffer et al., 1995; Steinlein et al., 2000). Some patients have long seizure-free intervals, which can last several years or even decades. Sometimes puberty, pregnancy or menarche marks the beginning or end of such a seizure-free interval, but again no intrafamilial consistencies are observed. Some patients enter total remission at some point, and in those with persisting seizures the frequency of nocturnal attacks tends to decrease with advanced age.

Relatives of patients often report gasps, grunts, or some vocalizations shortly before the seizures start. The aura, which sometimes precedes the seizure onset, includes a wide variety of phenomena, like somatosensory, sensory or psychic sensations. The seizures are characterized by thrashing hyperkinetic activity or tonic stiffening of arms and/or legs, often with superimposed clonic jerking. Sometimes, the hyperkinetic activity is pronounced enough to cause a fall from the bed and/or a minor injury. As in most focal epilepsies, secondary generalization with loss of consciousness can occur, but it happens rather infrequently. Interictal EEG abnormalities are rare, and even during a seizure EEG, recording often fails to show abnormal neuronal activity. ADNFLE can therefore be misinterpreted as nocturnal motor activity, benign nocturnal parasomnia, night terror, hyperactivity, or hysteria (Scheffer et al., 1994, 1995; Oldani et al., 1996; Hayman et al., 1997). Video-polysomnography is often helpful to confirm the diagnosis of nocturnal epilepsy, while a careful pedigree analysis helps to reveal the familial course of the disease and to differentiate ADNFLE from the more common forms of sporadic frontal lobe epilepsy.

Identification of the first ADNFLE gene

Linkage analysis assigned the first ADNFLE locus to the chromosomal region 20q13.2-q13.3, which is located close to the telomeric end of the long arm of chromosome 20. This first locus assignment was found by analyzing a large Australian ADNFLE pedigree, which included 27 affected individuals in six generations (Phillips et al., 1995). Comparative analysis of the human and mouse genome had previously helped to locate and clone a candidate gene in this region, CHRNA4, coding for the α4-subunit gene of the nAChR (Steinlein et al., 1994). As shown later the CHRNA4 gene is located close to the KCNQ2 gene, which is also associated with idiopathic epilepsy. The voltage-gated potassium channel KCNQ2 was identified as a major cause of BFNC (Biervert et al., 1998; Singh et al., 1998). KCNQ2 is separated from CHRNA4 by approximately 30 kb, and both genes are orientated in the same transcriptional direction. So far, it is unknown if they both share the same transcription control elements, or if their colocalization is due to coincidence by chance (Xiao et al., 2001).

Analysis of the CHRNA4 gene in the Australian ADNFLE family identified a mutation within the second transmembrane region (TM2). A cytosine to thymine nucleotide transition caused the replacement of a neutral serine by a complex aromatic phenylalanine in position 248 of the predicted CHRNA4 protein (α4-Ser248Phe mutation, numbering according to the *Torpedo* αAChR-subunit). According to the model proposed by Unwin (1993), the mutated Ser248 would be located at the most constricted part of the closed ion channel pore.

Since 1998, two more ADNFLE mutations (α4-Ser252Leu, α4-776ins3) were found in the CHRNA4 gene (Fig. 1). Like α4-Ser248Phe, both mutations are located within the TM2 region (Steinlein et al., 1995, 1997; Hirose et al., 1999; Saenz et al., 1999; Phillips et al., 2000; Steinlein et al., 2000). Although the ADNFLE patients from different families share most of their clinical features, some mutation-specific differences might be present. The Norwegian α4-776ins3 family (Steinlein et al., 1997) showed an unusual high rate of schizophrenia-like symptoms and other psychiatric disorders (Magnusson, A., Stordal, E., Brodtkorb, E., Steinlein, O., personal communication), whereas both known α4-Ser252Leu families had several members with mild to moderate mental retardation (Hirose et al., 1999; Cho, Y.-W., Motamedi, K.M., Laufenberg, I., Lesser, R.P., Sohn, S.-I., Lim, J.-G., Lee, H., Yi, S.-D., Lee, J.-H., Kim, D.-K., Steinlein, O.K., personal communication).

CHRNA4 (20q13.3)   CHRNB2 (1q21-q22)

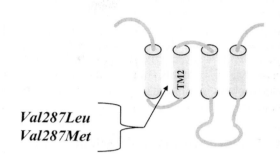

Ser248Phe
Ser252Leu
776ins3

Val287Leu
Val287Met

Fig. 1. Schematic view of the α4- and the β2-nAChR subtype. The positions of the five known ADNFLE mutations within the second transmembrane domains are indicated.

It will be interesting to identify additional ADNFLE families to see if the suspected genotype-phenotype relations can be confirmed.

Founder effect or independent occurrence?

From the three known mutations within the CHRNA4 gene, at least two have been found in different families. The α4-Ser248Phe mutation, first described in the Australian ADNFLE family (Steinlein et al., 1995), was later found again in a Spanish (Saenz et al., 1999) as well as in a Norwegian ADNFLE family (Steinlein et al., 2000). This raised the question if these families shared a common founder, or if the mutation had occurred independently in the different families. The α4-Ser248Phe mutation is located within the large exon 5 of the CHRNA4 gene, which contains several noncoding single nucleotide polymorphisms (SNPs). By sequencing through the exon, the haplotypes of the two CHRNA4 genes could be determined in patients from the Australian and the Norwegian ADNFLE families. The analysis showed that the mutation-carrying genes from both families did not share the same SNP-alleles. Thus, the haplotype analysis in both families excluded the possibility of a shared ancestral mutation (Steinlein et al., 2000). That ADNFLE mutations can be due to independent mutational events was also shown for another

CHRNA4 mutation. α4-Ser252Leu was found in a Japanese family (Hirose et al., 1999) as well as in a sporadic patient of Lebanese origin. The latter patient did not inherit the mutation from one of her parents, but had a de novo mutation (Phillips et al., 2000). The localization of all so far known mutations within TM2, together with the independent origin of some of these mutations in different families suggests that the character and position of the mutation within the CHRNA4 gene is critical for the molecular aetiology of ADNFLE.

Electrophysiological characteristics of CHRNA4 mutations

Each mutation in the highly conserved pore-lining TM2 region of α4 exhibits its own specific electrophysiological properties. The α4-Ser248Phe mutation has profound effects on the function of the receptor. Electrophysiological experiments demonstrated an increase of the desensitization rate, followed by a prolonged resensitization period; both effects are probably due to a destabilization of the ion channels open configuration (Weiland et al., 1996). The second ADNFLE mutation, α4-776ins3, increases the receptors apparent affinity for acetylcholine, and reduces the $Ca2+$-permeability, but has only minor effects on the receptor desensitization (Weiland et al., 1996; Steinlein et al., 1997; Bertrand et al., 1998; Weiland et al., 2000). Despite these differences, the main effect of both mutations is a reduced Ca^{2+} influx, which could be interpreted as a loss-of-function effect. However, an additional gain-of-function effect cannot be excluded. The mutated receptors are still expressed on the cell surface and are competing for space with the wild type receptors. They therefore represent a nAChR-subpopulation with altered electrophysiological and pharmacological characteristics in the brain.

CHRNA4 mutations and antiepileptic drugs

In most ADNFLE patients, antiepileptic drugs are able to markedly reduce the seizure frequency or even to help the patient to become seizure-free. There are some indications for the existence of mutation-specific differences regarding the response towards certain drugs. In the reported families with the α4-Ser248Phe mutation, most if not all patients seemed to benefit from drug therapy, especially if carbamazepine, or (in some patients) phenytoine, is given (Phillips et al., 1995; Scheffer et al., 1995; Steinlein et al., 1995; Saenz et al., 1999; Steinlein et al., 2000). The drug response seems to be less favourable in the two known families with the α4-Ser252Leu mutation. In the Japanese α4-Ser252Leu-family, unresponsiveness to antiepileptic drug treatment has been described in affected children, but not in adults (Hirose et al., 1999). In five out of seven affected members of a recently identified Korean α4-Ser252Leu-family, antiepileptic drugs led to no or only moderate decrease of seizure frequency, suggesting that this specific mutation might be associated with reduced drug responsiveness (Cho, Y.-W., Motamedi, K.M., Laufenberg, I., Lesser, R.P., Sohn, S.-I., Lim, J.-G., Lee, H., Yi, S.-D., Lee, J.-H., Kim, D.-K., Steinlein, O.K., personal communication). So far, the number of families with known mutations is too low to decide if these observations are indeed due to genotype-specific differences in drug-response.

Carbamazepin and ADNFLE

One of the most effective drugs in the treatment of nocturnal frontal lobe seizures is carbamazepine, a well-known antiepileptic drug. Although widely used,

the mechanisms by which carbamazepine reduces seizure frequency remain to be elucidated. It has been shown that carbamazepine, like many other antiepileptic drugs, acts on several different voltage-gated channel proteins and on at least some ligand-gated receptors. Electrophysiological studies demonstrated that voltage-gated sodium and calcium channels, NMDA-(N-methyl-D-aspartate), GABA-(γ-aminobutyric acid) receptors, and nAChR's are affected by administration of carbamazepine (Granger et al., 1995; Yoshimura et al., 1995; Hough et al., 1996). It is therefore probably safe to assume that carbamazepine decreases seizure activity by altering not only one but several different membrane proteins involved in the regulation of neuronal excitability. A generalized downregulation of neuronal excitability by unspecific interaction of different channels might therefore be part of the antiepileptic effect of carbamazepine. However, the differences in therapeutic effectiveness observed for different antiepileptic drugs raise questions about the mechanisms underlying such drug-specific profiles. The detection of an association between specific nAChR mutations and ADNFLE provided the possibility to simulate the drug-receptor interaction in an easy-to-study model outside the patients brain.

Xenopus oocytes or cultured eukaryotic cells such as HEK-cells are well known tools to study the electrophysiological and pharmacological properties of ligand-gated ion channels. To analyze the effect of ADNFLE mutations on the action of carbamazepine, α4-Ser248Phe and α4-776ins3 were introduced into specific expression vectors (Weiland et al., 1996). Since CHRNA4 proteins alone are not able to build functional AChRs, the mutation constructs were coexpressed with wild type β2-nAChR (Picard et al., 1999). The heteromeric α4-Ser248Phe/β2-wt and α4-776ins3/β2-wt AChRs were reconstituted in *Xenopus* oocytes and analyzed for their response to carbamazepine. The first set of experiments was designed to investigate the mechanism by which carbamazepine interacts with nAChRs. It was shown that an increase in concentration of the agonist ACh was not sufficient to overcome the receptors inhibition caused by carbamazepine. Thus, carbamazepine obviously does not interact with the agonists binding site formed mostly by parts of the extracellular regions of the nAChR α-subunits. ACh-evoked currents recorded in the presence of carbamazepine showed an increased apparent desensitization, compatible with a progressive blockade of the ion channel pore. These results suggest that carbamazepine enters the ionic pore and most probably acts as an open-channel-blocker (Picard et al., 1999). Another common antiepileptic drug, valproate, was also applied to *Xenopus* oocytes expressing wildtype or mutant nACHR, but few or no reactions were seen even for concentrations 10 fold higher than the upper antiepileptic plasma level in humans. Thus, valproate seems not to interact with nAChRs, which is consistent with the observation that this drug is not effective in controlling seizures in ADNFLE patients. Compared to the wild type receptor, the mutated nAChRs showed a threefold increased sensitivity towards carbamazepine (Picard et al., 1999).

ADNFLE can also be caused by CHRNB2 mutations

The nAChRs are pentameric ion channels made up of various hetero- or homologous combinations of eight α-subunits ($\alpha2$-$\alpha7$, $\alpha9$-$\alpha10$) and three β-subunits ($\beta2$-$\beta4$). The α-subunits, which are providing the agonist-binding sites, are characterized by two adjacent cysteines at positions 192 and 193 (numbering according to the $\alpha1$ muscle type subunit) (Galzi and Changeux, 1995). The β-subunits are thought to have a structural role, and they probably contribute to the pharmacological specificity of the receptor, too (Luetje and Patrick, 1991). Together with the $\beta2$-subunit, the $\alpha4$-subunit builds one of the most abundant neuronal nicotinic acetylcholine receptor subtypes in brain (Whiting et al., 1992). When it became obvious that only a few ADNFLE families could be linked to the CHRNA4-containing region on chromosome 20q13.3, the gene coding for the $\beta2$-subunit, CHRNB2, became the most likely candidate for further studies.

The CHRNB2 gene was previously assigned to chromosome 1q21-q22 (Rempel et al., 1998), but linkage to this region was only found several years later. This delay was probably due to the fact that CHRNB2-linked families are even more rare than CHRNA4-linked ones. To date only two ADNFLE families have been identified in which the

nocturnal seizures are caused by mutations in the CHRNB2 gene. The first linked family was a three-generation ADNFLE pedigree from southern Italy (Gambardella et al., 2000). Screening approaches in ADNFLE families subsequently identified two different missense mutations affecting the same amino acid residue, $\beta2$-Val287Leu and $\beta2$-Val287Met (Fusco et al., 2000; Phillips et al., 2001) (Fig. 2). The mutated valine residue is located near the extracellular end of TM2, and faces the pore of the ion channel in both the open and the closed state (Devillers-Thiery et al., 1993). Functional reconstitution experiments in *Xenopus* oocytes showed that, compared to the wild-type $\alpha4/\beta2$, the $\alpha4/\beta2$-Val287Met receptor exhibited a considerable slower desensitization after nicotine application (Fusco et al., 2000). The electrophysiological properties of $\beta2$-Val287Met containing receptors differ from those with the $\beta2$-Val287Leu mutation, although both mutations are affecting the same amino acid residue. The $\beta2$-Val287Met mutation causes a higher apparent acetylcholine affinity of the receptor, but no significant alteration of the desensitization properties (Phillips et al., 2000). Thus, the alteration of the receptor caused by the $\beta2$-Val287Met mutation is similar to that observed for the $\alpha4$-776ins3 mutation, although these mutations are affecting different subunits (Phillips et al., 2000).

CHRNA4 and CHRNB2 might not be the only ADNFLE genes

Despite the identification of mutations in two nAChR subunit genes, CHRNA4 and CHRNB2, the genetic basis of ADNFLE is still not completely known. There are several ADNFLE families in which neither CHRNA4 nor CHRNB2 mutations can be found, meaning that at least a third gene must exist which can cause this rare familial epilepsy. The most likely candidates are other genes coding for nAChR subunits. Several of them have already been localized within the human genome. The genes coding for subunits $\alpha2$ (CHRNA2) and $\alpha6$ (CHRNA6) are located on chromosome 8p21 and 8p22, respectively (Anand and Lindstrom, 1992), while the $\alpha7$ gene and the recently discovered $\alpha10$ gene, (CHRNA7 and CHRNA10) were assigned to chromosome 15q14 and 11p15.5, respectively (Chini et al., 1994; Elliott et al., 1996; Lustig et al., 2001). The gene for the structural subunit $\beta3$, CHRNB3, was mapped to chromosome 8p11.2 (Anand and Lindstrom, 1992). The chromosomal region 15q24 harbours a cluster of three nAChR genes (CHRNA3, CHRNA5 and CHRNB4) (Eng et al., 1991). Several attempts have been made to analyze ADNFLE families not linked to either CHRNA4 or CHRNB2. Phillips et al. (1998) studied six ADNFLE families and found significant lod scores with polymorphic markers from the chromosome 15-region containing the cluster of three nAChR genes in one of the families. However, subsequent analysis of these three genes failed to detect any mutation.

Besides the obvious candidate genes from the subfamily of nAChR genes, other ion channel gene families also have to be taken into consideration. There are examples for idiopathic epilepsies, which can be caused by mutations from different types of ion channel genes. Mutations in genes coding for several subunits from the family of the voltage-gated sodium channels (Wallace et al., 1998; Escayg et al., 2000; Sugawara et al., 2001), as well as in the GABRG2 gene (Baulac et al., 2001; Wallace et al., 2001), can cause GEFS$^+$, a syndrome characterized by febrile seizures, febrile seizures after the age of 6 years (therefore named 'febrile seizures plus'), and different types of generalized or even partial seizures (Scheffer and Berkovic, 1997). Recently, it has been shown that idiopathic epilepsies are not always due to mutations in ion channels. A leucine-rich repeat containing gene LGI1 on chromosome 15q24 was found to be associated with the partial epilepsy syndrome of autosomal dominant lateral temporal lobe epilepsy (ADLTE) (Kalachikov et al., 2002). Thus the number of likely candidates genes for ADFLE and other idiopathic epilepsies is ever increasing. Large, well-characterized families are needed to identify a region which might contain the third ADNFLE gene.

The possible pathomechanisms

There are several hypotheses about the patho-mechanisms, that might lead from nAChRs with impaired function to epileptic seizures. The nAChRs are thought to be involved in signal transduction by fast synaptic transmission, axo-axonic transmission, as well as in the modulation of presynaptic transmitter release (Zhang et al., 1993; Futami et al., 1995; Role and Berg, 1996; Lena and Changeux, 1997; Wonnacott, 1997; Hefft et al., 1999). The first hypothesis is based on the observation that nAChRs are located on inhibitory interneurons (Jones and Yakel, 1999). The mutations found in ADNFLE families might therefore promote seizures by reduction of tonic and/or phasic inhibition in the brain. In view of the high Ca^{2+} permeability of the nAChRs, another mechanism might also be important with respect to the epileptogenic effect. Activation of presynaptically located nAChRs would cause a Ca^{2+} influx, which itself could activate Ca^{2+}-dependent Cl^- or K^+ currents and second messenger systems (Fieber and Adams, 1991; Bertrand et al., 1993; Seguela et al., 1993). This could subsequently lead to an enhanced rate of integration of neurotransmitter-carrying vesicles into the presynaptic cell membrane. Another possibility is that the Ca^{2+} influx might activate voltage-gated calcium channels via the depolarization of the presynaptic cell membrane, thus indirectly enhancing presynaptic transmitter release. Depending on which presynaptic transmitters are released, and on the physiological role of the postsynaptic neurons affected by these transmitters, nAChRs can have either an excitatory or a inhibitory effect. Presynaptic nAChRs with impaired function could therefore cause an increased excitability of some neurons. Such a potentially unbalanced state, triggered by some unknown factors, could then lead to episodic hypersynchronization of larger neuronal networks.

The second hypothesis about the pathomechanisms underlying the epileptic seizures caused by mutated nAChRs is based on the observed expression pattern of the receptor during brain development. Evidence of functional nicotinic acetylcholine receptor channels in fetal mouse cerebral cortex has been found as early as embryonic day 10, when the cortex consists of dividing stem and progenitor cells. The detectable amount of nAChR expression declined with embryonic age (Atluri et al., 2001). Considerable amounts of CHRNA4 mRNA can be detected in fetal brain, especially in the thalamus, cortex, pyramidal layer of Ammon's horn, and granular layer of dentate gyrus (Agulhon et al., 1999a). Such an early expression was also described for CHRNA3 (Atluri et al., 2001) and CHRNA7 (Agulhon et al., 1999b); the fetal expression patterns of other nAChR subunits have not been reported so far. The early expression of nAChRs, together with the decline of this expression at later embryonic stages, points towards a possible function in the development of the fetal brain. Neurotransmitters and their receptors are thought to be involved in the complex signaling processes which direct cell migration and synaptic circuitry in the developing brain. Impaired nAChR function might results in structural abnormalities promoting hyperexcitability through shortcuts in neuronal networks.

Animal models for nAChR mutations

Gene targeting and homologous recombination experiments can be used to create mice carrying mutations in genes associated with different diseases in humans. Several mice strains lacking individual nAChR subunits have already been cloned and analyzed. For most nAChR subunits, mice homozygous for a null mutation (knockout mice) were viable and did not exhibit any gross neurological or behavioural deficits (Picciotto et al., 1995; Orr-Urtreger et al., 1997; Marubio et al., 1999; Cordero-Erausquin et al., 2000; Ross et al., 2000). Only in α3-knockout mice were the phenotypic findings more severe, including unexplained lethality within the first week of life, postnatal growth deficiency, megacystis and mydriasis (Xu et al., 1999). These abnormalities are probably due to disturbances of the autonomic control of peripheral organs, and the severity of the phenotype shows that the function of the α3-subunit cannot be fully substituted by other nAChR subunits. For most of the other nAChR subunits, redundancy seems to be sufficient to prevent any severe phenotype. In particular, mice missing the two subunits mutated in ADNFLE patients, CHRNA4 or CHRNB2, did not show spontaneous seizures or an increased rate of induced seizures (Cordero-Erausquin et al., 2000).

However, knockout mice are not the optimal animal models to study ADNFLE. The patients do not have a complete loss-of-function of specific

nAChR subtypes, because the mutated genes are still expressed, translated, and their proteins integrated into a nAChR. Experiments on reconstituted receptors have demonstrated that nAChRs carrying ADNFLE mutations result in altered electrophysiological and pharmacological profile (Weiland et al., 1996; Steinlein et al., 1997; Fusco et al., 2000; Phillips et al., 2001). To study such a gain-of-function effect, knockin animals carrying nAChR alleles with point mutations are better suited. So far, no such animal models exist for the five known ADNFLE mutations. The only known mouse model for a gain-of-function nAChR mutation is the CHRNA7-L250T mouse, which carries a missense mutation within TM2 of the α7 nAChR subunit. This mutation has not been found in humans yet, but is known to increase the current amplitude and to decrease the desensitization rate of the receptor. Mice homozygous for the L270T mutation die shortly after birth, while heterozygous animals are viable but more sensitive to the convulsant effects of nicotine (Broide et al., 2002; Gil et al., 2002). CHRNA7-knockout mice showed no changed sensitivity for nicotine-induced seizures. Thus, the altered properties rather than the absence of the nAChR subunit might be important for the susceptibility towards nicotine-induced seizures (Franceschini et al., 2002). If the same is true for the pathophysiological mechanisms underlying ADNFLE, nAChR knockout mice cannot be expected to mimic the human phenotype.

The challenges of future research

Idiopathic epilepsies are a heterogeneous group of disorders with distinct age-dependency and clinical expression. Up to 2% of the general population are affected by idiopathic epilepsy by the age of 80 years, making it one of the most common neurological conditions. Idiopathic epilepsies are not caused by any kind of brain damage, and presumably depend on a suspected or known genetic predisposition. In most forms, the mode of inheritance is thought to be oligo- or polygenic rather than monogenic. Nevertheless, the rare single gene epilepsies like ADNFLE can serve as models to investigate the pathomechanisms of idiopathic epilepsies. For future approaches they can give us some ideas about which gene families should be regarded as candidate genes, and which are less likely to be involved in epileptogenesis. The most important lesson we have learned so far is that there are probably many more genes associated with epilepsy than previously suspected. For most of the single gene epilepsies with a known genetic etiology, at least two, sometimes more genes have been identified. This strongly suggests that the number of genes involved in the common epilepsies with complex inheritance is quite high. But considering the complexity of brain functions and the fact that approximately 50% of all genes are at least transiently expressed in the brain, this should not be surprising.

References

Agulhon, C., Abitbol, M., Bertrand, D. and Malafosse, A. (1999a) Localization of mRNA for CHRNA7 in human fetal brain. NeuroReport, 10: 2223–2237.

Agulhon, C., Blanchet, P., Kobetz, A., Marchant, D., Faucon, N., Sarda, P., Moraine, C., Sittler, A., Biancalana, V., Malafosse, A. and Abitbol, M. (1999b) Expression of FMR1, FXR1, and FXR2 genes in human prenatal tissues. J. Neuropathol. Exp. Neurol., 58: 867–880.

Anand, R. and Lindstrom, J. (1992) Chromosomal localisation of seven neuronal nicotinic acetylcholine receptor subunit genes in humans. Genomics, 13: 962–967.

Atluri, P., Fleck, M.W., Shen, Q., Mah, S.J., Stadfelt, D., Barnes, W., Goderie, S.K., Temple, S. and Schneider, A.S. (2001) Functional nicotinic acetylcholine receptor expression in stem and progenitor cells of the early embryonic mouse cerebral cortex. Dev. Biol., 240: 143–156.

Baulac, S., Huberfeld, G., Gourfinkel-An, I., Mitropoulou, G., Beranger, A., Prud'homme, J.F., Baulac, M., Brice, A., Bruzzone, R. and LeGuern, E. (2001) First genetic evidence of GABA(A) receptor dysfunction in epilepsy: a mutation in the gamma2-subunit gene. Nature Genet., 28: 46–48.

Bertrand, D., Galzi, J.L., Devillers-Thiery, A., Bertrand, S. and Changeux, J.P. (1993) Mutations at two distinct sites within the channel domain M2 alter calcium permeability of neuronal alpha 7 nicotinic receptor. Proc. Natl. Acad. Sci. USA, 90: 6971–6975.

Bertrand, S., Weiland, S., Berkovic, S.F., Steinlein, O.K. and Bertrand, D. (1998) Properties of neuronal nicotinic acetylcholine receptor mutants from humans suffering from autosomal dominant nocturnal frontal lobe epilepsy. Br. J. Pharmacol., 125: 751–760.

Biervert, C., Schroeder, B.C., Kubisch, C., Berkovic, S.F., Propping, P., Jentsch, T.J. and Steinlein, O.K. (1998) A potassium channel mutation in neonatal human epilepsy. Science, 279: 403–406.

Broide, R.S., Salas, R., Ji, D., Paylor, R., Patrick, J.W., Dani, J.A. and De Biasi, M. (2002) Increased sensitivity to nicotine-induced seizures in mice expressing the L250T alpha 7 nicotinic acetylcholine receptor mutation. Mol. Pharmacol., 61: 695–705.

Charlier, C., Singh, N.A., Ryan, S.G., Lewis, T.B., Reus, B.E., Leach, R.J. and Leppert, M. (1998) A pore mutation in a novel KQT-like potassium channel gene in an idiopathic epilepsy family. Nature Genet., 18: 53–55.

Chini, B., Raimond, E., Elgoyhen, A.B., Moralli, D., Balzaretti, M. and Heinemann, S. (1994) Molecular cloning and chromosomal localization of the human alpha 7-nicotinic receptor subunit gene (CHRNA7). Genomics, 19: 379–381.

Cordero-Erausquin, M., Marubio, L.M., Klink, R. and Changeux, J.P. (2000) Nicotinic receptor function: new perspectives from knockout mice. Trends Pharmacol. Sci., 21: 211–217.

Devillers-Thiery, A., Galzi, J.L., Eisele, J.L., Bertrand, S., Bertrand, D. and Changeux, J.P. (1993) Functional architecture of the nicotinic acetylcholine receptor: a prototype of ligand-gated ion channels. J. Membr. Biol., 136: 97–112.

Elliott, K.J., Ellis, S.B., Berckhan, K.J., Urrutia, A., Chavez-Noriega, L.E., Johnson, E.C., Velicelebi, G. and Harpold, M.M. (1996) Comparative structure of human neuronal alpha 2-alpha 7 and beta 2-beta 4 nicotinic acetylcholine receptor subunits and functional expression of the alpha 2, alpha 3, alpha 4, alpha 7, beta 2, and beta 4 subunits. J. Mol. Neurosci., 7: 217–228.

Eng, C.M., Kozak, C.A., Beaudet, A.L. and Zoghbi, H.Y. (1991) Mapping of multiple subunits of the neuronal nicotinic acetylcholine receptor to chromosome 15 in man and chromosome 9 in mouse. Genomics, 9: 278–282.

Escayg, A., MacDonald, B.T., Meisler, M.H., Baulac, S., Huberfeld, G., An-Gourfinkel, I., Brice, A., LeGuern, E., Moulard, B., Chaigne, D., Buresi, C. and Malafosse, A. (2000) Mutations of SCN1A, encoding a neuronal sodium channel, in two families with GEFS+2. Nature Genet., 24: 343–345.

Fieber, L.A. and Adams, D.J. (1991) Acetylcholine-evoked currents in cultured neurones dissociated from rat parasympathetic cardiac ganglia. J. Physiol., 434: 215–237.

Franceschini, D., Paylor, R., Broide, R., Salas, R., Bassetto, L., Gotti, C. and De Biasi, M. (2002) Absence of alpha7-containing neuronal nicotinic acetylcholine receptors does not prevent nicotine-induced seizures. Brain Res. Mol. Brain Res., 98: 29–40.

Fusco, M.D., Becchetti, A., Patrignani, A., Annesi, G., Gambardella, A., Quattrone, A., Ballabio, A., Wanke, E. and Casari, G. (2000) The nicotinic receptor beta2 subunit is mutant in nocturnal frontal lobe epilepsy. Nature Genet., 26: 275–276.

Futami, T., Takakusaki, K. and Kitai, S.T. (1995) Glutamatergic and cholinergic inputs from the pedunculo-pontine tegmental nucleus to dopamine neurons in the substantia nigra pars compacta. Neurosci. Res., 21: 331–342.

Galzi, J.L. and Changeux, J.P. (1995) Neuronal nicotinic receptors: molecular organization and regulations. Neuropharmacol., 34: 563–582.

Gambardella, A., Annesi, G., De Fusco, M., Patrignani, A., Aguglia, U., Annesi, F., Pasqua, A.A., Spadafora, P., Oliveri, R.L., Valentino, P., Zappia, M., Ballabio, A., Casari, G. and Quattrone, A. (2000) A new locus for autosomal dominant nocturnal frontal lobe epilepsy maps to chromosome 1. Neurology, 55: 1467–1471.

Gil, Z., Sack, R.A., Kedmi, M., Harmelin, A. and Orr-Urtreger, A. (2002) Increased sensitivity to nicotine-induced seizures in mice heterozygous for the L250T mutation in the alpha7 nicotinic acetylcholine receptor. NeuroReport, 13: 191–196.

Giupponi, M., Rivier, F., Vigevano, F., Beck, C., Crespel, A., Echenne, B., Lucchini, P., Sebastianelli, R., Baldy-Moulinier, M. and Malafosse, A. (1997) Linkage mapping of benign familial infantile convulsions (BFIC) to chromosome 19q. Hum. Mol. Genet., 6: 473–477.

Granger, P., Biton, B., Faure, C., Vige, X., Depoortere, H., Graham, D., Langer, S.Z., Scatton, B. and Avenet, P. (1995) Modulation of the gamma-aminobutyric acid type A receptor by the antiepileptic drugs carbamazepine and phenytoin. Mol. Pharmacol., 47: 1189–1196.

Hayman, M., Scheffer, I.E., Chinvarun, Y., Berlangieri, S.U. and Berkovic, S.F. (1997) Autosomal dominant nocturnal frontal lobe epilepsy: demonstration of focal frontal onset and intrafamilial variation. Neurology, 49: 969–975.

Hefft, S., Hulo, S., Bertrand, D. and Muller, D. (1999) Synaptic transmission at nicotinic acetylcholine receptors in rat hippocampal organotypic cultures and slices. J. Physiol., 515: 769–776.

Hirose, S., Iwata, H., Akiyoshi, H., Kobayashi, K., Ito, M., Wada, K., Kaneko, S. and Mitsudome, A. (1999) A novel mutation of CHRNA4 responsible for autosomal dominant nocturnal frontal lobe epilepsy. Neurology, 53: 1749–1753.

Hough, C.J., Irwin, R.P., Gao, X.M., Rogawski, M.A. and Chuang, D.M. (1996) Carbamazepine inhibition of N-methyl-D-aspartate-evoked calcium influx in rat cerebellar granule cells. J. Pharmacol. Exp. Ther., 276: 143–149.

Kalachikov, S., Evgrafov, O., Ross, B., Winawer, M., Barker-Cummings, C., Boneschi, F.M., Choi, C., Morozov, P., Das, K., Teplitskaya, E., Yu, A., Cayanis, E., Penchaszadeh, G., Kottmann, A.H., Pedley, T.A., Hauser, W.A., Ottman, R. and Gilliam, T.C. (2002) Mutations in LGI1 cause autosomal-dominant partial epilepsy with auditory features. Nature Genet., 3: 335–341.

Lena, C. and Changeux, J.P. (1997) Role of Ca2+ ions in nicotinic facilitation of GABA release in mouse thalamus. J. Neurosci., 17: 576–585.

Luetje, C.W. and Patrick, J. (1991) Both alpha- and beta-subunits contribute to the agonist sensitivity of neuronal nicotinic acetylcholine receptors. J. Neurosci., 11: 837–845.

Lustig, L.R., Peng, H., Hiel, H., Yamamoto, T. and Fuchs, P.A. (2001) Molecular cloning and mapping of the human nicotinic acetylcholine receptor alpha10 (CHRNA10). Genomics, 73: 272–283.

Malacarne, M., Gennaro, E., Madia, F., Pozzi, S., Vacca, D., Barone, B., dalla Bernardina, B., Bianchi, A., Bonanni, P., De Marco, P., Gambardella, A., Giordano, L., Lispi, M.L., Romeo, A., Santorum, E., Vanadia, F., Vecchi, M., Veggiotti, P., Vigevano, F., Viri, F., Bricarelli, F.D. and Zara, F. (2001) Benign familial infantile convulsions: mapping of a novel locus on chromosome 2q24 and evidence for genetic heterogeneity. Am. J. Hum. Genet., 68: 1521–1526.

Marubio, L.M., del Mar Arroyo-Jimenez, M., Cordero-Erausquin, M., Lena, C., Le Novere, N., de Kerchove d'Exaerde, A., Huchet, M., Damaj, M.I. and Changeux, J.P. (1999) Reduced antinociception in mice lacking neuronal nicotinic receptor subunits. Nature, 398: 805–810.

Neubauer, B.A., Fiedler, B., Himmelein, B., Kampfer, F., Lassker, U., Schwabe, G., Spanier, I., Tams, D., Bretscher, C., Moldenhauer, K., Kurlemann, G., Weise, S., Tedroff, K., Eeg-Olofsson, O., Wadelius, C. and Stephani, U. (1998) Centrotemporal spikes in families with rolandic epilepsy: linkage to chromosome 15q14. Neurology, 51: 1608–1612.

Oldani, A., Zucconi, M., Ferini-Strambi, L., Bizzozero, D. and Smirne, S. (1996) Autosomal dominant nocturnal frontal lobe epilepsy: electroclinical picture. Epilepsia, 37: 964–976.

Orr-Urtreger, A., Goldner, F.M., Saeki, M., Lorenzo, I., Goldberg, L., De Biasi, M., Dani, J.A., Patrick, J.W. and Beaudet, A.L. (1997) Mice deficient in the alpha7 neuronal nicotinic acetylcholine receptor lack alpha-bungarotoxin binding sites and hippocampal fast nicotinic currents. J. Neurosci., 17: 9165–9171.

Ottman, R., Risch, N., Hauser, W.A., Pedley, T.A., Lee, J.H., Barker-Cummings, C., Lustenberger, A., Nagle, K.J., Lee, K.S. and Scheuer, M.L. (1995) Localization of a gene for partial epilepsy to chromosome 10q. Nature Genet., 10: 56–60.

Phillips, H.A., Scheffer, I.E., Berkovic, S.F., Hollway, G.E., Sutherland, G.R. and Mulley, J.C. (1995) Localization of a gene for autosomal dominant nocturnal frontal lobe epilepsy to chromosome 20q13.2. Nature Genet., 10: 117–118.

Phillips, H.A., Scheffer, I.E., Crossland, K.M., Bhatia, K.P., Fish, D.R., Marsden, C.D., Howell, S.J., Stephenson, J.B., Tolmie, J., Plazzi, G., Eeg-Olofsson, O., Singh, R., Lopes-Cendes, I., Andermann, E., Andermann, F., Berkovic, S.F. and Mulley, J.C. (1998) Autosomal dominant nocturnal frontal-lobe epilepsy: genetic heterogeneity and evidence for a second locus at 15q24. Am. J. Hum. Genet., 63: 1108–1116.

Phillips, H.A., Marini, C., Scheffer, I.E., Sutherland, G.R., Mulley, J.C. and Berkovic, S.F. (2000) A de novo mutation in sporadic nocturnal frontal lobe epilepsy. Ann. Neurol., 48: 264–267.

Phillips, H.A., Favre, I., Kirkpatrick, M., Zuberi, S.M., Goudie, D., Heron, S.E., Scheffer, I.E., Sutherland, G.R., Berkovic, S.F., Bertrand, D. and Mulley, J.C. (2001) CHRNB2 is the second acetylcholine receptor subunit associated with autosomal dominant nocturnal frontal lobe epilepsy. Am. J. Hum. Genet., 68: 225–231.

Picard, F., Bertrand, S., Steinlein, O.K. and Bertrand, D. (1999) Mutated nicotinic receptors responsible for autosomal dominant nocturnal frontal lobe epilepsy are more sensitive to carbamazepine. Epilepsia, 40: 1198–1209.

Picciotto, M.R., Zoli, M., Lena, C., Bessis, A., Lallemand, Y., LeNovere, N., Vincent, P., Pich, E.M., Brulet, P. and Changeux, J.P. (1995) Abnormal avoidance learning in mice lacking functional high-affinity nicotine receptor in the brain. Nature, 374: 65–67.

Rempel, N., Heyers, S., Engels, H., Sleegers, E. and Steinlein, O.K. (1998) The structures of the human neuronal nicotinic acetylcholine receptor beta2- and alpha3-subunit genes (CHRNB2 and CHRNA3). Hum. Genet., 103: 645–653.

Role, L.W. and Berg, D.K. (1996) Nicotinic receptors in the development and modulation of CNS synapses. Neuron, 16: 1077–1085.

Ross, S.A., Wong, J.Y., Clifford, J.J., Kinsella, A., Massalas, J.S., Horne, M.K., Scheffer, I.E., Kola, I., Waddington, J.L., Berkovic, S.F. and Drago, J. (2000) Phenotypic characterization of an alpha 4 neuronal nicotinic acetylcholine receptor subunit knock-out mouse. J. Neurosci., 20: 6431–6441.

Saenz, A., Galan, J., Caloustian, C., Lorenzo, F., Marquez, C., Rodriguez, N., Jimenez, M.D., Poza, J.J., Cobo, A.M., Grid, D., Prud'homme, J.F. and Lopez de Munain, A. (1999) Autosomal dominant nocturnal frontal lobe epilepsy in a Spanish family with a Ser252Phe mutation in the CHRNA4 gene. Arch. Neurol., 56: 1004–1009.

Scheffer, I.E. and Berkovic, S.F. (1997) Generalized epilepsy with febrile seizures plus. A genetic disorder with heterogeneous clinical phenotypes. Brain, 120: 479–490.

Scheffer, I.E., Bhatia, K.P., Lopes-Cendes, I., Fish, D.R., Marsden, C.D., Andermann, F., Andermann, E., Desbiens, R., Cendes, F. and Manson, J.I. (1994) Autosomal dominant frontal epilepsy misdiagnosed as sleep disorder. Lancet, 26: 515–517.

Scheffer, I.E., Bhatia, K.P., Lopes-Cendes, I., Fish, D.R., Marsden, C.D., Andermann, E., Andermann, F., Desbiens, R., Keene, D., Cendes, F. and Berkovic, S.F. (1995) Autosomal dominant nocturnal frontal lobe epilepsy. A distinctive clinical disorder. Brain, 118: 61–73.

Scheffer, I.E., Phillips, H.A., O'Brien, C.E., Saling, M.M., Wrennall, J.A., Wallace, R.H., Mulley, J.C. and Berkovic, S.F. (1998) Familial partial epilepsy with variable foci: a new partial epilepsy syndrome with suggestion of linkage to chromosome 2. Ann. Neurol., 44: 890–899.

Seguela, P., Wadiche, J., Dineley-Miller, K., Dani, J.A. and Patrick, J.W. (1993) Molecular cloning, functional properties, and distribution of rat brain alpha 7: a nicotinic cation channel highly permeable to calcium. J. Neurosci., 13: 596–604.

Singh, N.A., Charlier, C., Stauffer, D., DuPont, B.R., Leach, R.J., Melis, R., Ronen, G.M., Bjerre, I., Quattlebaum, T., Murphy, J.V., McHarg, M.L., Gagnon, D., Rosales, T.O., Peiffer, A., Anderson, V.E. and Leppert, M. (1998) A novel potassium channel gene, KCNQ2, is mutated in an inherited epilepsy of newborns. Nature Genet., 18: 25–29.

Steinlein, O., Smigrodzki, R., Lindstrom, J., Anand, R., Kohler, M., Tocharoentanaphol, C. and Vogel, F. (1994) Refinement of the localization of the gene for neuronal nicotinic acetylcholine receptor α4 subunit (CHRNA4) to human chromosome 20q13.2-13.3. Genomics, 22: 493–495.

Steinlein, O., Mulley, J.C., Propping, P., Wallace, R.H., Phillips, H.A., Sutherland, G., Scheffer, I.E. and Berkovic, S.F. (1995) A missense mutation in the neuronal nicotinic acetylcholine receptor α4 subunit is associated with autosomal dominant nocturnal frontal lobe epilepsy. Nature Genet., 11: 201–203.

Steinlein, O., Magnusson, A., Stoodt, J., Bertrand, S., Weiland, S., Berkovic, S.F., Nakken, K.O., Propping, P. and Bertrand, D. (1997) An insertion mutation of the CHRNA4 gene in a family with autosomal dominant nocturnal frontal lobe epilepsy. Hum. Mol. Genet., 6: 943–947.

Steinlein, O.K., Stoodt, J., Mulley, J., Berkovic, S., Scheffer, I.E. and Brodtkorb, E. (2000) Independent occurrence of the CHRNA4 Ser248Phe mutation in a Norwegian family with nocturnal frontal lobe epilepsy. Epilepsia, 41: 529–535.

Sugawara, T., Tsurubuchi, Y., Agarwala, K.L., Ito, M., Fukuma, G., Mazaki-Miyazaki, E., Nagafuji, H., Noda, M., Imoto, K., Wada, K., Mitsudome, A., Kaneko, S., Montal, M., Nagata, K., Hirose, S. and Yamakawa, K. (2001) A missense mutation of the Na + channel alpha II subunit gene Na(v)1.2 in a patient with febrile and afebrile seizures causes channel dysfunction. Proc. Natl. Acad. Sci. USA, 98: 6384–6389.

Szepetowski, P., Rochette, J., Berquin, P., Piussan, C., Lathrop, G.M. and Monaco, A.P. (1997) Familial infantile convulsions and paroxysmal choreoathetosis: a new neurological syndrome linked to the pericentromeric region of human chromosome 16. Am. J. Hum. Genet., 61: 889–898.

Unwin, N. (1993) Nicotinic acetylcholine receptor at 9 A resolution. J. Mol. Biol., 229: 1101–1124.

Wallace, R.H., Wang, D.W., Singh, R., Scheffer, I.E., George, A.L. Jr., Phillips, H.A., Saar, K., Reis, A., Johnson, E.W., Sutherland, G.R., Berkovic, S.F. and Mulley, J.C. (1998) Febrile seizures and generalized epilepsy associated with a mutation in the Na + -channel beta1 subunit gene SCN1B. Nature Genet., 19: 366–370.

Wallace, R.H., Marini, C., Petrou, S., Harkin, L.A., Bowser, D.N., Panchal, R.G., Williams, D.A., Sutherland, G.R., Mulley, J.C., Scheffer, I.E. and Berkovic, S.F. (2001) Mutant GABA(A) receptor gamma2-subunit in childhood absence epilepsy and febrile seizures. Nature Genet., 28: 49–52.

Weiland, S., Witzemann, V., Villarroel, A., Propping, P. and Steinlein, O. (1996) An amino acid exchange in the second transmembrane segment of a neuronal nicotinic receptor causes partial epilepsy by altering its desensitization kinetics. FEBS Letters, 298: 91–96.

Weiland, S., Bertrand, D. and Leonard, S. (2000) Neuronal nicotinic acetylcholine receptors: from the gene to the disease. Behav. Brain Res., 113: 43–56.

Whiting, P., Schoepfer, R., Lindstrom, J. and Priestly, T. (1992) Structural and pharmacological characterization of the major brain nicotine acetylcholine receptor subtype stably expressed in mouse fibroblasts. Mol. Pharmacol., 40: 463–472.

Wonnacott, S. (1997) Presynaptic nicotinic ACh receptors. Trends Neurosci., 20: 92–98.

Xiao, J.F., Fischer, C. and Steinlein, O.K. (2001) Cloning and mutation analysis of the human potassium channel KCNQ2 gene promoter. NeuroReport, 12: 3733–3739.

Xiong, L., Labuda, M., Li, D.S., Hudson, T.J., Desbiens, R., Patry, G., Verret, S., Langevin, P., Mercho, S., Seni, M.H., Scheffer, I., Dubeau, F., Berkovic, S.F., Andermann, F., Andermann, E. and Pandolfo, M. (1999) Mapping of a gene determining familial partial epilepsy with variable foci to chromosome 22q11-q12. Am. J. Hum. Genet., 65: 1698–1710.

Xu, W., Gelber, S., Orr-Urtreger, A., Armstrong, D., Lewis, R.A., Ou, C.N., Patrick, J., Role, L., De Biasi, M. and Beaudet, A.L. (1999) Megacystis, mydriasis, and ion channel defect in mice lacking the alpha3 neuronal nicotinic acetylcholine receptor. Proc. Natl. Acad. Sci. USA, 96: 5746–5751.

Jones, S. and Yakel, J.L. (1999) Inhibitory interneurons in hippocampus. Cell. Biochem. Biophys., 31: 207–218.

Yoshimura, R., Yanagihara, N., Terao, T., Minami, K., Abe, K. and Izumi, F. (1995) Inhibition by carbamazepine of various ion channels-mediated catecholamine secretion in cultured bovine adrenal medullary cells. Naunyn Schmiedebergs Arch. Pharmacol., 352: 297–303.

Zhang, M., Wang, Y.T., Vyas, D.M., Neuman, R.S. and Bieger, D. (1993) Nicotinic cholinoceptor-mediated excitatory postsynaptic potentials in rat nucleus ambiguus. Exp. Brain Res., 96: 83–88.

Progress in Brain Research, Vol. 145
ISSN 0079-6123

CHAPTER 20

Neurochemistry of consciousness: cholinergic pathologies in the human brain

Elaine K. Perry* and Robert H. Perry

*Development in Clinical Brain Ageing: MRC Building, Newcastle General Hospital,
Westgate Road, Newcastle upon Tyne NE4 6BE, UK*

Introduction

While consciousness is often considered to be beyond the realms of current scientific paradigms, the subject is increasingly part of the neurobiological literature. A satisfactory definition is still elusive, but the challenge of identifying neural correlates of consciousness (NCC) is being met in relation to neuroanatomy, functional imaging, neurophysiology and neurochemistry. With respect to neurotransmitter systems, acetylcholine (ACh) has been the focus of a number of recent articles, providing the evidence that this transmitter may be a prime candidate for a neural correlate of consciousness (NCC) (Woolf, 1997; Perry et al., 1999, 2002). Disorders of the human brain which involve disturbances in consciousness and cholinergic neuropathology provide a basis for exploring potential cholinergic NCC.

Cholinergic systems and consciousness

Cholinergic systems are more numerous than the other modulatory systems in the brain and distributed in a wide range of areas (reviewed Perry et al., 1999; Perry and Young, 2002). Of particular interest in relation to consciousness, discrete nuclei situated rostrally and caudally provide both widely divergent and, in some key areas such as cortex and thalamus,

also convergent projections. In the basal forebrain, septal neurons innervate hippocampus, cingulate and retrosplenial cortex; diagonal band neurons innervate entorhinal and olfactory cortex; and nucleus of Meynert the remainder of the cortex, amygdala and select thalamic nuclei. Functions of these and brainstem cholinergic nuclei include arousal, selective attention and REM sleep.

According to Mesulam (1995), the size of nucleus basalis cholinergic projections to the cortex indicates that 'this pathway is likely to constitute the single most substantial regulatory afferent system of the cerebral cortex'. Based on the maintenance of cortical activation during REM sleep, in the absence of monoaminergic (e.g., noradrenergic and 5-HT) activity, but continued firing of cholinergic nucleus basalis neurons, Buzsaki et al. (1988) concluded that 'the ascending cholinergic system alone is capable of keeping the neocortex in its operative mode'. The consensus view on the role of these cholinergic projections is that they control selective attention. Since Delacour (1995) has speculated that selective attention and consciousness overlap, and Baars (1998) has highlighted the importance of selective attention in a recent model of consciousness, the two processes may share a common neural basis.

Basal forebrain cholinergic neurons also project to select thalamic nuclei, including the reticular nucleus, which with topographical cortical input is also implicated in selective attention and consciousness (McAlonan and Brown, 2002). 85–95% of brainstem afferents to most thalamic nuclei, including both

---

*Corresponding author: Tel.: 0191-273-5251;
Fax: 0191-272-5291; E-mail: e.k.perry@ncl.ac.uk

DOI: 10.1016/S0079-6123(03)45020-6

specific relay and reticular nuclei, originate in the region of the rostral midbrain core where cholinergic pedunculopontine and lateral dorsal tegmental nuclei are maximally developed. Amongst thalamic nuclei, the intralaminar, also implicated in conscious awareness (Bogen, 1995), receive the highest density of brainstem cholinergic afferents. These thalamic inputs are excitatory both via direct, early nicotinic and slower muscarinic depolarization and via indirect hyperpolarization of GABAergic reticular neurons. Coactivation of rostrally projecting brainstem and forebrain cholinergic neurons, such as occurs in both wakefulness and REM sleep, provides thalamo-cortical circuitry with an integrative modulation that could represent a component mechanism of conscious awareness.

Electrophysiologically, acetylcholine is involved in both tonic and phasic neurotransmission. The basal forebrain cholinergic system has been implicated in the generation of an attending potential known as the P_{300}, thought to be a component of conscious awareness (reviewed Perry et al., 1999). Human studies have demonstrated P_{300} latency increases and amplitude reductions with the administration of scopolamine, reversed by physostigmine. These findings are consistent with animal studies which have additionally shown that physostigmine alone increases the P_{300} amplitude. Lesions of cholinergic basal forebrain neurons result in P_{300} latency delay and amplitude reductions. This effect is reversed with vagal implants, restoring P_{300} characteristics which correlate with the restoration of cortical levels of the cholinergic enzyme choline acetyltransferase (ChAT). High frequency (40 Hz) synchronizations considered to be a key neuronal correlate of consciousness, can be induced in vitro by acetylcholine (Fisahn et al., 1998), an effect blocked by atropine. Interestingly scopolamine administration in man is reported to increase performance in tasks which depend on implicit/nonconscious processing (Callaway and Band, 1958).

One of the most compelling arguments in favor of a primary role for forebrain cholinergic neurons in conscious awareness is that, unlike other transmitter systems, acetylcholine release in the cortex correlates with the level of awareness from high during waking, low (but not abolished) during slow wave sleep, and increased again during REM sleep (below, also reviewed Perry and Piggott, 2000). It can be argued that this variation relates to the extent of conscious awareness, if it is accepted that consciousness exists (albeit with different characteristics) during REM sleep and dreaming. Recent evidence indicates that dreaming occurs not only during REM but also during nonREM or slow wave sleep (Solms, 2000). There is then a low level of conscious awareness associated with the less complex and vivid dreaming during nonREM sleep, which could relate to minimal forebrain cholinergic activity in this state. In contrast to the basal forebrain, brainstem (e.g., pedunculo-pontine) cholinergic activity is evident during REM sleep and waking, but silent during nonREM sleep. A 'cholinergic NCC hypothesis', based on relative levels of consciousness during the sleep–wake cycle, would thus focus primarily on the basal forebrain cholinergic system.

If cholinergic rostral projecting neurons are centrally involved in generating conscious awareness, the question of whether any particular receptor subtype is involved arises. Muscarinic metabotropic receptors are concentrated in cortex and striatum, the predominant subtypes M1 and M4 being found to the greatest extent in these regions. M3 is predominant in the thalamus and M2 in cerebellum. Nicotinic ionotropic receptor subtype distributions are still being established, although much is known of the distribution of the two principal receptor subtypes (Court et al., 2001a,b). The high affinity receptor (predominantly $\alpha 4\beta 2$) is widely distributed and concentrated in thalamus (particularly lateral geni-culate nucleus), midbrain (e.g., substantia nigra), striatum and limbic cortex (e.g., subiculum and entorhinal cortex). The low affinity nicotinic receptor ($\alpha 7$) is found in hippocampus, substantia nigra and notably in the reticular nucleus of the thalamus. Neither subtype has however so far been specifically localized in thalamic intralaminar nuclei.

If acetylcholine functions to enlist and integrate the activity of target neurons into the networks associated with consciousness, such neurons obviously need to be responsive, either directly or indirectly to the transmitter. In the cortex for example, muscarinic and nicotinic receptors are widely distributed on pyramidal and nonpyramidal neurons. In many cortical neurons, acetylcholine is not directly excitatory, but enhances neuronal responses to other transmitters glutamate. Reinforcing the actions of executive transmitter such as glutamate and GABA, acetylcholine could increase competition between excitatory and inhibitory states, so restricting the neuronal numbers depolarising at any one time to those most relevant to the immediate external or internal environment. In rat brain, cortical activation reduces intralaminar inhibition through nicotinic receptors and promotes intracolumnar inhibition through muscarinic receptors (Xiang et al., 1998), so changing the direction of information flow within cortical circuits. In this chapter, evidence is provided that the nicotinic $\alpha_4\beta_2$ receptor may be of particular importance in potential cholinergic mechanisms of consciousness.

Role of acetylcholine in disorders of consciousness

The cholinergic system is implicated in a broad range of disorders of the human brain, including those associated with development (Rett's syndrome, autism and schizophrenia for example), and in old age (Alzheimer's and Parkinson's disease, dementia with Lewy bodies (DLB), and progressive supranuclear palsy, for example). Amongst these, autism and DLB are particular examples of brain disorders, which include as central feature disturbances in consciousness.

Autism can be considered as a disorder of consciousness. In this disease, there is not generally evidence of reflection and imagination that characterize human consciousness. In a recent analysis of self-referent memory tasks, performance in high functioning autism was lower compared to a normal group, indicative of deficits in self consciousness (Toichi et al., 2002). Evidence of cholinergic dysfunction in autism includes neuropathological abnormalities originally reported in the basal forebrain cholinergic nuclei. In relation to neurochemical activities, there are significant and extensive reductions in the $\alpha_4\beta_2$ nicotinic receptor subtype in the cerebral cortex (Perry et al., 1991). Therapeutic intervention based on cholinergic receptor modulation is currently being explored and results so far indicate positive and nonreversible improvements in function, especially language, using cholinesterase inhibitors.

Dementia with Lewy bodies (DLB) is the second most common degenerative dementia in old age, after Alzheimer's disease (Perry et al., 1990a). In DLB, there are frequent fluctuations in the level of awareness and attention together with persistent visual hallucinations. Neurochemical data indicate an extensive loss of presynaptic cholinergic activity in the cerebral cortex and also reductions in the thalamus and striatum in DLB. In the cortex, these losses are associated with visual hallucinations. Also, as in autism, there are reductions in the $\alpha_4\beta_2$ nicotinic receptor, and in this disorder this may be related to the disturbances in consciousness.

This chapter provides details of these cholinergic pathologies and explores how these may add to our understanding of the neurochemistry of consciousness.

Autism

Autism is a developmental disorder, involving abnormalities in social function, language and imagination. There is as yet no aetiology-based treatment or cure. Neurotransmitter signaling systems are relevant to symptom aetiology, treatment, and brain development. Transmitters so far implicated in autism include the monoamines-serotonin, dopamine and noradrenaline, together with acetylcholine, GABA, glutamate and several neuropeptides. Investigations have mainly relied on measurements in blood or cerebrospinal fluid or responses to pharmaceutical agents, and more recently on genetic linkage data and observations using human brain tissue.

The cholinergic system has received less attention in autism than the monoaminergic (e.g., 5-HT and dopamine) and neuropeptide (e.g., opioid) systems. It has been implicated on the basis of pathological abnormalities reported in the basal forebrain (septal) cholinergic neurons, such as increased and decreased size and numbers in, respectively, younger and adult individuals (Bauman and Kemper, 1994).

Cholinergic activities during development

Cholinergic afferents innervate the cerebral cortex during the most dynamic periods of neuronal differentiation and synapse formation, suggesting they play a regulatory role in these events (Hohmann and Berger-Sweeney, 1998). In the human cerebral cortex and hippocampus, cholinergic parameters including choline acetyltransferase activities (ChAT, the enzyme synthesizing the transmitter) and high affinity nicotinic receptor binding alter substantially postnatally, within the early postnatal period, when symptoms of autism first manifest (Court et al., 1993; Lainhart and Piven, 1995; Court et al., 1997). It is during the early postnatal period that human consciousness most likely develops, if it can be assumed that the absence of any recollection of experience during the first few years of life reflects a lack of conscious awareness.

Disruption of cholinergic innervation during early postnatal development (e.g., neonatal basal forebrain cholinergic lesions in rats) results in delayed cortical neuronal development, permanent changes in cortical cytoarchitecture, and impaired cognitive function (Hohmann and Berger-Sweeney, 1998). Abnormalities in cortical cytomorphology, including altered thalamic afferent distribution in layer IV (Hohmann et al., 1991a) are said to resemble pathologies associated with developmental disorders resulting in mental retardation (Huttenlocher, 1975).

Choline acetyltransferase in autism

The status of the cholinergic system in autism, based on basal forebrain and cortical indices in adults, is summarized in Table 1. No abnormalities are apparent in the activity of the transmitter synthesizing enzyme choline acetyltransferase (ChAT) in frontal or parietal cortex or the basal forebrain, although activity in the basal forebrain tends to be increased (Perry et al., 2001). In relation to the original neuropathological evidence of basal forebrain cholinergic dysfunction in autism (reduced neurons in adult autistics, Bauman and Kemper, 1994), the finding of normal ChAT both in frontal and parietal cortex and in the basal forebrain suggests the presynaptic cholinergic innervation of the cortex is structurally intact in autism. This indicates either that cholinergic neurons are not depleted in the cases examined or that compensatory

axonal sprouting has occurred in conjunction with cell loss. In another developmental disorder, Rett's syndrome, in which ChAT and the vesicular acetylcholine transporter (another presynaptic marker) are reduced in various areas including cortex (Wenk et al., 1991), disruption of cholinergic innervation is likely due to developmental or degenerative neuronal abnormalities occurring before or shortly after birth in the absence of compensation. The contrast with autism is striking and elevated BDNF in autism (below) may perhaps play a role in maintaining cortical cholinergic projections.

Cholinergic neurotrophins

Of two neurotrophins which control cholinergic neuronal development and maintenance, NGF levels are normal, but BDNF levels are significantly elevated (three-fold) in the basal forebrain in autism (Perry et al., 2001). The increase in BDNF is also apparent in the cerebral cortex where, although BDNF levels are normally much lower than in the basal forebrain, levels were over five-fold higher in autistic compared to control cases (Perry et al., unpublished).

The finding of increased BDNF in the basal forebrain and cortex can be interpreted variously. BDNF plays a role in sculpting synaptic connexions. Since BDNF is upregulated by cholinergic activity in developing rat hippocampus (da Penha Berzaghi et al., 1993), it is possible that the abnormality is due to a transient developmental cholinergic hypertrophy. Another possible explanation for the elevation, since neurotrophins influence the development and function of basal forebrain cholinergic neurons (Hohmann and Berger-Sweeney, 1998; Hashimoto et al., 1999), is that it reflects a regional compensatory mechanism. This may underlie the finding that cholinergic biomarkers, such as ChAT in the cortex and basal forebrain, are normal or even elevated. A further explanation is that overexpression of BDNF is an intrinsic component of the autism disease process. Recent findings of elevated blood levels of BDNF (amongst three other brain peptides or proteins including vasoactive intestinal peptide and calcitonin gene related peptide) in newborn autistic

Table 1. Basal forebrain cholinergic system in autism[a]

Neurons in septal nucleus	Increased numbers in children[b], decreased in adults
Choline acetyltransferase	Normal enzyme activity in frontal and parietal cortex
Acetylcholinesterase	Normal in cerebral cortex and basal forebrain
Nicotinic receptor, $\alpha_4\beta_2$	Extensive loss of receptor binding, immunoreactivity and mRNA in frontal, parietal and occipital cortex
Nicotinic receptor, α_7	Normal receptor binding, immunoreactivity and mRNA in frontal and parietal cortex
Muscarinic M1 receptor	Normal receptor binding, except for moderate loss in parietal cortex
Muscarinic M2 receptor	Normal receptor binding in frontal and parietal cerebral cortex
BDNF	Increased in parietal cortex and basal forebrain Elevated blood levels in neonates[b]
NGF	Normal in basal forebrain

[a]All of the findings reported, except[b], are based on adult cases.

and other mentally retarded individuals (Nelson et al., 2001) suggest an intrinsic rather than compensatory mechanism.

Muscarinic receptors

Binding to the M2 muscarinic receptor (which is located presynaptically on a variety of neuronal types) is normal in frontal and parietal cortex in autism (Perry et al., 2001). According to one report cortical muscarinic M1 receptor binding in the same areas is in contrast decreased (Perry et al., 2001). Moderate M1 receptor loss was apparent in frontal and parietal cortex and in both outer and inner cortical layers, and reached significance in parietal cortex. This neurochemical abnormality may be specific to autism, since it is not apparent in nonautistic mentally retarded individuals.

Reduced M1 muscarinic receptor binding in the parietal cortex in autism may indicate a specific abnormality in cholinoceptive function, since the M1 receptor is located postsynaptically. This could be related to epilepsy, which occurs in up to 40% of autistic individuals (Minshaw et al., 1997), since a loss has been reported in hippocampal sclerosis associated with temporal lobe epilepsy (Pennell et al., 1999). However, the finding of normal pirenzepine binding in a similar series of cases in the hippocampus (Blatt et al., 2001), an area particularly

susceptible in epilepsy, suggests another basis for the receptor loss such as dendritic dysfunction.

Nicotinic receptors

With respect to nicotinic receptor binding, there is no alteration of α-bungarotoxin (αBT) binding, immunoreactivity or mRNA in the cerebral cortex (Perry et al., 2001). In contrast to the α7 subtype, in almost all cortical areas so far examined (frontal, parietal and occipital but not cingulate), significant and extensive reductions of epibatidine binding (to 20–30% of the normal) are apparent throughout the different cortical layers. Epibatidine binding is also significantly reduced in nonautistic mentally retarded individuals but not in the Down's syndrome. It is unlikely that the receptor loss is due to differential use of tobacco (which results in increased receptor levels) between the groups since in the basal forebrain nicotine binding is normal.

Based on Western blotting, a selective loss of α4 and β2, but not α3 immunoreactivities has also been identified in autism (Martin-Ruiz et al., unpublished). Alpha 4 mRNA detected using RT-PCR in conjunction with a housekeeper mRNA (Martin-Ruiz et al., unpublished) is also reduced to the same extent as the receptor binding. This suggests that the abnormality exists at the level of gene expression and raises the question of whether there are any abnormalities of the alpha 4 subunit gene (CHRNA 4) or its promoter in autism. Such abnormalities are apparent in familial nocturnal epilepsy (Steinlein, 2000).

Since nicotinic receptors are thought to play a particular role in regulating synaptic/dendritic plasticity, it is likely that the receptor reduction in autism relates to this aspect of neuronal function. Knockout ($\beta 2$) mouse models, lacking high affinity agonist receptor binding, develop a degree of age-related cortical atrophy and neuronal loss in conjunction with cognitive impairment (Picciotto et al., 1999; Zoli et al., 1999). In the development of retinal ganglion cells, for example, exposure to the nicotinic antagonist curare from early embryonic stages aborts dendritic proliferation (Sernagor et al., 2001). Relationships between loss of the $\alpha 4$ nicotinic receptor subtype and synaptophysin, identified in the cerebral cortex in another cerebral disorder (Alzheimer's disease, Sabbagh et al., 1998), suggest that the receptor loss in autism may be associated with abnormal synaptic morphology and function, possibly involving overextensive synaptic pruning during development.

Loss of the high affinity nicotinic receptor agonist site not only in autism, but also in the nonautistic mental retardation group, could indicate a lack of specificity to autism, and that the receptor loss may be a consequence rather than cause of cortical dysfunction. However, developmental brain abnormalities occur in both groups, and an overlap in the processes involved would be expected. It will be important in the future to determine at what age the receptor abnormality in autism occurs. This could be investigated using novel nicotinic receptor ligands applicable to PET or SPECT in vivo imaging.

Clinical and therapeutic implications of the nicotinic receptor loss in autism

It is striking that amongst a range of cholinergic activities investigated in autism, and of other transmitter activities such as dopaminergic and 5-HT$_1$ also investigated in the same series (Perry et al., unpublished), the only consistent and extensive phenotypic

abnormality is the loss of the nicotinic $\alpha_4\beta_2$ receptor subtype. Although not yet established, the nicotinic $\alpha_4\beta_2$ receptor loss from the cerebral cortex could relate to clinical features of autism, such as attentional abnormalities, pain perception, anxiety, epilepsy or altered consciousness. So far, it does not appear that the receptor loss is associated with epilepsy (Perry et al., 2001) although other clinical correlates remain to be explored. The nicotinic receptor is implicated in attention (Mirza and Stolerman, 2000), and nicotine administered acutely in normal volunteers results in improvements in extended vigilance tasks, divided attention and rapid information processing (Wesnes, K., personal communication). The receptor loss may thus relate to attentional deficits in autism. Nicotinic agents are also analgesic (Bannon et al., 1998), and on the basis of a gene knockout model (Marubio et al., 1999, Table 1) the $\alpha 4$ subunit has been implicated in pain perception. Low levels of this receptor subtype in autism could thus relate to reduced pain reactivity evident in the disease (Tordjman et al., 1999).

There is no FDA-approved pharmacotherapy for autism, although neuroleptics, benzodiazepines, anticonvulsants and SSRI's are prescribed symptomatically. There is presently one report that cholinesterase inhibitors (donepezil and rivastigmine) administered to autistic children result in symptomatic improvement, especially language, in 70% of cases (Chez et al., 2001 and unpublished). Particularly encouraging, symptoms did not revert on cessation of drug treatment, consistent with the concept that restoring deficient cholinergic neurotransmission is not only of short term but also long term benefit, related perhaps to neurotrophic effects of chemical transmitters such as acetylcholine (Belluardo et al., 2000). Controlled clinical trials of currently available and developing cholinergic therapies in autism are warranted.

Therapeutic strategies could include nicotinic receptor agonists, such as nicotine provided via patches, gum or vaporisers. Nicotine is of clinical value in Tourettism (Sandberg et al., 1997) and Down's syndrome (Seidi et al., 2000). Such treatment may also be disease modifying. Nicotine administration in animal models or tobacco use in man results in an increase in nicotinic receptor levels (particularly $\alpha 4$, Martin-Ruiz et al., 1999). Nicotine significantly increases branching of both axons and dendrites in cultured hippocampal neurons (Audersirk and Cabell, 1999). In adolescent rat brain, nicotine exposure results in persistent upregulation of nicotinic receptors in a variety of brain areas including cortex (Trauth et al., 1999) and in adult mouse brain chronic nicotine exposure promotes retention of nicotinic receptors that decline in aging (Rogers et al., 1998). Newly available nicotinic drugs targeted to the $\alpha 4\beta 2$ would thus be worth testing in autism.

Dementia with Lewy bodies

Dementia with Lewy bodies (DLB) is a primary neurodegenerative dementia sharing several clinical and pathological characteristics with Parkinson's disease. It is characterized by a variable combination of fluctuating cognition/consciousness, neuropsychiatric disturbance, particularly visual hallucinations and parkinsonism (McKeith et al., 1996). The presynaptic protein α-synuclein is a major component of cortical and subcortical Lewy bodies and neurites that characterize the neuropathology of DLB (Spillantini et al., 1998).

The cholinergic system in DLB

A summary of cholinergic activities in DLB is provided in Table 2. The basal forebrain cholinergic system is profoundly affected. In addition to low

ChAT in the cortex, measured at autopsy (reductions to a less than one third of normal levels), ^{11}C-N-methyl-4-piperidyl acetate (a cholinesterase inhibitor) positron emission tomography imaging has identified cholinergic deficits in DLB in vivo (Shinotoh et al., 2001). Not only are neocortical presynaptic cholinergic activities affected, but there are also losses of ChAT in the thalamus (Perry et al., 1991) that may reflect not only basal forebrain input but also pathology of the pedunculopontine pathway.

Cortical muscarinic M1 receptor binding is elevated in DLB (Perry et al., 1990a), a finding recently confirmed by immunoabsorption studies. Cortical nicotinic receptor changes include a loss of the high affinity agonist binding site (reflecting the $\alpha 4\beta 2$ subtype), in many areas but there is no consistently reported change in alpha 7 subunit or alpha bungarotoxin binding sites. In the thalamus, by contrast, there is little change in nicotine binding, but a highly significant reduction in alpha bungarotoxin binding in the reticular nucleus (Court et al., 2001b). While loss of cortical high affinity nicotinic receptor binding has been related to synapse loss, as measured by synaptophysin levels in AD (Sabbagh et al., 1998), synaptophysin loss only occurs in DLB when the pathology includes that of the Alzheimer type.

Clinical correlates of cholinergic abnormalities

In contrast to the neurochemical measures in autism, cholinergic activities in DLB have been measured in

Table 2. Basal forebrain cholinergic system in dementia with lewy bodies

Neurons in nucleus of Meynert	Reduced numbers, α-synuclein positive Lewy bodies, neurites and inclusion bodies
Choline acetyltransferase	Reduced in all cerebral cortical areas examined, also in thalamus, including the reticular nucleus
Acetylcholinesterase	Reduced in all cortical areas examined
Nicotinic receptor, $\alpha_4\beta_2$	Reduced in cortical areas examined
Nicotinic receptor, α_7	Inconsistent reports: normal or reduced in cortex
Muscarinic M1 receptor	Increased in all cortical areas examined
Muscarinic M2 receptor	Reduced in cortex

patients prospectively assessed clinically, so permitting valid correlations between cerebral functions and symptoms of the disease. Cortical ChAT reductions in DLB correlate with cognitive impairments, independent of the extent of Alzheimer-type pathology, (Perry et al., 1990b; Samuel et al., 1997). The loss of thalamic cholinergic activity, which occurs particularly in the reticular formation, is likely to reflect loss of pedunculopontine neurones. Whether this relates to attentional dysfunction and/or disturbances in consciousness has not however yet been established.

Cortical ChAT reductions also relate to visual hallucinations, being more extensive in hallucinators compared to nonhallucinators (Ballard et al., 2000). The nicotinic receptor, measured using α-bungarotoxin binding (alpha-7 subunit) is also reported to be lower in hallucinating compared with nonhallucinating individuals (Court et al., 2001b). Hallucinations are principally visual, integrated images of people and animals. Deficits in visual perception and hallucinations measurable arise as a result of disruptions in the integration of visual perception and memory, secondary to defective cholinergic transmission. Interestingly, similar hallucinations are induced in normal individuals as a result of ingestion of muscarinic receptor antagonists (Perry et al., 1999).

The muscarinic M1 receptor is more elevated in individuals with delusions than those without (Ballard et al., 2000). This suggests that delusions may be associated with a loss of presynaptic cholinergic activity and a consequent, perhaps hypercompensatory, elevation in the M1 receptor subtype. Cholinergic therapy, which reduces delusions, is likely to be associated with a reduction in M1 receptors.

Disturbances in consciousness in DLB frequently include fluctuations in the level of awareness and less frequently periodic loss of consciousness. Variations can be detected over very short periods of time (on a second to second basis), suggesting that fluctuating consciousness arises from dysregulation of continuously active arousal systems. DLB is characterized by increased variability in performance on cognitive tasks, both within and between subjects and when compared to both age-matched controls and patients with Alzheimer's disease (AD) (Ballard et al., 2001). This variability is particularly evident in executive and attentional tasks.

Fluctuating cognition shows no correlation with the density of Lewy bodies (DLB) in either neocortical or paralimbic areas (Gomez-Tortosa et al., 1999).

Disturbances in consciousness in DLB, in contrast to hallucinations or delusions, do not relate to presynaptic cholinergic activities or the muscarinic M1 receptor, (or to a range of dopamine or 5-HT activities). However these are associated with abnormalities in the nicotinic receptor subtype binding agonists such as epibatidine with high affinity primarily $\alpha4\beta2$ in temporal association cortex (Ballard et al., 2002). This is the first report specifically relating disturbances in consciousness, assessed prospectively using verified clinical tools in a patient population, to an index of neurotransmitter functions, specifically the nicotinic $\alpha_4\beta_2$ receptor. Paradoxically, the receptor was reduced to a lesser extent in individuals with disturbances in consciousness, compared to those not experiencing fluctuating awareness. The receptor is located on a variety of cortical neurons (principally GABA but also glutamate) and on presynaptic terminals (cholinergic, 5-HT and dopamine for example). It is likely that different populations of the receptor are affected in the two DLB subgroups and that a more selective loss in the affected subgroup leads to an imbalance that critically affects the control of conscious awareness.

Cholinergic therapy in DLB

There are no disease-modifying pharmacological therapies for DLB, which might be based for example as an understanding of the neurobiology of synuclein. However, symptomatic treatments are available. Neuroleptics can control neuropsychiatric symptoms, but many patients who receive neuroleptics experience neuroleptic sensitivity (McKeith et al., 1992a). This is characterized by sedation, immobility, rigidity, postural instability, falls, increased confusion and unconsciousness.

A rational, effective and safer alternative for treating DLB appears to be to compensate for the cholinergic deficits described, by the use of cholinesterase inhibitors. Open studies have shown improvements in both cognitive and noncognitive symptoms with donepezil and rivastigmine, without significant deterioration in motor function (Shea et al., 1998). Apathy, anxiety, impaired attention, hallucinations, delusions, sleep disturbance, and cognitive test performance are the most frequently cited treatment responsive symptoms. Improvements are generally reported as greater than those achieved with similar doses of these drugs in AD, for which such drugs were originally designed, and a recent open label study found evidence of sustained efficacy after two years of treatment (Grace et al., 2001). A multi-center, randomized placebo-controlled trial of rivastigmine in DLB confirmed these observational studies (McKeith et al., 2000b). 120 DLB patients were treated with rivastigmine at daily doses up to 12 mg or placebo for 20 weeks. Subjects had a clinical diagnosis of probable and possible DLB and a MMSE score of greater than 10. Approximately twice as many patients on rivastigmine (63.4%) than on placebo (30.0%) showed at least a 30 percent improvement from baseline with regard to DLB-typical psychiatric symptoms. Patients also showed improvements on computer-based tests of attention. Adverse effects were similar to those observed with the drug in AD and to those reported for other cholinesterase inhibitors. Parkinsonian symptoms did not worsen on treatment, although emergent tremor was noted in a small number of treated patients.

In a further analysis of DLB patients treated with rivastigmine, Wesnes et al. (2002) discovered that responders were more likely to be those who experienced hallucinations and fluctuations in attention. This suggests, in keeping with the nicotinic receptor correlate described above, that disturbances/fluctuations in consciousness in DLB are associated not with an irreversible pathology but with the retention of retrievable function. Disturbances in consciousness could then be considered due to a functional imbalance between key neuronal populations receiving cholinergic input. Pursuing this concept in further clinico-pathological and treatment intervention studies may provide new insights into key mechanisms controlling conscious awareness.

New insights in to the neurochemistry of consciousness

In exploring a 'cholinergic hypothesis of consciousness', this chapter depends on assumptions and correlations which are admittedly highly speculative. In the first place, a broad range of human cerebral pathologies could be linked to consciousness, since so many human brain activities involves conscious awareness, from sensory perception to higher cognitive functions. Disorders such as blindsight, schizophrenia, narcolepsy, delirium, depression, Alzheimer's disease, frontotemporal dementia, epilepsy, stroke, and a range of developmental diseases associated with mental retardation have been or could be examined from this perspective. Secondly, correlations between any two variables, as in this instance between brain pathology and function do not necessarily imply any causative linkage. However, what is striking in the two diseases under discussion is that the prominent cholinergic abnormality in the one (autism) and neurochemical correlate of disturbed consciousness in the other (DLB) is the nicotinic $\alpha_4\beta_2$ receptor, not any of the other presynaptic, muscarinic or nicotinic receptors investigated. If this is a meaningful coincidence, it could provide a new lead in research into the neurobiology of consciousness. There is previous evidence that this particular nicotinic receptor is implicated in the mechanism of general anaesthesia. Neuronal nicotinic $\alpha4\beta2$ receptors are inhibited at clinically relevant concentrations by a range of general anaesthetic agents (Flood and Role, 1998; Tassonyi et al., 2002).

Although clinical pathological correlations are not available for the autism data, autism has been described as 'a disorder of consciousness and feelings' (Trevarthen, 1996), and restricted and repetitive behaviors have been interpreted as indicative of nonreflective thinking or absence of higher consciousness. There is no unitary agreement on the core cognitive characteristic in autism, but the finding of extensive nicotinic receptor abnormalities could be linked to consciousness and this hypothesis could be explored in future investigations. For example, the receptor could be imaged in vivo in affected individuals and related to performance in a variety of tasks dependant on explicit or implicit processing. Similarly responses to nicotinic drugs such as nicotine or more specific $\alpha_4\beta_2$ agonists could be correlated with such measures.

In DLB, tools for assessing disturbances in consciousness are already available including variations in attentional performance or EEG patterns. These indices could similarly be related to in vivo imaging or response to nicotinic drugs.

If the cortical $\alpha_4\beta_2$ receptor is particularly implicated in consciousness, it will also be worth considering other disorders in which abnormalities in this receptor have been implicated. These include Alzheimer's and Parkinson's disease and more controversially schizophrenia. In addition the receptor could be investigated in some of the other disorders mentioned above. Cognitive effects of nicotine in normal individuals could also be monitored using implicit and explicit tasks to discover if, as predicted, the latter are more affected. Whether $\alpha_4\beta_2$ knockout mouse models are affected in terms of consciousness is a question that can only be answered if it is accepted that such animals are conscious and if so, whether appropriate investigative tools can be devised.

In addition to nicotinic receptor localization, the balance between nicotinic and muscarinic receptors may be important. In DLB, there is an imbalance between lower $\alpha_4\beta_2$ nicotinic and elevated M1 receptors, and in autism the $\alpha_4\beta_2$ is lost to a much greater extent than the M1 receptor.

To meet the obvious challenge that consciousness is a highly complex integrative process that cannot be reduced to relatively simple molecular mechanisms, it is necessary to propose that a candidate molecular mechanism governs the key integrative process. Such possibilities for the $\alpha_4\beta_2$ nicotinic receptor include its localization on intrinsic GABA inhibitory neurons in the cerebral cortex and thalamus, linked globally by cholinergic input from a discrete nucleus in the basal forebrain. Since conscious awareness at any one moment involves only a minute proportion of cerebral activity (the vast majority of which is

conscious) it is likely that selective inhibition plays a major role. Nicotinic modulation of GABA neurons may be one potential mechanism worth exploring further.

Chemical neurotransmission may thus hold one of the keys to integrating conscious awareness. According to Hoffer and Osmond (1964), 'If any light is ever to be shed on the almost absolute darkness which envelopes these cerebral processes, then such light will only originate from chemistry and never from morphological research'. Such a viewpoint may not be so outdated.

Acknowledgments

The authors' research into autism and dementia with Lewy bodies is supported by respectively Cure Autism Now and the Medical Research Council. Thanks to Lorraine Hood for manuscript preparation. Sections of this chapter are based on previous publications (Perry and Young, 2002; Perry and Lee, 2002; Burn et al., 2002).

References

Audersirk, T. and Cabell, L. (1999) Nanomolar concentrations of nicotine and cotinine alter the development of cultured hippocampal neurons via non-acetylcholine receptor-mediated mechanisms. Neurotoxicol., 20: 639–646.

Baars, B.J. (1998) Metaphors of consciousness and attention in the brain. Trends Neurosci., 21: 58–62.

Ballard, C.G., Court, J.A., Piggott, M., Johnson, M., O'Brien, J., Mckeith, I., Holmes, C., Lantos, P., Jaros, E. and Perry, R. and Perry, E. (2002) Disturbances of consciousness in dementia with Lewy bodies associated with alteration in nicotinic receptor binding in the temporal cortex. J Consc. Cognition., in press.

Ballard, C., O'Brien, J., Gray, A., Cormack, F., Ayre, G., Rowan, E., Thompson, P., Bucks, R., McKeith, I., Walker, M. and Tovee, M. (2001) Attention and fluctuating attention in patients with dementia with Lewy bodies and Alzheimer's disease. Arch. Neurol., 58: 977–982.

Ballard, C., Piggott, M., Johnson, M., Cairns, N.J., Perry, R., McKeith, I., Jaros, E., O'Brien, J., Holmes, C. and Perry, E. (2000) Delusions associated with elevated muscarinic binding in dementia with Lewy bodies. Ann. Neurol., 48: 868–876.

Bannon, A.W., Decker, M.W., Holladay, M.W., Curzon, P., Donnelly-Roberts, D., Puttfarcken, P.S., Bitner, R.S.,

Diaz, A., Dickenson, A.H., Porsolt, R.D., Williams, M. and Arneric, S.P. (1998) Broad-spectrum, non-opioid analgesic activity by selective modulation of neuronal nicotinic acetylcholine receptors. Science, 279: 77–81.

Bauman, M.L. and Kemper, T.L. (1994) Neuro-anatomic observations of the brain in Autism. In: M.L. Bauman and T.L. Kemper (Eds.), The Neurobiology of Autism. Johns Hopkins University Press, Baltimore, pp. 119–145.

Belluardo, N., Mudo, G., Blum, M., Amato, G. and Fuxe, K. (2000) Neurotrophic effects of central nicotinic receptor activation. J. Neural. Trans. Suppl. 227–245.

Blatt, G.J., Fitzgerald, C.M., Guptill, J.T., Booker, A.B., Kemper, T.L. and Bauman, M.L. (2001) Density and distribution of hippocampal neurotransmitter receptors in autism: an autoradiographic study. J. Autism Dev. Disord., 537–543.

Bogen, J.E. (1995) On the neurophysiology of consciousness: I. An overview. Cognition, 4: 52–62.

Burn, D., Perry, E.K., O'Brien, Y.T., Collerton, D., Yaros, E., Perry, R., Piggott, M.A., Morris, C.M., Mclaren, A.

Buzsaki, G., Bickford, R.G., Armstrong, D.M., Ponomareff, G., Chen, K.S., Ruiz, R., Thal, L.J. and Gage, F.H. (1988) Electric activity in the neocortex of freely moving young and aged rats. J. Neurosci., 26: 735–744.

Callaway, E. and Band, R.I. (1958) J. Neurol. Psychiat., 79: 91–102.

Court, J.A., Lloyd, S., Johnson, M., Griffiths, M., Birdsall, N.J.M., Piggott, M.A., Oakley, A.E., Ince, P.G., Perry, E.K. and Perry, R.H. (1997) Nicotinic and muscarinic cholinergic receptor binding in the human hippocampal formation during development and aging. Dev. Brain Res., 101: 93–105.

Court, J.A., Perry, E.K., Johnson, M., Piggott, M.A., Kerwin, J.M., Perry, R.H. and Ince, P.G. (1993) Regional patterns of cholinergic and glutamate activity in the developing and aging human brain. Dev. Brain Res., 74: 73–82.

Court, J.A., Martin-Ruiz, C., Graham, A. and Perry, E. (2001) Nicotinic receptors in human brain: topography and pathology. J. Chem. Neuroanat., 20: 281–298.

Court, J.A., Ballard, C.G., Piggott, M.A., Johnson, M., O'Brien, J.T., Holmes, C., Cairns, N., Lantos, P., Perry, R.H., Jaros, E. and Perry, E.K. (2001) Visual hallucinations are associated with lower a-bungarotoxin binding in dementia with Lewy bodies., Pharmacol. Biochem. Behav., 70: 571–579.

da Penha Berzaghi, M., Cooper, J., Castren, E., Zafra, F., Sofroniew, M., Thoenen, H. and Lindholm, D. (1993) Cholinergic regulation of brain-derived neurotrophic factor (BDNF) and nerve growth factor (NGF) but not neurotrophin-3 (NT-3) mRNA levels in the developing rat hippocampus. J. Neurosci., 13: 3818–3826.

Delacour, J. (1995) An introduction to the biology of consciousness. Neuropsychologica, 33: 1061–1074.

Fisahn, A., Pike, F.G., Buhl, E.H. and Paulsen, O. (1998) Cholinergic induction of network oscillations at 40 Hz in the hippocampus in vitro. Nature, 394: 186–189.

Flood, P. and Role, L.W. (1998) Neuronal nicotinic acetylcholine receptor modulation by general anaesthetics. Toxicol. Lett., 100–101: 149–153.

Gomez-Tortosa, E., Newell, K., Irizarry, M.C., Albert, M., Growdon, J.H. and Hyman, B.T. (1999) Clinical and quantitative pathologic correlates of dementia with Lewy bodies. Neurology, 53: 1284–1291.

Grace, J., Daniel, S., Stevens, T., Shankar, K.K., Walker, Z., Byrne, E.J., Butler, S., Wilkinson, D., Woolford, J., Waite, J. and McKeith, I.G. (2001) Long-term use of rivastigmine in patients with dementia with Lewy bodies: An open label trial. Int. Psychogeriatr., 13: 199–205.

Hashimoto, Y., Abiru, Y., Nishio, C. and Hatanaka, H. (1999) Synergistic effects of brain-derived neurotrophic factor and ciliary neurotrophic factor on cultured basal forebrain cholinergic neurons from postnatal 2-week old rats. Brain Res. Devel. Brain Res., 115: 25–32.

Hoffer, A. and Osmond, H. (1964) The Hallucinogens. Academic Press, New York.

Hohmann, C.F., Kwiterovich, C., Oster-Granite, M.L. and Coyle, J.T. (1991a) Newborn basal forebrain lesions disrupt cortical cytodifferentiation as visualized by Rapid Golgi staining. Cerebral Cortex, 1: 143–157.

Hohmann, C.F. and Berger-Sweeney, J. (1998) Cholinergic regulation of cortical development and plasticity. New Twists to an old story. Perspect. Dev. Neurobiol., 5: 401–425.

Huttenlocher, P.R. (1975) Synaptic and dendritic development and mental defect. In: Brain Mechanism in Mental Retardation (M. Brazier, ed.) 123–139.

Lainhart, J.E. and Piven, J. (1995) Diagnosis, treatment and neurobiology of Autism in children. Curr. Opin. Paediatrics, 7: 392–400.

Martin-Ruiz, C., Court, J.A., Molnar, E., Lee, M., Gotti, C., Mamalaki, A., Tsouloufis, T., Tzartos, S., Ballard, C., Perry, R.H., Perry, E.K. (1999) Alpha4 but not alpha3 and alpha7 nicotinic acetylcholine receptor subunits are lost from the temporal cortex in Alzheimer's disease. J. Neurochem., 73: 1635–1640.

Marubio, L.M., del Mar Arroyo-Jimenez, M., Cordero-Erausquin, M., Lena, C., Le Novere, N., de Kerchove d'Exaerde, A., Huchet, M., Damaj, M.I. and Changeux, J.P. (1999) Reduced antinociception in mice lacking neuronal nicotinic receptor subunits. Nature, 398(6730): 805–810.

McAlonan, K. and Brown, V.J. (2002) The thalamic reticular nucleus: more than a sensory nucleus? Neuroscientist, 8: 302–305.

McKeith, I., Fairbairn, A., Perry, R., Thompson, P. and Perry, E. (1992a) Neuroleptic sensitivity in patients with senile dementia of Lewy body type. Brit. Med. J., 305: 673–678.

McKeith, I.G., Galasko, D., Kosaka, K. et al. (1996) Consensus guidelines for the clinical and pathological diagnosis of dementia with Lewy bodies (DLB): report of the consortium on DLB international workshop. Neurology, 47: 1113–1124.

McKeith, I.G., Spano, P.F., Del Ser, T., Emre, M., Wesnes, K., Anand, R., Cicin-Sain, A., Ferrara, R. and Spiegel, R. (2000b) Efficacy of rivastigmine in dementia with Lewy bodies: a randomised, double-blind, placebo-controlled international study. Lancet, 356: 2031–2036.

Mesulam, M.M. (1995) Cholinergic pathways and the ascending reticular activating system of the human brain. Ann. N.Y. Acad. Sci., 757: 169–179.

Minshaw, N.J., Sweeney, J.A. and Bauman, M. (1997) Neurological aspects of Autism. In: H. Cohen and F. Volkmar (Eds.), Handbook of Autism and Developmental Disorders. John Wiley & Sons, pp. 360–360.

Mirza, N.R. and Stolerman, I.P. (2000) The role of nicotinic and muscarinic acetylcholine receptors in attention. Psychopharmacology, (Berl), 148: 243–250.

Nelson, K.B., Grether, J.K., Croen, L.A., Dam brosia, J.M., Dickens, B.F., Jelliffe, L.L., Hansen, R.L. and Phillips, T.M. (2001) Neuropeptides and neurotrophins in neonatal blood of children with autism or mental retardation. Ann. Neurol., 49: 597–606.

Pennell, P.B., Burdette, D.E., Ross, D.A., Henry, T.R., Albin, R.L., Sackellares, J.C. and Frey, K.A. (1999) Muscarinic receptor loss and perservation of presynaptic cholinergic terminals in hippocampal sclerosis. Epilepsia., 40: 38–46.

Perry, E., Ashton, H. and Young, A. (Eds.), (2000) Neurochemistry of Consciousness. John Benjamins Oublishing Company, Amsterdam.

Perry, E.K. and Young, A.H. (2002) Neurotransmitter systems. In: E. Perry, H. Ashton and A. Young (Eds.), Neurochemistry of Consciousness. John Benjamins, Amsterdam, pp. 3–24.

Perry, E., Walker, M., Grace, J. and Perry, R. (1999) Acetylcholine in mind: a neurotransmitter correlate of consciousness? Trends Neurosci., 22: 273–280.

Perry, E.K., Lee, M., Martin-Ruiz, C.M., Court, J., Volsen, S., Iversen, P., Bauman, M., Perry, R. and Wenk, G. (2001) Cholinergic activities in autism: abnormalities in the cerebral cortex and basal forebrain. Am. J. Psychiat., 158: 1058–1066.

Perry, E.K., McKeith, I., Thompson, P., Marshall, E., Kerwin, J., Jabeen, S., Edwardson, J.A., Ince, P., Blessed, G. and Irving, D. (1991) Topography, extent, and clinical relevance of neurochemical deficits in dementia of Lewy body type, Parkinson's disease, and Alzheimer's disease. Ann. N.Y. Acad. Sci., 640: 197–202.

Perry, E.K., Smith, C.J., Court, J.A. and Perry, R.H. (1990) Cholinergic nicotinic and muscarinic receptors in dementia of Alzheimer, Parkinson and Lewy body types. J. Neural. Transm. Park. Dis. Dement. Sect., 2: 149–158.

Perry, E.K. and Piggott, M.A. (2000) Behavioural Brain Sciences.

Perry, R.H., Irving, D., Blessed, G., Fairbairn, A.F. and Perry, E.K. (1990) Senile dementia of the Lewy body type: a clinically and neuropathologically distinct form of Lewy body disease. J. Neurol. Sci., 95: 119–139.

Picciotto, M.R., Zoli, M. and Changeux, J.P. (1999) Use of knock-out mice to determine the molecular basis for the actions of nicotine. Nicotine Tob. Res., 2: 121–125.

Rogers, S.W., Gahring, L.C. and Collins, A.C. (1998) Marks M: Age related changes in neuronal nicotinic acetylcholine receptors subunit α4 expression are modified by long term nicotine administration. J. Neurosci., 18: 4825–4832.

Sabbagh, M.N., Reid, R.T., Corey-Bloom, J., Rao, T.S., Hansen, L.A., Alford, M., Masliah, E., Adem, A., Lloyd, G.K. and Thal, L.J. (1998) Correlation of nicotinic binding with neurochemical markers in Alzheimer's disease. J. Neural. Transm., 105: 709–717.

Samuel, W., Alford, M., Hofstetter, C.R. and Hansen, L. (1997) Dementia with Lewy bodies versus pure Alzheimer disease: differences in cognition, neuropathology, cholinergic dysfunction, and synapse density. J. Neuropathol. Exp. Neurol., 56: 499–508.

Sandberg, P.R., Silver, A.A., Shytle, R.D., Philipp, M.K., Cahill, D.W., Fogelson, H.M. and McConville, B.J. (1997) Nicotine for the treatment of Tourette's syndrome. Pharmacol. Ther., 74: 21–25.

Shea, C., MacKnight, C. and Rockwood, K. (1998) Donepezil for treatment of dementia with Lewy bodies: a case series of nine patients. Int. Psychogeriatr., 10: 229–238.

Shinotoh, H., Aotsuka, A., Ota, T., Tanaka, N., Fukushi, K., Nagatsuka, S., Tanada, S. and Ivie, T. (2001) Brain cholinergic function in dementia with Lewy bodies and Alzheimer's disease measured by PET. Abstracts, Proc 5th Int Conference on Progress in PD and AD, 74.

Solms, M. (2000) Dreaming and REM sleep are controlled by different brain mechanisms. Behav. Brain Sci., 23: 843–850.

Steinlein, O.K. (2000) Neuronal nicotinic receptors in human epilepsy. Eur. J. Pharmacol., 30, 393: 243–247.

Spillantini, M.G., Crowther, R.A., Jakes, R., Hasegawa, M. and Goedert, M. (1998) alpha-Synuclein in filamentous inclusions of Lewy bodies from Parkinson's disease and dementia with lewy bodies. Proc. Nat. Acad. Sci. USA., 95: 6469–6473.

Tassonyi, E., Charpantier, E., Muller, D., Dumont, L. and Bertrand, D. (2002) The role of nicotinic acetylcholine receptors in the mechanism of anaesthesia. Brain Res. Bul., 133–150.

Toichi, M., Kamio, Y., Okado, T., Sakihama, M., Youngstrom, E.A., Findling, R.L. and Yamamoto, K. (2002) A lack of self-consciousness in autism. Am. J. Psychiatry, 159: 1422–1424.

Tordjman, S., Antoine, C., Cohen, D.J., Gauvain-Piquard, A., Carlier, M., Roubertoux, P. and Ferrari, P. (1999) Study of the relationships between self-injurious behavior and pain reactivity in infantile Autism. Encephale., 25: 122–134.

Trauth, J.A., Seidler, F.J., McCook, E.C. and Slotkin, T.A. (1999) Adolescent nicotine exposure causes persistent upregulation of nicotinic cholinergic receptors in rat brain regions. Brain Res., 851: 9–19.

Trevarthen, C. 'Children with Autism', 1996.

Wenk, G.L., Naidu, S., Casanova, M.F., Kitt, C.A. and Moser, H. (1991) Altered neurochemical markers in Rett's syndrome. Neurology, 41: 1753–1756.

Wesnes, K.A., McKeith, I.G., Ferrara, R., Emre, M., Del Ser, T., Spano, P.F., Cicin-Sain, A., Anand, R. and Spiegel, R. (2002) Effects of rivastigmine on cognitive function in dementia with lewy bodies: a randomised placebo-controlled international study using the cognitive drug research computerised assessment system. Dement. Geriatr. Cogn. Disord., 13: 183–192.

Woolf, N.J. (1997) A possible role for cholinergic neurons of the basal forebrain and pontomesencephalon in consciousness. Conscious Cogn., 6: 574–596. Review.

Xiang, Z., Huguenard, J.R. and Prince, D.A. (1998) Cholinergic switching within neocortical inhibitory networks. Science, 281: 985–988.

Zoli, M., Picciotto, M.R., Ferrari, R., Cocchi, D. and Changeux, J.P. (1999) Increased neurodegeneration during ageing in mice lacking high-affinity nicotine receptors. EMBO J., 18: 1235–1244 Dlb.

Progress in Brain Research, Vol. 145
ISSN 0079-6123

CHAPTER 21

Functional studies of cholinergic activity in normal and Alzheimer disease states by imaging technique

Agneta Nordberg*

Karolinska Institute, Neurotec Department, Division of Molecular Neuropharmacology,
Huddinge University Hospital, S-141 86 Stockholm, Sweden

Introduction

Alzheimer's disease (AD) is the most common form of dementia. It is a devastating illness affecting not only the patient, but also his family and the society. AD is a neurodegenerative disease with a relentless course, starting during healthy ageing, when only minor changes in performances of cognitive tasks are seen compared to healthy young individuals, followed by a progressive change in cognitive, affective and behavioral parameters. A great demand is presently put on available functional neuroimaging methods, since they may reveal neuronal dysfunction before structural changes appear (Knopman et al., 2001). Positron emission tomography (PET), a noninvasive neuroimaging technique, allows for quantification and three-dimensional measures of distinct physiological variables, such as glucose metabolism, cerebral blood flow and neurotransmitter and receptor function (Nordberg, 1999). Regional deficits in cerebral glucose metabolism (CMRGlu) in the parieto-temporal region, assessed with [18F] 2-fluoro-2-deoxy-D-glucose (FDG) as tracer, have consistently been reported in AD. The fact that the metabolic impairment correlates to deficits in neuropsychological domains and increases with progression of the disease suggest that PET may also provide a sensitive means to assess disease progression and severity (Nordberg, 1993; Mielke

et al., 1994). As will be discussed in this chapter, recently available surrogate markers for cholinergic neuronal activity as well as for pathological processes, such as beta amyloid, will permit further in vivo studies of changes in functional mechanisms in AD brains.

With the conservative diagnostic criteria for AD, the diagnosis is often made when the clinical symptoms of the disease are quite evident. While the present AD therapies have symptomatic effects with some slowing down of disease progression (Nordberg and Svensson, 1998; Nordberg, 2000a; Doody et al., 2001; Giacobini, 2001), the drug therapies of the future are aiming to cure/prevent AD (Hardy and Selkoe, 2002). Early diagnosis at preclinical stages of the disease will be a prerequisite for successful AD therapy in the future. The clinical symptoms of AD are probably preceded by a period of unknown duration, during which neuropathological alterations may accumulate in the AD brain without detectable changes in cognition (Davis et al., 1999; Goldman et al., 2001) (Fig. 1). Neuropsychological studies suggest that mild cognitive impairment (MCI) is a condition characterized by subtle cognitive deficits, before functional impairments are evident in the patient. MCI might therefore represent an early stage of AD (Almkvist et al., 1998; Petersen et al., 1999; Arnáiz et al., 2000). Selective regional impairments of cortical glucose metabolism are found in MCI patients (Minoshima et al., 1997; Jelic and Nordberg, 2000). Presently, there is a great interest to investigate whether cerebral glucose metabolism or

*Corresponding author: Tel.: +46-8-58580000;
Fax: +46-8-6899210; E-mail: agneta.nordberg@neurotec.ki.s

DOI: 10.1016/S0079-6123(03)45021-8

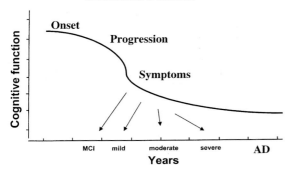

Fig. 1. Schematic representation of the time course of Alzheimer's disease (AD) with onset, progression and symptomatic period. MCI = mild cognitive impairment.

any other biological marker could be used as predictor for AD. There is a great need for surrogate markers. The present chapter deals with current developments based on novel technologies, such as PET, for directly imaging cholinergic neuroreceptors, enzymes or amyloid plaques, in addition to measuring cerebral blood flow and metabolic changes in the brain of AD patients.

Predictors of Alzheimer's disease

Mild cognitive impairment

How important is an early detection of MCI patients? How important is it to initiate drug treatment in MCI patients before they may convert to AD? Is there any possibility of predicting which MCI patients are likely to convert to AD? MCI is considered as an intermediate in the continuum from normalcy to dementia (Almkvist et al., 1998). The most commonly used criteria for MCI include memory complaints and memory impairment, but normal general cognitive function and daily living (Petersen et al., 1999). In a four year follow-up of 76 subjects selected for MCI, the annual conversion rate to AD was 12% compared to 1% in normal subjects (Petersen et al., 1999), whereas in a two year follow-up study, it was estimated to be 25% (Flicker et al., 1991). It has recently been estimated that, among subjects with MCI, the rate of progression to dementia or AD is 6–c25% per year (Petersen et al., 2001). We followed

27 subjects with MCI for 24 months and 26% developed AD (Jelic and Nordberg, 2000). All these MCI patients underwent PET studies of brain glucose metabolism before the longitudinal study started. Impairments in cerebral glucose metabolism were already present at the beginning of the follow-up period in the group of MCI patients who progressed to AD, with a 93% predictive power of converting to AD (Jelic and Nordberg, 2000; Nordberg et al., 2001). Similarly, in a group of 20 MCI patients followed for 36 months, 9 patients converted to AD while 11 patients remained clinically stable (and were classified as stable MCI). Neuropsychological assessments, such as Block Design, gave a correct classification of 65% while the combined measure of Block Design and cerebral glucose metabolism gave a correct classification of 90% (Arnáiz et al., 2001). It is apparent that early functional metabolic changes can be observed in subjects with MCI and that the degree of functional disturbances in glucose metabolism as measured by PET can be used as a predictive marker for conversion from MCI to AD, especially when combined with neuropsychological testing. CSF markers will probably also represent another valuable complement in a near future (Knopman et al., 2001).

Apolipoprotein E ε4 carriers

Two independent studies have reported that the APOE ε4 allele is not associated with specific alterations in glucose metabolism in patients with AD (Corder et al., 1997; Hirono et al., 1998). There is evidence that the presence of the ε4 allele could cause preclinical deficits similar to what can be observed in manifest disease (Small et al., 1996; Reiman et al., 1996). Small et al. (2000) also showed recently that, in nondemented APOE ε4 carriers followed longitudinally for two year, the memory performance scores did not decline significantly, while there was a significant decline in cortical metabolic ratio. Similar findings were also reported by Reiman et al. (2001), who found a significant decline in cerebral glucose metabolism in the temporal cortex, posterior cingulate cortex, prefrontal cortex, basal forebrain, parahippocampal gyri and thalamus of ε4 heterozygotes, in the absence of cognitive alterations during two years. Reiman et al. (2001) estimated that 50–150 cognitively normal ε4 heterozygotes would probably be a sufficient number of subjects for testing the outcome of a preventive AD therapy. Cerebral metabolic rates and genetic factors may provide a means for preclinical AD detection and importantly assist in the evaluation of the efficiency of preventive drug treatments in the future.

In vivo measurement of cholinergic activity in normal and Alzheimer brains

The cholinergic neurotransmitter system, which is involved in cognitive processes, is severely impaired in AD (Perry, 1986; Court et al., 2001; Nordberg, 2001). Functional imaging allows measurement of acetylcholinesterase (AchE) (Iyo et al., 1997; Kuhl et al., 1999) and butyrylcholinesterase (BuChE) activities (Snyder et al., 2001), nicotinic (Nordberg et al., 1995), muscarinic receptor binding (Dewey et al., 1990), and vesicular acetylcholine transporter (Kuhl et al., 1996) in normal subjects and AD patients (Table 1). These cholinergic surrogate markers are suggested to be more sensitive to early changes in brain than cerebral glucose metabolism (Nordberg et al., 1995; Kuhl et al., 1999). Fig. 2 illustrates multi-tracer studies, measuring AChE activity ([^{11}C]PMP), nicotinic receptors ([^{11}C]nicotine) and cerebral glucose metabolism ([^{18}F]FDG), performed in an AD patient. The differences in regional distribution and deficits in brain are illustrated for the PET markers. MCI patients show deficits in [^{11}C]nicotine binding in the

temporoparietal cortex, although the changes are more pronounced in AD patients (Fig. 3). An upregulation of the enzyme choline acetyltransferase (ChAT) in hippocampal postmortem brain tissue from MCI patients was recently reported by DeKosky et al. (2002). It was concluded from the findings that cholinergic deficits might represent a late phenomena in the course of AD disease, and that early treatment with cholinesterase inhibitors should not be effective. However, as also pointed out by DeKosky et al. (2002) and recently discussed by Sarter and Bruno (2002), ChAT is not the rate limiting step in the synthesis and release of acetylcholine. It is thus quite plausible that the enzyme becomes upregulated when the cholinergic system starts to be impaired in certain areas of the brain. We have observed an upregulation of the nicotinic receptors in mice overexpressing acetylcholinesterase and which are expected to have lower brain contents in acetylcholine (Svedberg et al., 2002). As pointed out in this chapter, deficits in [^{11}C]nicotine have been observed in MCI patients, and further studies with PET in MCI patients will show how the cholinergic system is impaired. It must also be underlined that MCI is a heterogeneous patient group, and that it is not possible to follow the patients and see whether they are converting to AD or not by studying autopsy material.

Recently, we reported the use of a [^{11}C]PIB for visualizing amyloid binding in vivo in AD patients (Engler et al., 2002). The possibility of detecting amyloid in vivo by imaging technique will allow to study the time course of amyloid deposition and its relationship with cholinergic deficits, glucose metabolism and drug effects.

Table 1. PET and SPECT ligands for visualization of cholinergic activity in human brain in vivo

Cholinergic parameter	Radioligand	Imaging technique
AChE	^{11}C-MP4A	PET
	^{11}C-PMP	PET
BuChE	^{11}C-nBMP	PET
	^{11}C-nVMP	PET
Cholinergic terminals	^{123}I-IBVM	SPECT
Nicotinic receptors	^{11}C-nicotine	PET
Muscarinic receptors	^{11}C-benzotropine	PET
	^{11}C-scopolamine	PET

Brain activation studies in Alzheimer patients

In order to explore the anatomical patterns of activation associated with cognitive processing in normal and diseased brain, PET and functional magnetic resonance imaging (fMRI) are presently used for functional imaging studies (Cabeza and Nyberg, 2000). With both techniques, changes in cerebral blood flow can be measured as an index of neural activity. Since episodic memory and attention are affected early in AD, functional brain imaging has

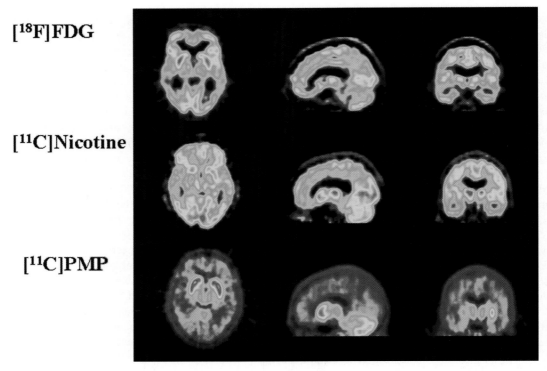

Fig. 2. Multi-tracer PET study with [18F]FDG, [11C]nicotine and [11C]PMP for the respective measurement of cerebral gluc metabolism, nicotinic receptor binding and acetylcholinesterase activity (AChE) in an Alzheimer patient. The color scale indica red = high, yellow = moderate and blue = low activity. (Photo: Uppsala PET Center, Uppsala and Huddinge University Hospit Stockholm).

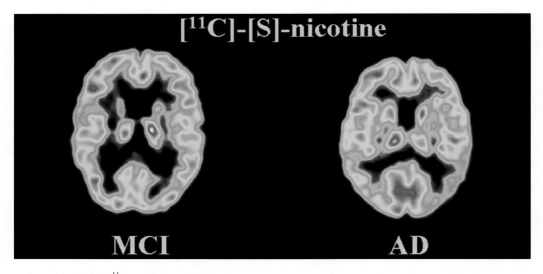

Fig. 3. Uptake and binding of [11C]nicotine in brain of a patients with mild cognitive impairment (MCI) and Alzheimer's disease (Al respectively. The color scale indicates red = high, yellow = moderate and blue = low activity. (Photo: Uppsala PET Center, Upps and Huddinge University Hospital, Stockholm).

focused on the regional brain activation patterns for these mental performances in AD patients. Studies in healthy elderly subjects showed activation in the left parietal cortex and left hippocampus during recall of words, while AD patients displayed increased activation in the left prefrontal cortex and left cerebellum, in addition to increased activation in a number of brain regions common to both controls and AD (Bäckman et al., 1999). When brain activation was studied in young and older adults during episodic retrieval, we found that both age groups showed blood flow increases in the prefrontal cortex; the younger adults also showed a selective increase in the left cerebellum (Bäckman et al., 1997). These findings suggest ongoing changes in brain plasticity during normal ageing. It is well known from studies of postmortem brain tissue that there are significant decreases in the number of nicotinic receptor binding sites in brain with normal ageing (Marutle et al., 2000). What impact such neuroreceptors changes may have on the functional activity of the normal brain remains to be determined.

The basal forebrain cholinergic system provides the major innervation of the cortex, including prefrontal cortex, thalamus and the parietal lobes (Mesulam and Geula, 1988). Impairment of the cholinergic pathways causes attentional deficits (Wesnes et al., 1988; Lawrence and Sahakian, 1995; Perry and Hodges, 1999). Studies on sustained attention in healthy young subjects with a Rapid Visual Information Processing Task (RVIP) revealed an activation of the parietal cortex and right frontal cortex (Coull et al., 1996). When a similar task was performed by AD and MCI patients, an increased activation was bilaterally observed in the parietal cortex, basal ganglia, and a decreased activation in the prefrontal regions (Almkvist et al., unpublished). During sustained attention, Johannsen et al. (1999) have observed a deactivation pattern in the frontal cortex of AD subjects, which was not seen in controls. PET may provide valuable insights into the early dynamic disturbances in brain function related to cognitive function in MCI and AD patients. The somewhat different activation patterns observed between AD and MCI patients may reflect compensatory efforts to keep up with task requirements, and the fact that networks normally used in these tasks are distressed or impaired by the disease process in AD. (Almkvist et al., unpublished). Brain imaging should now help to further explore the potential of drugs currently developed to influence these processes, and to elucidate underlying neurobiological mechanisms.

Functional effects of drug treatment

Cholinesterase inhibitors, such as tacrine, donepezil and rivastigmine, have been shown to increase cerebral blood flow and glucose utilization after some months of treatment in AD patients (Nordberg 2000b; Venneri et al. 2002), and to preserve cortical glucose metabolism following longer treatment periods (Stefanova et al., 2003). An improvement of nicotinic receptors has also been reported following treatment with cholinesterase inhibitors as well as with nerve growth factors (Nordberg, 2000b). PET studies have shown a reduced cortical AChE activity during donepezil treatment in AD patients (Kuhl et al., 2000; Shinotoh et al., 2001). The inhibition of AChE measured by PET in AD patients treated with donepezil was lower than that measured in the peripheral blood (Rogers et al., 1998). In CSF, an increase in AChE activity has been observed after long-term treatment with tacrine, donepezil or galantamine (Nordberg et al., 1999; Davidsson et al., 2001). Rivastigmine treatment, on the other hand, caused a persistent inhibition of AChE as well as BuChE in the CSF (Darreh-Shori et al., 2002). After 3–6 months of treatment with rivastigmine, there was a positive correlation between the inhibition of AChE and BuChE in the CSF and plasma and improvement in cognitive function (Darreh-Shori et al. 2002) (Fig. 4). AD patients treated with rivastigmine also showed increased nicotinic receptor binding in the temporal cortex, as measured by PET (Fig. 5). The improvement in [^{11}C]binding (decrease in K2* value) appeared to be more pronounced after rivastigmine treatment compared to those treated with tacrine (Fig. 5). These PET findings illustrate the influence of drug interventions on synaptic neuronal markers, and their relation to changes in enzyme activity and cognitive performance.

By modulating the cholinergic system in brain with drugs, we expect the improvement of memory.

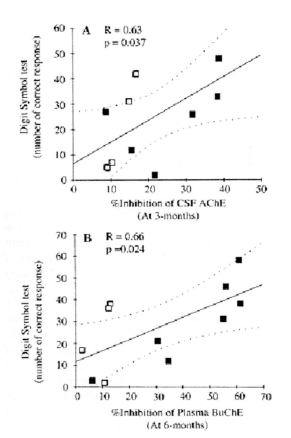

Fig. 4. Positive correlations between cognitive performance and cholinesterase inhibition in the CSF (A) and the plasma (B), as assessed using the attentional digit symbol test (attention DS) in AD patients treated with rivastigmine. The correlation between test score and CSF AChE inhibition after 3 months, and that between test score and plasma BuChE inhibition after 6 months of treatment. Filled squares = high dose group ($n = 7$); open squares = lower dose ($n = 4$). AChE, acetylcholinesterase; BuChE, butyrylcholinestesterase. Data from Darreh-Shori et al., 2002.

Furey et al. (1997) studied the changes in cerebral blood flow in healthy subjects receiving an infusion of physostigmine during the performances of a working memory task. They observed an increase in cerebral blood flow in the right prefrontal cortex during the task, which was less pronounced during the infusion of physostigmine. Since physostigmine improved the working memory test performance, they concluded that this was related to reduced task-related activity in the prefrontal cortex, suggesting that enhanced

[¹¹C]nicotine binding in temporal cortex

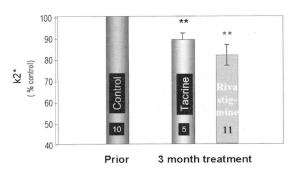

Fig. 5. The effect of three months of rivastigmine (9 mg daily) and tacrine (80 mg daily) treatment on the binding of [¹¹C]nicotine in the temporal cortex of AD patients. The [¹¹C]nicotine binding is expressed as $k2^*$, where a decrease in $k2^*$ value corresponds to an increase in [¹¹C]nicotine binding (for further details about the methodology for calculating $k2^*$, see Nordberg et al., 1995). The data are expressed as percentage of control $* p < 0.05$, $** p < 0.01$. N, number of patients.

cholinergic function in brain could ameliorate processing efficiency (Furey et al., 1997). Cholinergic enhancement with physostigmine has also been shown to cause selective increases in the perceptual processing processes during working memory test (Furey et al., 2000). In addition, Sperling et al. (2002), using functional MRI technique, recently observed that administration of scopolamine to healthy young subjects decreased the activation in hippocampus and inferior prefrontal cortex, but not in the striate cortex, during a face recognition test.

We have recently examined the effect of rivastigmine on the changes in frontal cortical blood flow associated with attentional task performances. Three months of rivastigmine treatment modified the frontal cortical activation pattern in AD patients, which then became more similar to that of MCI patients (Nordberg et al., unpublished). Functional PET activation studies performed during memory tasks should also provide valuable insights into the mechanisms of action of new drugs, how they interact with various brain processes, and how they can improve the efficacy of cognitive processes in AD patients.

Acknowledgments

This study was supported by grants from the Swedish Medical Research Council (project no 05817), Stiftelsen för Gamla Tjänarinnor, Stohne's Foundation.

References

Almkvist, O., Basun, H., Bäckman, L., Herlitz, A., Lannfelt, L., Small, B., Viitanen, M., Wahlund, L.O. and Winblad, B. (1998) Mild cognitive impairment—an early stage of Alzheimer's disease? J. Neural Transm., 54 (suppl.): 21–29.

Arnáiz, E., Blomberg, M., Fernaeus, S.E., Wahlund, L.O., Winblad, B. and Almkvist, O. (2000) Psychometric discrimination of Alzheimer's disease and mild cognitive impairment. Alzheimer Report, 2: 97–201.

Arnáiz, E., Jelic, V., Almkvist, O., Wahlund, L.O., Winblad, B., Valind, S. and Nordberg, A. (2001) Impaired cerebral metabolism and cognitive functioning predict deterioration in mild cognitive impairment. NeuroReport, 12: 851–855.

Bäckman, L., Almkvist, O., Andersson, J., Nordberg, A., Winblad, B., Reineck, R. and Långström, B. (1997) Brain activation in young and older adults during implicit and explicit retrieval. J. Cognitive Neurosci., 9: 378–391.

Bäckman, L., Andersson, J.L.R., Nyberg, L., Winblad, B., Nordberg, A. and Almkvist, O. (1999) Brain regions associated with episodic retrieval in normal aging and Alzheimer's disease. Neurology, 52: 1861–1870.

Cabeza, R. and Nyberg, L. (2000) Imaging cognition II. An empirical review of 275 PET and fMRI studies. J. Cognitive Neurosci., 12: 1–47.

Corder, E., Jelic, V., Basun, H., Lannfelt, L., Valind, S., Winblad, B. and Nordberg, A. (1997) No difference in cerebral glucose metabolism in patients with Alzheimer's disease and differing apolipoprotein E genotype. Arch. Neurol., 54: 273–277.

Coull, J.T., Frith, C.D., Frackowiak, R.S.J. and Grasby, P.M. (1996) A fronto-parietal network for rapid visual processing: a PET study of sustained attention and working memory. Neuropsychologia, 34: 1085–1095.

Court, J., Martin-Ruiz, C., Piggott, M., Spurden, D., Griffiths, M. and Perry, E.K. (2001) Nicotinic receptors abnormalities in Alzheimer's disease. Biol. Psychiatry, 49: 175–184.

Darreh-Shori, T., Almkvist, O., Guan, Z.Z., Garlind, A., Strandberg, B., Svensson, A.L., Soreq, M., Hellström-Lindahl, E. and Nordberg, A. (2002) Sustained cholinesterase inhibition in AD patients receiving rivastigmine for 12 months. Neurology, 59: 563–572.

Davidsson, P., Blennow, K., Andreasen, N., Erikssson, B., Minthon, L. and Hesse, C. (2001) Differential increase in cerebrospinal fluid acetylcholinesterase after treatment with acetylcholinesterase inhibitors in patients with Alzheimer's disease. Neurosci. Lett., 300: 157–160.

Davis, D.G., Schmitt, F.A., Wekstein, D.R. and Markesbery, W.R. (1999) Alzheimer's neuropathologic alterations in aged cognitive normal subjects. J. Neuropathol. Exp. Neurol., 58: 376–388.

DeKosky, S.T., Ikonomovic, M.D., Styren, S.D., Beckett, L., Wisniewski, S., Bennett, D.A., Cochran, E.J., Kordower, J.H. and Mufson, E.J. (2002) Upregulation of choline acetyltransferase activity in hippocampus and frontal cortex of elderly subjects with mild cognitive impairment. Ann. Neurol., 5: 145–155.

Dewey, S.L., Macgregor, R.R., Brodie, J.D., Bendriem, B., King, P.T., Volkow, N.D., Schlyer, D.J., Fowler, J.S., Wolf, A.P., Gatley, S.J., et al. (1990) Mapping muscarinic receptors in human and baboon using [N-^{11}C-methyl]-benztropine. Synapse, 5: 213–223.

Doody, R.S., Stevens, J.C., Beck, C., Dubinsky, R.M., Kaye, J.A., Gwyther, L., Mohs, R.C., Thal, L., Whitehouse, P., DeKosky, S.T. and Cummings, J.L. (2001) Practice parameter: standard subcommittee of the American Academy of Neurology. Neurology, 56: 1154–1166.

Engler, H., Nordberg, A., Blomqvist, G., Bergström, M., Estrada, H., Barletta, J., Sandell, J., Antoni, G., Långström, B., Klunk, B., Debnath, M., Holt, D., Wang, Y., Huang, G. and Mathis, C. (2002) First human study with a benzothiazole amyloid imaging agent in Alzheimer's disease and control subjects. Neurobiol. Aging, 231S: 1568.

Flicker, C., Ferris, S.H. and Reisberg, B. (1991) Mild cognitive impairment in the elderly: predictors of dementia. Neurology, 41: 1006–1009.

Furey, M.L., Pietrini, P., Haxby, J.V., Alexander, G.E., Lee, H.C., VanMete, J., Grady, C.L., Shetty, U., Rapoport, S.I., Schapiro, M.B. and Freo, U. (1997) Cholinergic stimulation alters performance and task-specific regional cerebral blood flow during working memory. Proc. Natl. Acad. USA, 94: 6512–6516.

Furey, M.L., Pietrini, P. and Haxby, J.V. (2000) Cholinergic enhancement and increased selectivity of perceptual processing during working memory. Science, 290: 2315–2319.

Giacobini, E. (2001) Do cholinesterase inhibitors have disease-modifying effects in Alzheimer's disease? CNS Drugs, 15: 85–91.

Goldman, W.P., Price, J.L., Storandt, M., Grant, E.A., McKee, D.W., Rubin, E.H. and Morris, J.C. (2001) Absence of cognitive impairment or decline in preclinical Alzheimer's disease. Neurology, 56: 361–367.

Hardy, J. and Selkoe, D.J. (2002) The amyloid hypothesis of Alzheimer's disease: progress and problems on the road to therapeutics. Science, 297: 353–356.

Hirono, N., Mori, E., Yasuda, M., Ishii, K., Ikejiri, Y., Shimomura, T., Hashimoto, M., Yamashita, H. and Sasali, M. (1998) Lack of association of apolipoprotein E epsilon4 allele dose with cerebral glucose metabolism in Alzheimer disease. Alzheimer Dis. Assoc. Disord., 12: 362–367.

Iyo, M., Namba, H., Fukushi, K., Shinotoh, H., Nagatsuka, S., Suhara, T., Sudo, Y., Suzuki, K. and Irie, T. (1997)

Measurement of acetylcholinesterase by positron emission tomography in the brain of healthy controls and patients with Alzheimer's disease. Lancet, 349: 1805–1809.

Jelic, V. and Nordberg, A. (2000) Early Diagnosis of Alzheimer's disease with positron emission tomography. Alzheimer Dis. Assoc. Disord., 14 (suppl.): S109–S113.

Johannsen, P., Jacobsen, J., Bruhn, P. and Gjedde, A. (1999) Cortical responses to sustained and divided attention in Alzheimer's disease. NeuroImage, 10: 269–281.

Knopman, D.S., DeKosky, S.T., Cummings, J.L., Corey-Bloom, J., Relkin, N., Small, G.W., Miller, B. and Stevens, J.C. (2001) Practice parameters: Diagnosis of dementia (an evidence-based review). Report of the quality standard subcommittee of the American Academy of Neurology. Neurology, 56: 1143–1153.

Kuhl, D.E., Minoshima, S., Fessler, J.A., Frey, K.A., Foster, N.L., Ficardo, E.P., Wiland, D.M. and Koeppe, R.A. (1996) In vivo mapping of cholinergic terminals in normal aging, Alzheimer's disease, and Parkinson's disease. Ann. Neurol., 40: 399–410.

Kuhl, D.E., Koeppe, R.A., Minoshima, S., Snyder, S.E., Ficardo, E.P., Foster, N.L., Frey, K.A. and Kilbourn, M.R. (1999) In vivo mapping of cerebral acetylcholinesterase activity in aging and Alzheimer's disease. Neurology, 52: 691–699.

Kuhl, D.E., Minoshima, S., Frey, K.A., Foster, N.L., Kilbourn, M.R. and Koeppe, R.A. (2000) Limited donepezil inhibition of acetylcholinesterase measured with positron emission tomography in living Alzheimer cerebral cortex. Ann. Neurol., 48: 391–395.

Lawrence, A.D. and Sahakian, B.J. (1995) Alzheimer disease, attention, and the cholinergic system. Alzheimer Dis. Assoc. Disord., (suppl.2): 43–49.

Marutle, A., Warpman, U., Bogdanovic, N. and Nordberg, A. (1998) Regional distribution of subtypes of nicotinic receptors in human brain and effect of aging studied by $\pm[^3\text{H}]$epibatidine. Brain Res., 801: 143–149.

Mesulam, M.-M. and Geula, C. (1988) Nucleus basalis (Ch4) and cortical cholinergic innervation in the human brain: observations based on the distribution of acetylcholinesterase and choline acetyltransferase. J. Comp. Neurol., 275: 216–240.

Mielke, R., Herholz, K., Grind, M., Kessler, J. and Heiss, W.D. (1994) Clinical deterioration in probable Alzheimer disease correlates with progressive metabolic impairment of association areas. Dementia, 5: 215–234.

Minoshima, S., Giordani, B., Berent, S., Frey, K.A., Foster, N.L. and Kuhl, D.E. (1997) Metabolic reduction in the posterior cingulate cortex in the very early Alzheimer's disease. Ann. Neurol., 42: 85–94.

Nordberg, A. (1993) Clinical studies in Alzheimer patients with positron emission tomography. Behav. Brain Res., 57: 215–234.

Nordberg, A. (1999) Functional brain imaging in Alzheimer's disease. In: S. Gauthier (Ed.), Clinical Diagnosis and Management of Alzheimer's Disease, 2nd Edition. Martin Dunitz, London, pp. 117–132.

Nordberg, A. (2000a) Neuroprotection in Alzheimer's disease—new strategies in treatment. Neurotoxicity Res., 2: 157–165.

Nordberg, A. (2000b) The effect of cholinesterase inhibitors studied with imaging. In: E. Giacobini (Ed.), Cholinesterases and Cholinesterase Inhibitors. Martin Dunitz, London, pp. 237–247.

Nordberg, A. (2001) Brain functional imaging in early and preclinical Alzheimers Disease. Nicotinic receptor abnormalities of Alzheimer's disease: Therapeutic implications. Biol. Psychiatry, 49: 200–210.

Nordberg, A. and Svensson, A.L. (1998) Cholinesterase inhibitors in the treatment of Alzheimer's disease. A comparison of tolerability and pharmacology. Drug Safety, 19: 465–480.

Nordberg, A., Lundqvist, H., Hartvig, P., Lilja, A. and Långström, B. (1995) Kinetic analysis of regional (S) (−) ^{11}C-nicotine binding in normal and Alzheimer brains—in vivo assessment using positron emission tomography. Alzheimer Dis. Assoc. Disord., 1: 21–27.

Nordberg, A., Hellström-Lindahl, E., Almkvist, O. and Meurling, L. (1999) Activities of actetylcholinesterase in CSF increase in Alzheimer patients after treatment with tacrine. Alzheimer Report, 2: 347–352.

Nordberg, A., Jelic, V., Arnáiz, E., Långström, B. and Almkvist, O. (2001) Title missing. In: K. Iqbal, S. Sisodia and B. Winblad (Eds.), Alzheimer's Disease: Advances in Etiology, Pathogensis and Therapeutics. John Wiley & Sons, Chichester, pp. 153–164.

Perry, E.K. (1986) The cholinergic hypothesis—ten years on. Br. Med. Bull., 42: 63–69.

Perry, R.J. and Hodges, J.R. (1999) Attention and executive deficits in Alzheimer's disease. A critical review. Brain, 122: 383–404.

Petersen, R.C., Smith, G.E., Waring, S.C., Ivnik, R.H.J., Tangalos, E.G. and Krokmen, E. (1999) Mild cognitive impairment: clinical characterisation and outcome. Arch. Neurol., 56: 303–308.

Petersen, R.C., Stevens, J.C., Ganguli, M., Tangalos, E.G., Cummings, J.L. and DeKosky, S.T. (2001) Practice parameter: Early detection of dementia: Mild cognitive impairment (an evidence-based review). Report of the quality standard subcommittee of the American Academy of Neurology. Neurology, 56: 1133–1142.

Reiman, E., Caselli, R.J., Yun, L.S., Chen, K., Bandy, D., Minoshima, S., Thibodeau, S.N. and Osborne, D. (1996) Preclinical evidence of Alzheimer's disease in persons homozygous for the epsilon4 allele for apolipoprotein E. New Engl. J. Med., 334: 752–758.

Reiman, E., Caselli, R.J., Chen, K., Alexander, G.E., Bandy, D. and Frost, J. (2001) Declining brain activity in cognitively normal apolipoprotein E epsilon4 heterozygotes: A foundation for using positron emission tomography to efficiently test treatments to prevent Alzheimer's disease. Proc. Natl. Acad. Sci. USA, 98: 3334–3339.

Rogers, S.L., Farlow, M.R., Doody, R.S., Mohs, R. and Friedhoff, L.T. (1998) A 24-week, double-blind, placebo-controlled trial of donepezil in patients with Alzheimers disease. Neurology, 50: 136–145.

Sarter, M. and Bruno, J.P. (2002) Mild cognitive impairment and the cholinergic hypothesis: a very different take on recent data. Ann. Neurol., 52: 384–385.

Shinotoh, H., Aotsuka, A., Fukushi, K., Nagatsuka, S., Tanaka, N., Ota, T., Tanada, S. and Irie, T. (2001) Effect of donepezil on brain acetylcholinesterase activity in patients with AD measured by PET. Neurology, 56: 408–410.

Small, G.W., Komo, S., La Rue, A., Saxena, S., Phelps, M.E., Mazziotta, J.C., Saunders, A.M., Haines, J.L., Pericak-Vance, M.A. and Roses, A.D. (1996) Early detection of Alzheimer's disease by combining apolipoprotein E and neuroimaging. Ann. N.Y. Acad. Sci., 802: 70–78.

Small, G.W., Ercoli, L.M., Silverman, D.H., Huang, S.C., Komo, S., Bookheimer, S.Y., Lavretsky, H., Miller, K., Siddarth, P., Rasgon, N.L., Mazziotta, J.C., Saxena, S., Wu, H.M., Mega, M.S., Cummings, J.L., Saunders, A.M., Pericak-Vance, M.A., Roses, A.D., Barrio, J.R. and Phelps, M.E. (2000) Cerebral metabolic and cognitive decline in persons at genetic risk for Alzheimer's disease. Proc. Natl. Acad. Sci. USA, 97: 6037–6042.

Snyder, S.E., Gunupudi, N., Sherman, P.S., Butch, E.R., Skaddab, M.B., Kilbourn, M.R., Koeppe, R.A. and Kuhl, D.E. (2001) Radiolabeled cholinesterase substrates: In vitro methods for determining structure-activity relationship and identification of a positron emission tomography radiopharmaceutical for in vivo measurement of butyrylcholinesterase activity. J. Cereb. Blood Flow, 21: 132–143.

Sperling, R., Greve, D., Dale, A., Killianty, R., Holmes, J., Rosas, D., Cocchiarella, A., Firth, P., Rosen, B., Lake, S., Lange, N., Routledge, C. and Albert, M. (2002) Functional MRI detection of pharmacologically induced memory impairment. Proc. Natl. Acad. Sci. USA, 99: 455–460.

Stefanova, E., Blennow, K., Almkvist, O., Hellström-Lindhal, E. and Nordberg, A. (2003) Cerebral glucose metabolism, cerebrospinal fluid-beta-amyloid(1–42) (CSF-Abeta42), tau and apolipoprotein E genotype in long-term rivastigmine and tacrine treated Alzheimer disease (AD) patients. Neurosci Lett., 338: 159–163.

Svedberg, M.M., Svensson, A.L., Johnson, M., Lee, M., Cohe, O., Court, J., Soreq, H., Perry, E.K. and Nordberg, A. (2002) Upregulation of neuronal nicotinic receptors subunits α4, β2 and α7 in transgenic mice overexpressing human acetylcholinesterase. J. Mol. Neurosci., 18: 211–222.

Venneri, A., Shanks, M.F., Staff, R.T., Pestell, S.J., Forbes, K.E., Gemmell, H.G. and Murray, A.D. (2002) Cerebral blood flow and cognitive responses to rivastigmine treatment in Alzheimer's disease. NeuroReport, 13: 83–87.

Wesnes, K., Simpson, P. and Kidd, A. (1988) An investigation of the range of cognitive impairments induced by scopolamine 0.6 mg sc. Human Psychopharmacol., 3: 27–41.

Progress in Brain Research, Vol. 145
ISSN 0079-6123

CHAPTER 22

Paying attention to acetylcholine: the key to wisdom and quality of life?

Peter J. Whitehouse*

Case Western Reserve University, 12200 Fairhill Road, Suite C357, Cleveland, OH 44120-1013, USA

Introduction

The scientific encounter on 'Acetylcholine in the Cerebral Cortex' gives me an opportunity to review the history and current state of the art, as well as, make some predictions about the future of the role of acetylcholine in cognitive enhancement. As the Montreal meeting convened so many contributors to our understanding of the role of acetylcholine in the cortex, it is a unique opportunity to discuss some of the clinical implications of these basic studies. I will address this topic in the context of the so-called cholinergic hypothesis of Alzheimer's disease (AD) and other related dementias. Although most attention has been paid to AD and the cholinergic pathology in this condition, we now know that many dementias are characterized by loss of cells in the cholinergic basal forebrain (Whitehouse et al., 1981, 1983). As a consequence persons with a variety of dementias may benefit from new treatments of this deficiency with drugs such as cholinesterase inhibitors. We are also increasingly aware that the clinical symptoms in these conditions do not just include memory problems, but also include alterations in attentional mechanisms (Patterson et al., 1996). For the most part, these dementias occur in older individuals. As such, age-related changes in the cholinergic basal forebrain provide the backdrop for our understanding of disease mechanisms. Moreover, as clinicians and basic scientists studying the biology

of these dementias have begun to recognize, studies of cholinergic mechanisms may have implications for understanding how to treat mild degrees of impairment occurring with age.

I will first focus on attention and its important role in cognition. I will discuss the relationship between attentional mechanisms and higher order cognitive processes such as frontal lobe functions in health and disease and speculate about the relationship between attention and wisdom. Second, I will briefly consider the cholinergic hypothesis particularly as it relates to the role of nicotine in attention. Third, I will review our current knowledge of the effects of cholinergic drugs in AD and other dementias. Finally, I will end with some new data on the effects of cholinergic drugs in normal individuals and discuss the implications for cholinergic cognitive enhancement quite generally (Whitehouse et al., 1997). In this admittedly speculative and personal perspective, I will focus on the research I have conducted with colleagues at The Johns Hopkins University and Case Western Reserve University/University Hospitals of Cleveland, as well as, clinical experiences gained at these institutions over the past 20 years.

Attention

The most fundamental aspect of cognitive function is arousal. All other cognitive functions, be they language, visual spatial abilities or memory, depend on appropriate functioning of subcortical and cortical mechanisms to permit the awake organism

*Tel.: (216) 844-6448; Fax: (216) 844-6466;
E-mail: pjw3@po.cwru.edu

DOI: 10.1016/S0079-6123(03)45022-X

to attend to stimuli generated in the environment or internally. At this very basic level of cognition, we can barely differentiate cognitive and noncognitive systems. Arousal is intimately related to cognitive processing as well as motivation. The concept of apathy, which is used clinically to characterize patients who are not motivated or able to generate enough internal drive to interact with the world, has both cognitive and noncognitive components. When we speak of 'paying attention' to a task, we are communicating that the degree of cognitive effort can be modified by internal and external states. Above and beyond this general arousal is, however, an aspect of selectivity to our attention. The successful organism will attend to relevant stimuli in the environment and also avoid irrelevant distractions. The survival advantages of these attentional processes seem apparent, particularly in challenging environments.

So-called higher order cognitive abilities depend on attention. In particular intimately related to arousal and attention are executive functions, sometimes also called frontal lobe functions, although their biological substrates involve many brain regions. The executive functions, as the words imply, structure behaviors around goals and develop plans to achieve those goals in relationship to other organisms and the environment (Patterson et al., 1996). Frontal lobe functions include feedback mechanisms. A goal is established by the organism and a plan is created to achieve that goal. Monitoring of the success of the plan provides information to the organism to allow adaptive behavioral changes that maximize the likelihood that the goal will be achieved. Executive function involves an integration of motivation, i.e., the desire to achieve the goal, as well as, the necessary cognitive skills to implement a successful plan. A variety of clinical disorders affect executive functions, particularly those that involve damage to frontal and temporal structures, functions.

In my own experience and those of others, attention mechanisms are affected in a variety of neurological and psychiatric conditions including dementia, attention deficit disorder and schizophrenia. My own research began in schizophrenia and was motivated by clinical parallels between paranoid schizophrenia and amphetamine psychosis. I found differences between paranoid and nonparanoid

schizophrenics on tasks requiring attending to fields of numbers selectively and avoiding distractor items (Whitehouse, unpublished). Adult attention deficit disorder is frequently not accompanied by hyperactivity as in children, but does manifest as cognitive disorganization, poor judgment and inability to focus attention. Its phenomenology is similar to early frontal lobe dementias where a combination of dysfunctions of higher cognitive functions can be associated with behavior disinhibition. However, executive dysfunction can also occur in Alzheimer disease and other dementias at different stages.

Patients with disorders in attention have a difficult time dealing with life's normal challenges, especially those involving interactions with other people. They often demonstrate poor social judgment and problem-solving skills. They could be said to lack common sense or even a kind of practical wisdom. Some clues to understand the basics of the best in human thinking may be found in studying those unfortunates who have intellectual and affectual problems that accompany brain disease.

Although the topic of wisdom has not received much focus from neurobiologists or even psychologists, we can question about the relationships between attentional mechanisms and this higher order integrative skill. Wisdom implies the ability to solve practical life-problems by integrating both cognitive information about a particular situation, and appropriate values, in order to achieve a certain desired outcome. Fundamental to wisdom is the recognition of limits, and with that, an attitude of humility. There are things that cannot be known or attended to adequately. There is an intrinsic uncertainty in facing problems in the real world. In many ways, wisdom represents higher order attentional processing. People who are wise know what to attend to and what to ignore. Thus, it is actually possible by better understanding the neurobiology of attention that might contribute in improving the highest aspects of human cognitive abilities, executive functions and wisdom.

The topic of wisdom is also intimately related to quality of life (Whitehouse et al., 2001). The concept of quality of life is a multi-component, integrative concept that is important to human survival and thriving. In medicine, the topic of quality of life is receiving attention, including in the area of dementia, because it should be the ultimate goal for our therapeutic interventions. Quality of life includes cognitive ability, mental health, social relationships, and financial well being, as well as other subjective and even spiritual domains (Whitehouse and Rabins, 1992; Mack and Whitehouse, 2001). Is it possible that understanding the neurobiology of attention can contribute to our ability to enhance the quality of lives of people who are affected by diseases that have damaged these mechanisms? The achievement of a high quality of life does include an element of wisdom. Both wisdom and quality of life depend on having realistic expectations about what is possible to achieve. The achievement of wisdom and quality of life both involve integrating cognitive and emotional data and living one's life in a manner that optimizes one's own life and the lives of others around them.

My colleagues and I have suggested that there is an intimate relationship between wisdom and quality of life (see Fig. 1, Whitehouse et al., 2001). The oriental yin and yang symbol captures the interactions through time between wisdom and quality of life. As we age the processes of learning and adaptation allow us opportunities for gaining wisdom and improving our own quality of life and that of those around us. Wisdom is a manifestation of the subjective aspects of quality of life, just as enjoying a high quality of life is external evidence of a subjective state of wisdom. Addressing values and motivations is as important, or more important than addressing just facts and thoughts in constructing the wisdoms

Human experience through the life course

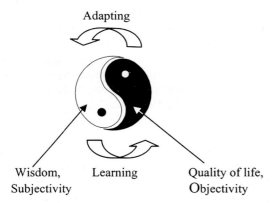

Fig. 1. An integrated spacial model of aging, wisdom, and quality of life. From Whitehouse Ballenger and Katz (2001).

that we will need for the future. The very survival of our species and of life itself depends on knowing what to value for the long-term and attending wisely to relevant issues (like ecological sustainability today). Once again, I say that these thoughts are speculative, but the connections amongst arousal, selective attention, motivation, executive functions, wisdom, and quality of life raise the stakes for understanding some of the basic mechanisms of attention and how diseases disrupt these mechanisms. Others frame the issues a bit differently but ask similar provocative questions about the relationship between acetylcholine and consciousness (Perry et al., 2002).

Cholinergic hypothesis

The history of the role of acetylcholine in cognition dates back centuries when human beings discovered that some drugs could impair cognition and also caused somatic side effects that we now know are consistent with cholinergic blockade. It was not until the early 1900s that scientific understanding of some of these mechanisms of action began to emerge. The identification of acetylcholine as a neurotransmitter was followed rapidly by the elaboration of the theory of receptors and the identification of muscarinic and nicotinic receptors. The distribution of cholinergic markers in the brain confirmed an important role for this neurotransmitter in a variety of systems related to cognition. At the same time, the development of brain psychiatry led to an understanding of the importance of brain mechanisms in diseases. It was not, however, until the development of neurotransmitter chemistry in the beginning of the second half of the twentieth century that the cholinergic hypothesis of dementia emerged.

The so-called cholinergic hypothesis was often not stated in a rigorous scientific way and was more often used as a political statement. The strong form of the cholinergic hypothesis was that loss of the cholinergic system markers explained all the symptoms of AD, whereas, its weak form implied that damage to this system had no role to play in the clinical picture. Clearly the truth lies in between. The phrase cholinergic hypothesis was used in discussions about where to appropriate resources for research. Currently, the so-called amyloid hypothesis has become the dominant way of framing scientific issues in AD. Particular aspects of the cholinergic hypothesis could be stated as scientific hypotheses, for example, proposing that levels of choline acetyl transferase activity relate to scores on specific clinical cognitive tests.

The relative decline of neuropathology and neurochemistry as fields of study, and the ascendancy of molecular biology and genetics, parallels the dominance of amyloid over acetylcholine as a brain substance of the decade. It also illustrates the delays in development of medications based on basic biological research. Cholinomimetic drugs to improve cognition were not approved by regulatory bodies until decades after the descriptions linking cholinergic pathology to symptoms of disease appeared.

Our own work focused initially on the cholinergic basal forebrain in AD, where cell shrinkage and loss occurs, and is particularly severe in young onset cases. We also found some evidence for cell loss in aged controls when compared to younger individuals. We extended these findings by exploring muscarinic and nicotinic receptors in AD and discovered that a major neurotransmitter receptor loss in the cortex was in nicotinic-binding sites (Whitehouse et al., 1986).

After describing our results in AD and normal aging, we explored other dementias. We rediscovered the original observation that pathology (Lewy bodies) occurred in the basal forebrain in Parkinson's disease (Whitehouse et al., 1988). Now it is clear that cholinergic dysfunction occurs not only in Alzheimer's and Parkinson's disease, but also in variants such as Lewy body and vascular dementias. Many disease states characterized by cognitive impairment now appear to have a degree of cholinergic pathology. In fact, it is only in the rare degenerative dementias, such as Pick's disease and Huntington's disease, that cholinergic cortical pathologies are not a feature. Dysfunction in cholinergic mechanisms have also been claimed to occur in attention deficit disorder, schizophrenia, delirium and various amnesic states.

Cholinergic therapies

After a gap of over a decade, from the original neurochemical observations in medicines were eventually approved to treat the condition based on

an understanding of the cholinergic pathology (Davis et al., 1992; Mayeux and Sano, 1999). Initial attempts at using precursor loading therapy with choline and lethicin failed. Several cholinesterase inhibitors are now available around the world, and are of relatively comparable efficacy although they vary somewhat in their side effects. These drugs have been approved by regulatory bodies because they have been demonstrated to improve cognition on objective tests and to have also achieved a modest degree of clinical meaningfulness as judged by clinicians' ratings and/or scores of caregiver ratings of activities of daily living. Clearly, these drugs are only first step efforts to improve the cognitive function in dementias but they do represent some modest success for rational drug development.

Efforts to improve the symptomatic effects of cholinesterase inhibitors have taken several directions. Initial attempts at receptor agonist therapies focused on muscarinic cholinergic receptors because they are found in high concentrations in the central nervous system and are relatively intact in postmortem samples of brains of persons with AD. However various agents failed in the clinic because of lack of efficacy and/or side effects. Probably most of the energy is being expended now on nicotinic agonists. Although much of the focus has been on the improvement of memory, many people think that cholinesterase inhibitors might actually work by improving attention. Drugs that act on nicotinic receptors, particularly a subtype of receptors such as the alpha 4 beta 2 or alpha 7, may be more efficacious and also may avoid some of the side effects of the cholinesterase inhibitors. We know from animal and human studies that nicotine and nicotine-related compounds can improve cognition. (Witte et al., 1997; Levin et al., 1998; Waters and Sutton, 2000; Phillips et al., 2000; Woodruff-Pak and Gould, 2002). Sorting out the relevant receptor actions of these agents, their effects in peripheral and central nervous system and their specific effects of cognition and mood has proven challenging.

The latest drug to be approved for AD, galantamine, is claimed to have both cholinesterase inhibitory and nicotinic allosteric receptor modulating properties. It remains to be seen whether this additional property has clinical meaningfulness. Interestingly, the cholinesterase inhibitors have been shown to improve more than just memory, attention and cognitive abilities. They appear to have mild positive effects on behavioral disturbances. Improvements in apathy have been consistently reported, which is interesting given the intimate relationship of apathy to arousal and attention. Claims have also been made that cholinergic drugs may modify the course of the disease by altering basic pathogenesis of amyloid plaques, for example. Nicotinic agonists are protective in a variety of experimental models of cell death. However, there is as yet no clear clinical evidence supporting disease modifying effects of any of these drugs (Woodruff-Pak and Gould, 2002).

Clinical issues in dementia and related conditions

It is possible that more effective approaches will be developed for the treatment of dementias based on our increasing understanding of the role of acetylcholine in cerebral cortex. Currently, much energy is being spent on the prevention of AD and related dementias. In order to prevent the condition, one needs to identify people at risk for becoming demented. Currently, the concept of Mild Cognitive Impairment (MCI) is attracting attention in this regard. People with MCI have features of AD but not severely enough to warrant the diagnosis of dementia. Logically, all degenerative dementias must pass through a preclinical phase where the symptoms are present but subtle enough to escape attention or of insufficient magnitude to warrant a diagnosis of dementia. However, clinically this label is problematic. If someone is labeled as having MCI does it mean that he (she) has AD, does not have AD or may get AD? Thus, although this concept may be helpful for research, it has significant limitations when used in a clinical context. In fact, the concept of MCI challenges the often made statement that AD is not related to normal aging. It is possible that we would all develop AD if we lived long enough. Other concepts have been offered in the literature prior to MCI, including Benign Senile Forgetfulness, Aging-Associated Memory Impairment and Aging-related cognitive decline (Crook et al.,

1986). Having been involved in many of these efforts to develop 'diagnostic' categories I can attest to the influences of the motivation to improve pharmacological treatments in developing these labels. Yet each one of them is an arbitrary category superimposed on a continuum. Whether we consider basal forebrain cell loss, cholinergic neurochemical markers, densities of plaques and tangles or cognitive performance, the processes we call AD occur gradually. Any diagnostic labeling necessarily involves an arbitrary threshold being established.

The implications of this continuum of pathology are profound for clinical diagnosis, especially the search for biological markers. Much political energy is put into defending that AD is in fact a disease and not normal aging and promoting that finding a specific diagnostic neuroimaging, CSF or blood biological marker is essential for progress to be made in the field. Yet, where we set the threshold will determine how many people get labeled as diseased. Drug companies love to have more customers. Experts have a bias towards making their favorite categories more common to gather grants and clinical reimbursements. From my point of view AD will remain a clinical diagnosis and as such a sociomarker. We must remember the power of labels to alter people's lives. Positive effects include helping them get access to medical services and providing an explanation for symptoms they may experience. Perhaps more ignored, labels have the power to negatively affect people's lives. In creating new labels on the continuum of cognitive aging, have we created a terror of AD beyond what it might otherwise be (Whitehouse, 2002)?

The focus on cholinergic diseases such as AD as discrete categories distorts other processes as well. Perhaps we should look to approve drugs for cognition more like pain killers than disease-specific treatments. The FDA and other regulatory bodies permit drugs to be approved in two categories: category specific, e.g., anti-dementia or better anti-AD, and noncategory specific, symptomatic treatments, e.g., pain relievers. Since many conditions that affect cognition involve cholinergic mechanisms, perhaps we should approve cognitive enhancers as symptomatic drugs rather than disease specific ones. Similar discussions have been held at the FDA about drugs to improve behavioral symptoms in dementia. Should we approve drugs for categories, e.g., the psychosis of AD, or as symptomatic agents e.g., agitation and hallucinations in AD?

Categories of disease seem to isolate those with the label from the rest of us, where as situating symptoms on a continuum means that we can all share them, despite being at a different point along the trajectory. Does it help to medicalize cognitive aging and diagnose people, or should we all relate to the effects of aging on our thinking and feeling abilities? One of the problems of the medical process is that it focuses on the negative aspects of cognitive aging. Perhaps we should focus on enhancing wisdom as we age, and not just treating dementia. How can we think more deeply about cognitive aging in all its aspects as members of societies facing more older people, and fewer younger ones in our communities? Could we actually develop a drug that would help us attend to the critical aspects of these issues, avoid distracting simplifications and become collectively wiser?

In persons with AD cholinesterase inhibitors have been shown to improve activities of daily living. What about such effects in normal people? In order to push the limits of the cholinergic hypothesis of cognition we administered donepezil, the leading cholinesterase inhibitor for treatment of AD, to 18 airplane pilots with an average age of 53 years (Yesavage et al., 2002). They were tested in a flight simulator, which is an ecologically valid means of assessing aviation skills, and is representative of complicated activities of daily living. After one month of daily donepezil treatment (5 mgs), the pilots in the active condition appeared to improve their performance when compared to those on placebo. Interestingly, the two areas of enhanced performance were landing, and handling emergencies, both of which are attention-intensive tasks. Thus, in the future, the question may be asked as to whether the concept of AD itself has outlived its usefulness and if we should examine the neurobiology of memory and aging on a continuum. This raises many complex issues about the potential use of cognitive enhancement drugs in normal people (Whitehouse et al., 1997).

Having been involved in cholinergic research for over 20 years in both the laboratory and clinic, I was interested in the effects of cholinesterase drugs on my own cognition. Having provided scientific evidence that they might work in normal people of my own age, and having published about the ethical issues of such cognitive enhancement, I decided to pursue a systematically evaluated course of donepezil treatment by prescribing it to myself. Extensive pretesting included cognition, activities of daily living and quality of life. The clinical treatment plan is still in progress but initial observations of the effects on cognition were affected by the appearance of some gastrointestinal symptoms (mild nausea) which limited the length of the first trial period.

It is clear to me from this initial work, however, that the wisest way to enhance my own cognition is not through drugs, even cholinergic ones. Our own clinical and research programs designed to improve higher order cognitive abilities in people with a wide range of memory and attention issues include an NIH funded randomized controlled study of book reading, electronic reminiscence therapy using multimedia DVDs of family archived material, a university course on wisdom and various forms of story telling, including some using the theatre arts (http://www.timeslips.org). The most integrative program is The Intergenerational School founded by my wife, Catherine Whitehouse, myself and Stephanie FallCreek (Whitehouse et al., 2000). This is the world's first public learning community organized to create educational opportunities for people of all ages and abilities (including those with cholinergic pathologies such as attention deficit disorder, MCI and AD). In this environmentally responsible, value driven, and experiential learning oriented school, people work to improve themselves, their fellow students and their community.

As the world struggles to address the new and evolving terrors that the future can bring, we must learn to enhance our collective wisdom in as many ways as possible. Understanding the role of acetylcholine in the cerebral cortex has contributed in a small way to such learning. Only integrated biopsychosocial approaches to considering age-related cognitive changes will offer us much hope for improving our quality of lives, however. Such approaches will need to include the wisdom that the investigation of biology alone can take us only so far.

Acknowledgments

Special acknowledgement for support of this manuscript goes to the following research grants from the National Institutes of Health Enhancement Ethics and the Molecular Genetics of Aging: Issues for Families, Health Professionals, and Society R01 HG01446-01, Shiego and Megumi Takayama Foundation and the National Institutes of Health National Institute on Aging Medical Goals in Dementia: Ethics and Quality of Life R01 AG/HS17511-01A1.

References

Crook, T., Bartus, R.T., Ferris, S.H., Whitehouse, P., Cohen, G.D. and Gershon, S. (1986) Age-associated memory impairment: Proposed diagnostic criteria and measures of clinical change—Report of a National Institute of Mental Health Work Group. Develop. Neuropsych., 2: 261–276.

Davis, K.L., Thal, L.J., Gamzu, E.R., Davis, C.S., Woolson, R.F., Gracon, S.I., Drachman, D.A., Schneider, L.S., Whitehouse, P.J., Hoover, T.M. (1992) A double-blind, placebo-controlled multicenter study tacrine in Alzheimer's disease. The Tacrine Collaborative Study Group. New Eng. J. Med., 327: 1253–1259.

Mack, J.L. and Whitehouse, P.J. (2001) Quality of life in dementia: state of the art—report of the International Working Group for Harmonization of Dementia Drug Guidelines and the Alzheimer's Society satellite meeting. Alzheimer Dis. Assoc. Disord., 15: 69–71.

Mayeux, R. and Sano, M. (1999) Treatment of Alzheimer's disease. New Eng. J. Med., 341: 1670–1679.

Levin, E.D., Conners, C.K., Silva, D., Hinton, S.C., Meck, W.H., March, J. and Rose, J.E. (1998) Transdermal nicotine effects on attention. Psychopharm., 140: 135–141.

Patterson, M.B., Mack, J.L., Geldmacher, D. and Whitehouse, P.J. (1996) Executive functions and Alzheimer's disease: problems and prospects. Eur. J. Neurol., 3: 5–15.

(2002). In Perry E., Ashton H. and Young A. (Eds.), Neurochemistry of consciousness. John Benjamins, The Netherlands.

Phillips, J.M., McAlonan, K., Robb, W.G.K. and Brown, V.J. (2000) Cholinergic neurotransmission influences covert orientation of visuospatial attention in the rat. Psychopharm., 150: 112–116.

Waters, A.J. and Sutton, S.R. (2000) Direct and indirect effects of nicotine/smoking on cognition in humans. Addictive Behav., 25: 29–43.

Whitehouse, P.J. (1974) Attentional mechanisms in paranoid and non paranoid schizophrenics. Unpublished Master's thesis, the Johns Hopkins University.

Whitehouse, P.J. (2002) Book review: Losing my mind: an intimate look at life with Alzheimer's by Thomas DeBaggio. New Eng. J. Med., 347: 861.

Whitehouse, P.J. and Rabins, P.V. (1992) Quality of life and dementia. Alzheimer Dis. Assoc. Disord., 6: 135–138.

Whitehouse, P.J., Price, D.L., Clark, A.W., Coyle, J.T. and DeLong, M.R. (1981) Alzheimer disease: evidence for selective loss of cholinergic neurons in the nucleus basalis. Ann. Neurol., 10: 122–126.

Whitehouse, P.J., Hedreen, J.C., White, C.L., III and Price, D.L. (1983) Basal forebrain neurons in the dementia of Parkinson disease. Ann. Neurol., 13: 243–248.

Whitehouse, P.J., Martino, A.M., Antuono, P.G., Lowenstein, P.R., Coyle, J.T., Price, D.L. and Kellar, K.J. (1986) Nicotinic acetylcholine binding sites in Alzheimer's disease. Brain Res., 371: 146–151.

Whitehouse, P.J., Martino, A.M., Marcus, K.A., Zweig, R.M., Singer, H.S., Price, D.L. and Kellar, K.J. (1988) Reductions in acetylcholine and nicotine binding in several degenerative diseases. Arch. Neurol., 45: 722–724.

Whitehouse, P.J., Juengst, E.T., Mehlman, M. and Murray, T. (1997) Enhancing cognition in the intellectually intact: possibilities and pitfalls. Hastings Cent. Rep., 3: 14–22.

Whitehouse, P.J., Bendezu, E., FallCreek, S. and Whitehouse, C. (2000) Intergenerational community schools: a new practice for a new time. Educ. Gerontol., 26: 761–770.

Whitehouse, P.J., Ballenger, J. and Katz, S. (2001). How we think (deeply but with limits) about quality of life. In: Weisstub D.N., Thomasma D.C., Gauthier S. and Tomossy G.F. (Eds.), Aging: decisions at the end of life. Kluwer, The Netherlands, pp. 1–19.

Witte, E.A., Davidson, M.C. and Marrocco, R.T. (1997) Effects of altering brain cholinergic activity on covert orienting of attention: comparison of monkey and human performance. Psychopharm., 132: 324–334.

Woodruff-Pak, D.S. and Gould, T.J. (2002) Neuronal nicotinic acetylcholine receptors: involvement in Alzheimer's disease and schizophrenia. Behav. Cog. Neurosci. Rev., 1: 5–20.

Yesavage, J.A., Mumenthaler, M.S., Taylor, J.L., Friedman, L., O'Hara, R., Sheikh, J., Tinklenberg, J. and Whitehouse, P.J. (2002) Donepezil and flight simulator performance: Effects on retention of complex skills. Neurol., 59: 123–125.

Subject Index

Note: references to whole chapters are in **bold**, and references to figures and tables in *italics*.

322